Ethnomedicinal Plant Use and Practice in Traditional Medicine

Akash
Gurukul Kangri University, India

Navneet
Gurukul Kangri University, India

B.S. Bhandari
Garhwal University, India

A volume in the Advances in Medical Diagnosis, Treatment, and Care (AMDTC) Book Series

IGI Global
DISSEMINATOR OF KNOWLEDGE

Published in the United States of America by
IGI Global
Medical Information Science Reference (an imprint of IGI Global)
701 E. Chocolate Avenue
Hershey PA, USA 17033
Tel: 717-533-8845
Fax: 717-533-8661
E-mail: cust@igi-global.com
Web site: http://www.igi-global.com

Library of Congress Cataloging-in-Publication Data

Names: Saklani, Akash, 1988- editor. | Navneet, 1963- editor. | Bhandari, B. S., 1969- editor.
Title: Ethnomedicinal plant use and practice in traditional medicine / Akash, Navneet, B.S. Bhandari, editors.
Description: Hershey, PA : Medical Information Science Reference, [2020] | Includes bibliographical references and index. | Summary: “”This book explores the use and applications of natural-based resources and products from ethnomedicinal plants”-- Provided by publisher.
Identifiers: LCCN 2019028360 (print) | LCCN 2019028361 (ebook) | ISBN 9781799813200 (h/c) | ISBN 9781799813217 (eISBN)
Subjects: MESH: Plants, Medicinal | Medicine, Traditional | Ethnobotany | Ethnopharmacology | Plant Extracts--therapeutic use
Classification: LCC RM666.H33 (print) | LCC RM666.H33 (ebook) | NLM QV 766 | DDC 615.3/21--dc23
LC record available at https://lccn.loc.gov/2019028360
LC ebook record available at https://lccn.loc.gov/2019028361

This book is published in the IGI Global book series Advances in Medical Diagnosis, Treatment, and Care (AMDTC) (ISSN: 2475-6628; eISSN: 2475-6636)

British Cataloguing in Publication Data
A Cataloguing in Publication record for this book is available from the British Library.

Advances in Medical Diagnosis, Treatment, and Care (AMDTC) Book Series

ISSN:2475-6628
EISSN:2475-6636

Mission

Advancements in medicine have prolonged the life expectancy of individuals all over the world. Once life-threatening conditions have become significantly easier to treat and even cure in many cases. Continued research in the medical field will further improve the quality of life, longevity, and wellbeing of individuals.

The **Advances in Medical Diagnosis, Treatment, and Care (AMDTC)** book series seeks to highlight publications on innovative treatment methodologies, diagnosis tools and techniques, and best practices for patient care. Comprised of comprehensive resources aimed to assist professionals in the medical field apply the latest innovations in the identification and management of medical conditions as well as patient care and interaction, the books within the AMDTC series are relevant to the research and practical needs of medical practitioners, researchers, students, and hospital administrators.

Coverage

- Disease Management
- Internal Medicine
- Medical Testing
- Critical Care
- Patient-Centered Care
- Diagnostic Medicine
- Experimental Medicine
- Chronic Conditions
- Patient Interaction
- Emergency Medicine

The Advances in Medical Diagnosis, Treatment, and Care (AMDTC) Book Series (ISSN 2475-6628) is published by IGI Global, 701 E. Chocolate Avenue, Hershey, PA 17033-1240, USA, www.igi-global.com. This series is composed of titles available for purchase individually; each title is edited to be contextually exclusive from any other title within the series. For pricing and ordering information please visit http://www.igi-global.com/book-series/advances-medical-diagnosis-treatment-care/129618. Postmaster: Send all address changes to above address.

Titles in this Series

For a list of additional titles in this series, please visit: www.igi-global.com/book-series

Advanced Pharmacological Uses of Medicinal Plants and Natural Products
Ajeet Singh (National Dairy Research Institute, India) Padam Singh (Vaccine and Infectious Disease Research Centre, Translational Health Science and Technology Institute, India) and Navneet Bithel (Gurukul Kangri Universiy, India)
Medical Information Science Reference • ©2020 • 540pp • H/C (ISBN: 9781799820949) • US $425.00

Quality Control of Cellular Protein in Neurodegenerativ Disorders
Md. Sahab Uddin (Southeast University, Bangladesh) and Ghulam Md. Ashraf (King Abdulaziz University, Saudi Arabia)
Medical Information Science Reference • ©2020 • 260pp • H/C (ISBN: 9781799813170) • US $295.00

Multidimensional Perspectives and Global Analysis of Universal Health Coverage
Yeter Demir Uslu (Istanbul Medipol University, Turkey) Hasan Dinçer (Istanbul Medipol University, Turkey) and Serhat Yüksel (İstanbul Medipol University, Turkey)
Medical Information Science Reference • ©2020 • 499pp • H/C (ISBN: 9781799823292) • US $265.00

Examining Biological Foundations of Human Behavior
Barre Vijaya Prasad (Dharwad Institute of Mental Health and Neurosciences, India)
Medical Information Science Reference • ©2020 • 266pp • H/C (ISBN: 9781799828600) • US $285.00

Handbook of Research on Medical Interpreting
Izabel E.T. de V. Souza (Osaka University, Japan) and Effrossyni (Effie) Fragkou (National and Kapodistrian University of Athens, Greece)
Medical Information Science Reference • ©2020 • 511pp • H/C (ISBN: 9781522593089) • US $345.00

Quality Assurance in the Era of Individualized Medicine
Anastasius S. Moumtzoglou (P. & A. Kyriakou Children's Hospital, Greece)
Medical Information Science Reference • ©2020 • 346pp • H/C (ISBN: 9781799823902) • US $285.00

Diagnosing and Managing Hashimoto's Disease Emerging Research and Opportunities
Mahmoud Sakr (Alexandria University, Egypt)
Medical Information Science Reference • ©2020 • 263pp • H/C (ISBN: 9781522596554) • US $245.00

Using Narrative Writing to Enhance Healing
Jennifer Lynne Bird (Oxbridge Academy, USA)
Medical Information Science Reference • ©2020 • 355pp • H/C (ISBN: 9781799819318) • US $275.00

IGI Global
DISSEMINATOR OF KNOWLEDGE

701 East Chocolate Avenue, Hershey, PA 17033, USA
Tel: 717-533-8845 x100 • Fax: 717-533-8661
E-Mail: cust@igi-global.com • www.igi-global.com

Table of Contents

Section 2
Pharmacology of Ethnobotanical Plants

Section 3
Traditional Practices of Ethnobotanical Plants

Detailed Table of Contents

Ethnobotanical and economic studies endorse Allium stracheyi Baker, of the Alliaceae family, for future food and health security. The species is important to the socio-cultural, spiritual, and medicinal lives of the indigenous Himalayan people and those in the surrounding urban and peri-urban areas. The species is used by the indigenous Bhotiya people as a flavoring, spice/condiment (Jambu Faran), and a remedy for colds/coughs, jaundice, stomachaches, and various other ailments. A perennial, medicinal, and aromatic plant (MAP) species, A. stracheyi is harvested two or three times annually and is categorized as an endemic and endangered species. It has a high value and represents an important cultural element in the lives of indigenous Bhotiya folk living in the mountains of Central Himalaya, India. Among medicinal and aromatic plants, A. stracheyi is an economic indicator of the plant-based economy and tradition of Bhotiya ethnic groups in the Niti and Mana Valleys of Uttarakhand, India.

The use of ethnomedicinal plants of home gardens has played an important role in treating minor illnesses related to physical and psychological wellbeing among rural folks. Local healers often use plants from their home garden or nearby forest. In this chapter, the traditional uses of medicinal plants using various tools with sample data have been used to facilitate ethnobotanical research on home gardens. Samples measuring the relative importance of species through quantitative analysis are tabulated. These tools, if applied during in-situ data collection, would help generate reliable information on characteristics of home gardens and medicinal plants. Conservation of medicinal plants from the perspective of home gardening is important. To document how medicinal plants are used to treat different diseases, their use value and ethical considerations would be significant for bioprospecting of medicinal plants and protecting intellectual right of the associated traditional knowledge.

Traditional medicines and natural products from ethnomedicinal plants have great significance in recent time. Various forms of medicines like Ayurveda, traditional Chinese medicine, kampo, Unani, have been plasticising in recent days due to their effectiveness against various human ailments and also have blossomed into the regulated systems of traditional medicine. This chapter reviews the relationship of plants and humans, along with their cultural relationship and role of the traditional medicines, by exploring the methodologies and various concepts for the discovery of various drugs. Further, this will also illustrate traditional medicines that have their incomparable advantages over the modern allopathic medicines.

In recent decades, species extinction has increased sharply. Most species in the world are discovered in tropical forests – covering more than 10% of the planet. Sustainable management and conservation efforts must include local communities and their traditional knowledge. The traditional use of forest resources, especially non-timber products such as medicinal plants, not only has deep roots in indigenous populations, but is also practiced in a broad segment of culture. Using medicinal herbs is often a costly alternative to modern medicine that is economically inevitable. This traditional use's basic knowledge is carried from one generation to the next. Medical use in particular reflects an extremely vibrant, ever-evolving process in which fresh knowledge is constantly acquired and connected to traditional methods. An instance from West Bengal's West Midnapore district in India is used to illustrate the impacts of an integrated strategy to preserving biodiversity and culture.

Section 2
Pharmacology of Ethnobotanical Plants

Cyperus rotundus (nut grass in English) is a perennial erect sedge plant and is distributed in over 90 countries of the world, where it has been mostly classified as a highly invasive weed. Despite this classification, the plant has been considered from traditional times to be medicinally important. The traditional uses of the plant in various countries include uses against various gastrointestinal tract disorders, skin diseases, leprosy, fever, and neurological disorders. Evaluation of the plant and especially its rhizomes in a scientific manner has revealed the presence of numerous phytochemicals and wide-ranging pharmacological activities, which include anti-microbial, gastrointestinal, wound healing, anti-diabetic, anti-cancer, anti-malarial, anti-obesity, hepatoprotective, and anti-pyretic activity. The scientific validation of a number of traditional uses strongly indicates that the plant may prove useful in the discovery of a number of lead compounds and novel drugs.

This is an overview of plant use for medicinal applications, a practice from old civilizations still used around the world. According to WHO, nearly 80% of people use herbal medicine plant extracts as their primary health solution. Ethnobotany emerges as a research field to document and understand the traditional knowledge about plants and their roles in society. Diseases like diabetes, hypertension, hypercholesterolemia, and cancer are the predominant challenges to global health, chronic diseases accounting for two-thirds of deaths worldwide. This document discusses ethnobotanical studies on many medicinal plants affecting these diseases.

Genotoxicity is considered as a potential cause of various diseases including cancer. During the last decade, herbal extracts attained a great deal of attention due to its safe and effective applications against various DNA damaging agents. However, the mechanism of DNA strand breaks by various mutagens and genotoxins is often correlated with the generation of Reactive Oxygen Species (ROS). Herbal extracts constitute a number of phytochemicals and those are reported to have considerable antioxidant properties, which are in turn capable of neutralizing ROS mediated DNA damage. The botanical family Zingiberaceae is reported to have significant antioxidant and antigenotoxic potential by various researchers. Among a number of species belonging to this family, six species, namely Alpinia galanga, A. zerumbet, Curcuma amada, C. caesia, Zingiber officinale, and Z. zerumbet, attract notable attention due to their remarkable ethnobotanical and medicinal importance. This chapter deals with phytochemical composition, antioxidant, and antigenotoxic properties of these six Zingiberaceous plant extracts.

Hemidesmus indicus (L.) R. Br. is commonly known as Indian Sarsaparilla or Anantmula. Traditionally, it has been utilised as a vital herb for the treatment of several disorders. Indian Sarsaparilla is rich in a wide range of phytoconstituents such as pregnane glycosides, steroids, terpenoids, aromatic aldehydes, lignans, saponins, flavonoids and aliphatic acids which may further contribute to its pharmacological properties. This chapter gathers and compiles the traditional ethnobotanical and ayurvedic aspects of H. indicus and recently updated knowledge regarding the pharmacology, phytochemistry, adulteration, and current trends of this medicinally important herb in the field of modern phytomedicine. It also presents the ayurvedic pharmacology of this herb and summarizes the biomedical researches in as much as it helps glean a better understanding of H. indicus safety and effectiveness in humans, and describes the various natural products and polyherbal medicines containing H. indicus.

Aloe vera is a perennial, drought-resisting, succulent plant belonging to the Asphodelaceae family that has a long history of having many tremendous medicinal and anti-inflammatory as well as anti-bacterial properties. This chapter highlights the great potentials of the aloe vera species. Desk study and document analysis guided the systematic and extensive review of both published and unpublished resources on the Aloe vera Species. The chapter contends that horticulturists and plant scientists in the field of agriculture and medicine must pay attention to the Aloe vera as a medicinal plant that possesses the most powerful antibacterial and anti-inflammatory properties, which protects the body against constipation through enhancing the body metabolism, skin, and worm infestations.

From ancient time, plants have been utilized as a great source of medicinal products for several types of diseases and disorders. Traditional knowledge is an important source for the development of new drugs. Several studies revealed that traditional knowledge of medicinal plants is being practised among several tribes throughout the world. Many researchers have been evaluated the authenticity of this information. Family Asteraceae got an important place among this medicinal heritage. This is one of the widely distributed families and large numbers of plants have been utilized in various skin-related problems. This chapter highlights the ethnopharmacological properties of this family.

Humans have known the therapeutic properties and plant origin poisons of many plant species since ancient times. Ancient people who had no real and scientific knowledge on diseases and treatments tried to find solutions to these problems by the opportunities they had and the theories they could produce. Plants are the most important source of traditional folk medicine. People around the world have benefited from plants grown in their environment for this purpose. In these drugs, whole plant or various plant organs or secondary substances were used. Over the years, several therapies have emerged to treat herniated disc, ranging from conservative treatment to minimally invasive and percutaneous techniques to open surgical methods. This chapter examines the recent studies to form a compilation on ethnobotanical uses of medicinal plants for the treatment of herniated disc and their local names, parts, and usage forms in Turkish traditional medicine.

The number of plants used for treatments has been steadily increasing since ancient times. There were around 250 herbal drugs used during the Mesopotamian civilization. During the time of the Greeks, 60 medicinal plants were known. During the Arab-Persian civilization, this number increased to about 4,000. At the beginning of the 19th century, the number of known medicinal plants reached 13,000. These herbal drugs are used to treat many diseases, from diabetes to cancer. Back pain and related symptoms rank as the second most common affliction of humankind. One of the proposed methods to control and treat back pain symptoms is traditional folk medicine. This chapter examines the recent ethnobotanical studies to form a compilation on medicinal plants used as Turkish herbal medicine in the treatment of back pain and their local names, parts, and usage forms.

Leucorrhoea is a common disorder that is prevalent in females. Ethnic communities still treat leucorrhoea with plants that grow in their surroundings due to lack of sufficient medical facilities and/or costly modern treatment. Therefore, the author tried to look for plants used in traditional ethno-medicine and Ayurvedic systems of medicine for treatment of leucorrhoea in India. This review reveals the use of 345 plants by various folk communities and 31 unique plants from Ayurveda, which gives a figure of 376 plant species prevalent in Indian traditional medicine for treatment of leucorrhoea. Further analysis revealed maximum use of roots followed by the leaves and the stem bark of these plants. The chapter indicates 29 highly credible ethnomedicinal plant species with a rating of 5. This review shall prove helpful in screening of potential ethnomedicinal plant species for carrying out further scientific validation studies to assess their efficacy in leucorrhoea, which may contribute to the development of novel phyto-therapeutic molecules for effective treatment.

This chapter deals with scientific information about the wild socioeconomic plants used by the local inhabitants of the Champawat district of Uttarakhand for the enhancement of their livelihood. During this entire study, it was observed and documented that poor rural people of this area are educationally and financially backward, and they do not have any permanent source of income; due to this reason they are unable to fulfill their necessary requirements. Therefore, they are totally dependent on various forest ecosystem services, viz., selling prepared herbal medicines, wild edible fruits, and wild ornamental plants, etc., which shares approximately 45-50% of their total annual income and plays a vital role for the upliftment of their livelihood. During this research work, a total of 63 plant species belonging to 58 genera under 46 families have been documented, which are being used as livelihood resources by the locals of this area.

Villages in the Indian Himalayan Region have a rich repository of the indigenous knowledge and practices. Documentation of such knowledge is urgently required in view of the advent of modernization. Therefore, an attempt is made to document the indigenous uses and practices of the plants utilized by the local inhabitants of the Gulmarg Wildlife Sanctuary, Jammu and Kashmir., India. A total of 48 plant species belonging to 46 genera and 25 families are used traditionally to cure various diseases. Various plant parts of these species are used to cure cold, cough, fever, liver disorder, kidney stones, skin diseases, and eye and ear complaints. Such kinds of studies will help in developing a comprehensive database of the plants used traditionally, strengthening the health care system in the villages, and conserving traditional knowledge for the future.

Preface

From ancient times, human societies have always relied on plants and their products for curing ailments. Indigenous groups developed knowledge of ethnomedicinal plants and plant-based products through their skills to solve various problems they faced during their routine activities and through practical experience. In certain areas of the world, the plant-based prescription is endemic, especially in developing nations.

In recent years, the World Health Organization (WHO) has focused on the traditional knowledge associated with plant-based herbal medicine as a large mainstream of rural and folk groups in various developing countries are still using herbal medicine as first safety for curing various diseases. WHO observed that approximately 80% of the world population depends on plant-based medicine for curing primary health. Sometimes, the commercialization of these plants and excessive dependency on the products from plants led to the rapid extinction of medicinal plants. The global market for herbal medicine and medicinal plants is estimated to be worth US$800 billion annually. Traditional medicines are continuously increasing due to features like safety and efficacy against diseases.

In the past few years, special attention has been paid to alternative medicine for curing ailments because of their fast reaction, but at the same time, alternative medicine or allopathic medicine have caused side effects. So, WHO has encouraged people through various programs to interact with Western and indigenous people with a view on developing new bioactive compounds that could address non-microbial origins and microbial infections and offer better, safer remedies. At the same time, the documentation of ethnomedicinal knowledge and herbal preparation is an essential approach for the use of various medicinal plant species to treat different ailments and also for conservation of ethnic sources. The documentation of traditional knowledge through ethnobotanical surveys and studies is essential for the conservation of cultural and biological diversity.

The research of natural products is based on ethnobotanical knowledge as various drugs today are developed from plants and the repository source of medicinal plants. All the primary information in drug discovery comes from the traditional source which is generally gathered from the indigenous communities. Even more essential, traditional communities, which have a lot of information to contribute to better human health, are now studied, as a central task of modern ethnobotanical surveys and research. A major portion of ethnobotanical research in recent years has been focused on a better understanding of the pharmacological effects of each of the plants separately which systematically have explored the human health effects of the ethnopharmacopoeia. Current studies examine the status of ethnomedicinal plants in phytochemical and pharmacological terms, but less frequent are studies that combine cultural and traditional approaches.

Ethnobotanical knowledge has two important aspects, (a) indigenous ethnobotanical knowledge and (b) sources of useful economically important plants. The first category will help in spreading traditional knowledge associated with plants from one to another generation whereas the second category can help in the documentation of plants through bioprospecting. The present edited book of ethnobotany in different regions of world tries to bring together all of the knowledge of ethnobotany in one place. For this purpose, competent authors have been selected to summarize the information regarding the different aspects of ethnobotany, such as ethnoecology, ethnopharamacology, ethnoconservation, bioprospectiong of ethnobotanical knowledge and its protection. Further this book provided the details information on ethnomedicinal plants used in different ways along with their essentials properties which will be helpful to the people interested in traditional ethnobotany.

The first chapter provides an ethnobotanical assessment of one of the most important but endangered plants, *Allium stracheyi* of the Central Himalayan region of India. This plant has a great position in socio-cultural, spiritual, and medicinal areas of indigenous lives of the Central Himalaya region, thus further prompting urban and peri-urban areas. It has important cultural value for the lives of the indigenous Bhotiya community. *A. stracheyi* is an indicator of the plant-based economy and traditions of Bhotiya ethnic group from Niti and Mana Valley of Uttarakhand Himalaya. Further, this species is used as a flavoring, condiment, and as medicine for cold and cough, jaundice, stomach ache, and various other ailments in Himalayan region.

The second chapter examines the use of ethnomedicinal plants of home gardens of Bhutan, which played an important role in treating major illnesses related to physical and psychological well-being in rural areas. Conservation of medicinal plants is important from the perspective of home gardening. To document how medicinal plants are used to treat different diseases, their use-value and ethical considerations are significant for bioprospecting of medicinal plants and protecting intellectual rights regarding the associated traditional knowledge. Local healers often use plants that are either available within their home garden or in nearby forests. Further, this chapter also focuses on traditional uses of medicinal plants using various tools as sample data, which have been used to facilitate ethnobotanical research on home gardens.

The third chapter explains traditional knowledge associated with plant and plant-based products for curing various human ailments and sufferings. The traditional use of forest resources, especially non-timber products such as medicinal plants, not only has deep roots in indigenous populations but is also practiced in a broad segment of culture. Different forms of medicines like Ayurveda, traditional Chinese medicine, Kampo, Unani have been practicing in recent days due to their effectiveness against various human ailments and also have blossomed into the regulated systems of traditional medicine. Further, the chapter shows the cultural relationship between plants and humans, along with the role of traditional medicines, by exploring the methodologies and various concepts for the discovery of various drugs.

The fourth chapter discusses the role of indigenous knowledge of plants and the different social institution involved in the conservation of plant-based traditional knowledge. Species extinction has increased sharply in recent decades. Most species in the world are discovered in tropical forests, which cover just 10% of the surface of the planet. Sustainable management and conservation efforts must include local communities and along with their traditional knowledge. Traditional use includes basic knowledge carried from one generation to the next. Medical use, in particular, reflects an extremely vibrant, ever-evolving process in which fresh knowledge is constantly acquired and connected to traditional methods.

The fifth chapter explains ethnomedicinal uses of *Cyperus rotundus,* one of the most important ethnomedicinal plants used in a variety of human ailments. *Cyperus rotundus* is a perennial erect sedge plant, and it is distributed in over 90 countries of the world, where it has been mostly classified as a highly invasive weed. The plant has been considered to be medicinally important since traditional times. The traditional uses of the plant in various countries include uses against various gastrointestinal tract disorders, skin diseases, leprosy, fever, and neurological disorders. Scientific evaluation of the plant, especially its rhizomes, has revealed the presence of numerous phytochemicals and wide-ranging pharmacological activities, which include anti-microbial, gastrointestinal, wound healing, anti-diabetic, anti-cancer, anti-malarial, anti-obesity, hepatoprotective and antipyretic activity.

The sixth chapter considers the therapeutic potential of ethnomedicinal plants used in different diseases. Despite little scholarly attention to the use of plants for medicinal applications, an old practice of old civilizations, they are still used around the world. According to WHO, nearly 80% of people use herbal medicine plant extracts as their primary source of health. Thereby, the ethnobotany emerges as a research field to document and understand people's knowledge of the plants and their roles in society. As a result, the contribution of the botanical field for medicine throughout evolution has been helpful and is an important alternative to use and manage plants to treat diseases in the world. Diseases like diabetes, hypertension, hypercholesterolemia, and cancer are the predominant challenges to global health, the chronic diseases accounting two-thirds of deaths worldwide. This chapter also emphasizes ethnobotanical studies on multiple medicinal plants affecting chronic diseases like diabetes, hypertension, hypercholesterolemia, obesity, cancers among others.

The seventh chapter discusses antigenotoxic and phytochemistry of six important medicinal plants of family zingiberaceae. *Alpinia galanga, A. zerumbet, Curcuma amada, C. caesia, Zingiber officinale,* and *Z. zerumbet* attracted notable attention, due to their remarkable ethnobotanical and medicinal importance, as it is well known that genotoxicity is considered a potential cause of various diseases including cancer. Herbal extracts attained a great deal of attention due to its safe and effective applications against various DNA damaging agents. However, the mechanism of DNA strand breaks by various mutagens and genotoxins is often correlated with the generation of Reactive Oxygen Species (ROS). Herbal extracts constitute a number of phytochemicals, and these are reported to have considerable antioxidant properties capable of neutralizing ROS mediated DNA damage. The family Zingiberaceae has great potential in curing various ailments due to its eminent plant species.

The eighth chapter explains the ethnobotanical, phytochemical and pharmacological investigation of medicinally important shrub *Hemidesmus indicus*. This plant has a great role in curing various diseases and has been utilized for nearly thousands of years as a vital medicine, which is also mentioned in Ayurveda. Further, its use in several ailments such as skin diseases (e.g., eczema and psoriasis), liver diseases, arthritis, asthma, epileptic seizures, bronchitis, syphilis, high blood pressure, chronic nervous diseases, rheumatism, impotence, and immune disorders is discussed. The root of *Hemidesmus indicus* has diaphoretic, tonic, depurative, alterative, and diuretic properties. Recent experimental data proved that this shrub also has antihyperlipidaemic, anticancer, renoprotective, antidiarrheal, hepatoprotective, antiulcerogenic, antileprotic, antiatherogenic, antioxidant, analgesic, anti-inflammatory, antipyretic, immunomodulatory, and antimicrobial properties. *Hemidesmus indicus* is rich in a wide range of biologically active phytoconstituents, such as pregnane glycosides, steroids, terpenoids, aromatic aldehydes, lignans, saponins, flavonoids, and aliphatic acids, which play vital roles in its pharmacological properties.

The ninth chapter focuses on the chemical constituents and anti-inflammatory properties, along with the ethnomedicinal aspects of *Aloe barbadensis. Aloe vera* is a perennial, drought-resisting, succulent plant belonging to the Asphodelaceae family that has a long history of having many tremendous medicinal and anti-inflammatory as well as anti-bacterial properties. This chapter highlights the great potential of the *Aloe vera* species. The study also contends that horticulturists and plant scientists in agriculture and medicine must pay attention to the *Aloe vera* as a medicinal plant that possesses the most powerful antibacterial and anti-inflammatory properties. It also protects the body against skin and worm infestations, as well as constipation, by enhancing the metabolism.

The tenth chapter focuses on the ethnopharmacological properties of family Asteraceae. Asteraceae is one of the widely distributed family as large numbers of plants occur throughout the world. All of these plants have been utilized in various skin-related problems. *Ageratum conyzoides, Artemisia absinthium, Bidens pilosa, Calendula officinalis, Eclipta alba, Grangea maderaspatana, Helianthus annus, Parthenium hysterophorus, Sochus oleraceus, Sphaeranthus indicus, Tridax procumbens, Tagetes erecta,* and *Vicoa indica* have provided incentive for the medication advancement in the present ordinary restorative framework. Moreover, the treatment of epilepsy, gastric problems, enlargement of spleen, urinary disorders, wound healing, chronic fever, swellings, and inflammation of the liver is based on traditional knowledge provided by studies of an organized survey of each plant.

The eleventh chapter describes important ethnomedicinal plants regarding herniated discs in Turkey, a major issue there with a very costly treatment. Turkey has a rich diversity of ethnomedicinal plants, so the therapeutic properties of these plant species have been known and used by humans since ancient times. Ancient people had no real scientific knowledge on diseases and treatments and tried to find solutions to these problems through experience and theories about their experience. Over time, several therapies emerged to treat a herniated disc, ranging from conservative treatment to minimally invasive and percutaneous techniques to open surgical methods. This chapter examines recent studies and compiles ethnobotanical uses for the treatment of herniated discs along with their local names, parts, and usage forms in Turkish traditional medicine.

Chapter twelve discusses important aspects of ethnomedicinal plants in the treatment of back pain. Around 250 herbal drugs were used during in early Mesopotamian civilization. In ancient Greece, 60 medicinal plants were known; at the time of Arab-Persian civilization, this amount increased to about 4,000. At the beginning of the 19th century, the number of known medicinal plants reached 13,000. These herbal drugs are used in the treatment of diseases ranging from diabetes to cancer. Back pain and related symptoms rank among the second most common affliction of humankind. One proposed treatment method to control and treat back pain symptoms is traditional folk medicine. This chapter examines recent ethnobotanical studies and compiles medicinal plants used as Turkish herbal treatments for back pain.

Chapter thirteen of this book explains traditional uses of plants in the treatment of Leucorrhoea. Leucorrhoea is a common disorder prevalent in females. Ethnic communities still treat it by plants that grow in their surroundings, due to a lack of sufficient medical facilities or costly modern treatment. Therefore, an attempt was made to look for plants that are employed in traditional ethnomedicine and the Ayurveda system of medicine for treatmenting leucorrhoea in India. This chapter reveals the use of 345 plants by various folk communities and 31 unique plants from Ayurveda, which gives a figure of 376 plant species prevalent in Indian traditional medicine for treatmenting leucorrhoea. This chapter is also helpful in screening potential ethnomedicinal plant species for carrying out further scientific validation studies to assess their efficacy in leucorrhoea, which may lead to the development of novel phytotherapeutic molecules for effective treatment.

Chapter fourteen deals with wild plants used by the local inhabitants of Champawat district of Uttarakhand to livelihoods and sustainability. Local inhabitants are very poor and are backward in terms of education. They do not have any permanent source of income. Therefore, they are totally dependent on various forest and plant-based forest products as well as various ecosystem services (selling prepared herbal medicines, wild edible fruits, etc.), approximately 45–50% of their total income per year. So, ethnomedicinal plants are playing a vital role in improving their livelihoods. This chapter focuses on the status of locals along with their socioeconomic status and dependency on plants for their food security.

Chapter fifteen deals with plant species used as ethnomedicine by the inhabitants of Gulmarg area of Kashmir Himalaya. The Kashmir Himalaya is an important Indian State which has a great diversity of medicinal plants. Further the area also serves greatly to the local inhabitants both in terms of food security and ecosystem services. This chapter also documents the indigenous knowledge of the area associated with ethnomedicinal plants. As the indigenous knowledge of plants is being eroded in younger generations, this present chapter could serve in filling this gap.

Akash
Gurukul Kangri University, India

Navneet
Gurukul Kangri University, India

B. S. Bhandari
HNB Garhwal University, India

Section 1

Traditional Bases of Ethnobotany

Chapter 1
Ethnobotanical Assessment of Endangered *Allium stracheyi* (Baker):
A Lesser Known Species of the Nanda Devi Biosphere Reserve, Central Himalaya, India

Kusum Payal
https://orcid.org/0000-0001-8518-5128
Guru Gobind Singh Indraprastha University, New Delhi, India

Kottapalli S. Rao
Department of Botany, University of Delhi, Delhi, India

R. K. Maikhuri
G. B. Pant National Institute of Himalayan Environment and Sustainable Development, Srinagar Grahwal, India

L. S. Kandari
Graphic Era Hill University, Dehradun, India

ABSTRACT

Ethnobotanical and economic studies endorse Allium stracheyi Baker, of the Alliaceae family, for future food and health security. The species is important to the socio-cultural, spiritual, and medicinal lives of the indigenous Himalayan people and those in the surrounding urban and peri-urban areas. The species is used by the indigenous Bhotiya people as a flavoring, spice/condiment (Jambu Faran), and a remedy for colds/coughs, jaundice, stomachaches, and various other ailments. A perennial, medicinal, and

DOI: 10.4018/978-1-7998-1320-0.ch001

aromatic plant (MAP) species, A. stracheyi is harvested two or three times annually and is categorized as an endemic and endangered species. It has a high value and represents an important cultural element in the lives of indigenous Bhotiya folk living in the mountains of Central Himalaya, India. Among medicinal and aromatic plants, A. stracheyi is an economic indicator of the plant-based economy and tradition of Bhotiya ethnic groups in the Niti and Mana Valleys of Uttarakhand, India.

INTRODUCTION

The high-altitudes of Himalaya are extremely rich in plant life, abounding with a genetic diversity of medicinal plants, a wild relative of crop plants, and other economically-important species. These fragile ecosystems are marked by rareness of topography, climate, soil, vegetation, and ethno-culture (Hajra & Jain, 1983; Negi Tiwari, & Gaur, 1987; Rikhari, Negi, & Singh, 1993; Sundriyal & Sharma, 2016). Uttarakhand state lies in the Central Himalaya, which is rich in biodiversity, as this area exhibits a large number of plants and animals (Rau, 1975). Besides providing essential commodities to the large population living in the vicinity, including wildlife and livestock, a large number of plants have been fulfilling the demands of various industries based on natural products, such as dyes, cosmetics, tannins, resins, and medicinal plants, etc. The high altitude is a rich repository of medicinal plants being used by the indigenous people and the pharmaceutical industry in the form of advanced drugs. While there are innumerable medicinal plants in these ecosystems, *Allium stracheyi* Baker plays a significant role in human life, in general, and among indigenous people, in particular. All major pharmacopoeias of Asia recognize the role of *A. stracheyi* Baker in their holistic medicine system (Lama, Ghimire, & Aumeeruddy, 2001).

Although the wild *A. stracheyi* Baker, according to the Red Data Book, is a threatened species that grows in the alpine areas of the Himalaya (Nayar & Shastry, 1990), it has also has been brought under cultivation, but, only on a small scale. The Bhotiya indigenous people of the Central Himalaya use it as a substitute for onions and garlic. Locally, it is known as *Jambu Faran,* which is used to cure a variety of respiratory and gastrointestinal diseases. The plant is used as a green leafy vegetable and its hammered, dry, leaves are used as a spice and condiment in the Himalayan region (Nautiyal, Maikhuri, Semwal, & Rao, 1998). The pungent sulphur smell of the leaves repel insects and is used as an insect repellent in the region. The demand for *A. stracheyi*-based spices has also amplified with the increased human population. The plant is used as a green leafy vegetable in soups and curries. The dry, processed, leaves are the key spice/condiment for Bhotiya's indigenous people's culinary use, and the commercial product is known as *Jambu Faran*. The increased demand has resulted in loss of natural habitat of the plant species, however, the environmental and climatic conditions of this region are very suitable for several lesser known *Allium* species of high economic value. The area is encircled by India's second highest peak, 'Nanda Devi,' situated in the Nanda Devi Biosphere Reserve (NDBR).

The indigenous Bhotiya people are the real custodians of these Himalayan alpine resources. The Bhotiya people of the Garhwal Himalaya inhabit the Chamoli district of the mountainous Uttarakhand state in India. The Buffer zone of NDBR, particularly in the Niti Valley, located 2,000 to 4,000 meters above sea level (m a.s.l.), represents an underprivileged population with small land holdings and a very low (43.29%) literacy count (Nautiyal et al., 1998). The Bhotiya community is comprised of three small sub-communities: Tolcha, Marcha and Jadh. These subcommunities have two dwellings and practice seasonal migration every year from the higher Himalayan zone (2,000-3,600 m asl) to the lower Himalayan elevation (400-1,400 m a.s.l.) in winters and again back to higher Himalaya during summers. During seasonal migration local people practices small business for livelihood options and *A. stracheyi* is sold among customers with local name *Jambu Faran* as a spice and condiment (Nautiyal et al., 1998; Nautiyal, Maikhuri, Rao, & Saxena, 2001). The Marcha sub-community and Tolcha sub-community are semi-agrarians (practice agriculture in May through October) whereas Jadhs sub-community is semi pastoral (practice animal rearing in alpine grasslands May through September). The semi-agrarian communities grow lentils May through October and, in parallel to this, some family members collect leafy vegetables, fuel, fodder, spices, edible fungi, and medicinal herbs from the wild for the rest of their household needs throughout the winter. These fresh products are mostly dried in the sun in low atmospheric temperatures (16-25°C) so that the food value and fungal and bacterial infection-fighting properties are not altered too much. All three of the underprivileged communities are dependent on nearby forests to sustain their households and they collect a variety of non-timber forest produce (NTFPs), such as *Betula utilis, Hippophae rahmnoides, Morchella esculenta, Prunus persica, Juglans regia, Corylus cornuta, Diplazium esculentum, Taxus wallichiana,* and *Viburnum grandiflorum*. All collected plants, spice, fodder, medicines, and mushrooms are dried and stored until early winter arrives and are consumed by the Bhotiya people when the whole region is snow-covered.

These products also play a vital role as a means of income and gift material to kinsfolks. All this provides a supplementary income source to the Bhotiya indigenous folk during the summer season (May through October) following barter system and monetary purchases, as per business prospects. The indigenous people exchange spices and other forest products for edibles (pulses, rice, wheat, and oils, etc.) from other parts of the state. In this system, the *A. stracheyi*-based spice, *Jambu Faran,* is popular, high in demand, and is presently playing a key role in their economy. *A. stracheyi* has been cultivated in the Niti Valley for 70 years, initially by low volume in comparatively lower altitudes (2,650 m a.s.l.) than its natural habitat (Nautiyal, et al., 1998). The natural alpine habitat of *A. stracheyi* is around 2,900-3,000 m a.s.l. in the Niti Valley in which the plants can be acclimatized, approximately 2,300 to 2,800 m a.s.l. in the nearby terraces of the village (Payal, 2007).

The Nanda Devi Biosphere Reserve (NDBR) is a complex and unique habitat of medicinal and aromatic plant species having high economic value. These medicinal plants play a significant role in the traditional system of medicines, and some of the plant species among them are in the list of rare and endangered species, that includes *Podophyllum hexandrum*, *Aconitum* spp., *Dactylorhiza hatagirea, Saussurea obvallata*, and *Nardostachys grandiflora* (Nair & Sastry, 1987). Besides the aforementioned herbaceous species, some tree species, such as *Taxus baccata* and *Betula utilis,* locally known as *Thuner* and *Bhojpatra,* respectively, are also under severe threat of depletion due to on-going anthropogenic pressure. *T. baccata* is used for local beverages and for its medicinal value in treating tumours and cancers. On the other hand, *B. utilis* is used for religious and aesthetic purposes and is an economically important species (Purohit, Maikhuri, Rao, & Nautiyal, 2001).

Most of the areas under *A. stracheyi* cultivation are generally planted on steep slopes in small pockets. As the demand is increasing in national and local markets, it may have serious implications on the survival of the *A. stracheyi* species belonging to the family Alliaceae, which is already in on the endangered plant list. The price of spice is 300 to 500 Rs/Kg ($4,18 to $6.95 per kilogram in US dollars), whereas during times of scarcity prices go up to 600 to 700 Rs/kg. ($8.36 to $9.75 per kilograms) Locally people purchase three pathi, a local measure weighing around 850 grams of dry spice for an average size family of four for a year's consumption (Payal, 2007). It is reported that *A. stracheyi* Baker is also distributed across the Indian Himalaya, including Jammu, Kashmir, Himachal Pradesh, Uttarakhand, Sikkim, and the high altitudes of the north-eastern region (Negi et al., 1987). There are also records of *A. stracheyi* being distributed in the Nanda Devi Biosphere Reserve (NDBR) and adjoining areas (Nautiyal, et al., 1998; Payal, 2007; Payal, Maikhuri, Rao, & Kandari, 2014). All the parts (flower, shoot, leaves, seeds, and roots) of the *A, stracheyi* are eaten as a vegetable and also have multi-uses (medicine, spice, condiment, and food). In addition, *A. stracheyi* is a rich source of carbohydrates, vitamins, and minerals. Not only the leaves, but also the fibrous roots provide energy; Bhotiya indigenous folk eat the raw bulbs during trekking.

STATUS OF *ALLIUM* AND MAPS IN ALPINE AREAS

Allium plants have been traded throughout the Himalayan region and the indigenous Bhotiya people have been collecting *Jambu Faran* throughout the alpine zones for centuries, because of the great trade value the plant provides (Atkinson, 1882). Maikhuri, Nautiyal, Rao, and Saxena (1998) and Payal (2007) reported that the area cultivating the *Allium* species was higher than any other medicinal plant species cultivation by the Bhotiya people in the Niti Valley. In this context, Silori and Badola (2000) reviewed the cultivation of *A. stracheyi* in five villages in the Pithoragarh district of Uttarakhand. In these villages, 90% of the families were found to cultivate medicinal plants, including *Aconitum heterophyllum, Angelica glauca, Picrorhiza kurrooa,* and *Pleurospermum angelicoides,* which are the leaves and edible parts of *Allium stracheyi, Rheum australe* (Kandari et al. 2012), *R. moorcroftianum.* Also, the fruits of the *Podophyllum hexandrum* are used in the high Himalayan region (Samant & Dhar, 2000; Dhar, et al., 2002).

Medicinal plant cultivation is emerging more and more with time and will continue to be the most profitable business in the future for the Himalayan states. The maximum productivity per unit area of higher altitude was recorded for *Angelica glauca* at 593 kg/ha/year, locally known as *Gandhrayan,* while the minimumrecorded for *Allium carolinianum* was 79 kg/ha/year. *P. kurrooa* and *R. emodi* were observed for economic viability compared to lower altitudes and found suitable for cultivation up to 1,800 m a.s.l. (Nautiyal & Nautiyal, 2004).

There are two species of *Allium, A. stracheyi* and *A. humile,* locally called *Jambu Faran* or *Tibetan Faran.* In the region, nearly 76% families cultivate these species to fulfill medicinal and other economic interests. The flavor of *Jambu Faran,* which is made from *A. starcheyi,* has a pleasing aroma and is very much liked and endorsed by the indigenous Bhotiya folk. Apart from anthropogenic pressure, other factors, such as natural disasters, pathogens, lack of pollinators, and invasion of exotic and aggressive species, affect the population dynamics of these valuable endemic Himalayan regions (Dhar, 2000; Maikhuri, et al. 1998).

Cultivation of these *Allium* species is the only viable option. Still, in the majority of the villages, most of the collection of all of these medicinal and aromatic plant species is from the wild (Maikhuri et al., 1998), which may cause the complete eradication of these *Allium* species (which are only 4 to 5 in number) from the region, which would be an irreversible loss of valuable MAPs resources now and forever (Nautiyal et al., 1998; Negi et al., 1987; Payal, 2007). There were no appropriate agro-techniques available for massive propagation and cultivation of these species until year 2000, but now large-scale cultivation packages are available as a result of in depth research work and medicinal plant-based projects that were carried out in the region by government agencies using appropriate propagation techniques via seed, bulbs, and plantlets, under polyhouses and shade houses in the alpine regions (Payal, 2007).

GENESIS, CULTIVATION EFFORTS, INDIGENOUS SOCIO-ECONOMY AND CULTURAL BACKGROUND

The middle Asiatic countries are considered the primary centers of the origin of genus *Allium*. In India, evidence by references, is found in the "*Charak Samhita,*" one of the ancient texts of the medical tradition in India since time immemorial. It has been in cultivation for over 5,000 years and use of its wild form is uncertain., In the Niti Valley, most of the villages cultivate various medicinal plants, but buffer zone villages, namely Kosa and Mahergaon, depend completely on the alpine regions for plant collection due to failed cultivation (See Figure 1). In the higher altitudes of the Himalayas, the wild plant is used by the locals as a substitute for onions and garlic, locally referred to as *Jambu Faran*. Bhotiya indigenous folk are known for their herbal medicines and spices; MAPs play a key role in their economy. The most popular species with the Bhotiya is *A. stracheyi* vern *Jambu Faran* due to its pleasing aroma and taste. Traditional *Jambu Faran* growers have been cultivating this particular species as a kitchen garden crop since 1945. Into the beginning of 1940s, the occurrence of *Jambu Faran* was rare with the inhabitants of the Niti Valley and was considered an exotic spice/condiment, especially if imported from the Khampas/Dulbas communities of Tibet, and was a medium of barter.

The pastoral communities of the Garpak and Kaga villages of the Niti Valley, who live in the alpines rearing sheep, were the first to cultivate *A. stracheyi*. The initial experimental plants were collected from the Sumna Valley and planted in the growers' kitchen gardens for better plant management. Although the growers could not address species conservation on a large-scale due to a lack of suitable conditions, such as altitude, temperature, and soil conditions (Payal, 2007). These growers introduced the people of the valley to *Jambu Faran* cultivation and encouraged the local people to identify seven varieties of wild *Allium* bulbs in their adjacent high alpine altitude regions of Mallari, Kailashpur, Himtoli, and Suki. It

Figure 1. Percent share of Allium spp. with other MAP species for different surveyed villages in Niti Valley, Nanda Devi Biosphere Reserve, India.

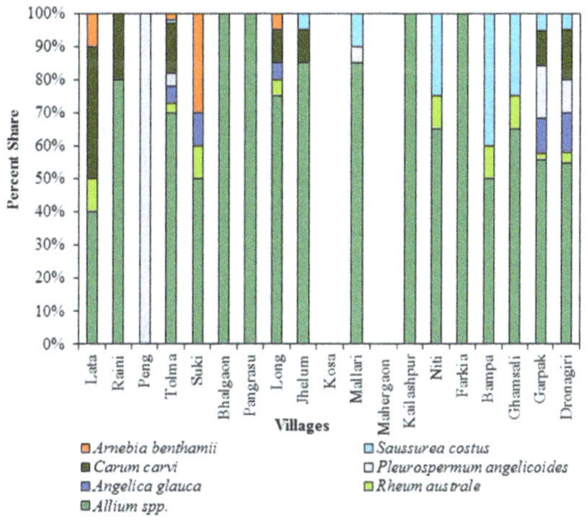

was a very slow process of cultivation for independent growers of *Jambu Faran*, who often achieved less success. The farmers from the of villages such as Tolma, Garpak, Dronagiri, Jhelum, Long, and Suki, made continuous efforts in this direction achieved success to some extent by bulb propagations.

Some villages are blessed with naturally suitable soil conditions for *Jambu Faran* cultivation. The 98% collection of MAPs was totally from wild habitat of the species. Until the 1970s, the cereal-dominated agriculture environment slowly started being transformed into a high-value, cash-crop-based, commercial agriculture with the increased production of kidney beans, potatoes, oil seeds, and horticulturally based crops. A revolution came in the form of herbal spices like *Allium stracheyi*, *angelica glauca, pleurospermum angelicoides, and rehum austral,* etc., after scientific inputs from government agencies for mountain economy development (Raghupathy, 2001; Maikhuri et al., 2003). Traditional crops, such as *Eleusine coracana, Fagopyrum esculentum,* and *Panicum miliaceum,* etc., have been reduced 25% to 50% under cultivation for the last three decades, which has resulted in an intensive cultivation of MAPs, especially *Allium stracheyi,* through their traditional practices.

Ethnobotanical Survey and Data Collection

The ethno-botanical facts presented in this chapter were collected during the years 2003 to 2010 at an interval of six months. Community wise, small meetings for talks on multipurpose medicinal plants were organized through outreach programmes for all species-related validation. In the region, inhabitants have knowledge of the edible and medicinal properties of the region's plants, and every household indulges in collection and cultivation of these medicinal plants. Herbarium catalogues and photographic images of all plants were prepared to identify species. Informants were interviewed twice using structured and semi-structured questionnaires. This helped to develop a checklist of ethno-botany purposes and their ethno-medicinal uses.

Climate And Soils Of The Region

The climate of the area is typically temperate, monsoon dependent, and divisible into three seasons, namely long winter (Oct.-April), short summer (April-June), and rainy (July-Sept.). The buffer zone of the Nanda Devi Biosphere (NDBR) is situated between 30°17' N and 30°41' N latitude, and 79°40' E and 80°05¢ E longitude, and covers an area of a 624.62 km^2 core zone and a 1612.12 km^2 buffer zone in the Garhwal region of the Central Himalayas in the Chamoli district of Uttarakhand, India (Purohit, Maikhuri, Rao, & Nautiyal, 2001). The total area of the NDBR, by adding the Valley of Flower National Park as the second core zone, has been given the status of a World Heritage Site by the United Nations Educational, Scientific, and Cultural Organization (UNESCO) as having cultural, historical, scientific or other form of significance, and which is legally protected by international treaties (Maikhuri et. al., 1998). The whole area is a rich repository of *A. stracheyi* and other wild medicinal species. People harvest *A. stracheyi* from alpine to subalpine area and use it in spices, foods, and medicines. Now, the total area of the NDBR has increased from 2236.74 km^2 to 630.33 Km^2. A total of 47 villages are situated in the buffer zone of the NDBR of which 34 villages, including the Badrinath shrine, lie in the Garhwal region (Chamoli district) and 13 villages in Pithoragarh and Bageswar districts of Kumaun region of Uttarakhand (Figure 2). The human settlements in the Chamoli district of the NDBR are located between the 2,200-3,600 m a.s.l. elevations (Maikhuri et al., 1998). The north of the Higher Himalaya region is known as the cold and arid zone of Trans Himalaya. The microclimatic conditions typically differ from

Figure 2. Map (not to scale) A stracheyi cultivation in the Nanda Devi Biosphere Reserve, India

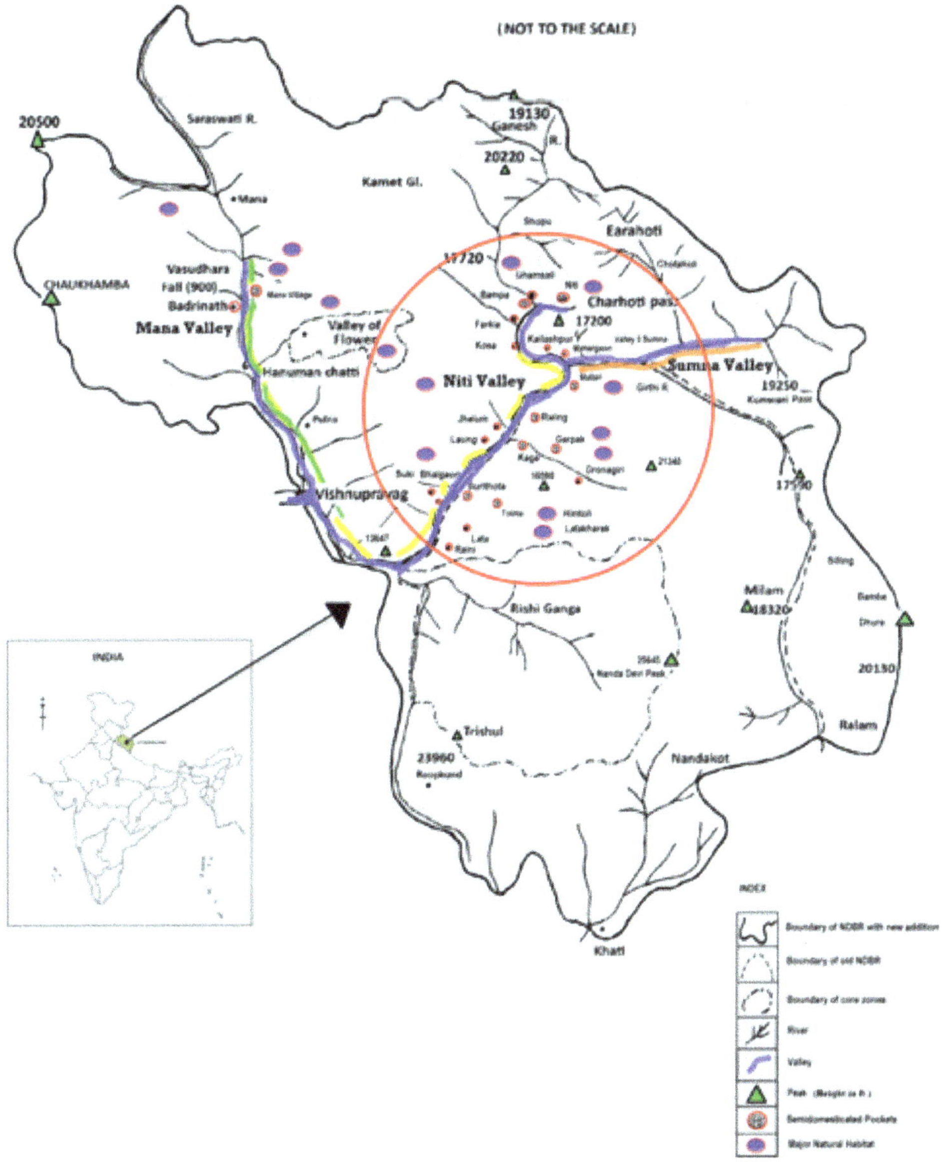

valley to valley and locality-to-locality according to: 1). Direction; 2). Degree of slope; 3). Aspect; 4). Forest vegetation and cover strength and; 5). Distance from glaciers. Parent rock material is crystalline rock, which includes granetiferous mica schists, garnet mica quartz schists, and mica quartzite. The soils in general are deep in the agricultural lands, black in color, loam to sandy loam, and drainage is well to excessive.

The Custodians of Traditional Knowledge

The target group of Bhotiya indigenous folk of the Garhwal Himalaya live in the northern most mountain border areas of the Chamoli district in the Nanda Devi Biosphere Reserve (NDBR), particularly in the Niti valley (2,000 to 4,000 m a.s.l.) representing a primarily poor populace of 10,069 people with small land holding and a very low literacy percentage (43.29%). The Bhotiya community is comprised of three small subcommunities: Tolcha, Marcha, and Jadh. These subcommunities have two dwellings and practice a transhumant life migrating every year from the higher Himalayan zone (2,000-3,600 m a.s.l.) to the lower Himalayan zone (400-1,400 m a.s.l.) in winters, then back to the higher Himalaya regions during the summers. The Marcha and Tolcha subcommunities are semi-agrarians, whereas the Jadhs subcommunity is semi-pastoral, and all the three communities are wholly dependent on the surrounding forests for their daily necessities. Most of the area agricultural areas are rain-fed and generally cultivated on steep slopes, and very limited areas are under irrigated agriculture. The main occupation

Table 1. A. stracheyi and its elevation-wise occurrence, domestication statistics, harvesting technique, and Bhotiya indigenous folk in the buffer zone villages of the Niti Valley in the NDBR.

Parameters	Lower Areas	Higher Areas
Altitude (m above sea level)	1900-2700	2700-3600
Transhumance	Not Practiced	Practiced
Cropping pattern	3 crops/ 2 years	1 crop/ year
Main occupation	Agriculture	Agriculture
Subsidiary occupation	Animal husbandry, Non Timber Forest Products collection, Woolen carpet production	Animal husbandry, Non Timber Forest Products collection, Woolen carpet production
Villages (Practicing Transhumance)	Tolma, Lata, Reni, Surithota, Peng, Bhalgaon, Suki, Phagati, Parsari, Markuda and Badagaon,	Mallari, Dronagiri, Jhelum, Kaga, Garpak, Ghamsali Bampa, Niti, Long
Cultivation of *A. stracheyi*	Lata, Raini, Bhalgaon, Tolma,	Mallari, Dronagiri, Jhelum, Kaga, Garpak, Rewing, Ghamsali Bampa, Niti, Long
Harvesting frequency	3-4 harvest	2 harvest
Harvesting Months	May, July-August, October	June-July, September

of inhabitants in the NDBR is livestock rearing and practicing marginal agriculture, in addition to a few opportunities as daily wage labor in government's professed development schemes (Table 1).

Concept of Sacred Land and Community Conservation

The entire area is considered Nanda Dev Bhoomi (*The land of Nanda Devi*) and is treated like sacred grooves around Nanda Devi, the second highest peak of India, which is also considered a goddess among the indigenous Bhotiya folk. Local organic farm yard manure (FYM) is prepared (cow dung + soil + leaf litter + water) and is transported to tiny crop fields located in far-flung hill slopes. However, some of the new generation's progressive farmers make organic compost pits near the cattle sheds and transfer the cattle dung and fodder residue to the compost pit occasionally for a period of one year (Decomposition is slow due to low atmospheric temperature.). Decomposed FYM is transferred to the crop fields and disseminated at the time of plantation bed preparation.

Ethnobotany and Botanical Attribute of *A. stracheyi*

A. stracheyi is a herbaceous plant and its dark green leaves are linear in shape (20-45 cm. in length). Its flowers are purplish-pink in color and its seeds are tiny and black (Table 2). The leaves and flowers are used as spices, medicine, and green vegetables. The bulbs are small and robust and can multiply. Its leaves are believed to assist healing and are regularly used to treat patients with Jaundice (Table 3). The plant is also used as a cough and cold decoction. The most important pharmacological functions attributed to *A. stracheyi* are antimicrobial activity and pain relief when taken as a decoction. The ethnomedicinal value of this endemic plant is very high and promising. The traditional use knowledge of *A. stracheyi* among the Bhotiya people in the Himalaya has unique dietetic-use categories, which include using the leafy vegetable as food (curry and soups), a condiment, and a spice (Table 4).

Propagation Techniques Using Plant Parts

Bulbs, bulbils, and the seeds of the *A. stracheyi* plant are all used. *A. stracheyi* bulbs (single or two pieces in a vertical section) and bulbils are normally transplanted/directly seeded in the spring (April-July). They require a moist and shady place for the first two months. Irrigation once or twice a week is also required for its cultivation. Moisture holding soil is considered good for propagation of the bulbs, bulbils, seeds, and cuttings. The seeds should be soaked either in cold or hot water (warmed up to 80°C) and GA_3 (up to 100 ppm). Pre-soaking in a 100 mg/l GA_3 solution and incubated at 20°C under light conditions enhances the germination significantly (Payal et al., 2014). The plantlets grow best at a depth of 10-15cm, in well-drained, sandy loam soil, with ample organic matter. It thrives best at a pH of 6-7. Before planting the roots in the rainy season, the fields are well prepared by ploughing until the soil is soft. A larger quantity of bio-compost (250 q/ha) duly prepared by using dry leaves, soil, and cow dung, is required to prepare the land for planting (Payal, 2007).

Rotation and Transfer of Bulbs of *A. stracheyi*

The *A. stracheyi* field should be rotated after 20 years. The bulbs of *A. stracheyi* can be propagated in soil during late February and March. Appropriate care of nutrient and soil moisture is a must during

Table 2. A. stracheyi growth and harvest status in the buffer zone villages of the NDBR situated at lower and higher altitudes

Attribute of *A. stracheyi*	Value of Parameters (± SE)
Composite flower/ mature plant	1.00 (±0.00)
Flowers/ composite flower	37.25 (±9.98)
Max seeds formation /capsule	6.00 (±0.00)
Occurrence of mature seeds/capsule	4.18 (±0.75)
Seed maturation time	Sep.-Oct.
10 Seed Weight in grams	0.022 (±.0009)
% Moisture (Dried at 60° C temperature)	67.6 (±0.52)

propagation to prevent thin bulbs from wilting. In a prepared planting bed, the transfer of bulbs should be done approximately three inches deep, and spaced four to six inches apart during the first year. The roots should be buried and just the stem of the bulb should be above the surface. The spacing between plants should be at least four to six inches so that moisture retention and shade will be available to all of the plants. after one to two years, the bulbs can be transferred to a 30 centimeter spacing. Planting bulbs at the proper depth is important for bulb survival. If space is limited, clusters of 5 to 10 plants can be assembled together in the soil. Mulch the planting bed with biocompost in at least a two to three-inch layer to prevent moisture loss and nutrient status improvement (Payal, 2007).

Package of Practices With New Dimensions

A. stracheyi requires dry, cold, climatic conditions and prefers growing in alpine to sub-alpine regions. Agro-techniques are available for massive commercialization of these species (Payal, 2007). Presently, there is a mismatch between a growing demand and a diminishing supply of *A. stracheyi*-based spices in the region. Forest societies, in a majority of cases, adapted to isolation to preserve their culture, a tendency very uncommon with urban people. Moreover, wild alternatives of some edible plants have medicinal value and proved to be income-generating for the rural and tribal populace. The role of the *A. stracheyi* species in the local healthcare system, food habits, and socio-economic development of the indigenous people is very important. Low-cost, scientific intervention, to traditional agronomic practices via bulb or seed propagation, strengthens the market and encourages farmers towards large-scale cultivation. All the cultivation efforts help in the conservation of the species in situ and ex-situ farm conditions.

Table 3. Ethnomedicinal value of species A. stracheyi in Nanda Devi Biosphere Reserve, India

A. stracheyi Baker	Ethnomedicinal Value
Part used	Green linear Leaves and Roots
Disease Cured	Asthma, Flu fever (Influenza), Cold and cough (Allergic), Gastric disorders, Antiseptic, Cancer, Appetizer, Blood Purifier, jaundice
Medicinal Properties	Leaves have sweet and acrid taste with warm potency, used in wind disease (Lama, 2000). Stimulate the appetite and cures gastric ailments, lung and nasal infection controls cold, heals wound, checks bleeding, purify the impure blood. The roots have highly expectorant properties and cures acute asthma when taken daily.
Indigenous Use Method	The decoction of green leaves nourishes the blood and body, cures lung infection, and daily leaves decoction or soup intake is prescribed in Flu, Cold and cough, Bronchitis, Asthma and jaundice. Local leaves juice application on cuts and wounds gives antiseptic. Anti-inflammatory and Analgesic Potential of Leaf Extract (Ranjan et al, 2010)
Veterinary purposes and Uses Method	Used in some veterinary diseases like Diarrhoea or loose Dung, Flu conditions, Stomach pain Lung and Liver disease. In stomachache long boiled warm leaves (Plantlet) are used externally on stomach of livestock and for flu conditions leaves cut in to pieces and boiled in water with cattle food mixture.

The cultivation *of Jambu Faran* was remarkably started in the 1960's, while some earlier evidence was also found in the region. Until the 1980's, *Allium* spp. cultivation was on small-scale basis by the Bhotiya tribes. In this regard, through scientific efforts, different low-cost agro-techniques were introduced, such as seed germination, vegetative propagation, habitat study, and information about the impact of different microclimatic conditions on this species were introduced to the region. Now, it has grown to a larger scale by the people in the villages, namely Tolma, Garpak, Kaga, and Dronagiri of the Niti Valley. Percent consumption and percent of total families cultivating *A. stracheyi* with other MAP species in the Niti Valley, NDBR is provided in Figure 3.

CONCLUSION AND RECOMMENDATIONS

Even before and after the creation of the Uttarakhand state, medicinal plant cultivation and conservation has clearly been identified as a potential resource for elevating the state's economy. The hill state of

Table 4. Ethnobotanical and traditional use knowledge of A. stracheyi by means of various dietetic use categories (food, spice) among the indigenous Bhotiya folk in the Himalaya

A. stracheyi	**Food value and dietetic use method**
Part Used	Leaves, flower, Seeds, Green Vegetable,
Use Category	Soup, Culinary spice, Spice, Condiment, Curries
Major Dietetic Use	Green leaves cooked as vegetable with salt and pepper. Sundried leaves are reckoned in oil for strong aroma and alluring flavour to prepare cuisine viz. fried rice, curries and seasoning ingredient to curd preparations. Dry flowers also used for sweet and strong aroma and known as *Kaseri Pharan*.
Spice processing method	For condiment green leaves are sundried and simultaneously hammered to flatten and processed and left for slow drying in shade.
Other usage	Black small seeds are used in pickles. A distinct type of salt seasoning is made from green leaves of *A. stracheyi* fusion with other herbs. The plantlets of *A. strachyi* is grinded with common salt, green/red chillies, seeds of *Carum carvi*, and dry root powder of two other medicinal herbs from the region known as *Angelica glauca* and *Pleurospermum angelicoides*. The preparation of flavoured salt is known as Dongcha.

Uttarakhand was identified for medicinal plants cultivation and it is marketed as an important aspect of the state's economic sector. Considering biodiversity and its importance in the Central Himalaya, conservation of all of the sectors that are directly or indirectly related to the life of the communities residing in the mountainous regions has been recommended. Forest diversity is the primary source of livelihood of the people living in Central Himalaya. For the past few decades, due to population pressure and over exploitation, forest density has diminished, which is a serious issue (Sundriyal & Sharma, 2016). The theme of medicinal plant cultivation has been prioritized since its origin as a state on November 9, 2000. The resources developed and programs designed for the application and diffusion of agro-technologies need to be augmented. Efforts to develop planting materials, processing, and semi-processing facilities at important collection and cultivation points to add value to the raw materials, proper and better marketing opportunities through farmer cooperatives, and appropriate insurance schemes to reduce the farmers' risks in cultivation need to be promoted. However, the previous studies carried out, particularly on the

Figure 3. Percent consumption and total families (percent) cultivating A. stracheyi with other MAP species in Niti Valley, Nanda Devi Biosphere Reserve, India.

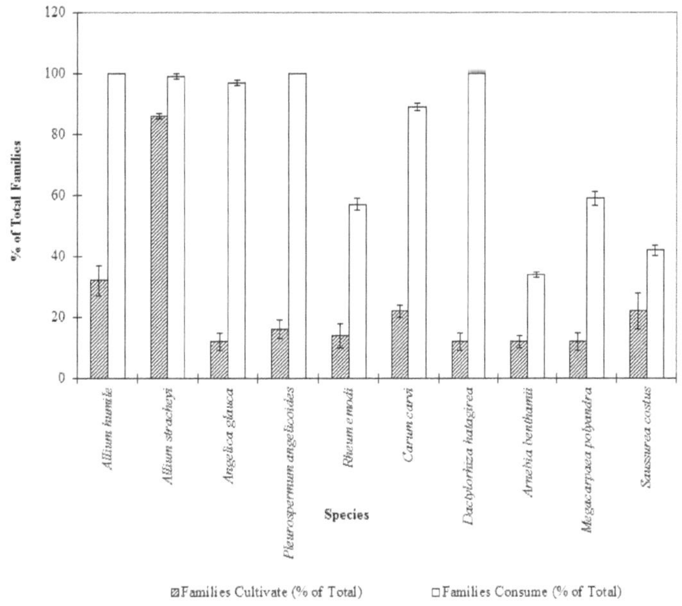

medicinal and aromatic plants (MAPs) in the region, suggested *Allium* spp. as a potential MAP suitable for high altitude areas for large-scale domestication/cultivation.

Appropriate agrotechniques are now available for the large-scale cultivation of these species, and with realization as to the impact of species cultivation on the socio-economic development of the local communities,' attempts were made to develop appropriate packages of practices based on in-depth scientific experimentation for large-scale cultivation. Initiation of large-scale cultivation of the valuable *A. stracheyi* that are a high-value, low-volume crop, will not only improve the socio-economic conditions of the Bhotiya tribe, but will also help with conservation of these species in their natural habitat (Payal, 2007, 2014).

The government's policies have so far had a marginal impact on the cultivation and marketing of medicinal plant produce and have, thus, indirectly, favoured illegal collection from the forests. Effective mechanisms need to be identified to promote and sustain institutional collaboration and networking. A cross-sectoral approach needs to include mechanisms for cultivation, conservation, management, marketing, etc., which can strengthen the economy of the people while conserving natural resources. However, the majority of the farmers/local people are currently facing problems/risks for undertaking the large-scale cultivation of MAPs due to following reasons:

- Lack of consistent support in terms of technical knowledge.
- Little experience of capacity building for market and resource management.
- Problems in getting planting materials (seeds/seedlings) to start nursery raising and cultivation.
- Lack of support for the purchase of planting materials, land improvement, etc.
- Lack of promotional activities to familiarize farmers with potential of MAPs as cash crops.

Therefore, there is a need to develop location-specific technologies to maximize the use of local resources. Value addition to *A. stracheyi* can upgrade the socioeconomic status of the indigenous Bhotiya people and can become a significant livelihood option in the geographically tough terrain of the Himalaya. This will help farmers improve their livelihoods and social life, future food securit,y and simultaneously conserve the natural biodiversity, as well as ethic culture and indigenous knowledge.

The major challenge is to bring indigenous knowledge, science, and technology inputs together under one umbrella to meet the growing demand of livelihood and food security. The policies governing medicinal plant cultivation/development and the government's role e.g., ground rules for medicinal and aromatic plant cultivators/farmers, market and marketing, timely availability of good quality of seeds, planting materials, etc., need to be given due importance. Certain policy recommendations are required to be considered, examined, and implemented, to fully utilize *A. stracheyi* and other medicinal plants to improve the socioeconomic development of the region.

ACKNOWLEDGMENT

Special acknowledgement goes to the Director of the G.B. Pant National Institute of Himalayan Environment and Sustainable Development, Garhwal Research Centre, Srinagar, Uttarakhand, India, for experimental site, facility, design, and data collection. All the research work was carried out in the research sites of the G.B. Pant National Institute of Himalayan Environment and Sustainable Development, Garhwal Research Centre, and the High-Altitude Plant Physiology Research Centre (HAPPRC) of H.N.B. Garhwal Central University, Srinagar Garhwal, Uttarakhand, India. Special thanks to the Department of Biotechnology (DBT), Ministry of Science and Technology (Government of India), and the Department of Botany, University of Delhi. The authors are also grateful to the indigenous Bhotiyafolks of high Himalaya for assistance during this research work.

References

Atkinson, E. T. (1882). *The Himalayan gazetteer*. New Delhi, India: Cosmo.

Bisht, A. K., Bhatt, A., & Dhar, U. (2006). Prioritization and conservation of Himalayan medicinal plants: Angelica glauca as a case study. [Not cited in text.]. *Ethnobotany Research and Applications*, *4*, 11–23. doi:10.17348/era.4.0.11-24

Dhar, U. (2000). Prioritization of conservation sites in the timberline zone of west Himalaya. In S. Singh, A. R. K. Sastry, R. Mehta, & V. Uppal (Eds.), *Setting biodiversity conservation priorities for India* (pp. 193–211). New Delhi, India: WWF India.

Dhar, U., Manjkhola, S., Joshi, M., Bhatt, A., Bisht, A. K., & Joshi, M. (2002). Current status and future strategy for development of medicinal plants sector in Uttaranchal, India. *Current Science*, *83*, 956–964.

Hajra, P. K., & Jain, K. K. (1983). *A contribution to the botany of Nanda Devi National Park*. Howrah: Botanical Survey of India.

Kandari, L. S., Phondani, P. C., Payal, K. C., Rao, K. S., & Maikhuri, R. K. (2012). Ethnobotanical study towards conservation of medicinal and aromatic plants in upper catchments of Dhauli Ganga in the Central Himalaya. *Journal of Mountain Science*, *9*(2), 286–296. doi:10.100711629-012-2049-7

Lama, Y. C., Ghimire, S. K., & Aumeeruddy, T. Y. (2001). *Medicinal plants of Dolpo: Amchis' knowledge and conservation*. Kathmandu: WWF Nepal Program.

Maikhuri, R. K., Nautiyal, S., Rao, K. S., & Saxena, K. G. (1998). Medicinal plant cultivation and biosphere reserve management: A case study from the Nanda Devi Biosphere Reserve, Himalaya. *Current Science*, *74*(2), 157–163.

Maikhuri, R. K., Rao, K. S., Chauhan, K., Kandari, L. S., Prasad, P., & Rajasekaran, C. (2003). Development of marketing of medicinal plants and other forest products: Can it be a pathway for effective management and conservation? *Indian Forester*, *129*, 169–178.

Nautiyal, M. C., & Nautiyal, B. P. (2004). Collaboration between farmers, research institutions and industry: Experiences of Picrorhiza kurrooa cultivation at Gheshe village in Chamoli district, Uttaranchal. In G. Alam, & J. Belt (Eds.) Searching synergy: stakeholder views on developing a sustainable medicinal plant chain in Uttaranchal, India. Amsterdam, The Netherlands: KIT Publishers, KIT (Bulletin no. 359 pp. 63–71).

Nautiyal, S., Maikhuri, R. K., Rao, K. S., & Saxena, K. G. (2001). Medicinal plant resources in Nanda Devi Biosphere Reserve in the Central Himalaya, India. *Herbs Spices & Medicinal Plants*, *8*(4), 47–64. doi:10.1300/J044v08n04_06

Nautiyal, S., Maikhuri, R. K., Semwal, R. L., & Rao, K. S. (1998). Conservation through cultivation: A case study of medicinal plants in buffer zone village of NDBR. In Research for mountain development: Some initiatives and accomplishments. Gyanodaya Prakashan, Nainital: Himvikas Publication. (vol. 12. pp. 357–374).

Nayar, M. P., & Shastry, A. R. K. (1990). *Red data book of Indian plants* (Vol. I). Calcutta, India: Botanical Survey of India.

Negi, K. S., Tiwari, J. K., & Gaur, R. D. (1987). A contribution to the flora of Khatling glacier in the Garhwal Himalaya (District-Tehri). *U. P. J. Bombay Nat. Hist. Soc*, *84*, 585–598.

Payal, K. (2007). Eco-physiological and socio-economic studies of some multipurpose Allium species of higher Himalaya. (Ph.D. Thesis), HAPPRC, H.N.B. Garhwal University. Srinagar Garhwal Central University, India.

Payal, K., Maikhuri, R. K., Rao, K. S., & Kandari, L. S. (2014). Effect of temperature, photoperiod, gibberellic acid and water based pre-soaking treatments on seeds germination of *Allium stracheyi* Baker: An endangered alpine species of Central Himalaya, India. *Journal Plant Biosystems—An International Journal Dealing with all Aspects of Plant Biology, 148*(6), 1075–1084.

Purohit, A., Maikhuri, R. K., Rao, K. S., & Nautiyal, S. (2001). Impact of bark removal on Taxus baccata L. (Himalayan Yew) in Nanda Devi Biosphere Reserve, Garhwal Himalaya, India. *Current Science*, *81*(5), 586–590.

Raghupathy, L. (2001). Conservation and sustainable use of medicinal plants: Current issues. In S. Samant, U. Dhar, & L. M. S. Palni (Eds.), Himalayan medicinal plants: Potential and prospects (415–426). Nainital: Gyanodaya Prakashan.

Ranjan, S., Jadon, V. S., Sharma, N., Singh, K., Parcha, V., Gupta, S., & Bhatt, J. P. (2010). Anti-inflammatory and analgesic potential of leaf extract of *Allium stracheyi.* [Not cited in text.]. *Journal of Applied Sciences Research*, *6*(2), 139–143.

Rau, M. A. (1975). *High altitude flowering plants of West Himalaya.* Howrah: Botanical Survey of India - Indian Botanic Garden.

Rikhari, H. C., Negi, G. C. S., & Singh, S. P. (1993). Species and community diversity pattern in an alpine meadow of central Himalaya. In: US. Dhar (Ed.), Himalayan biodiversity: Conservation strategies. (pp. 205–218). Kosi, Almora: G. B. Pant Institute of Himalayan Environment and Development.

Samant, S. S., & Dhar, U. (1997). Diversity, endemism and economic potential of wild edible plants of Indian Himalaya. *International Journal of Sustainable Development and World Ecology*, *4*(3), 179–191. doi:10.1080/13504509709469953

Silori, C. S., & Badola, R. (2000). Medicinal plant cultivation and sustainable development: A case study in the buffer zone of the Nanda Devi Biosphere Reserve, Western Himalaya, India. *Mountain Research and Development*, *20*(3), 272–279. doi:10.1659/0276-4741(2000)020[0272:MPCASD]2.0.CO;2

Sundriyal, M., & Sharma, B. (2016). Status of biodiversity in Central Himalaya. *Applied Ecology and Environmental Sciences*, *4*(2), 37–43.

Chapter 2
Ethnobotanical Wealth of Home Gardens

Bimal K. Chetri
https://orcid.org/0000-0002-2154-7676
Department of Environment and Life Science, Sherubtse College, Royal University of Bhutan, Bhutan

Namgay Rinchen
Department of Environment and Life Science, Sherubtse College, Royal University of Bhutan, Bhutan

Karma Wangchuck
https://orcid.org/0000-0002-6513-3467
Department of Environment and Life Science, Sherubtse College, Royal University of Bhutan, Bhutan

ABSTRACT

The use of ethnomedicinal plants of home gardens has played an important role in treating minor illnesses related to physical and psychological wellbeing among rural folks. Local healers often use plants from their home garden or nearby forest. In this chapter, the traditional uses of medicinal plants using various tools with sample data have been used to facilitate ethnobotanical research on home gardens. Samples measuring the relative importance of species through quantitative analysis are tabulated. These tools, if applied during in-situ data collection, would help generate reliable information on characteristics of home gardens and medicinal plants. Conservation of medicinal plants from the perspective of home gardening is important. To document how medicinal plants are used to treat different diseases, their use value and ethical considerations would be significant for bioprospecting of medicinal plants and protecting intellectual right of the associated traditional knowledge.

INTRODUCTION

Home gardens are limited, terraced lands available for sustenance farming to every house-hold in the village of the Himalayan region. They usually vary in size (Agnihotri et al., 2004; Ali, 2005; Barbhuiya et al., 2016), with rich floral composition (Bargali, 2016; Rana et al., 2016; Singh et al., 2014; Yang et al.,

DOI: 10.4018/978-1-7998-1320-0.ch002

2014). Medicinal plants and their products are frequently used by the local healers or by elderly people in a family. Also, for many healers, this practice is part of their livelihoods and community service. Using medicinal plants and their products available within their home garden or from nearby forest against minor illnesses during rituals are common among people living along the foothills of Bhutan. Similar to home gardening is the culture of kitchen gardening, common in Nepal (Gautam et al., 2009; Shrestha et al., 2001; Subedi et al., 2006), Bhutan (Chetri et al., 2018; Tobgay, 2006) and many other parts of the world (Galluzzi et al., 2010; High & Shackleton, 2000; Inta et al., 2008; Kehlenbeck & Maass, 2004; Reyes-García et al., 2014; Yusuf et al., 2007).

The *home garden* is a land-use system that combines trees and crops, along with animals (Fernandes & Nair, 1986; Paembonan et al., 2018). It is an important site for *in situ* conservation (Trinh et al., 2003). Home gardens are found to have complex structures (Mesfin et al., 2009), traditionally with multifunctional roles and considered hotspots with regard to their importance in conserving genetic resources (Das & Das, 2005; Galluzzi et al., 2010). A traditional home garden is a typical practice of agriculture used by indigenous people for their own sustainability in a small area (Huai & Hamilton, 2009). Home gardens within a small area constitute annual crops while the wider lands have various species, including medicinal plants.

They contribute to income generation influenced by the mixture of plant species and the area of land. The economic contribution income ranges from 0% to 50% annually (Alam, 2011), 43.27% to 49.06% (Paembonan et al., 2018), and up to 54% (Trinh et al., 2003). According to Clarke et al. (2014), rural people with low income tend to choose garden plant species that they are knowledgeable with regard to edibility and medicinal uses as they have minimum access to urban market. Some of these plants are ornamental (Blanckaert et al., 2004). Home gardens are an important source of traditional medicines. The plants found in rural home gardens carry time-tested medicinal values and are used to treat ailments ranging from simple cuts to heart diseases (Rahman et al., 2013). Such an ethnobotanical wealth of home gardens practiced and passed down by elderly people and local healers from generation to generation must be documented with a sense of urgency.

While considerable research has been carried out on ethnobotany of home gardens in other parts of the world (Agelet et al., 2000; Alam, 2011; Chetri, 2019; Das & Das, 2005; Fernandes & Nair, 1986; Huia & Hamilton, 2009; Kala, 2010; Kala & Ratajc, 2012; Saikia & Khan, 2014; Shrivastava & Heine, 2005; Sthapit et al., 2006; Uprety et al., 2010; Vlkova et al., 2011; Vogel et al., 2004; Zimik et al., 2012), researching the socio-economic benefits of home gardens in enhancing local food production would be a significant contribution to food security among rural populations. These home gardens maintain the biological diversity of native and exotic as well as managed or wild species, and play an important role in improving the quality of life, economic and social welfare of people (Gautam et al., 2006; Rahman et al., 2013; Trinh et al., 2003) and food security (Marsh, 1998). As global food prices soar, there has been an increased emphasis on building local food systems and a need for research on the cost-benefit analysis of home gardening (Galhena et al., 2013).

The People, Place, and Plants of the Home Garden

The home garden is the most productive agroecosystem, with soil organic carbon and nutrient concentrations significantly higher than all other forest or agricultural land uses (Singh et al., 2008). In Kumaun Himalaya common locations for gardens in relation to house are backyards, front yards, side yards, and those that almost encircle the house (Bargali, 2015).

Traditional knowledge (TK) associated with ethnomedicines of home gardens comes from rural folk mostly from elderly people. This knowledge has religious and regional origins. There are papers (Agelet et al., 2000; Ban & Coomes, 2004; Kala, 2010, and references therein) on the origins of traditional knowledge among the people within the region and cross-boundary transfers of knowledge by exchanging plant materials in improving and maintaining plant diversity (see Table 1). Households with greater diversity in their home gardens have a higher potential for exchanging plant materials.

Ethnobotanically important plant species are cultivated within small areas based on traditional knowledge of the plants and their uses. Plant species are selected based on their environmental and ecological requirements as preferred by the local people. Some cultivated plants are used by specialized herbal practitioners and healers in curing specific diseases as and when required. Home gardens are small fragmented land available to farmers for subsistence farming too. This type of farming system is an ecosystem in itself. Often home gardens are more than agroforestry (Ffolliott, 1998; Lamont et al., 1999). Home gardens often conserve rich medicinal flora (see Table 2).

The Need to Study Home Garden Ethnobotany

The use of medicinal plants based on traditional knowledge practiced almost on daily basis by the local healers has contributed much to health care of women and children in rural villages. Therefore, it is urgent to conduct studies on limited resource utilization and ecosystem services provided by the home gardens in terms of food security and ethnobotanical wealth. The use of medicinal plants from home gardens by the local healers and women is a common practice for children's and women's healthcare systems in rural places. This invaluable contribution made by local healers and women to their community has been passed down to generations requires proper documentation. Kabir and Webb (2008) reported 419 species (59% native), including six on the IUCN Red List. Pala et al. (2019) reported 260 plant species, including 53 ethnomedicinal ones from the home gardens of the study area used in health-care system.

The study of the structural and functional dynamics of home gardens also offers the opportunity to understand the trends in socioeconomic sustainability and how these relate to ecological sustainability (Peyre et al., 2006). There are many reasons to conduct research on socio-economics (Abebe et al., 2010; Hoogerbrugge & Fresco, 1993; Linger, 2014; Montagnini, 2006; Rugalema et al., 1994; Soemarwoto, 1987), eco-services (Calvet-Mir et al., 2012; Eyzaguirre et al., 2002; Kehlenbeck et al., 2007), and medicinal plants (Agelet et al., 2000; Díaz-Reviriego et al., 2016; High & Shackleton, 2000) available within home gardens. Comprehensive collections of data engaging the local leaders, local healers, elderly people, and women during including PRA (participatory rural appraisal) and interviews will help generate sufficient sustainability data to inform policy and planning at the village block level.

Tools and Methods for Data Collection

Interviews are ethnobotanists' principal tool for obtaining ethnobotanical information (Albuquerque et al., 2014; Thomas et al., 2007) PRA when applied to ethnobiological issues (Sieber et al., 2014) regarding home gardens. Multiple approaches are useful in collecting data for the study on botanical inventories, collection of plant specimens, and both structured and informal interviews with local healers (Vogl et al., 2004), including freelisting, preference ranking, priority ranking, and direct matrix ranking (Cotton & Wilkie, 1996; Martin, 2010). The use of colored photographs and voucher specimens are added advantages in identify plant species during *in situ* data collection involving local healers (Thomas et al., 2007).

Table 1. Comparison of the use of common medicinal plant species from home gardens (NR = not reported)

Species	Sarpang Bhutan	S/jhonkhar Bhutan	Trashigang Bhutan	South Sikkim	Assam	North-East India
Achyranthes aspera L.	Leaves and roots to treat pneumonia	Leaves to treat pneumonia	NR	NR	Stem and leaf juice used to treat jaundice and menstrual problems	NR
Acorus calamus L.	Rhizome is mixed with shilajit, lacca, and shemriin for piles	NR	Powdered rhizomes as insecticide; Root paste is used to cure scabies and wounds	Paste from dried rhizomes is applied on forehead for fever and orally for fever, bronchitis, diarrhea, and dyspepsia	NR	NR
Ageratina adenophora (Spreng.) R.M.King & H.Rob.	Leaves and stem as poultice on fresh cuts	Leaf poultice used as a blood coagulant	NR	NR	NR	NR
Ageratum conyzoides (L.) L	Leaves are crushed and applied directly on cuts to control bleeding; pneumonia	Leaf juice and poultice helps withhemostasis	NR	NR	Leaves juice applied to fresh cuts and plant juice taken orally daily to treat Jaundice	Crushed leaves applied to cuts
Artemisia vulgaris L.	Leaves to treat postpartum, skin diseases, fresh cuts bleeding, and to disinfect	Leaves for skin diseases, insect repellents	Leaves applied on cuts, in cough, cold and piles	Freshly crushed leaves kept in nose to stop bleeding	NR	Fresh leaves on cuts, headaches, and nose bleeds
Bombax ceiba L.	Decoction mixed with sugar for vaginal bleeding	Bokrey (Cattle disease)	NR	NR	Seed used for liver and stomach troubles	NR
Caltrops gigantean (L.) Dryland.	Baked and warmed leaves directly applied to swollen parts for joint pain and sprains	NR	Latex used for leprosy, powdered root mixed with goat's milk for ear problems and boils	NR	Warm leaves are applied externally for rheumatism and flower paste for piles	One teaspoon leaf extract orally 3 times daily for malaria
Chromolana odorata (L.) R.M.King & H.Rob	Leaves are crushed and applied externally on deep cuts	Leaf poultice on deep cuts to stop bleeding	NR	NR	Leaf juice on fresh cuts and wounds; young leaf paste for headache	Whole plant used for fish poison
Clerodendrum infortunatum L.	Fresh leaf extract and paste directly applied to wounds	Bokrey (cattle disease)		NR	Young leaves used to treat diabetes and dysentery and mixed with rice as anthelmintic	Leaves tied on belly for digestion
Curcuma caesia Roxb.	NR	Crushed roots are dried and chewed; used for body aches and used as strong appetizer	NR	After segregation, roots crushed into powder, administered in small amount 3 times per day with water for gastritis	NR	NR
Drymaria cordata (L.) Willd. ex Schult.	Baked leaves sniffed by wrapping in cloth to treat sinusitis and nasal congestion	Baked leaves are sniffed by wrapping in cloth to treat sinusitis and nasal congestion	Used in cold, fever and throat problems	Whole herb rolled in a large banana leaf, heated on cinders, crushed yields juice used for infant fever and pneumonia. For sinusitis, heated plant material is inhaled from a thin cotton cloth.	NR	Used as a vegetable
Entada rheedii Spreng.	Boiled fruit is crushed, and endosperm is applied to head to treat dandruff and scalp diseases	Boiled fruit is crushed, and endosperm is applied to head to treat dandruff and scalp diseases	Powdered kernels mixed with spices used by women after delivery	NR	NR	NR
Euphorbia royleana Boiss. (Seuri)	Appropriate dose of latex as an appetizer	Appetizer	NR	NR	NR	NR
Euphorbia tithymaloides L. (Chetu)	Appropriate dose of latex as an appetizer	Appetizer	NR	NR	NR	NR
Ficus racemosa L.	NR	Apply latex on the area swollen portion of neck	Bark crushed into a fine powder is taken orally with warm water in the morning to treat piles	NR	NR	NR
Gmelina arborea Roxb.	Peeled bark given to cattle for foot and mouth disease	Leaves and peeled bark fed to cattle against bokrey	NR	NR	NR	NR
Justicia adhatoda L.	NR	Leaves rubbed externally; used for in-flammation; flowers eaten as a vegetable	Decoction of leaves for bronchial trouble; antihelminticand haemostatic	NR	Leaf juice used to treat cough	Used as vegetable

continued on following page

Table 1. Continued

Species	Sarpang Bhutan	S/jhonkhar Bhutan	Trashigang Bhutan	South Sikkim	Assam	North-East India
Morus macroura Miq.	Roots decoction mixed with alum to treat jaundice	Roots decoction mixed with alum to treat jaundice	NR	NR	Fruits are used in cough and root is used to treat jaundice	NR
Oroxylum indicum (L.) Kurz	Fruit and ash used on deep cuts	Fruit and ash used on deep cuts	Decoction of bark for fever and jaundice	Flower ash applied to burns. Seeds used for pneumonia, fever, and throat problems.	NR	NR
Paederia foetida L.	Leaves eaten as vegetable, to cool stomach	Poultice from leaves and stem applied to swollen nodes to treat *kalo bagay*	NR	Plant juice orally for ulcers and jaundice	NR	NR
Phlogacanthus thyrsiformis (Roxb. ex Hardw.) Mabb.	NR	Decoction from flower used for lowering high blood pressure	NA	Flowers are soaked in warm water then sun-dried and cooked with rice to regulate blood pressure	NA	NR
Pteris biaurita L.	NA	NA	NA	Fresh stem crushed, its juice applied on cuts and wounds	NA	NA
Rauvolfia serpentina (L.) Benth. ex Kurz	NR	Decoction from whole plant used to treat malaria	NR	NR	NR	NR
Ricinus communis L.	Seeds are crushed into paste and applied externally on affected part; chicken pox	NR	NR	Seed oil for rheumatism, leaf extract for head and body aches	NR	NR
Rumex nepalensis Spreng.	NR	Rubbed with garlic and applied on skin for ringworm	NR	Fresh or dried roots extract for wounds and hair loss	NR	
Sansevieria trifasciata Prain	Leaf juice orally and applied on affected body parts	Leaf juice orally and applied on affected body parts	NR	NR	NR	NR
Tagetes minuta L.	Decoction for pneumonia	NR	Infusion of leaf and root for fever	NR	Leaf juice for fresh cuts and plant juice for dysentery	NR
Thysanolaena latifolia (Roxb. ex Hornem.) Honda	Crushed root applied on boil to draw pus to the surface			Boiled fresh root extract to treat boils and burns. Roots (100 gm) are crushed and mixed with water and used to induce abortion after 3–4 months of pregnancy.	Broomstick	NR
Viscum album L.	Juice of stem for pneumonia, applied to fractured bones, joints pain	NR	NR	Whole plant is crushed and mixed with a teaspoon of water for body pain and bone fractures	NR	NR
Source	Chetri et al., 2016	Chetri, 2019	Sadruddin, 1997	Idrisi et al., 2010	Das et al., 2008	Gangwar & Ramakrishnan, 1990; Maikhuri & Gangwar, 1993

Prior to field visits, we prepared a template (Figure 1) for quickly collecting notes from respondents in the field, also useful for semi-structured interviews at informants' homes (Figure 2). Species richness, species density, and data were analyzed per standard statistical methods (Magurran, 2013). With objectives to study about rich agrodiversity and ethnomedicinal plants of home gardens, both *in situ* and *ex situ*, a diachronous study interviewing senior citizens was useful to understand pattern changes in terms of how family members used each medicinal plant on a daily basis.

Studying the socioeconomic values of home gardens to understand their function and nature is useful in rural areas where food sufficiency is a priority. It is important to select suitable home gardens for the study based on personal interest and data collection objectives. The authors to be careful about the type of study area (home gardens) based on their personal interests, budget, time, research questions and hypothesis. Vogl-Lukasser & Vogl (2016) have recommended both quantitative and qualitative research

Table 2. Estimated medicinal plant species per garden

Authors	Species
Agelet et al. (2000)	30–60
Chetri et al. (2018)	40–50
Das & Das (2005)	20
Eichemberg et al. (2009)	10–12
Panyadee et al. (2019)	20–59 (villages)
Saikia & Khan (2011)	96 (27 villages)

methods to get rigorous results. Ethnobotany covers a variety of traditional and cultural uses of plants for food, medicine, in material culture as a part of their spiritual beliefs, which will help in the conservation of biodiversity. Plants were used as medicines since ancient times. Some of the quantitative techniques are the freelisting (FL), preference ranking (PR), direct matrix ranking (DMR), use value (UV), and relative importance (RI). Preference ranking is one of the simplest analytical tools (Martin, 2010), in which key informants are asked to think of all medicinal plant species potentially useful for a particular treatment, and each informant ranks plant species according to their personal preference and experience using them. Each rank is given integer value (see Table 3).

Freelisting is an open-ended, straightforward, and precise method to gather information from an expansive test of people by oral interviews, which helps in collecting information from experienced respondents (Berlin & Berlin, 2005). It is a fast approach to gather information from a large set of people. The ease and accuracy of freelist interviewing make it ideal for collecting data on local knowledge from relatively large samples (Quinlan, 2005). Freelisting is an established ethnographic strategy based on the order of familiarity, basis of more knowledge, and respondents on locally prominent items being used (Table 4). It allows researchers to gain knowledge through subjects' experiences and knowledge.

If the information given by the local plant expert is not true or not practically revealed, however, the researcher might consider similar plants as medicinal, which becomes risky. Direct matrix ranking is

Figure 1. Data collection template

Royal University of Bhutan: Sherubtse College Flora of Bhutan

1. Voucher No: Date (___/___/20___)
2. Gewog/Village/Locality______________________.
3. Habitat (Herb/Shrub/Tree)[✓]
4. Local people/Ethnic group (Nga/Sha/Khe/Lho)[✓]
5. Local Name/ Vernacular name .
6. Family___________. Type (Domesticated/Wild)[✓]
7. Botanical name ___________________________.
8. Description of the plant
9. Parts used (Leaves/Roots/Rhizomes/Bark/Flower/Seeds)[✓]
10. Ethno (Uses-Prep)_________________________.
11. Name of the informant ____________________.
12. Collected by_________________________________.

Figure 2. Home garden, in-situ interview, and plant sample collections

Note: Top left panel: Typical home garden the field. Top right panel: Interviewing local healer in the field. Bottom left panel: Plant identification. Bottom right panel: Rauwolfia collected plant sample.

the next level of preference ranking (Martin, 2010). It is simple, quick, and informative, which helps researchers and respondents move quickly from early discussion and questioning to recording respondents' views directly in a table or matrix (Table 5).

The *use value (UV)* is the frequency of usage and the number of people that mention the plant; it is widely used in the ethnobotanical community to designate that certain species are considered important. The technique of Relative Importance (RI) is the method developed to measure the utility of medicinal plants. The value of a species increases as it treats more illnesses (Albuquerque et al., 2005). Use value demonstrates the relative importance of plants known locally (Parthiban et al., 2016). The use values of *Gmelina arborea* Roxb. and *Tinospora cordifolia* (Willd.) Miers are calculated based on different diseases and their local importance and perception (Table 6). The limitation of this technique is an emphasis on species that have many uses, even if it is used by only few people, whereas RI solely emphasizes plants

Table 3. PR of the use of medicinal plants for treating diseases by 7 informants (a)

List of medicinal plants	i1	i2	i3	i4	i5	i6	i7	Total	Rank
Vitex negundo L.	4	4	1	1	1	1	2	14	3rd
Curcuma caesia Roxb.	2	2	5	4	5	4	3	25	2nd
Euphorbia royleana Boiss.	1	1	2	2	2	2	1	11	4th
Euphorbia tithymaloides L.	3	5	3	5	5	5	5	31	1st
Researcher's name: Bimal K. Chetri							05/04/2019		

Note: The highest number is the most preferred; the lowest number is least preferred). Rank is determined based on the total score of each species. Top-ranked species indicate the most effective plants for the respective treatments. Source: Chetri (2019).

that have the greatest absolute number of uses. Thus, we cannot conclude that a plant species' importance is not merely because of its multiple uses.

Table 4. Freelist of medicinal plants done by 7 informants (a) in their local language for specific treatments

Name of the disease: Lantana poisoning (Bokrey) in cattle	
Gmelina arborea Roxb.	Bark concoction with leaves fed to cattle in a bamboo cup
Boehmeria penduliflora Wedd. ex D.G.Long	Root Infusion
Bombax ceiba L.	Bark rubbed on body
Clerodendrum infortunatum L.	Root and leaves extract fed in a bamboo made cup
Citrus reticulata Blanco	Bark concoction with leaves fed to cattle in a bamboo cup
Informant's name: OP Sharma	Date: 17/01/2019

Note. Most common diseases treated by informants in the 2 village blocks were identified and listed the medicinal plants used against each disease.

Table 5. Direct matrix ranking of multipurpose uses of medicinal plants

Species	*Bombax ceiba* **L.**	*Gmelina arborea* **Roxb.**	*Mallotus philippensis* **(Lam.) Müll. Arg. (Sidurey)**	*Oroxylum indicum* **(L.) Kurz**
Medicines	☑	☑	☑	☑
Timber	☑	☑	☒	☒
Articrafts	☑	☑	☒	☑
Rituals	☒	☑	☒	☑
Firewood	☒	☑	☑	☑
Fodder	☑	☑	☑	☑
Total	4	6	3	5
Rank	C	A	D	B

Note. The total score of three popular informants, indicating each species as ☑ (used) or ☒ (not used), accordingly. The species with the most tick marks is the most preferred species and is ranked as A.

Table 6. Estimating UV based on local perceptions

Name of the Informant: Mitra Lal Dhungyal		UV_{is}
Name of the species: *Tinospora cordifolia* **(Willd.) Miers**		
Sl.No.	Uses: Treatment	
1	Ulcer	2
2	Hypertension	
3	Diabetes	
4	Gastritis	
Total Uses	1+1+2	
No. of events	1+1	

Note. UVis *stands for* use value (UV) *for a given species* (s) *by one informant* (a) *during number of events* (n; *Cotton & Wilkie, 1996; Martin, 2010). Source: Chetri (2019)*

Figure 3. Plant parts used for ethnomedicinal preparations

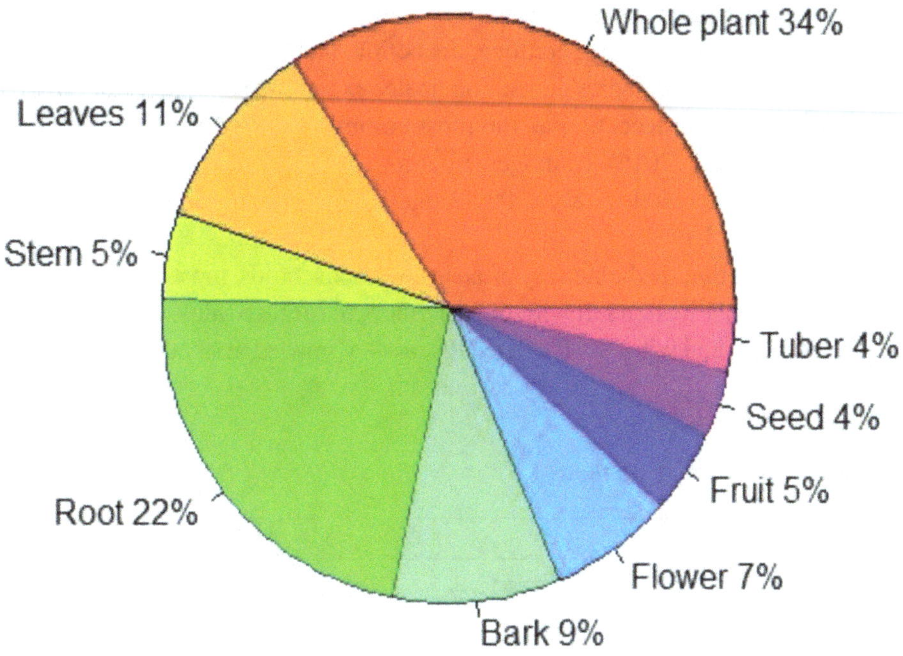

Table 7. Home remedies

Species	Preparation	Treatment
Crinum	Poultice: Sliced rhizome, baked and applied on affected part twice per day	Bone fracture
Rauwolfia	Decoction: 5 gm of leaves or flowers with one liter of water in the morning	Malaria
Rumex	Fresh leaves:1–2 gm fresh leaves, chewed	Piles

Relative cultural importance (RCI) indices are analytical tools used to measure plant species preferences in a particular area (Hoffman & Gallaher, 2007; Phillips & Gentry, 1993). They calculate value per folk or biological plant taxon, a growing trend in quantitative ethnobotanical research; however, the authors claim that only few attempts to compile or standardize divergent methods were made. Also, appropriate purposive sampling methods using structured, semi-structured, and minimally structured face-to-face interviews could be useful in order to reduce bias in findings (Tongco, 2007).

COMMON PLANT PARTS USED AND ETHNOMEDICINAL PREPARATION

An important way to identify plant parts with active properties is to consider what parts are used in healers' practices. Figure 3 shows the percentages of parts used for ethnomedicinal preparations from medicinal plants of home gardens. Mussarat et al. (2014) reported the leaves and fruit (31%) are the most frequently used, followed by the whole plant and its roots, barks, tubers, seeds, and stems. Bekalo et al. (2009) reported leaves being used the most for ethnomedicinal preparations (34.2%), followed by roots (30.9%), barks (8.2%), and other parts (19.8%).

Often preparations include pounding, poultice, decoction, and concoction based on healers' experience in treating patients. Home remedies depend on healer experience and the right dose (see Table 7). Mussarat et al. (2014) reported decoction as the most common method used for preparation, used in 43% of treatments (Sabran et al., 2016). Homogenizing in water (53.3%) is another common preparation, followed by crushing (20.1%; Bekalo et al., 2009). Table 8 shows the most popular preparation used by local healers from home gardens.

Elderly people practice ethnomedicine very often in their households, most commonly, elderly women and mothers. Since children are often left with their mothers in rural settings, women take care of minor illnesses, including injuries and witch-craft related diseases. Knowledge of ethnomedicine passed down through generations becomes an emergency means of treatment. Elderly people often bring rare and economically important plants to their home gardens for regular use, unknowingly helping to conserve threatened wild species.

Table 8. Frequency of preparation methods

Method	%	Treatment
Powder	6	6.5
Crushing or pounding	32	34.4
Chewing	4	4.3
Concoction	5	5.4
Decoction	10	10.8
Juices	8	8.6
Poultice	4	4.3
Roast	4	4.3
Boiled or as a soup	4	4.3
Rubbed on body	9	9.7
Infusions	7	7.5
Total	93	100

Home gardens are self-sustaining agricultural practices still prevalent among people from developing countries specifically from the Himalayan region, where traditional cropping system, rearing livestock, is common. Livestock is very often seen as extended family members. This adds another means to practice ethnoveterinary medicine among farmers, which is an integral part of home gardening (Tadesse et al., 2014). Among tribes, ethnoveterinary medicine very often becomes a usual practice to treat common diseases among cattle (Giday & Ameni, 2003; Kebede et al., 2017; Rajkumari et al., 2014; Ramana, 2008; Tekle, 2014). For example, ethnoveterinary use of *Helleborus* ssp. has been described to relieve pain in animals (Papp et al., 2013). Bharati & Sharma (2012) reported 37 plant species used in 15 broad categories of treating livestock ailments in Sikkim Himalaya.

Table 9 shows preference ranking based on total score. Firewood collection and unsustainable use are indicated as the greatest potential threats, and illegal trade of medicinal plants is the lowest threat, as perceived by the local healers.

EXAMPLES OF PREPARATION

Two examples of treatments are provided below. The most effective treatment against any disease uses a concoction of multiple plants (e.g., both treatments below), as reported by one of local healer Tikka Devi Adhikari. She was quite popular in treating jaundice with concoction of *Morus* spp. root, alum powder, and *Citrus* spp., along with knowledgeably administering appropriate doses and giving proper advice during treatment.

Hypertension Treatment

Informant: Timsina Jhakri (Shaman).

Preparation:

- *Zea mays* L. flour
- *Eleusine coracana* (L.) flour
- *Urtica parviflora* Roxb. tender leaves

Maize and finger millet flours from are sprinkled on the tender leaves of *Urtica parviflora* Roxb. and are boiled for half an hour until a slurry soup is formed. The soup is taken for morning breakfast and dinner.

Jaundice Treatment

Informant: Tika Devi Adhikari (a healer popular for treating jaundice in Beldara, Pemathang)

Table 9. Preference ranking of factors perceived as threats to medicinal plants

Factors	i1	i2	i3	i4	i5	i6	i7	Total	%	Rank
Forest fire	☒	☒	☒	☒	☒	☒	☑	1	14.28	6th
Developmental activities	☑	☒	☑	☒	☒	☒	☒	2	28.57	5th
Grazing	☑	☑	☒	☑	☑	☑	☑	6	85.71	2nd
Agricultural expansion	☑	☑	☑	☒	☒	☑	☒	4	57.14	4th
Fire wood	☑	☑	☑	☒	☑	☑	☒	5	71.42	3rd
Illegal trade	☒	☒	☒	☒	☒	☒	☒	0	0	7th
Migration	☒	☑	☑	☒	☑	☒	☒	3	42.85	5th
Unsustainable use	☑	☑	☑	☑	☑	☑	☑	7	100	1st

Note. The first rank represents the greatest potential threat. The symbols ☑ and ☒ represent potential threat *and* no threat, *respectively, to medicinal plants. Source: Chetri (2019)*

Preparation:

- One-third teaspoon alum (*fetkeri*)
- 200g *Morus alba* L. (sano kimbu) root
- 200g *Citrus medica* L. (bimira) root

Ingredients are mixed, and the concoction is administered orally with an empty stomach for four days. During this time, the patient should refrain from consuming oil, cholesterol-rich foods, chili peppers, turmeric, other spices, and excessive salt.

DISCUSSION

Conservation and Protection of Home Garden Medicinal Plants

The domestication of wild medicinal plants in home gardens plays an important role in their conservation and protection. Genetic sources already face threats at different degrees in the wild, including over-harvesting and illegal trade. Although domestication is a viable option (Canter et al., 2005) to challenge such threats, there are reports on factors affecting the medicinal properties of plants due to changes in habitat (Gairola et al., 2010; Szakiel et al., 2011). Local healers domesticate medicinal plants for their regular practices, which are their livelihoods and health care systems (Hamilton, 2004), although there are implications for biodiversity (Schippmann et al., 2002). Ethnobotany also plays roles in conservation medicinal plant use by both healers and farmers (Arora, 1987; Yang et al., 2014).

Plants are often used in ritual practices. It has been observed during the study that plants used in religious purposes are also therapeutically potential. *Achyranthes aspera* L., *Cynodon dactylon* (L.) Pers and *Ocimum sanctum* are a few Hindu examples. Such plants are often used in religious rituals or during treatment as ingredients or as part of chanting a mantra. The species of plants or parts of plants used for religious activities and socio-cultural activities or events are based on traditional beliefs (Kala, 2010). For example, *Rosmarinus officinalis*, *Paeonia officinalis* ssp., and *Salix* sp. have special significance and symbolic value in religious events (Vogl & Vogl-Lukasser, 2003).

Transfer of Traditional Values and Ethical Considerations

Traditional knowledge in the form of stories, songs, folklore, proverbs, cultural values, beliefs, rituals, community laws, local language, and agricultural practices, including the development of plant species and animal breeds is orally transmitted. It is the knowledge, innovations and practices of local people developed from experience gained over the centuries and adapted to the local culture and environment. This valuable knowledge about medicinal plants and traditional healers is depleting because of over-exploited resources and changes in land-use patterns. Article 7, which deals with access to traditional knowledge of the Nagoya Protocol associated with genetic resources, says that:

Each Party shall take measures, as appropriate, with the aim of ensuring that traditional knowledge associated with genetic resources that is held by indigenous and local communities is accessed with the prior and informed consent or approval and involvement of these indigenous and local communities, and that mutually agreed terms have been established. (Buck & Hamilton, 2011)

The Access and Benefit Sharing Policy of a country should ensure obtaining Prior Informed Consent (PIC) from the Competent Authority for fair and equitable sharing of benefits arising from research and commercial utilization of genetic resources, which are associated with traditional knowledge. The participation of informants depends on their willingness. Prior permission must be sought from the head or office concerned, regarding the purpose of the visits and the study, for example, consent of the local healer and identified informants, field visits, sampling, and data observations on-site to be carried out. The identities of the informants and their traditional knowledge must be acknowledged in the study report. Additionally, the information collected needed to be properly credited to the informants involved.

CONCLUSION AND FUTURE PROSPECTS

Home gardens will continue to be an important land-use system for small-scale farmers to cultivate different cultivars and medicinal plants. It is interesting to note that the medicinal plants identified and used by the local healers without any scientific knowledge have been proven to have positive effects in treatments. Home gardens are rich genetic resources. They are potential sources for bioprospection of bioactive compounds, which could cure deadly human diseases like cancer, diabetes, and HIV. Therefore, documenting traditional knowledge of ethnomedicines and chanting mantras used by local healers could be useful in the traditional medicine system.

Key informants during the research field visits and ethnobotanical data collections were eager to share their traditional knowledge of medicinal plants. Spending more time with them would allow collecting more data than the tools mentioned in this chapter could have captured. These informants' enthusiasm and honor in sharing their traditional knowledge on ethnobotanical studies indicates good ethnobotanical knowledge is still prevailing in the home gardens. The good ethnomedicinal knowledge still prevailing in traditional societies, particularly among the traditional healers and elderly persons, will keep traditions alive.

However, although people still partially depend on traditional medicine for treating human diseases as well as treating livestock, with our changing society, there is a high risk of losing this traditional knowledge. Therefore, digital documentation is necessary. With free and easy access to modern health care and medicines, people have switched to modern government health care systems. A comprehensive study focusing on sound objectives would reveal enormous ethnobotanical information from the home gardens that could be useful for socio-economic analysis of home gardens and the people surviving because of it.

REFERENCES

Abebe, T., Wiersum, K., & Bongers, F. (2010). Spatial and temporal variation in crop diversity in agroforestry homegardens of southern Ethiopia. *Agroforestry Systems*, *78*(3), 309–322. doi:10.100710457-009-9246-6

Agelet, A., Bonet, M. À., & Vallès, J. (2000). Homegardens and their role as a main source of medicinal plants in mountain regions of Catalonia (Iberian Peninsula). *Economic Botany*, *54*(3), 295–309. doi:10.1007/BF02864783

Agnihotri, R., Sharma, S., Joshi, M., & Palni, L. (2004). Crop diversity in the home gardens of the Kumaun region of central Himalaya, India. *Plant Genetic Resources Newsletter*.

Alam, M. (2011). Tropical homegardens in Bangladesh: Characteristics and sustainability. In Alternative Farming Systems, Biotechnology, Drought Stress, and Ecological Fertilisation (pp. 245–262). Springer.

Albuquerque, U. P., Andrade, L., & Caballero, J. (2005). Structure and floristics of homegardens in northeastern Brazil. *Journal of Arid Environments, 62*(3), 491–506. doi:10.1016/j.jaridenv.2005.01.003

Albuquerque, U. P., Ramos, M. A., de Lucena, R. F. P., & Alencar, N. L. (2014). Methods and techniques used to collect ethnobiological data. In *Methods and techniques in Ethnobiology and Ethnoecology* (pp. 15–37). Springer. doi:10.1007/978-1-4614-8636-7_2

Ali, A. M. S. (2005). Homegardens in smallholder farming systems: Examples from Bangladesh. *Human Ecology, 33*(2), 245–270. doi:10.100710745-005-2434-8

Arora, R. (1987). Ethnobotany and its role in the domestication and conservation of native plant genetic resources. In A Manual of Ethnobotany (pp. 94–102).

Ban, N., & Coomes, O. T. (2004). Home gardens in Amazonian Peru: Diversity and exchange of planting material. *Geographical Review, 94*(3), 348–367. doi:10.1111/j.1931-0846.2004.tb00177.x

Barbhuiya, A. R., Sahoo, U. K., & Upadhyaya, K. (2016). Plant diversity in the indigenous home gardens in the eastern Himalayan region of Mizoram, northeast India. *Economic Botany, 70*(2), 115–131. doi:10.100712231-016-9349-8

Bargali, K. (2015). Comparative participation of rural women in agroforestry home gardens in Kumaun Himalaya, Uttarakhand, India. *Asian Journal of Agricultural Extension. Economia e Sociologia (Evora, Portugal), 6*(1), 16–22.

Bargali, K. (2016). Traditional homegardens as a sustainable ecosystem for maintenance of biodiversity: A case study from Kumaun Himalaya, India. *Journal of Biodiversity, 7*(2), 88–100. doi:10.1080/0976 6901.2016.11884761

Bekalo, T. H., Woodmatas, S. D., & Woldemariam, Z. A. (2009). An ethnobotanical study of medicinal plants used by local people in the lowlands of Konta Special Woreda, southern nations, nationalities and peoples regional state, Ethiopia. *Journal of Ethnobiology and Ethnomedicine, 5*(1), 26. doi:10.1186/1746-4269-5-26 PMID:19775482

Berlin, E. A., & Berlin, B. (2005). Some Field Methods in Medical Ethnobiology. *Field Methods, 17*(3), 235–268. doi:10.1177/1525822X05277532

Bharati, K. A., & Sharma, B. (2012). Plants used as ethnoveterinary medicines in Sikkim Himalayas. *Ethnobotany Research and Applications, 10*, 339–356. doi:10.17348/era.10.0.339-356

Blanckaert, I., Swennen, R. L., Flores, M. P., López, R. R., & Saade, R. L. (2004). Floristic composition, plant uses and management practices in homegardens of San Rafael Coxcatlán, valley of Tehuacán-Cuicatlán, Mexico. *Journal of Arid Environments, 57*(2), 179–202. doi:10.1016/S0140-1963(03)00100-9

Buck, M., & Hamilton, C. (2011). The Nagoya protocol on access to genetic resources and the fair and equitable sharing of benefits arising from their utilization to the convention on biological diversity. *Review of European Community & International Environmental Law*, *20*(1), 47–61. doi:10.1111/j.1467-9388.2011.00703.x

Calvet-Mir, L., Gómez-Baggethun, E., & Reyes-García, V. (2012). Beyond food production: Ecosystem services provided by home gardens. A case study in Vall Fosca, Catalan Pyrenees, northeastern Spain. *Ecological Economics*, *74*, 153–160. doi:10.1016/j.ecolecon.2011.12.011

Canter, P. H., Thomas, H., & Ernst, E. (2005). Bringing medicinal plants into cultivation: Opportunities and challenges for biotechnology. *Trends in Biotechnology*, *23*(4), 180–185. doi:10.1016/j.tibtech.2005.02.002 PMID:15780709

Chetri, B. K. (2019). *Ethnobotanical Study of South Eastern Foothills of Bhutan*. Asian Plant Research Journal.

Chetri, B. K., Ghalley, L., Penjor, D., Dechen, K., & Gyeltshen, T. (2018). Ethnobotanical study on wealth of homegardens in Gosiling Gewog of Tsirang District. *Asian Journal of Research in Botany*.

Chetri, K., Zangpo, T., Tshering, P., Rai, S., Tenzin, P., & Roy, S. (2016). Ethnobotanical study of the southern foothills of the Himalayas in Bhutan. Shrub Deonte. *The Research Journal of Sherubtse College*, *12*(1 & 2), 63–70.

Clarke, L. W., Li, L., Jenerette, G. D., & Yu, Z. (2014). Drivers of plant biodiversity and ecosystem service production in home gardens across the Beijing municipality of china. *Urban Ecosystems*, *17*(3), 741–760. doi:10.100711252-014-0351-6

Cotton, C. M., & Wilkie, P. (1996). *Ethnobotany: Principles and applications*. Chichester, UK: John Wiley & Sons.

Das, A. K., Dutta, B., & Sharma, G. (2008). *Medicinal plants used by different tribes of Cachar district*. Assam.

Das, T., & Das, A. K. (2005). Inventorying plant biodiversity in homegardens: A case study in Barak valley, Assam, north east India. *Current Science*, 155–163.

Díaz-Reviriego, I., González-Segura, L., Fernández-Llamazares, Á., Howard, P. L., Molina, J., & Reyes-García, V. (2016). Social organization influences the exchange and species richness of medicinal plants in Amazonian home gardens. *Ecology and Society*, *21*(1), 1. doi:10.5751/ES-07944-210101 PMID:27668001

Eichemberg, M. T., Amorozo, M. C. M., & Moura, L. C. (2009). Species composition and plant use in old urban homegardens in Rio Claro, Southeast of Brazil. *Acta Botanica Brasílica*, *23*(4), 1057–1075. doi:10.1590/S0102-33062009000400016

Eyzaguirre, P., & Watson, J. (2002). Home gardens and agrobiodiversity: An overview across regions. *Home gardens and in Situ Conservation of Plant Genetic Resources in Farming Systems, 10*.

Fernandes, E. C., & Nair, P. R. (1986). An evaluation of the structure and function of tropical homegardens. *Agricultural Systems*, *21*(4), 279–310. doi:10.1016/0308-521X(86)90104-6

Ffolliott, P. (1998). Multiple benefits of arid land agroforestry home gardens and riparian ecosystems. In *North American Conference on Enterprise Development Through Agroforestry*.

Gairola, S., Shariff, N. M., Bhatt, A., & .... (2010). Influence of climate change on production of secondary chemicals in high altitude medicinal plants: Issues needs immediate attention. *Journal of Medicinal Plants Research*, *4*(18), 1825–1829.

Galhena, D. H., Freed, R., & Maredia, K. M. (2013). Home gardens: A promising approach to enhance household food security and wellbeing. *Agriculture & Food Security*, *2*(1), 8. doi:10.1186/2048-7010-2-8

Galluzzi, G., Eyzaguirre, P., & Negri, V. (2010). Home gardens: Neglected hotspots of agro-biodiversity and cultural diversity. *Biodiversity and Conservation*, *19*(13), 3635–3654. doi:10.100710531-010-9919-5

Gangwar, A., & Ramakrishnan, P. (1990). Ethnobiological notes on some tribes of Arunachal Pradesh, northeastern India. *Economic Botany*, *44*(1), 94–105. doi:10.1007/BF02861071

Gautam, R., Sthapit, B., & Shrestha, P. (2006). *Home gardens in Nepal: Proceeding of a workshop on enhancing the contribution of home garden to on-farm management of plant genetic resources and to improve the livelihoods of Nepalese farmers: Lessons learned and policy implications,* August 6–7, 2004. Pokhara, Nepal: LI-BIRD, Biodiversity International and SDC.

Gautam, R., Sthapit, B., Subedi, A., Poudel, D., Shrestha, P., & Eyzaguirre, P. (2009). Home gardens management of key species in Nepal: A way to maximize the use of useful diversity for the well-being of poor farmers. *Plant Genetic Resources*, *7*(2), 142–153. doi:10.1017/S1479262108110930

Giday, M., & Ameni, G. (2003). An ethnobotanical survey of plants of veterinary importance in two woredas of southern Tigray, northern Ethiopia. *Sinet. An Ethiopian Journal of Science*, *26*(2), 123–136.

Hamilton, A. C. (2004). Medicinal plants, conservation and livelihoods. *Biodiversity and Conservation*, *13*(8), 1477–1517. doi:10.1023/B:BIOC.0000021333.23413.42

High, C., & Shackleton, C. M. (2000). The comparative value of wild and domestic plants in home gardens of a South African rural village. *Agroforestry Systems*, *48*(2), 141–156. doi:10.1023/A:1006247614579

Hoffman, B., & Gallaher, T. (2007). Importance indices in ethnobotany. *Ethnobotany Research and Applications*, *5*, 201–218. doi:10.17348/era.5.0.201-218

Hoogerbrugge, I., & Fresco, L. O. (1993). Homegarden systems: Agricultural characteristics and challenges. *IIED, 39*.

Huai, H., & Hamilton, A. (2009). Characteristics and functions of traditional homegardens: A review. *Frontiers of Biology in China*, *4*(2), 151–157. doi:10.100711515-008-0103-1

Idrisi, M. S., Badola, H. K., & Singh, R. (2010). Indigenous knowledge and medicinal use of plants by local communities in Rangit Valley, south Sikkim, India. *NeBio*, *1*, 34–45.

Inta, A., Shengji, P., Balslev, H., Wangpakapattanawong, P., & Trisonthi, C. (2008). A comparative study on medicinal plants used in Akha's traditional medicine in China and Thailand: Cultural coherence or ecological divergence? *Journal of Ethnopharmacology*, *116*(3), 508–517. doi:10.1016/j.jep.2007.12.015 PMID:18280071

Kabir, M. E., & Webb, E. L. (2008). Can homegardens conserve biodiversity in Bangladesh? *Biotropica*, *40*(1), 95–103.

Kala, C. P. (2010). Home gardens and management of key species in the Pachmarhi biosphere reserve of India. *Journal of Biodiversity*, *1*(2), 111–117. doi:10.1080/09766901.2010.11884722

Kala, C. P., & Ratajc, P. (2012). High altitude biodiversity of the Alps and the Himalayas: Ethnobotany, plant distribution and conservation perspective. *Biodiversity and Conservation*, *21*(4), 1115–1126. doi:10.100710531-012-0246-x

Kebede, A., Ayalew, S., Mesfin, A., & Mulualem, G. (2017). An ethnoveterinary study of medicinal plants used for the management of livestock ailments in selected kebeles of Dire Dawa administration, eastern Ethiopia. *Journal of Plant Sciences*, *5*(1), 34–42.

Kehlenbeck, K., Arifin, H. S., & Maass, B. L. (2007). Plant diversity in homegardens in a socio-economic and agro-ecological context. In *Stability of Tropical Rainforest Margins* (pp. 295–317). Springer. doi:10.1007/978-3-540-30290-2_15

Kehlenbeck, K., & Maass, B. L. (2004). Crop diversity and classification of homegardens in central Sulawesi, Indonesia. *Agroforestry Systems*, *63*(1), 53–62. doi:10.1023/B:AGFO.0000049433.95038.25

Lamont, S. R., Eshbaugh, W. H., & Greenberg, A. M. (1999). Species composition, diversity, and use of homegardens among three Amazonian villages. *Economic Botany*, *53*(3), 312–326. doi:10.1007/BF02866644

Linger, E. (2014). Agro-ecosystem and socio-economic role of homegarden agroforestry in Jabithenan district, north-western Ethiopia: Implication for climate change adaptation. *SpringerPlus*, *3*(1), 154. doi:10.1186/2193-1801-3-154 PMID:24790810

Magurran, A. E. (2013). *Measuring biological diversity*. John Wiley & Sons.

Maikhuri, R., & Gangwar, A. (1993). Ethnobiological notes on the Khasi and Garo tribes of Meghalaya, northeast India. *Economic Botany*, *47*(4), 345–357. doi:10.1007/BF02907348

Marsh, R. (1998). Building on traditional gardening to improve household food security. *Food, Nutrition and Agriculture*, 4–14.

Martin, G. J. (2010). *Ethnobotany: A methods manual*. Routledge. doi:10.4324/9781849775854

Mesfin, F., Demissew, S., & Teklehaymanot, T. (2009). An ethnobotanical study of medicinal plants in Wonago Woreda, SNNPR, Ethiopia. *Journal of Ethnobiology and Ethnomedicine*, *5*(1), 28. doi:10.1186/1746-4269-5-28 PMID:19821994

Montagnini, F. (2006). Homegardens of Mesoamerica: biodiversity, food security, and nutrient management. In *Tropical Homegardens* (pp. 61–84). Springer. doi:10.1007/978-1-4020-4948-4_5

Mussarat, S. (2014). AbdEl-Salam, N. M., Tariq, A., Wazir, S. M., Ullah, R., & Adnan, M. (2014). Use of ethnomedicinal plants by the people living around Indus River. *Evidence-Based Complementary and Alternative Medicine*, 212634–212634. PMID:24778701

Paembonan, S., Millang, S., Dassir, M., & Ridwan, M. (2018). Species variation in home garden agroforestry system in south Sulawesi, Indonesia and its contribution to farmer's income. In *IOP Conference Series: Earth and Environmental Science* (Vol. 157, pp. 012004). IOP Publishing. 10.1088/1755-1315/157/1/012004

Pala, N. A., Sarkar, B. C., Shukla, G., Chettri, N., Deb, S., Bhat, J. A., & Chakravarty, S. (2019). Floristic composition and utilization of ethnomedicinal plant species in home gardens of the eastern Himalaya. *Journal of Ethnobiology and Ethnomedicine*, *15*(1), 14. doi:10.118613002-019-0293-4 PMID:30782184

Panyadee, P., Balslev, H., Wangpakapattanawong, P., & Inta, A. (2019). Medicinal plants in homegardens of four ethnic groups in Thailand. *Journal of Ethnopharmacology*, *239*, 111927. doi:10.1016/j.jep.2019.111927 PMID:31055002

Papp, N., Birkás-Frendl, K., Farkas, Á., & Pieroni, A. (2013). An ethnobotanical study on home gardens in a Transylvanian Hungarian csángó resources village (Romania). *Genetic and Crop Evolution*, *60*(4), 1423–1432. doi:10.100710722-012-9930-7

Parthiban, R., Vijayakumar, S., Prabhu, S., & Yabesh, J. G. E. M. (2016). Quantitative traditional knowledge of medicinal plants used to treat livestock diseases from Kudavasal Taluk of Thiruvarur district, Tamil Nadu, India. [Brazilian Journal of Pharmacognosy]. *Revista Brasileira de Farmacognosia*, *26*(1), 109–121. doi:10.1016/j.bjp.2015.07.016

Peyre, A., Guidal, A., Wiersum, K. F., & Bongers, F. (2006). Dynamics of homegarden structure and function in Kerala, India. *Agroforestry Systems*, *66*(2), 101–115. doi:10.100710457-005-2919-x

Phillips, O., & Gentry, A. H. (1993). The useful plants of Tambopata, Peru: I. Statistical hypotheses tests with a new quantitative technique. *Economic Botany*, *47*(1), 15–32. doi:10.1007/BF02862203

Quinlan, M. (2005). Considerations for collecting freelists in the field: Examples from ethnobotany. *Field Methods*, *17*(3), 219–234. doi:10.1177/1525822X05277460

Rahman, M. (2013). A study on exploration of ethnobotanical knowledge of rural community in Bangladesh: basis for biodiversity conservation. *ISRN Biodiversity*, 2013.

Rajkumari, R., Nirmala, R., Singh, P., Das, A. K., Dutta, B., & Pinokiyo, A. (2014). Ethnoveterinary plants used by the chiru tribes of Manipur, northeast India.

Ramana, M. V. (2008). Ethnomedicinal and ethnoveterinary plants from Boath, Adilabad district, Andhra Pradesh, India. *Ethnobotanical Leaflets, 2008*(1), 46.

Rana, P., Tewari, S. K., Kumar, V., & Kumar, A. (2016). Floristic structure, composition and functional characteristics of homegardens in Garhwal region, Uttarakhand, India. *International Journal of Agriculture Environment and Biotechnology*, *9*(6), 1045. doi:10.5958/2230-732X.2016.00133.9

Reyes-García, V., Aceituno-Mata, L., Calvet-Mir, L., Garnatje, T., Gómez-Baggethun, E., Lastra, J. J., ... Pardo-de-Santayana, M. (2014). Resilience of traditional knowledge systems: The case of agricultural knowledge in home gardens of the Iberian Peninsula. *Global Environmental Change*, *24*, 223–231. doi:10.1016/j.gloenvcha.2013.11.022

Rugalema, G., Johnsen, F., Okting'Ati, A., & Minjas, A. (1994). The homegarden agro-forestry system of Bukoba district, north-western Tanzania. 3. An economic appraisal of possible solutions to falling productivity. *Agroforestry Systems, 28*(3), 227–236. doi:10.1007/BF00704758

Sabran, S. F., Mohamed, M., & Bakar, M. F. A. (2016). Ethnomedical knowledge of plants used for the treatment of tuberculosis in Johor, Malaysia. *Evidence-Based Complementary and Alternative Medicine, 2016*, 2850845–2850845. doi:10.1155/2016/2850845 PMID:26881002

Sadruddin. (1997). Medicinal plants of Bhutan: A conspectus. *Sherub Doenme, 3*(1 & 2), 32–55.

Saikia, P., & Khan, M. (2014). Homegardens of upper Assam, northeast India: A typical example of on farm conservation of agarwood (*Aquilaria malaccensis* Lam.). *The International Journal of Biodiversity Science, Ecosystem Services & Management, 10*(4), 262–269. doi:10.1080/21513732.2014.973449

Saikia, P., & Khan, M. L. (2011). Diversity of medicinal plants and their uses in home-gardens of upper Assam, northeast India. [AJPBR]. *Asian Journal of Pharmaceutical & Biological Research, 1*(3).

Schippmann, U., Leaman, D. J., & Cunningham, A. (2002). *Impact of cultivation and gathering of medicinal plants on biodiversity: Global trends and issues*. Biodiversity and the Ecosystem Approach in Agriculture, Forestry and Fisheries.

Shrestha, P., Gautam, R., Rana, R. B., & Sthapit, B. (2001). Home gardens in Nepal: Status and scope for research and development. *Home gardens and in situ conservation of plant genetic resources in farming systems*, 17–19.

Shrivastava, R. J., & Heinen, J. T. (2005). Migration and home gardens in the Brahmaputra Valley, Assam, India. *Journal of Ecological Anthropology, 9*(1), 20–34. doi:10.5038/2162-4593.9.1.2

Sieber, S. S., da Silva, T. C., Campos, L. Z. O., Zank, S., & Albuquerque, U. P. (2014). *Participatory methods in ethnobiological and ethnoecological research* (pp. 39–58). New York, NY: Springer New York. doi:10.1007/978-1-4614-8636-7_3

Singh, A. K., Arunachalam, A., Ngachan, S. V., Mohapatra, K. P., & Dagar, J. C. (2014). *From shifting cultivation to integrating farming: Experience of agroforestry development in the northeastern Himalayan Region* (pp. 57–86). New Delhi, India: Springer India.

Singh, K., Maikhuri, R., Rao, K., & Saxena, K. (2008). Characterizing land-use diversity in village landscapes for sustainable mountain development: A case study from Indian Himalaya. *The Environmentalist, 28*(4), 429–445. doi:10.100710669-008-9164-6

Soemarwoto, O. (1987). Homegardens: A traditional agroforestry system with a promising future. *Agroforestry: A decade of development*, 157–170.

Sthapit, B., Gautam, R., & Eyzaguirre, P. (2006). The value of home gardens to small farmers. In *Home gardens in Nepal: Proceedings of the workshop on enhancing the contribution of home garden to on-farm management of plant genetic resources and to improve the livelihoods of Nepalese farmers: Lessons learned and policy implications,* August 6–7, 2004, Pokhara, Nepal. LI-BIRD, Bioversity International and SDC. Local Initiatives for Biodiversity, 324, 8.

Subedi, A., Suwal, R., Gautam, R., Sunwar, S., & Shrestha, P. (2006). Status and composition of plant genetic diversity in Nepalese home gardens. In *Home gardens in Nepal: Proceedings of the workshop on enhancing the contribution of home garden to on-farm management of plant genetic resources and to improve the livelihoods of Nepalese farmers: Lessons learned and policy implications, August 6–7, 2004, Pokhara, Nepal. LI-BIRD, Bioversity International and SDC. Local Initiatives for Biodiversity, 324*, 7.

Szakiel, A., Paczkowski, C., & Henry, M. (2011). Influence of environmental abiotic factors on the content of saponins in plants. *Phytochemistry Reviews, 10*(4), 471–491. doi:10.100711101-010-9177-x

Tadesse, B., Mulugeta, G., Fikadu, G., Sultan, A., & Nekemte, E. (2014). Survey on ethno-veterinary medicinal plants in selected woredas of east Wollega zone, western Ethiopia. *Journal of Biology. Agriculture and Healthcare, 4*(17), 97–105.

Tekle, Y. (2014). An ethno-veterinary botanical survey of medicinal plants in Kochore district of Gedeo zone, Southern Nations Nationalities and Peoples Regional State (SNNPRS). *Journal of Scientific and Innovative Research, 3*(4), 433–445.

Thomas, E., Vandebroek, I., & Van Damme, P. (2007). What works in the field? A comparison of different interviewing methods in ethnobotany with special reference to the use of photographs. *Economic Botany, 61*(4), 376–384. doi:10.1663/0013-0001(2007)61[376:WWITFA]2.0.CO;2

Tobgay, S. (2006). Agriculture diversification in Bhutan.

Tongco, M. D. C. (2007). Purposive sampling as a tool for informant selection. *Ethnobotany Research and Applications, 5*, 147–158. doi:10.17348/era.5.0.147-158

Trinh, L. N., Watson, J. W., Hue, N. N., De, N. N., Minh, N., Chu, P., ... Eyzaguirre, P. B. (2003). Agrobiodiversity conservation and development in Vietnamese home gardens. *Agriculture, Ecosystems, & Environment, 97*(1–3), 317–344. doi:10.1016/S0167-8809(02)00228-1

Uprety, Y., Asselin, H., Boon, E. K., Yadav, S., & Shrestha, K. K. (2010). Indigenous use and bio-efficacy of medicinal plants in the Rasuwa district, central Nepal. *Journal of Ethnobiology and Ethnomedicine, 6*(1), 3. doi:10.1186/1746-4269-6-3 PMID:20102631

Vlkova, M., Polesny, Z., Verner, V., Banout, J., Dvorak, M., Havlik, J., ... Krausova, J. (2011). Ethnobotanical knowledge and agrobiodiversity in subsistence farming: Case study of home gardens in Phong My commune, central Vietnam. *Genetic Resources and Crop Evolution, 58*(5), 629–644. doi:10.100710722-010-9603-3

Vogl, C. R., & Vogl-Lukasser, B. (2003). Tradition, dynamics and sustainability of plant species composition and management in homegardens on organic and non-organic smallscale farms in alpine eastern Tyrol, Austria. *Biological Agriculture and Horticulture, 21*(4), 349–366. doi:10.1080/01448765.2003.9755278

Vogl, C. R., Vogl-Lukasser, B., & Puri, R. K. (2004). Tools and methods for data collection in ethnobotanical studies of homegardens. *Field Methods, 16*(3), 285–306. doi:10.1177/1525822X04266844

Vogl-Lukasser, B., & Vogl, C. R. (2016). No Title. *Ethnobotany Research and Applications.*

Yang, L., Ahmed, S., Stepp, J. R., Mi, K., Zhao, Y., Ma, J., ... Xue, D. (2014). Comparative homegarden medical ethnobotany of Naxi healers and farmers in northwestern Yunnan, China. *Journal of Ethnobiology and Ethnomedicine*, *10*(1), 6. doi:10.1186/1746-4269-10-6 PMID:24410825

Yusuf, M., Wahab, M., Yousuf, M., Chowdhury, J. U., & Begum, J. (2007). Some tribal medicinal plants of Chittagong hill tracts, Bangladesh. *Bangladesh Journal of Plant Taxonomy*, *14*(2), 117–128. doi:10.3329/bjpt.v14i2.531

Zimik, L., Saikia, P., & Khan, M. (2012). Comparative study on homegardens of Assam and Arunachal Pradesh in terms of species diversity and plant utilization pattern. *Research Journal of Agricultural Science*, *3*(3), 611–618.

Chapter 3
Ethnobotany:
The Traditional Medical Science for Alleviating Human Ailments and Suffering

Akash
Gurukul Kangri University, India

Navneet
Gurukul Kangri University, India

Bhupendra Singh Bhandari
HNB Garhwal University, India

Surendra Singh Bisht
HNB Garhwal University, India

Dalip Kumar Mansotra
Gurukul Kangri University, India

ABSTRACT

Traditional medicines and natural products from ethnomedicinal plants have great significance in recent time. Various forms of medicines like Ayurveda, traditional Chinese medicine, kampo, Unani, have been plasticising in recent days due to their effectiveness against various human ailments and also have blossomed into the regulated systems of traditional medicine. This chapter reviews the relationship of plants and humans, along with their cultural relationship and role of the traditional medicines, by exploring the methodologies and various concepts for the discovery of various drugs. Further, this will also illustrate traditional medicines that have their incomparable advantages over the modern allopathic medicines.

DOI: 10.4018/978-1-7998-1320-0.ch003

INTRODUCTION

Plants and the products made from have been utilized for medicinal purposes since ancient times (Samuelsson, 2004). These medicinal products have taken the form of drugs used in various herbal formulations (Samuelsson, 2004). Different plants have been used to treat a variety of disorders and ailments, and specific parts of various plants can be used to treat specific ailments. Many of these early remedies were passed down through oral history.

In recent years, it was observed that the utilization of plants and plant-based herbal products as medicines has involved the isolation of the active components from the plants in different forms, just like the isolation of morphine from the opium plant (Kinghorn, 2001; Samuelsson, 2004). The discovery of medicinal plants further led to the isolation of various drugs, including quinine, digitoxin, codeine, etc., and some are still in the development phase today (Butler, 2004; Samuelsson, 2004). Moreover, the techniques of drug discovery have recently been applied to the standardization of drugs and to elucidate analytical marker compounds. The process of obtaining medicines and plant-based herbal formulations typically started with traditional botanists, ecologists, ethnobotanists, and ethnopharmacologists. The ecologist collects basic data from the study area to provide information about how animals and plants interact with the environment. This also includes gathering information on the various aspects of the plants, such as the medicinal aspects, food and fodder-related values, and traditional remedies associated with the plants and people in the environment.

The next task is photochemistry wherein the phytochemist prepares the plant extract to test it for biological screening in pharmacologically related assays and to characterize the isolated compounds through bioassay-based fractionation. The practice of pharmacognosy was first implemented about 200 years ago (Samuelsson, 2004), when the herbal-based products analysis progressed through the formulation of crude drugs by isolating the active compounds. In a broad spectrum, pharmacognosy deals with natural products obtained from various sources, such as plants, fungi, bacteria, etc. Pharmacognosy also involves the study of dietary supplements of botany and herbal medicinal remedies (Cardellina, 2002), as well as the search for new compounds that lead to the development of new drugs.

The development of ethnobotany in recent years has challenged the prevailing trend in various academic studies in disciplinary specialization. Ethnobotany reflects congruence with the efforts of humans for determining the place of human beings in the world. Further, ethnobotany cooperates with other interdisciplinary fields, such as environmental ethics, cultural ecology, environmental history, ecological economics, conservation biology, and ecology. Ethnobotany includes the study of life forms traditionally, but is no longer considered for the study of lichens, algae, and fungi. The study of traditional science could lead to a fulfilling and fascinating career within different institutions and organizations. There is ample space for people and communities who can enjoy both statistical and analytical methods, as well as for those who might be interested in the utilization of multiple-method approaches of ethnobotany. Although some ethnobotanical surveys can also be conducted in archaeological, botanical, a computer laboratory, or in a herbarium, it is working collaboratively with the traditional people, forest communities, and with those who spend enough time with the traditional healers and old communities that is essential in ethnobotanical research.

ETHNOBOTANY: PLANTS, PEOPLE, AND CULTURE

Ethnobotany is the study of the relationships between plants, people, and culture. It is well known that if plants did not exist on earth, human life would not be possible. Human beings, animals, and plants are interdependent in different ways. Plants play an essential role in ecosystem functioning and in the operation of human societies. Plants are responsible for the transference of energy that all of the organisms on earth utilize. Ethnobotany has developed into a scientific discipline that reviews the relationships between of plant-people in a multi-disciplinary manner not through anthropology and botany but also the other branches like ecology, pharmacology, public policy, economics and other disciplines (Balick & Cox, 1996). In the last two decades, the interest of quantitative ethnobotany has boosted up in the field of medicines and traditional remedial methods for different human ailments. Further scientists also have developed different applied tools and quantitative methods to prove or test the various hypotheses about the relationship of plants, people and environment (Reyes-García, Huanca, Vadez, Leonard, & Wilkie, 2006). Authors like Phillips and Gentry (1993) and Prance, Baleé, Boom, & Carneiro (1987) tried to develop different indices in the past, which measured the significance of culture and provided a quantitative estimation of ethnomedicinal plants. In ethnobotany, it is now a trend to measure the relationship between plants, people, cultures, and the environment quantitatively. In quantitative ethnobotanical analysis, various indices are used to measure ethnomedicinal plants, for use in veterinary and human-grade medicines (Upadhyay, Singh, & Kumar, 2011), as food (Pieroni, 2001), as a disease

Table: 1. Some of the important branches of ethnobotanical research

Research Branches of Ethnobotany	Key Points	Source
Food ways and Nutritional ethnobotany	Description and identification of nutritional elements of ethnomedicinal plants in medicine and in the human diet	Etkin, 2006; Pieroni & Price, 2006
Symbolic ethnobotany	Examination of ethnomedicinal plants by ritual in ceremonial and folkloristics healing	Vildarich, 2007
Ethnobotanical system of classification	Inventing the universal naming system and categorizing living things in terms of calibrating folk and scientific issues	Brown, 1984
Paleoethnobotany	Discovery of past cultures, including traditional systems of plant resources management	Lepofsky, Heyerdahl, Lertzman, Schaepe, & Mierendorf, 2003; Fritz, 2005
Interpretive ethnobotany	Emphasizing the traditional wisdom and philosophies, and the highlights of Indigenous teachings and narratives of plants	Turner, Marshall, & Thompson, 2008
Cognitive ethnobotany	Observing the distribution and forms of plant knowledge and its transmission	Ingold, 2004; Zarger & Stepp, 2004
Ethnobotany and conservation	Identifying and safeguarding of the traditional knowledge	Minnis, 2000
Ethnobotany and agrodiversity	Investigating the knowledge and conservation of the associated knowledge through various tools and techniques	Veteto & Skarbo, 2009
Historic migrations and ethnobotany	Observing the movements of ethnobotanical cultural	Ramirez-Sosa, 2006

* Data are based on the published research from different Journals of Ethnobotany

remedial drug (Kim & Song, 2013), as fodder, or for economically related reasons (Reyes-García et al., 2006). While the goals, study scope, and study subjects vary, the goals of quantitative ethnobotanical analysis remain the same and include documenting and comparing the knowledge of plants grown and/or used by indigenous and non-native cultures, and to provide data on the number of species, number of individuals, and guilds of plants used by those groups.

Ethnobiology, ethnopharmacology, ethnoecology, and other disciplines of ethnobotany have played significant roles in improving and understanding the relationships and sustainability of humans, plants, animals, cultures, and the environment on earth. Ethnobotany documents the cultural and biological diversity among plants and animals, as well as the unique positions of various groups of peoples and cultures. This science plays a significant role as an effective tool that may vary from gathering data that provides local solutions to providing information that addresses many of the essentials issues that confront us as a species at global level. The most important and challenging issues at present are food security, deforestation, and pollution. These are the issues that directly link with the maintenance of human and animal health, and the quality of life. The practices and utilization of all of the knowledge about ethnobotanical plants can be linked to the words of Aldo Leopold, who remarked, "…a land ethic changes the role of Homo sapiens from conqueror of the land-community to plain member and citizen of it. It implies respect for his fellow-members, and also respect for the community as such."

Overall, ethnobotany can develop and strengthen human links to the environment. It is a science of great value for understanding all of the records and old experiences from that of human cultures toward the diverse areas, or further, by using all of the available records and experiences to meet the challenges of the future. Ethnobotany has made all the things from the past easy to learn and use through its study and analyses of the diverse practical approaches used by various human groups in utilizing the plants. In summary, ethnobotany is the science of preserving the diversity of medicinal flora by observing and utilizing the existing knowledge so that knowledge can be used sustainably to improve economies and effectively treat different ailments throughout the world.

The traditional cultures and practices of ethnomedicinal plant cultivation, as well as seasonal limitations have contributed to the conservation and management of these plants. Today, indigenous folk knowledge of ethnomedicinal plants is gradually decreasing due to numerous influences including an increase in modern culture, educational programmes lacking in conservation awareness, the disinterest of young age groups, secrecy within cultures, various anthropogenic pressures, the expansion of agriculture, increased deforestation, and other environmental factors.

Anthropogenic pressures, in the form of destructive practices in harvesting wherein farmers collect the whole plant, but only a few parts are used the rest discarded, is presently the most important factor threatening the survival of ethnomedicinal plants. It has been a common destructive practice to remove a whole plant in order to use only a small part of the plant for use in an ethnomedicinal study, which ultimately poses a serious threat to this category of plants. In recent times, forest communities were more familiar with the uses of ethnomedicinal plants than urban peoples. To collect medicinal plants, indigenous people travelled long distances from one area to another to locate appropriate plants. In contrast, many urban people cultivated only certain plants that they believed were useful to them as medicines, spices, fodder, etc., in their home gardens. If they were not familiar with a plant species or did not have a specific use for a plant species, they viewed it a weed and uprooted it from their gardens or fields.

Responsibility for conservation, through training, by sustainable development, or by cultivating threatened or multipurpose plants, should be regulated to indigenous traditional healers or villagers. Other important information should be made available to the community in different forms, such as through posters, leaflets, and media accessibility, in order to make conservation successful. All recommendations and suggestions should be implemented effectively so that the wealth of ethnomedicinal and indigenous knowledge can be conserved. Further, initiation of biological and pharmacological activity through the testing of the most essential and important ethnomedicinal plants should be undertaken at the community level. This could also be enhanced by establishing botanical gardens, training communities in the conservation of germplasm (seeds or tissues maintained for the purpose of animal and plant breeding, preservation, and other research uses), or through encouraging and organizing community-based conservation activities and practices.

ETHNOBOTANY AND PHARMACOLOGY

Ethnobotany is one of the most well defined sciences in the study of human culture and plants, and it can be defined as the "science of human's interaction with plants" (Turner, 1995). Ethnobotanists gather field information based on the cultural aspects of plants, and then when they return to the laboratory, they test the reliability of the uses of the plants in treating different diseases. While in the field researching, ethnobotanists study how native plants are used by the local people for cooking, healing, hunting, building, and wearing, as well as for religious and other ceremonial purposes. While doing their fieldwork, ethnobotanist researchers often collaborate with shamans, healers, and specialists, such as doctors trained in traditional and modern medicine.

The pharmacological activities of a medicinal plant are part of ethnobotanical research. And, while the fields of ethnobotany and pharmacology are related and compliment one another, the primary goal of the ethnobotanist is not to develop a new drug or bioactive compound, but to demonstrate the medicinal property of a plant in a general way. Initially, both of these branches of botany initially faced challenges in order to foster the benefits.

Pharmacologists study drug and medication actions and attempt to isolate bioactive compounds (those that exert a biochemical or physiological effect on cells, tissues, organs, or organisms), with the goal of developing safe and effective drugs that provide beneficial treatments for various ailments. The main differences between ethnobotany and pharmacology are the applications and methodologies used, which are designed for the utilization of a particular plant. Exploration of new products is the primary aim of pharmacologist, whereas, ethnobotany is generally focussed more on the cultural aspects and relationships among humans, plants, and the utilization of plants parts. It was observed that in the past, bioprospecting has guided ethnobotanists in carrying out research, but more recently, the main objective of ethnobotany shifted towards observing human cultures and exploring relationships among plants and people. On the other hand, modern ethnobotanical research largely focussed on human populations and benefits with associated with the environment (Cunningham, 2008).

There are many challenges for ethnobotanists and pharmacologists, related to both cultural and bioprospecting, and researchers must be aware of these as ignorance of these challenges does not lead to safeguarding bioprospecting. Therefore, pharmacological research and training can enhance ethnobotanical research, which never makes a collaboration resulting into the beneficial effects on the further pharmacological researchers for new drug discovery and isolation of new compounds. As ethnobotany stands at a crossroads among the biological, social, and cultural sciences, it is responsible for demonstrating a number of social, ethical, and cultural issues towards which scientists are working. Though the common goal of ethnobotany and pharmacology seems to be the same as far as curing diseases and discovering the healing aspects of plants, most of the pharmacological studies are being focussed on only one aspect of a particular disease. Pharmacological studies necessarily involve the ethnobotanical remedial system along with a biomedical approach and medical anthropological research framework (Etkin, 1993; Hahn, 1992).

ETHNOMEDICINAL PLANTS USED IN DIFFERENT HUMAN ALIMENTS

Traditional medicines and natural products are of great importance today. All of these forms of medicines, such as Ayurveda, Traditional Chinese Medicine (TCM), Kampo, Korean, and Unani have been practiced in different regions of the world since the existence of life on this planet. Various methods have been implemented to obtain an understanding of the different compounds of plants for the discovery of new drugs. This also includes methods to isolate chemicals for the treatment of various diseases, such as Alzheimer's disease (Pirttila, Wilcock, Truyen, & Damaraju, 2004). Nitisinon, a drug that was recently extracted from a plant, was found to be useful in treating inherited disease and tyrosinemia (Frantz & Smith, 2003). It is a modification of an herbicide named as "mesotrione" (Mitchell et al., 2001).

In ancient times, people used natural products, such as those obtained from animals, microorganisms, and plants, to alleviate different human diseases and disorders. Fossil records, which can be traced to about 60,000 years back, have shown the use of plants as medicine (Fabricant & Farnsworth, 2001; Shi, Li, Huo, Zhang, & Wang, 2010). The use of plants as medicines had been a great challenge to the human beings of the past. It is also known that when ancient people consumed certain plants that were not safely edible, it tended to lead to intestinal distress, vomiting, and other toxic reactions, and could even include death for the individual. So, despite all of the trial and error, humans were able to eventually develop traditional knowledge about natural medicines and edible materials (Gao, Zhang, Zhang, Guo, & Zhong, 2007).

Today, medicinal and aromatic plants (MAPs) are used in healthcare for treating both mental and physical illnesses. Indigenous groups of people, tribes, and nomads from many different areas used and continue to use plants in various healing methods to combat a numerous threatening diseases that affected both humans and animals. The uses for traditional formulas are very broad, as they offer an alternative source of medicine that has played a very essential role in many countries in alleviating different ailments (WHO, 2000). These MAP medicines are derived purely from natural products. During the past few years, considerable advances and various experiences have been developed and accumulated in the context of drug preparation from plants, utilization of medicinal plants, identification, and selection of these ethnomedicinal plants. Traditional medicinal practices offer large merits over other medicinal systems in various areas, such as the ability to test the activity of drugs and explore toxicological characteristics, physicochemical, as well as provide biochemical characterization. If any form of traditional medicine is successfully used, it may surprisingly help in the preparation of new medicines, resulting in many beneficial affects, such as a reduction in costs and greater availability.

Different plants and plant products had and have been used by human beings in curing various diseases. The leaves of plants like *eupatorium adenophorum and ageratum conyzoides* are used to treat cuts and wounds. Further, cuts, and sores, are covered with oil from the *pandanus plant* along with the latex from breadfruit trees. The fresh leaves of the *cordyline plant* and *musa paradisiaca* fruit are used to reduce the pain of scabies and other skin infections such as abscesses. The famous anti-malarial drug, "quinine" is obtained from the bark of the *cichona calisaya plant*. The drug "quinine" is valued as a

Table 2. Drugs prepared from different Ethnomedicinal plants

S.No.	Scientific Name	Family	Name of Medicine	Uses
1	*Filipendula ulmaria*	Rosaceae	Aspirin	Analgesic, inflammation
2	*Atropa belladonna*	Solanaceae	Atropine	Ophthalmology
3	*Rauvolfia* spp.	Apocynaceae	Ajmaline	Heart arrhythmia
4	*Styrax tonkinensis*	Styracaceae	Benzoin	Oral disinfectant
5	*Camellia sinensis*	Theaceae	Caffeine	Stimulant
6	*Erythroxylum coca*	Erythroxylaceae	Cocaine	Ophthalmologic anaesthetic
7	*Colchicum autumnale*	Liliaceae	Colchicinet	Gout
8	*Cinnamomum camphora*	Lauraceae	Camphor	Rheumatic pain
9	*Colchicum autumnale*	Liliaceae	Demecolcine	Leukemia, lymphomata
10	*Rhamnus purshiana*	Rhamnaceae	Cascara	Purgative
11	*Papaver somniferum*	Papaveraceae	Codeine	Analgesic, antitussive
12	*Melilotus officinalis*	Fabaceae	Dicoumarol	Thrombosis
13	*Digitalis purpnrea*	Scrophulariaceae	Digoxin	Atrial fibrillation
14	*Rauvolfia canescens*	Apocynaceae	Deserpidine	Hypertension
15	*Digitalis purpurea*	Scrophulariaceae	Digitoxin	Atrial fibrillation
16	*Ephedra sinica*	Ephedraceae	Ephedrine	Bronchodilator
17	*Hyoscyamus niger*	Solanaceae	Hyoscyamine	Anticholinergic
18	*Syzygium aromaticurn*	Myrtaceae	Eugenol	Toothache
19	*Hamamelis virginiana*	Hamamelidaceae	Gallotanins	Hemorrhoid suppository
20	*Cephaelis ipecachuanha*	Rubiaceae	Emetine	Amoebic dysentery
21	*Papaver somniferum*	Papaveraceae	Morphine	Analgesic
22	*Carica papaya*	Caricaceae	Papain	Attenuates mucus
23	*Papaver somniferum*	Papaveraceae	Papaverine	Antispasmodic
24	*Papaver somniferum*	Papaveraceae	Noscapine	Antitussive
25	*Cephaelis ipecacuanha*	Rubiaceae	Ipecac	Emetic
26	*Hyoscyamus niger*	Solanaceae	Ipratropium	Bronchodilator
27	*Anamirta cocculus*	Menispermaceae	Picrotoxin	Barbiturate antidote
28	*Pilocarpus jaborandi*	Rutaceae	Pilocarpine	Glaucoma
29	*Physostigma venenosum*	Fabaceae	Physostigmine	Glaucoma
30	*Drimia maritima*	Liliaceae	Proscillaridin	Cardiac malfunction
31	*Psoralea corylifolia*	Fabaceae	Psoralen	Vitiligo
32	*Cinchona pubescens*	Rubiaceae	Quinidine	Cardiac arrhythmia
33	*Veratrum album*	Liliaceae	Protoveratrine	Hypertension
34	*Podophyllum peltatum*	Berberidaceae	Podophyllotoxin	Condylomata acuminata
35	*Ephedra sinica*	Ephedraceae	Pseudoephedrine	Rhinitis
36	*Cinchona pubescens*	Rubiaceae	Quinine	Malaria prophylaxis
37	*Rauvolfia serpentina*	Apocynaceae	Reserpine	Hypertension
38	*Cassia angustifolia*	Caesalpiniaceae	Sennoside A,B	Laxative
39	*Datura stramonium*	Solanaceae	Scopolamine	Motion sickness
40	*Rauvoljia serpentina*	Apocynaceae	Rescinnamine	Hypertension
41	*Strophanthus gratus*	Apocynaceae	Strophanthin	Congestive heart failure
42	*Cannabis sativa*	Cannabaceae	THC	Antiemetic
43	*Podophyllum peltatum*	Berberidaceae	Teniposide	Bladder neoplasms
44	*Physostigma venenosum*	Fabaceae	Stigmasterol	Steroidal precursor
45	*Strychnos guianensis*	Loganiaceae	Toxiferine	Surgery, relaxant
46	*Camellia sinensis*	Theaceae	Theophylline	Diuretic, asthma
47	*Catharanthus roseus*	Apocynaceae	Vinblastine	Hodgkin's disease
48	*Chondrodendron tomentosum*	Menispermaceae	Tubocurarine	Muscle relaxant
49	*Catharanthus roseus*	Apocynaceae	Vincristine	Pediatric leukemia
50	*Ammi majus*	Apiaceae	Xanthotoxin	Vitiligo

* Data was collected from various published research depository and from text books

medicine worldwide and also as a flavouring agent in the United States to flavour tonic water. The fresh leaves of plants like *nicotiana tobacum, ipomoea batata,* as well as *piper betle* are used in healing severe infections. The leaves are heated, and then moistened and placed over the abscess. Various plants such as *areca* nut, fruit of *piper* inflorescence and the *spilanthes achmella* have been used to cure mouth, gum, and other oral disorders in addition to use as a mouth freshener. *Zingiber officinale* is another ancient and important ethnobotanical plant that is most widely used to alleviate the respiratory infection. The

Table 3. Some of the important ethnomedicinal plants used in different human ailments

Name	Name of plant	Family	Ethnomedicinal Uses	Source
Adusa	*Adhatoda vesica*	Acanthaceae	Roots - amoebic dysentery	Bhatt & Negi, 2006
White Ghoosefoot	*Chenopodium album*	Amaranthaceae	The leaves - bladder stone.	Gaur, 1999
Wolfsbane	*Aconitum balfourii*	Ranunculaceae	Dry root - Rheumatism	Kumar, Bussmann, Mukesh, & Kumar, 2011
Alpine Onion	*Allium humile*	Amaryllidaceae	The leaves - cold and cough	Semwal et al., 2010
Maxican poppy	*Argemone mexicana*	Papaveraceae	Powder - dysentery	Senthilkumar, Gurumoorthi, & Janardhanan, 2006
Snooth Angelica	*Angelica glauca*	Apiaceae	Powder - stomach disorder, diarrhoea	Kandari, 2005; Anonymous, 1985
Red Hogweed	*Boerhavia diffusa*	Nyctaginaceae	Root- bloody dysentery	Mitra & Gupta, 1997
Japenese Wormwood	*Artemisia japonica*	Asteracaeae	Juice - stomach disorder	Bhattacharjee, & Kumar, 2000; Kandari & Maikhuri, 2009
Himalayan Birch	*Betula utilis*	Butulacaea	Bark - muscular pain and swelling	Nautiyal, Rajan, & Shibasaki, 2004
Hairy Bergenia	*Bergenia ciliata*	Sexifragaceae	Root -wound and cut	Negi & Chauhan, 2009
Bermuda Grass	*Cynodon dactylon*	Poaceaea	Whole - diarrhoea and dysentery	Negi, Maikhuri, & Vashishtha, 2011
Mustard	*Brassica compestris*	Brassicaceae	Oil - wound	Kadel & Jain, 2008; Semwal, Saradhi, Kala, & Sajwan, 2010
Indian Rosewood	*Dalbergia sissoo*	Fabaceae	Leaves- blood dysentery, diarrhoea	Shukla, Srivastava, & Rawat, 2010
Himalayan Cedar	*Cedrus deodara*	Pinaceae	Bark Oil- skin diseases	Kala, 2009
Silver Cockscomb	*Celosia argentea*	Amaranthaceae	Decoction of seed - diarrhoea	Raju & Reddy, 2005
Cluster Fig Tree	*Ficus racemosa*	Moraceae	The latex -dysentery	Singh, & Rawat, 2011
Indian laburnum	*Cassia fistula*	Fabaceae	The bark of the plant is crushed with pepper grains - dysentery	Raju & Reddy, 2005
Buck Weed	*Fagopyrum esculentum*	Polygonaceae	The root- in urinary disorder	Nautiyal et al., 2004; Pant, Samant, & Arya, 2009; Negi & Chauhan, 2009)
Holy Basil	*Ocimum sanctum*	Lamiaceae	Leaf paste along with black pepper - diarrhoea	Kala, 2009; Semwal et al., 2010
Indian fig Tree	*Ficus auriculata*	Moraceae	Milky latex is poured - diarrhoea	Sen & Behera, 2008
Prickly Amaranth	*Amaranthus paniculatus*	Amaranthaceae	Fried seed - liver worm in children	Pant et al., 2009
Gum Arabia	*Acacia nilotica*	Fabaceae	Bark - treat dysentery	Gaur, 1999
Flame of Forest	*Butea monosperma*	Fabaceae	Seed, gum and flowers- dysentery	Uniyal & Shiva, 2005
Indian banyan Tree	*Ficus benghalensis*	Moraceae	Bark infusion- diarrhoea	Tiwari, Ballabha, & Tiwari, 2010
Screw Tree	*Helicteres isora*	Sterculaceae	Powder of fruit -diarrhoea, dysentery	Gaur, 1999; Tiwari et al., 2010
Sensitive Plant	*Mimosa pudica*	Mimosaceae	Whole plant-dysentery	Gaur & Sharma, 2011
Indrajao	*Holarrhena pubescens*	Apocynaceae	The bark -dysentery	Kumar et al., 2011
Mahwa	*Madhuca longifolia*	Sapotaceae	The infusion of flowers- diarrhoea	Tiwari et al., 2010
Broken bone Plant	*Oroxylum indicum*	Bignoniaceae	Root bark and fruit - diarrhoea and dysentery	Gaur & Sharma, 2011
Myrobalan	*Terminalia chebula*	Combretaceae	The seeds - Asthma and indigestion	Tiwari, 2010
Monkey face Tree	*Mallotus phillipensis*	Euphorbiaceae	The red fruit -constipation	Tiwari et al., 2010
Indian Goosberry	*Phyllanthus emblica*	Phyllanthaceae	Fruit and leaves- hair falling, asthma and indigestion	Pant et al., 2009
Bolly Beech	*Listea chinensis*	Lauraceae	The bark and hard wood- fractured bones	Sharma, Gairola, Gaur, & Painuli, 2011
Daar	*Boemeria rugulosa*	Urticaceae	Stem-liver disorder	Gaur, 1999
Flame fire Bush	*Woodfordia fruticosa*	Lythraceae	Red flower- piles	Kumar, 2011
Chaff Flower	*Achyranthus aspera*	Amaranthaceae	The leaves-muscular pain	Tiwari et al., 2010
Nettle Fruit Tree	*Celtis australis*	Cannabaceae	Bark - fractured bones and seed - constipation	Sharma, 2011,
Creeping Wood sorrel	*Oxalis corniculata*	Oxalidaceae	Leaves - fever	Singh & Rawat, 2011
Amaranth	*Amaranthus spinosus*	Amaranthaceae	Leaves - scorpion bite	Gaur, 1999
Asiatic Pennywort	*Centella asiatica*	Apiaceae	Leaves - memory enhancer	Gaur, 1999
Chick Weed	*Ageratum conyzoides*	Asteraceae	Leaves- cuts and wound dressing	Gaur, 1999
Wormwood	*Artemisia annua*	Asteraceae	Leaves - cosmetics medicinal	Gaur, 1999
Black Jack	*Bidens pilosa*	Asteraceae	The paste of leaf - skin diseases.	Chandra, Sachan, & Pal,2015
False Daisy	*Eclipta alba*	Asteraceae	Leaves - liver problems	Adhikari, Babu, Saklani, & Rawat, 2010)
Famine weed	*Parthenium hysterophorus*	Asteraceae	Leaves -toothaches, insect bites and infertility	Gaur, 1999

continued on following page

Table 3. Continued

Name	Name of plant	Family	Ethnomedicinal Uses	Source
Coat Buttons	*Tridax procumbens*	Asteraceae	Plant leaves -bronchial catarrh and diarrhoea	Gaur, 1999
Clotbur	*Xanthium strumarium*	Asteraceae	Leaves - tooth infection	Adhikari et al., 2010
Moonseed	*Tinospora cordifolia*	Menispermaceae	The whole plant decoction chronic dysentery	Umapriya, Rajendran, Aravindhan, Thomas, & Maharajan, 2011
Indian Plum	*Ziziphus mauritiana*	Rhamnaceae	Decoction -diarrhoea	Tiwari et al., 2010
Chaste Tree	*Vitex negundo*	Lamiaceae	The powdered of flowers - diarrhoea	Umapriya et al., 2011

**Herbal product prepared with the help of different ethnomedicinal plants

leaves and stem of the *zingiber officinale plant* is chewed and spat into a wound to reduce infection. Additionally, the leaves are also used in dressings. In ancient times, people chewed ginger when they got cuts on their hands or legs from axes or knives. Whenever, someone had an inflammation in the eyes, a little bit of *zingiber officinale* was masticated. Another very universal use of this plant was to treat colds and coughs (Ayensu, 1981). Recently the effects of *Zingiber officinale* has been tested and applied in various pharmacological and antimicrobial activities.

The list of various herbal products prepared by different companies is shown in Table 4. The following information was collected from the published database of Scopus, Web of Sciences, Science Citation Index, Indian citation index, Google Scholar, Research Gate, and Wikipedia, and then confirmed from the official portals of the companies.

CHALLENGES IN DRUG DISCOVERY FROM MEIDICINAL PLANTS

The future endeavours will face many challenges despite the success of drug discovery from ethnomedicinal plants. The phytochemists, pharmacognosists, and other ethnobotanists need to improve the quality and quantity of the new drug development phase to ensure that it keeps pace with the efforts of other discovery (Butler, 2004). It has been observed that the process of new drug discovery from medicinal and ethnomedicinal plants generally takes about 10 years and costs approximately 800 million dollars (Dickson & Gagnon, 2004). During the drug discovery process, lots of money is spent on the isolation of new bioactive compounds. Further, it has been recorded from the study that approximately 5,000 lead compounds will be estimated with the help of clinical trials, and then they will be approved. The discovery of various drugs from medicinal plants has been traditionally lengthier and more complicated than the other methods. Different pharmaceutical companies have scaled down or eliminated their natural products (Koehn & Carter, 2005). The discovery of drugs from ethnomedicinal plants has evolved various fields and different methods of study. The process of drug discovery usually begins with a plant ecologist, ethnobotanist, or ethnopharmacologist, who collects the relevant information. The collection of the plant species may be known by the collector for its biological activity or it may gathered randomly for its large screening programmes.

Improving Collections and Processes of MAPs for Drug Discovery

It should be kept in mind that the intellectual property rights belong to the country from where the plants are collected (Baker et al., 1995). Additionally, pharmaceutical industries and natural scientists need to develop and improve the quality and quantity of compounds that enter into the new drug development

Table 4. Ingredients of Boro-plus antiseptic cream (Emami International Company)

S.No	Name	Botanical Name	Family	Part Used
1. 1	Spike zinger Lily	*Hedychium spicatum*	Zingiberaceae	Root/Shoot
2.	Indian Sandalwood	*Santalum album*	Santalaceae	Stem
3.	Aloe	*Aloe barbadensis*	Xanthorrhoeaceae	Leaves
4.	Indian lilac	*Azadirachta indica*	Maliaceae	Leaves/Bark
5.	Fragrant Swamp Mallow	*Paonia odorata*	Malvaceae	Leaves
6.	Holy Basil	*Ocimum sanctum*	Lamiaceae	Leaves
7.	Turmeric	*Curcuma longa*	Zingiberaceae	Rhizome/Leaves
8.	Arabic Gum	*Acacia catechu*	Fabaceae	Stem/ Bark

Table 5. Ingredients of Thuja cream (Homeo Indian Pvt.Ltd.)

S.No	Name	Botanical Name	Family	Part Used
1	White cedar	*Thuja occidentalis*	Cupressaceae	Leaves/Stem

Table 6. Ingredients of nourishing skin cream (Himalaya Drug Company)

S.No	Name	Botanical Name	Family	Part Used
1	Alow	*Aloe barbadensis*	Xanthorrhoeaceae	Leaves
2	Poison Gooseberry	*Withania somnifera*	Solanaceae	Bark/Stem
3	Indian Kono Tree	*Pterocarpus marsupium*	Fabaceae	Stem
4	Asiatic pennywort	*Centella asiatica*	Apiaceae	Stem

phase and discoveries. Various approaches for plant-based drug discovery were considered, which included chemical screening by random selection, biological assays, traditional uses of medicinal plants, use of ethnomedicinal plants in different forms, such as decoctions powders, and enriched and standard-

Table 7. Ingredients of Patanjali Tejus cream (Patanjali Ayurveda)

S.No	Name	Botanical Name	Family	Part Used
1	Aloe	*Aloe barbadensis*	Xanthorrhoeaceae	Leaves
2	Indian Sarsaparilla	*Hemidesmus indicus*	Apocynaceae	Stem/Leaves
3	Indian Madder	*Rubia cordifolia*	Rubiaceae	Stem
4	Sandalwood	*Santalum album*	Santalaceae	Stem
5	Turmeric	*Curcuma longa*	Zingiberaceae	Rhizome/Leaves
6	Holy Basil	*Ocimum tenuiflorum*	Lamiaceae	Stem/Leaves

Table 8. Ingredients of Himalaya antiseptic cream (Himalaya Drug Company)

S.No	Name	Botanical Name	Family	Part Used
1	Chaste Tree	*Vitex negundo*	Verbenaceae	Leaves
2	Indian Madder	*Rubia cordifolia*	Rubiaceae	Leaves /Stem
3	Aloe	*Aloe barbadensis*	Xanthorrhoeaceae	Leaves
4	Almond	*Prunus dulcis*	Rosaceae	Seed

Table 9. Ingredients of Himalaya foot care (Himalaya Drug Company

S.No	Name	Botanical Name	Family	Part Used
1	Turmeric	*Curcuma longa*	Zingiberaceae	Rhizome/Leaves
2	Salasara	*Shorea robusta*	Dipterocarpaceae	Stem
3	Ginger	*Zingibera officinale*	Zingiberaceae	Rhizome/Leaves
4	Fenugreek	*Trigonella foenum-graeceum*	Fabaceae	Seed

Table 10. Ingredients of Patanjali Boro safe antiseptic cream (Patanjali Ayurveda)

S.No	Name	Botanical Name	Family	Part Used
1	Aloe	*Aloe barbadensis*	Xanthorrhoeaceae	Leaves
2	Wheat	*Triticum sativum*	Poaceae	Oil
3	Indrajao	*Holarrhena pubescens*	Apocynaceae	Whole Plant
4	Indian madder	*Rubia cordifolia*	Rubiaceae	Stem/Flower
5	Holy Basil	*Ocimum tenuiflorum*	Lamiaceae	Stem/Leaves

ized extracts of plants. These methods are all applied on different levels in different ailments and often have to face many challenges wherever they are applied, in terms of practical utility. However, these approaches can all be used in the isolation of new bioactive compounds, which have a novel shape, size, or mechanism of action. The discovery of new drugs from plants has been time consuming, so the best methodologies should be implemented, which are compound isolation, bioassay screening, etc. (Koehn et

Table 11. Ingredient of Nomarks skin cream (Bajaj Corp Company)

S.No	Name	Botanical Name	Family	Part Used
1	Indian Gooseberry	*Emblica officinalis*	Phyllanthaceae	Whole plant
2	Dhritumari	*Aloe barbadensis*	Xanthorrhoeaceae	Leaves
3	Lemon	*Citrus limon*	Rutaceae	Whole plant
4	Rose	*Rosa damascus*	Rosaceae	Flower
5	Mentha	*Mentha arvensis*	Lamiaceae	Leaves
6	Camphor	*Cinnamomum camphora*	Lauraceae	Root

Table 12. Ingredients of Eraser Antimars cream (IPSA Labs)

S.No	Name	Botanical Name	Family	Part Used
1	Lemon	*Citrus limon*	Rutaceae	Whole Plant
2	Turmeric	*Curcuma longa*	Zingiberaceae	Rhizome/Leaves
3	Holy Basil	*Ocimum tenuiflorum*	Lamiaceae	Stem/Leaves
4	Sandalwood	*Santalum album*	Santalaceae	Stem
5	Indian madder	*Rubia cordifolia*	Rubiaceae	Stem/Flower
6	Saffron	*Crocus sativus*	Iridaceae	Whole Plant

Table 13. Ingredients of Vicco turmeric cream (VICCO - Vishnu Industrial Chemical Company)

S.No	Name	Botanical Name	Family	Part Used
1	Turmeric	*Curcuma longa*	Zinziberaceae	Rhizome
2	Sandalwood	*Santalum album*	Santalaceae	Stem

Table 14. Ingredients of anti-dandruff shampoo (M.P. Business Company)

S.No	Name	Botanical Name	Family	Part Used
1	Aloe	*Aloe barbedensis*	Asphodelaceae	Leaves, Stem
2	Grape	*Vitis vinifera*	Vitaceae	Fruit
3	Indian Sarsaparilla	*Hemidesmus indicus*	Apocynaceae	Roots
4	Tea	*Camellia sinensis*	Theaceae	Leaves

Table 15. Ingredients of blood cleaner (Herbal Forever Company)

S.No	Name	Botanical Name	Family	Part Used
1	Indian Lilac	*Azadirachta indica*	Meliaceae	Leaves, Stem
2	Picorhiza	*Picrorhiza kurroa*	Scrophulaniaceae	Rhizome
3	Indian Sarsaparilla	*Hemidesmus indicus*	Apocynaceae	Roots
4	Holy Basil	*Ocimum sanctum*	Lamiaceae	Leaves
5	Buttermilk Root	*Asparagus racemosus*	Asparagaceae	Leaves
6	Babchi	*Psoralea corylifolia*	Fabaceae	Leaves
7	Asian Barberry	*Barberis asiatica*	Berberidiceae	Roots, Fruit
8	Gandhak	-	-	-
9	Indian Fumitory	*Fumaria indica*	Fumariaceae	Leaves
10	Purple Flebane	*Centratherum anthelminticum*	Asteraceae	Seeds

Table 16. Ingredients of fair skin cream (Dhathri Drug Company)

S.No	Name	Botanical Name	Family	Part Used
1	Turmeric	*Curcuma longa*	Zingiberaceae	Rhizome
2	Lotus	*Nelumbo nucifera*	Nelumbonaceae	Flowers
3	Holy basil	*Ocimum sanctum*	Lamiaceae	Leaves
4	Indian sarsaparilla	*Hemidesmus indicus*	Apocynaceae	Roots
5	Jasmine	*Jasminum officinale*	Oleaceae	Flowers

al., 2005). Further, new and better innovative techniques are needed for the processing of plant collection with regard to legal issues that fulfil benefit-sharing agreements (Rosenthal, 2002; Soejarto et al., 2004)

At the present time, different crude extracts or plant exudates are directly formulated in the form of capsules and tablets and, to some extent, they are formulated as liquid preparations. These different forms of medicinal dosage are sometimes not counted as successful due to problems encountered with their therapeutic efficacy, absorption, or poor compliance. Further, crude extracts faced difficulty in formulating solid forms due to their varying poor solubility, hygroscopic nature, and stickiness. The capsule form requires a powder of the crude herb, but the size of the powder particle affects the process of compression, blending, and filling of the capsule. In addition to this, homogeneity is hard to achieve due to high moisture content and difficulties in the handling of huge quantities.

Drug development from plants has faced a variety of challenges from time to time because of the nature of diseases, types of microorganisms, etc. The considerable rate of the increasing human population leads to the different kinds of diseases, so fast relief and high demands for drugs are needed at the same time. In a general way, natural products are isolated in very low quantities, which are generally not sufficient to lead development, optimization, or clinical trials. Thus, there should be various strategies to develop collaborations between medicinal chemists and chemists working with synthetics to explore the possibilities of combining efforts (Rosenthal, 2002; Soejarto, 2004). The level of natural compounds can be enhanced by developing natural product libraries that can combine combinatorial chemistry methods with natural product methods. The approaches to new drug development are associated with various problems including implementation of appropriate processes, design, determination, and clinical difficulties executing these high-level processes (Rosenthal, 2002; Soejarto, 2004). The problem of solubility and screening of herbal extract libraries could pose other problems during the screening of herbal extracts, but various new techniques, such as the pre-fractionation of herbal extracts, could alleviate some of these challenges (Butler, 2004; Soejarto, 2004).)

Table 17. Ingredients of Kava stress (Yogii Drug Company)

S.No	Name	Botanical Name	Family	Part Used
1	Indian sarsaparilla	*Hemidesmus indicus*	Apocynaceae	Roots
2	Camphor	*Cinamomum camphora*	Lauraceae	Bark
3	Cardamom	*Elettaria cardamomum*	Zingiberaceae	Seed
4	Ginger	*Zingiber officinale*	Zingiberaceae	Roots
5	Stevia	*Stevia rebaudiana*	Asteraceae	Leaves

HARASS OF ETHNOBOTANICAL KNOWLEDGE

At the beginning of the science of ethnobotany, ethnobotanists was initially concerned with the uses of plants and their relationships to human beings and their traditions. Ethnomedicinal work was mostly centred on the need to catalogue traditional knowledge about ethnomedicinal plants and their uses. However, researchers now find themselves engaged in a race to gather this valuable, historical information due to the fast-moving threat of losing natural resources and indigenous sources this historical knowledgeable about these plants. It is universally known that there is a close relationship between the destruction of the natural forests and the decreasing of biological diversity. Traditional knowledge is being lost throughout the world primarily due to the world's population quickly increasing, natural resources not being adequately protected. However, one of the primary threats to obtaining and preserving this knowledge is the fact that a great number of villagers and indigenous tribal communities are moving to more modern urban areas, and indigenous knowledge is not being transferred from the parental communities to the next generation due to this modernization and urbanization trend.

Large numbers of tropical forests, which were once the repositories for many ethnomedicinal plants have now been converted into shopping malls and housing complexes, resulting in a severe loss of biodiversity. In developing countries, the primary cause of deforestation is urbanization, as more and more parts of tropical forests are giving way to manicured tourists attractions and to the housing developments. The cultural-based traditional knowledge of biodiversity is widely being lost globally and at the individual or community level despite the significance of ethnomedicinal plants in improving the health of humans and their ability to survive in developed and developing countries (Samuelsson, 2004; Kinghorn, 2001).

The continuing loss of traditional knowledge has been contributed to by various factors, and major factors include cultural homogenization, increasingly globalized communities and societies, the desire for modernization, as well as the declining knowledge of plants (Cragg, Schepartz, Suffness, & Grever, 1993; Lombardino & Lowe, 2004). Further, the knowledge of ethnomedicinal plants has been identified as vulnerable to loss in all over the world (Cragg & Boyd, 1996; Newman, Cragg, & Snader, 2003). Additional factors that now threaten biodiversity and the preservation of MAP information is the increased reliance on the biomedical healthcare system, devaluation of old herbal practitioners by new generations, lack of cultural and moral support, and various pressures put upon citizens by governmental programs encouraging the adoption of modernization (Van Agtmael, Eggelte, & van Boxtel 1999; Graul, 2001; Heinrich & Teoh, 2004; Prittila, Wilcock, Truyen, & Damaraju, 2004). The permanent settlement of communities from one area to another and from one country to another, and international migrations are other aspects of globalization that are also contributing to the loss of traditional knowledge (Mundy & Kirkpatrick, 2004). Hence, a new branch of ethnobotany arises that is referred to as "Urban ethnobotany," which studies the relationship of plants, people, and the environment in a modern way (Yu, Suzuki, Xie, Morris-Natschke & Lee, 2003). Urban ethnobotany offers ample opportunities to evaluate the data of traditional knowledge based on plant and plant-based products within a greater transnational framework by comparing the knowledge of plants and uses by the same cultural group in its new situation or environment.

CONCLUSION AND FUTURE PERSPECTIVES

It is evident from this study that nature offers a great combination of medicine and plant-based medicinal products that may be turn out to be remedies that can cure all the ailments of mankind. Various products from ethnomedicinal plants have been discovered to date, which are essentially needed clinically. In spite of the variety of problems and challenges encountered in the drug discovery from ethomedicinal plants, these plants still play a significant role in the isolation and prospects of finding new compounds in the new era of medicine. Proper utilization of these ethnomedicinal plants and plant-based products could be helpful in discovering novel compounds by employing new tools and techniques of drug discovery and through developing coordinated efforts. Further, researchers, governments, and communities all need to be aware of and care about the development activities that take humanity away from its traditional knowledge, unique cultures, and individual rights, as well as to the importance of preserving biodiversity and the earth's natural resources. Hence, using development in a well-planned, positive manner to empower traditional communities, which can be benefitted through an interest in ethnobotany. If diverse ethnobotanical knowledge survives, there will be people to exchange traditional and indigenous knowledge. The documentation of ethnomedicinal plants and their uses shows a rich tradition and one that can benefit humanity in the future, but the traditional knowledge of plants is deteriorating due to the modernization and low knowledge of medicinal plants by younger generations, which could be detrimental for future generations, thus, ethnobotany is a field of science that is critically important today and one that may be able to save the future.

References

Adhikari, B. S., Babu, M. M., Saklani, P. L., & Rawat, G. S. (2010). Medicinal plants diversity and their conservation status in Wildlife Institute of India (WII) campus, Dehradun. *Ethnobotanical Leaflets, 1*, 6–9.

Anonymous. (1885). The wealth of India: A dictionary of Indian raw materials and industrial products. New Delhi, India: A. Publication and Information Directorate CSIR. *Raw Materials, 1*.

Ayensu, E. S. (1981). Medicinal plants of the West Indies. Reference Publications, Inc.

Baker, J. T., Borris, R. P., Carté, B., Cordell, G. A., Soejarto, D. D., Cragg, G. M., & Tyler, V. E. (1995). Natural product drug discovery and development: New perspectives on international collaboration. *Journal of Natural Products, 58*(9), 1325–1357. doi:10.1021/np50123a003 PMID:7494142

Balick, M. J., & Cox, P. A. (1966). *Plants, people, and culture: The science of ethnobotany*. NewYork: Scientific American Library, W. H. Freeman & Co.

Bhatt, V. P., & Negi, G. C. S. (2006). Ethnomedicinal plant resources of Jaunsari tribe of Garhwal Himalaya, Uttaranchal. *Indian Journal of Traditional Knowledge, 5*(3), 331–335.

Bhattacharjee, K., & Kumar, S. (2000). *Handbook of medicinal plants*. India: Aavishkar Publishers.

Brown, C. H. (1984). *Language and living things: Uniformities in folk classification and naming*. New Brunswick, NJ: Rutgers University Press.

Butler, M. S. (2004). The role of natural product chemistry in drug discovery. *Journal of Natural Products*, *67*(12), 2141–2153. doi:10.1021/np040106y PMID:15620274

Cardellina, J. H. (2002). Challenges and opportunities confronting the botanical dietary supplement industry. *Journal of Natural Products*, *65*(7), 1073–1084. doi:10.1021/np0200515 PMID:12141880

Chandra, P., Sachan, N., & Pal, D. (2015). Protective effect of *Dalbergia sissoo* Roxb. ex DC. (Family: Fabaceae): Leaves against experimentally induced diarrhoea and peristalsis in mice. *Toxicology and Industrial Health*, *3*(12), 1229–1235. doi:10.1177/0748233713491815 PMID:23788395

Cragg, G. M., & Boyd, M. (1996). Drug discovery and development at the National Cancer Institute: The role of natural products of plant origin. In M. J. Balick, E. Elisabetsky, & S. A. Laird (Eds.), Medicinal Plant Resources of the Tropical Forest (101–136). New York: Columbia University Press.

Cragg, G. M., Schepartz, S. A., Suffness, M., & Grever, M. R. (1993). The taxol supply crisis: New NCI policies for handling the large-scale production of novel natural product anticancer and anti-HIV agents. *Journal of Natural Products*, *56*(10), 1657–1668. doi:10.1021/np50100a001 PMID:7903979

Cunningham, A. B. (2008). People's science: Southern African ethnobotany in global perspective. *South African Journal of Botany*, *2*(74), 357–359. doi:10.1016/j.sajb.2008.01.014

Dickson, M., & Gagnon, J. P. (2004). Key factors in the rising cost of new drug discovery and development. *Nature Reviews. Drug Discovery*, *3*(5), 417–429. doi:10.1038/nrd1382 PMID:15136789

Etkin, N. L. (1993). Anthropological methods in ethnopharmacology. *Journal of Ethnopharmacology*, *38*(2-3), 93–104. doi:10.1016/0378-8741(93)90003-N PMID:8510473

Etkin, N. L. (2006). *Edible medicines: An ethnopharmacology of food.* Tucson, AZ: The University of Arizona Press.

Fabricant, D. S., & Farnsworth, N. R. (2001). The value of plants used in traditional medicine for drug discovery. *Environmental Health Perspectives, 109*(1), 69-75.

Frantz, S., & Smith, A. (2003). New drug approvals for 2002. *Nature Reviews. Drug Discovery*, *2*(2), 95–96. doi:10.1038/nrd1014 PMID:12572538

Fritz, G. J. (2005). Paleoethnobotanical methods and applications. H. D. G. Maschner, & C. Chippindale (Eds.). Handbook of archaeological methods (pp. 771–832). Walnut Creek, CA: Altamira Press.

Gao, X. M., Zhang, T. M., Zhang, J. R., Guo, J. S., & Zhong, G. S. (2007). Chinese material medica. *Beijing: China Press of Traditional Chinese Medicine, 60, 323*–324.

Gaur, R. D. (1999). *Flora of the District Garhwal, North West Himalaya.* Transmedia.

Gaur, R. D., & Sharma, J. (2011). Indigenous knowledge on the utilization of medicinal plant diversity in the Siwalik region of Garhwal Himalaya, Uttarakhand. *Journal of Forest and Environmental Science*, *27*(1), 23–31.

Graul, A. I. (2001). The year's new drugs. *Drug News & Perspectives*, *14*(1), 12–31. PMID:12819805

Hahn, G. (1992). *Hypericum perforatum* (St. John's wort): A medicinal herb used in antiquity and still of interest today. *Journal of Naturopathic Medicines, 3*, 94–96.

Heinrich, M., & Teoh, H. L. (2004). Galanthamine from snowdrop: The development of a modern drug against Alzheimer's disease from local Caucasian knowledge. *Journal of Ethnopharmacology, 92*(2-3), 147–162. doi:10.1016/j.jep.2004.02.012 PMID:15137996

Kadel, C., & Jain, A. K. (2008). Folklore claims on snakebite among some tribal communities of central India. *Indian Journal of Traditional Knowledge, 7*(2), 296–299.

Kala, C. P. (2009). Aboriginal uses and management of ethnobotanical species in deciduous forests of Chhattisgarh state in India. *Journal of Ethnobiology and Ethnomedicine, 5*(1), 20–25. doi:10.1186/1746-4269-5-20 PMID:19653889

Kandari, L. S. (2005). *Eco-physiological and socio-economic studies of some rhizomatous medicinal and aromatic plants species,* 1–153. (Doctoral dissertation). HNB Garhwal University, India.

Kandari L. S., & Maikhuri, R. K. (2009). *Artemisia maritime (Purchu): Option to improve local economy of higher Himalaya. A report* (46–47). Kurukshetra, Hariyana: Publisher Name.

Kim, H. & Song, M. J. (2013). Ethnomedicinal practices for treating liver disorders of local communities in the southern regions of Korea and alternative medicine. doi:10.1155/2013/869176

Kinghorn, A. D. (2001). Pharmacognosy in the 21st century. *Journal of Pharma and Pharmacology, 53*, 135–148.

Koehn, F. E., & Carter, G. T. (2005). The evolving role of natural products in drug discovery. *Nature Reviews. Drug Discovery, 4*(3), 206–220. doi:10.1038/nrd1657 PMID:15729362

Kumar, M., Bussmann, R. W., Mukesh, J., & Kumar, P. (2011). Ethnomedicinal uses of plants close to rural habitation in Garhwal Himalaya, India. *Journal of Medicinal Plants Research, 5*(11), 2252–2260.

Lepofsky, D., Heyerdahl, E., Lertzman, K., Schaepe, D., & Mierendorf, B. (2003). Historical meadow dynamics in southwest British Columbia: A multidisciplinary analysis. *Conservation Ecology, 7*(3), 5–9. doi:10.5751/ES-00559-070305

Lombardino, J. G., & Lowe, J. A. III. (2004). A guide to drug discovery: The role of the medicinal chemist in drug discovery—then and now. *Nature Reviews. Drug Discovery, 3*(10), 853–862. doi:10.1038/nrd1523 PMID:15459676

Minnis, P. E. (2000). Ethnobotany: A reader, (1–325). Norman, OK: University of Oklahoma Press.

Mitra, R., & Gupta, R. C. (1997). Punarnava: An Ayurvedic drug of repute. *Applied Botany Abstracts, 17*(3), 209–227.

Mundy, C., & Kirkpatrick, P. (2004). Tiotropium bromide. *Nature Reviews. Drug Discovery, 3*(8), 643–645. doi:10.1038/nrd1472 PMID:15317149

Nautiyal, S., Rajan, K. S., & Shibasaki, R. (2004). Environmental conservation vs. compensation: Explorations from the Uttaranchal Himalaya. *Environmental Informatics Archives, 2*, 24–35.

Negi, V. M., & Chauhan, N. S. (2009). Medicinal and aromatic plants wealth of a tribal district Kinnaur in Himachal Himalayas. *Indian Forester*, *135*(6), 838–852.

Negi, V. S., Maikhuri, R. K., & Vashishtha, D. P. (2011). Traditional healthcare practices among the villages of Rawain Valley, Uttarkashi, Uttarakhand, India. *Indian Journal of Traditional Knowledge*, *10*(3), 533–537.

Newman, D. J., Cragg, G. M., & Snader, K. M. (2003). Natural products as sources of new drugs over the period 1981–2002. *Journal of Natural Products*, *66*(7), 1022–1037. doi:10.1021/np030096l PMID:12880330

Pant, S., Samant, S. S., & Arya, S. C. (2009). Diversity and indigenous household remedies of the inhabitants surrounding Mornaula reserve forest in West Himalaya. *Indian Journal of Traditional Knowledge*, *8*(4), 606–610.

Phillips, O., & Gentry, A. H. (1993). The useful plants of Tambopata, Peru: Statistical hypotheses tests with a new quantitative technique. *Economic Botany*, *47*(1), 15–32. doi:10.1007/BF02862203

Pieroni, A. (2001). Evaluation of the cultural significance of wild food botanicals traditionally consumed in Northwestern Tuscany. *Italy. Journal of Ethnobiology*, *21*(1), 89–104.

Pieroni, A., & Price, L. (2006). Eating and healing: Traditional food as medicine (1–237). Binghamton, NY: Haworth Press.

Pirttilä, T., Wilcock, G., Truyen, L., & Damaraju, C. V. (2004). Long-term efficacy and safety of galantamine in patients with mild-to-moderate Alzheimer's disease: Multicenter trial. *European Journal of Neurology*, *11*(11), 734–741. doi:10.1111/j.1468-1331.2004.00885.x PMID:15525294

Prance, G. T., Baleé, W., Boom, B. M., & Carneiro, R. L. (1987). Quantitative ethnobotany and the case for conservation in Ammonia. *Conservation Biology*, *1*(4), 296–310. doi:10.1111/j.1523-1739.1987.tb00050.x

Raju, V. S., & Reddy, K. N. (2005). Ethnomedicine for dysentery and diarrhoea from Khammam district of Andhra Pradesh. *Indian Journal of Traditional Knowledge*, *4*(4), 443–447.

Ramirez-Sosa, C. R. (2006). Quantitative ethnobotany in El Salvador, Central America: A model to study ethnobotanical knowledge dynamics. (2005). In *Proceedings of the 4th International Congress of Ethnobotany (ICEB 2005)*, (557–564). Central America.

Reyes-García, V., Huanca, T., Vadez, V., Leonard, W., & Wilkie, D. (2006). Cultural, practical, and economic value of wild plants: A quantitative study in the Bolivian Amazon. *Economic Botany*, *60*(1), 62–74. doi:10.1663/0013-0001(2006)60[62:CPAEVO]2.0.CO;2

Rosenthal, J. (2002). Curtain has fallen on hopes of legal bioprospecting. *Nature*, *416*(6876), 15–19. doi:10.1038/416015a PMID:11882865

Samuelsson, G. (2004). Drugs of natural origin: A textbook of pharmacognosy, (1–350). Stockholm, Sweden: Swedish Pharmaceutical Press.

Semwal, D. P., Saradhi, P. P., Kala, C. P., & Sajwan, B. S. (2010). Medicinal plants used by local Vaidyas in Ukhimath block, Uttarakhand. *Indian Journal of Traditional Knowledge*, *9*(3), 480–485.

Sen, S. K., & Behera, L. M. (2008). Ethnomedicinal plants used by the tribals of Bargarh district to cure diarrhoea and dysentery. *Indian Journal of Traditional Knowledge*, *7*(3), 425–428.

Senthilkumar, M., Gurumoorthi, P., & Janardhanan, K. (2006). Some medicinal plants used by Irular, the tribal people of Marudhamalai hills, Coimbatore, Tamil Nadu. *Natural Product Radiance*, *5*(5), 382–388.

Sharma, J., Gairola, S., Gaur, R. D., & Painuli, R. M. (2011). Medicinal plants used for primary healthcare by Tharu tribe of Udham Singh Nagar, Uttarakhand, India. *International Journal of Medicinal and Aromatic Plants*, *1*(3), 228–233.

Shi, Q., Li, L., Huo, C., Zhang, M., & Wang, Y. (2010). Study on natural medicinal chemistry and new drug development. *Zhongcaoyao-Chinese Traditional and Herbal Drugs*, *41*(10), 1583–1589.

Shukla, A. N., Srivastava, S., & Rawat, A. K. S. (2010). An ethnobotanical study of medicinal plants of Rewa district, Madhya Pradesh. *Indian Journal of Traditional Knowledge, 9*(1) 191–202. Singh, G. & Rawat, G. S. (2011). Ethnomedicinal survey of Kedarnath Wildlife Sanctuary in Western Himalaya, India. *Indian Journal of Fundamental and Applied Life Sciences*, *1*(1), 35–46.

Soejarto, D. D., Gyllenhaal, C., Fong, H. H. S., Xuan, L. T., Hiep, N. T., Hung, N. V., ... Pezzuto, J. M. (2004). The UIC ICBG (University of Illinois at Chicago International Cooperative Biodiversity Group) Memorandum of Agreement: A model of benefit-sharing arrangement in natural products drug discovery and development. *Journal of Natural Products*, *67*(2), 294–299. doi:10.1021/np0304363 PMID:14987071

Tiwari, J. K., Ballabha, R., & Tiwari, P. (2010). Diversity and present status of medicinal plants in and around Srinagar hydroelectric power project in Garhwal Himalaya, India: Needs for conservation. *Researcher*, *2*(2), 50–60.

Turner, N. J. (1995). Ethnobotany today in Northwestern North America Portland. In R. E. Schultes, & R. S. Von (Eds.), Ethnobotany: Evolution of a discipline (pp. 264–283). Portland, OR: Dioscorides Press.

Turner, N. J., Marshall, A., & Thompson, J. C. (2008). Ebb and flow- transmitting environmental knowledge in a contemporary aboriginal community. In J. S. Lutz, & B. Neis (Eds.), *Making and moving knowledge. Interdisciplinary and community-based research in a world on the edge* (pp. 45–63). Montreal, Canada: McGill-Queen's University Press.

Umapriya, T., Rajendran, A., Aravindhan, V., Thomas, B., & Maharajan, M. (2011). Ethnobotany of Irular tribe in Palamalai Hills, Coimbatore, Tamil Nadu. *Natural Product Research*, *2*(2), 250–255.

Uniyal, B., & Shiva, V. (2005). Traditional knowledge on medicinal plants among rural women of the Garhwal Himalaya, Uttaranchal. *Indian Journal of Traditional Knowledge*, *4*(3), 259–266.

Upadhyay, B., Singh, K. P., & Kumar, A. (2011). Ethno-veterinary uses and informants consensus factor of medicinal plants of Sariska region, Rajasthan, India. *Journal of Ethnopharmacology*, *133*(1), 14–25. doi:10.1016/j.jep.2010.08.054 PMID:20817085

Van Agtmael, M. A., Eggelte, T. A., & van Boxtel, C. J. (1999). Artemisinin drugs in the treatment of malaria from medicinal herb to registered medication. *Trends in Pharmacological Sciences*, *20*(5), 199–205. doi:10.1016/S0165-6147(99)01302-4 PMID:10354615

Veteto, J. R., & Skarbø, K. (2009). Sowing the seeds: Anthropological contributions to agrobiodiversity studies. *Culture & Agriculture*, *31*(2), 73–87. doi:10.1111/j.1556-486X.2009.01022.x

Viladrich, A. (2007). Between bellyaches and lucky charms: Revealing Latinos' plant-healing knowledge and practices in New York City. In A. Pieroni, & I. Vandebroek (Eds.), Travelling cultures and plants: The ethnobiology and ethnopharmacy of migrations, (64–85). New York: Berghahn Books.

World Health Organisation (WHO). (2000). *General guidelines for methodologies on research and evaluation of traditional medicine*. Geneva, Switzerland: World Health Organisation.

Yu, D., Suzuki, M., Xie, L., Morris-Natschke, S. L., & Lee, K. H. (2003). Recent progress in the development of coumarin derivatives as potent anti-HIV agents. *Medicinal Research Reviews*, *23*(3), 322–345. doi:10.1002/med.10034 PMID:12647313

Zarger, R., & Stepp, J. (2004). Persistence of botanical knowledge among Tzeltal Maya children. *Current Anthropology*, *45*(3), 413–418. doi:10.1086/420908

Chapter 4
Role of Traditional Ethnobotanical Knowledge:
Culture and Indigenous Institutions in Medicinal Plant Conservation

Uday Kumar Sen
https://orcid.org/0000-0002-1901-3069
Department of Botany and Forestry, Vidyasagar University, West Bengal, India

Ram Kumar Bhakat
Ecology and Taxonomy Laboratory, Department of Botany and Forestry, Vidyasagar University, Midnapore, India

ABSTRACT

In recent decades, species extinction has increased sharply. Most species in the world are discovered in tropical forests – covering more than 10% of the planet. Sustainable management and conservation efforts must include local communities and their traditional knowledge. The traditional use of forest resources, especially non-timber products such as medicinal plants, not only has deep roots in indigenous populations, but is also practiced in a broad segment of culture. Using medicinal herbs is often a costly alternative to modern medicine that is economically inevitable. This traditional use's basic knowledge is carried from one generation to the next. Medical use in particular reflects an extremely vibrant, ever-evolving process in which fresh knowledge is constantly acquired and connected to traditional methods. An instance from West Bengal's West Midnapore district in India is used to illustrate the impacts of an integrated strategy to preserving biodiversity and culture.

INTRODUCTION

As a manner of understanding, traditional wisdom is comparable to contemporary science because it is based on an accumulation of observations, but in some basic respects, it is distinct from science (Berkes, 1993). Traditional knowledge, as a way of knowing, is comparable to modern science because it is based

DOI: 10.4018/978-1-7998-1320-0.ch004

on an accumulation of observations (Berkes, Colding, & Folke, 2000). These two ways of knowing are two parallel modes of acquiring knowledge about the universe; the two sciences are fundamentally distinct. Interest in Traditional Ecological Knowledge (TEK) has grown in latest years, partially because such understanding can contribute to biodiversity conservation (Gadgil, Berkes, & Folke, 1993), rare species (Colding, 1998), protected areas (Johannes, 1998), ecological processes (Alcorn, 1989), and to sustainable resource use in general (Schmink, Redford, & Padoch, 1992; Berkes, 1999). For science, social or economic factors, conservation biologists, ecological anthropologists, ethnobiologists, other scientists, and the pharmaceutical industry share an interest in traditional knowledge. In the course of the field grows, the research of Traditional Ecological Knowledge started with the research of species identification and classification (ethnobiology), taking into account the understanding of ecological processes by populations and their relationship with the setting (Williams & Baines, 1993; Berkes, 1999).

On a worldwide scale, accessible evidence points to a direction in which Traditional Ethnobotanical Knowledge (TEK) is increasingly relevant as an invaluable, under-used and undocumented knowledge pool (Camara-Leret, Paniagua-Zambrana, Balslev, & Macia, 2014). This provides a strong tool for many advanced and developing countries, especially in Asia, to tackle the problems of plant resource conservation (Chepkosgey & Jerotich, 2016). In 1992, the Convention on Biological Diversity (CBD) was the first to create policies linked to the conservation and sustainable use of biodiversity for the use and preservation of traditional knowledge. Countries that remain are anticipated to (1) Promoting the use of Indigenous Knowledge (IK) technologies in the management of natural resources, (2) Increase and scale up the use of indigenous knowledge and (3) encourage equity and access to benefits from the use of indigenous knowledge technologies (CBD, 1992). In their draft papers, the United Nations Scientific Conference Organization (UNESCO) and the International Council for Science Union (ICSU) recognize IK's position and beg for its implementation in all types of human commitment (UNESCO, 2002). In defining TEK, different writers concentrate on the characteristics of local communities' perception, management and use of plant resources (Balick & Cox, 1996; Ugulu, 2011). Specifically, TEK study focuses on 'how individuals classify, recognize and connect to plant resources, examine plant and human interactions, taxonomic identification of chosen plants and biological and chemical analysis of their components' (Ugulu, 2011). In other words, TEK includes plant research as used in indigenous cultures for food, medicine, rituals, construction, household appliances, firewood, pesticides, clothing, shelter and other useful purposes (Amsalu, Bezie, Fentahun, Alemayehu, & Amsalu, 2018). We postulate that the management and use of plant resources by individuals is based on the understanding, priorities and perceptions of the concerned natural environmental resources and ecological processes. We postulate that the management and use of plant resources by individuals is based on the understanding, priorities and perceptions of the concerned natural environmental resources and ecological processes. The study identifies the plant resources that are perceived by people as Indigenous People resources and undertakes an evaluation of these resources. It documents and assesses the TEK associated with the utilization of plant resources and examines how the resources are defined by use and culture.

1. TEK in Perspective

In a vibrant equilibrium based on cycles and fluctuations, ecosystems maintain themselves, which are nonlinear processes. Thus, environmental awareness will only emerge when we combine our rational understanding with an intuition of the non-linear nature of our surroundings. Such intuitive wisdom is characteristic of traditional, non-literate societies, where life has been structured around an extremely sophisticated environmental awareness. TEK reflects knowledge gained from direct human interaction with the setting over thousands of years. Despite the extensive use of the word TEK in the 1980s, TEK practice is as old as ancient societies of hunter-gatherers. In several areas, besides ecology, the study of traditional knowledge is appreciated. For instance, study on traditional knowledge has a wealthy background in agriculture, pharmacology and ethnobotany. Indeed, the research of indigenous knowledge in ecology is comparatively recent compared to these areas.

2. Practical Significance of TEK

It follows from these factors that for social and cultural reasons the conservation of TEK is essential. TEK is a concrete element of a manner of life that can be deemed useful for the group in question. There are also concrete and practical reasons for TEK to be so essential for the remainder of the globe, quite apart from the ethical imperative of cultural diversity preservation. The following list is tailored from the IUCN Traditional Conservation Knowledge Program.

a) Traditional knowledge for new biological and ecological insights. New scientific knowledge can be derived from perceptive investigations of traditional environmental knowledge systems.
b) Traditional resource management expertise. For modern natural resource management, a lot of traditional knowledge is important.
c) Traditional knowledge for protected areas and education in conservation. Protected areas can be established so that resident groups can continue their traditional lifestyles with the advantages of conservation.
d) Traditional knowledge for the planning of growth. Using traditional knowledge can help development agencies deliver more realistic environmental assessments.
e) Traditional environmental evaluation knowledge. People who depend for their livelihood on local resources are often better prepared to evaluate the real expenses and advantages of growth than any external evaluator.

3. Key Socio-Cultural Role of TEK Transmission

Besides its practical elements of ensuring sustainable resource management, traditional knowledge transmission is of basic socio-cultural significance to any community. Social institutions are gradually crystallized during information transmission over several generations; routine or usual methods of doing stuff gradually become the usual way things are accomplished. For children, the customary way of a community eventually becomes the social world received, an analog of the biological-physical world with which it overlaps. The transmitter's sense of truth is reinforced in the process of transferring understanding to a fresh generation. During transmission, the social world, embodied in traditional knowledge, is expanded. But, of course, every fresh generation of information recipients knows the history and context of the institutions of their society only through increasingly attenuated hearsay. Therefore, educators must provide the rationale behind custom, tradition, normative and real conduct and rules and regulations through coherent and thorough legitimating to educators (Hannan & Freeman, 1989; Haunss, 2013).

4. TEK and Biodiversity: The Collaboration of Formal and Non-Formal Institutions

The significance of biodiversity conservation and worldwide engagement is no longer in question. The accelerating loss rates of floral and faunal species and the predicted adverse human effects of this germplasm loss (Braatz, 1992; Dobson, 1992). Many global foundations, development agencies, and international agricultural research centres also add the strength of their collective interest and resolve to address the conditions leading to species loss (Horton & Mackay, 2003). While these papers concentrate on the instant and long-term adverse biological and financial effects of biodiversity loss, several bring the supplementary significance of cultural diversity that is often reflected in indigenous knowledge of the management of natural resources, including plants and animals (Grenier, 1998). For instance, the Global Biodiversity Strategy involves the principle of biodiversity as one of its ten principles "Cultural

Table 1. Social-ecological practices and mechanisms in traditional knowledge and practice (adapted from Folke, Berkes, & Colding, 1998).

1. Social mechanisms behind management practices		2. Management practices based on ecological Knowledge	
(A). Generation, accumulation and transmission of local ecological knowledge	a) Reinterpreting signals for learning b) Revival of local knowledge c) Folklore and knowledge carriers d) Integration of knowledge e) Intergenerational transmission of knowledge f) Geographical diffusion of knowledge	(A). Practices found both in conventional resource management and some local and traditional Societies	a) Monitoring resource abundance and change in ecosystems b) Total protection of certain species c) Protection of vulnerable life history stages d) Protection of specific habitats e) Temporal restrictions of harvest
(B). Structure and dynamics of institutions	a) Role of stewards/wise people b) Cross-scale institutions c) Community assessments d) Taboos and regulations e) Social and religious sanctions	(B). Practices largely abandoned by conventional resource management but still found in some local and traditional societies	a) Multiple species management b) Maintaining ecosystem structure and function c) Resource rotation d) Succession management
(C). Mechanisms for cultural internalization	a) Rituals, ceremonies and other traditions b) Cultural frameworks for resource management	(C). Practices related to the dynamics of complex systems, seldom found in conventional resource management but found in some traditional societies	a) Management of landscape patchiness b) Watershed-based management c) Managing ecological processes at multiple scales d) Responding to and managing pulses and surprises e) Nurturing sources of ecosystem renewal
(D). World view and cultural values	a) A world view that provides appropriate environmental ethics b) Cultural values of respect, sharing, reciprocity, humility and other deshi char-dwellers. Other examples include integrating farming and aquaculture. While common in traditional societies, such practices have largely been abandoned in conventional resource management (Berkes, Colding, & Folke, 2000).		

diversity is strongly related to biodiversity. Humanity's collective understanding of biodiversity and its use and management rests on cultural diversity; on the other hand, biodiversity conservation often contributes to strengthening cultural integrity and values" (World Resources Institute, 1992).

Progress in understanding the complementarity of cultural diversity and biodiversity has been produced over the previous century. *Ex-situ* germplasm conservation has been well created and a parallel set of organizations is now being developed that document local understanding about the environment, indigenous knowledge resource centres that could play a future part in *in-situ* conservation. Understanding the indigenous management policies of farmers and other rural people that promote diversity in domesticated and wild species can assist to develop domestic programs for *in-situ* germplasm conservation that complement existing *ex-situ* programs (Frankel, Brown, & Burdon, 1995; Padulosi, Leaman, & Quek, 2002). To gain an extensive knowledge base for the conservation of genetic resources, the genetic institution must acknowledge a mandate that deals not only with germplasm but also with the knowledge systems that generate it (Manicad, 1996). The role of colonialism and social science in ignoring and sometimes maligning indigenous knowledge was well documented (Warren, 1992). Studies that portrayed local groups and their expertise as primitive, easy and static are now countered by a quickly growing database produced by biological and social researchers that explains the complexity and sophistication of many native natural resource management structures (Robbins, 2011).

5. TEK, Social-Ecological Methods and Processes

a) Monitoring Resource Abundance and Change in Ecosystems

Monitoring is a key activity of conservation of biodiversity and preservation biology (Marsh & Trenham, 2008) and has been defined worldwide as a centrepiece of nature conservation (Schmeller, 2008). And yet, as is often stated, many surveillance programs do not have a sound ecological foundation, are poorly intended, do not result in measures or reactions to leadership, and are disconnected from decision-making. Monitoring is often provided low priority because the implementation can be hard and costly and monitoring programs are often poorly financed and poorly enforced (Danielsen et al., 2009). Monitoring essentially involves making credible observations of nature to detect, evaluate and draw conclusions as to how species and ecosystems change through time and space, either naturally or as a result of intentional or inadvertent human interference. It is implemented in many distinct ways – to monitor the status of threatened species, the spread of invasive species, ecosystem health, the efficiency of protected fields and other conservation activities, and more usually to evaluate biodiversity status and major trends through indices and tracking at domestic, regional and international level (Pringle, 2017).

b) Protection of Species

Species conservation relies on the preservation of their habitats. Habitat classifications can be based on structural or abiotic characteristics, but they are often based on syntax, conveying a summary of difficult to assess or model ecosystem characteristics. Preserving the diversity of existing plant communities is believed to protect the survival not only of plants, but also of livestock, fungi, and microorganisms of typical species, as well as the maintenance of present ecosystem processes (Naveh & Lieberman, 2013).

c) Protection of Vulnerable Taxa

For many centuries, much of the world's biodiversity has been in the hands of traditional peoples, hunting and gathering societies, herders, fishermen, farmers. Most of the earth's living resources have been used for a long time in history. The reality is that pre-scientific, traditional management systems were the primary means for societies to manage natural resources for thousands of years. The reality is that pre-scientific, traditional management systems were the primary means for societies to manage natural resources for thousands of years (Berkes, Folke, & Gadgil, 1994). Despite, and in some instances because of, these management schemes, biological diversity has continued so that we have to talk about any biodiversity today.

d) Protection of Specific Habitats

Habitat is the physical and biological environment in which organisms reside and other environmental elements are found (Krebs, 1985; Jones, 1987). Habitat is the physical and biological setting in which organisms live and in which the other components of the environment are encountered (Clements & Shelford, 1939); habitat is a basic requirement of all living organisms (McNaughton, 1989). Habitat, together with weather factors, nutrients and other interacting organisms, is one of the four elements of a species' setting (Andrewartha & Birch, 1954). To understand its complete function in the Endangered Species Act, the fact that habitat serves a variety of species is critical. Many species were not categorized or determined their status. Our species knowledge is too restricted and the species deserving of endangered or threatened status too countless to list all that might merit it in an appropriate timeframe for their protection. It is estimated that the amount of unclassified living species is two to ten times the amount recognized and named (Wilson, 1992). Because the job is so enormous and there are few experts for many plant taxa, prospects are scanty for a substantial, near-term shift in this information gap. Therefore, wide conservation measures based on ecosystems that are not dependent on taxonomic awareness or the determination of individual species listing status are also required to avoid species extinction.

6. Traditional Conservation of Plants

The use of herbal medicines has increased over time due to the belief that herbal remedies are secure, cheap and accessible to local populations in developing nations. Local individuals have gained significant traditional knowledge about their use of medicinal plants while interacting with the setting. Due to enhanced use and economic value of some crops, demand for medicinal plants has also risen. Past studies have shown that nearly one-third of medicinal plant species may extinct, in Asia and Africa (Hamilton, 2004; Njoroge, Kaibui, Njenga, & Odhiambo, 2010). Traditionally sustainable use of medicinal plants has been controlled by management procedures such as taboos, seasonal and social constraints on growing medicinal plants that have been used to restrict the consumption of medicinal plants. Cunningham (1993) and Mavi & Shava (1997) demonstrate that other factors limit species stress from over-exploitation:

i. Use of taboos in India which restricted menstruating women from collecting medicinal plants. This was thought to prevent a decrease in medicine's healing energy.
ii. Women in India, while males practised as herbalists, tended to practice as diviners, limiting the number of resource consumers.

iii. Traditional control of instruments used to remove some plants' barks to wooden sticks rather than metal instruments.

7. Taboos and Social Controls on the Harvesting of Medicinal Plants

In addition to harvesting checks, there have been taboos that have guided people's social behaviour in culture in harvesting traditional medicinal plants. Interviewed elderly heads of households and herbalists reported social controls and prohibitions on who can harvest or manage medicinal plants. Medicinal plants were shielded and preserved by the taboos and myths. For example, a female in her reproductive era was not permitted to practise community herbal medicine. In society, mature females who had previous menopause were permitted to practise herbal medicine. However, mothers with childbearing were only permitted to harvest components of medicinal plants for their kids to be treated. This limited the individuals who harvested medicinal plants to only mature individuals who were held accountable (Little, 1965; Monica, Mark, & Paul, 2016).

8. Social Institutions in Ecosystem Management and Biodiversity Conservation

Local organizations and management procedures linked to TEK schemes can provide hints in many environments around the globe for enhancing natural resource management. The main component in the growth of many traditional and local resource management methods appear to be ecological knowledge and understanding supplied through environmental monitoring. Local community management methods do not exist in a vacuum but are framed by a social context. They, therefore, tend to be linked and integrated with informal organizations and other kinds of social processes that are backed by a worldview and cultural values that do not separate individuals from their reliance on the natural system (Colding, Folke, & Elmqvist, 2003).

Successful resource management systems involve flexible social processes to adapt continuously to the dynamics of the environment. Therefore, institutional buildings are required to take into account environmental variability and ecological feedback with the ability to respond to such dynamics. In local groups, there are wide ranges of procedures with the ability to confer resilience in ecological systems. Informal organizations also exist with the ability to adapt to ecological feedback and the ability to preserve critical biota (Colding, Folke, & Elmqvist, 2003).

Table 2: Resource and habitat taboos and their nature conservation and resource management functions (Colding & Folke, 2001).

Category	Function
Segment taboos	Regulate resource withdrawal
Temporal taboos	Regulate access to resources in time
Method taboos	Regulate methods of withdrawal
Life-history taboos	Regulate withdrawal of vulnerable life history stages of species
Specific-species taboos	Total protection to species in time and space
Habitat taboos	Restrict access and use of resources in time and space

9. Cultural Internalization

Cultural theory often emphasizes internalized beliefs and values, but empirical research often focuses solely on attributing values and beliefs to leaders of organizations. Admittedly, there are advantageous synergies to be obtained from the game between cultural meanings and internalized private beliefs and values, especially with regard to commitment and inherent motivations (Sathe, 1985). These synergies to claim that the core of culture is sharing those norms, internalized values and beliefs that are more taken for granted.

10. Ethnobotanical Information on the Uses of Plants in Medicinal and Other Purposes

An ethnobotanical survey was carried out to collect information on the uses of plants in medicinal and other purposes by local people in West Midnapore district, West Bengal, India. Detailed information on medicinal plants was gathered from the medicine-men known as 'Gunin/Kabiraj' to whom the knowledge was passed on from their ancestors (Das & Pandey, 2007). Other ethnobotanical data were prepared, including the collection of information through folk, oral tradition, etc. and voucher specimens were made as per Jain & Rao (1977).

West Midnapore district was blessed with a rich and diverse heritage of cultural traditions. These traditions were associated with the use of wild plants as medicinal herbs. The use of medicinal herbs is still a tradition adopted by ethnic communities who are living in undulating areas and plains. The ethnic people of the district use an extensive range of wild plants for their health care by monoherbal (Table 3) and polyherbal (Table 4) formulation (Sen, 2016).

Drug formulation is based on two principles: Use as a single drug and use of more than one drug, in which the latter is known as polyherbal formulation. This key traditional therapeutic herbal strategy exploits the combining of several medicinal herbs to achieve extra therapeutic effectiveness, usually known as polyherbal. Due to synergism, polyherbal confers some benefits not available in the single herbal formulation. It is evident that a better therapeutic effect may be reached with a single multi-constituent formulation.

CONCLUSION

Attempts to preserve and handle biodiversity need to incorporate local communities opinions, needs and traditional knowledges. In this phase, the connection between researchers, non-governmental organizations and local individuals is inevitable, as each aspect of the network needs to contribute specifically to its achievement. At the academic level, pure ethnobotanical studies must be included in broader interdisciplinary research, bringing together as many natural and social science disciplines as possible to provide a sound basis for implementation. In order to be helpful, the primary findings of such studies must be made accessible in local languages. Only management and conservation actions based on such an integrated strategy are viable on a long-term basis.

The traditional medicine used in the region lacks phyto-therapeutic evidence. It is too necessary to perform photochemical or pharmacological studies to explore the potentiality of plants used for medicinal purpose. The unsustainable harvesting of such medicinal plants that are obtained from wild sources may

Table 3: Local use of monospecific medicinal plants to treat different diseases.

Sl. No.	Disease	Plants	Family	Parts used
1.	**Anaemia**	*Hygrophila auriculata* (Schumach.) Heine	Acanthaceae	Leaf decoction
2.	**Anti-allergic**	*Solanum americanum* Mill.	Solanaceae	Root paste
3.	**Anti-dandruff**	*Sapindus trifoliatus* L.	Sapindaceae	Fruit soaked in water
4.	***Strychnos nux-vomica* poisoning**	*Canna indica* L.	Cannaceae	Stalk boiled
5.	**Antiseptic**	*Bauhinia variegata* L.	Fabaceae	Leaf paste
		Centipeda minima (L.) A. Braun & Asch.	Compositae	Flower and leaf juice
		Citrus maxima (Burm.) Merr.	Rutaceae	Fruit juice
		Holoptelea integrifolia Planch.	Ulmaceae	Leaf paste
		Hydrolea zeylanica (L.) Vahl	Hydroleaceae	Leaf paste
		Limnophila heterophylla (Roxb.) Benth.	Plantaginaceae	Plant juice
		Melochia corchorifolia L.	Malvaceae	Root paste
		Ocimum tenuiflorum L.	Lamiaceae	Leaf juice
		Schoenoplectiella articulata (L.) Lye	Cyperaceae	Root juice
		Trianthema portulacastrum L.	Aizoaceae	Crushed root
6.	**Appetizer**	*Azadirachta indica* A.Juss.	Meliaceae	Cooked leaves
		Citrus medica L.	Rutaceae	Fruit
		Jasminum multiflorum (Burm.f.) Andrews	Oleaceae	Roasted
		Polycarpon prostratum (Forssk.) Asch. & Schweinf.	Caryophyllaceae	Cooked young twigs with leaves
		Trichosanthes dioica Roxb.	Cucurbitaceae	Cooked leaves
7.	**Astringent**	*Agave vivipara* L.	Asparagaceae	Root paste
		Laphangium luteoalbum (L.) Tzvelev	Compositae	Leaf extract
		Soymida febrifuga (Roxb.) A. Juss.	Meliaceae	Bark extract
8.	**Bite**			
A.	**Bee bite**	*Indigofera linnaei* Ali	Fabaceae	Leaf juice
B.	**Cat-fish bite**	*Calotropis gigantea* (L.) Dryand.	Apocynaceae	Latex
		Vitex negundo L.	Lamiaceae	Latex
C.	**Dog bite**	*Lygodium flexuosum* (L.) Sw.	Lygodiaceae	Leaf paste
		Ochna obtusata DC.	Ochnaceae	Root paste
D.	**Miscellaneous insect bite**	*Abutilon indicum* (L.) Sweet	Malvaceae	Crushed leaf, root juice
		Acalypha indica L.	Euphorbiaceae	Leaf juice
		Calotropis gigantea (L.) Dryand.	Apocynaceae	Latex
		Euphorbia nivulia Buch.-Ham.	Euphorbiaceae	Latex
		Ipomoea aquatica Forssk.	Convolvulaceae	Plant sap
		Nerium oleander L.	Apocynaceae	Root bark paste
		Vitex negundo L.	Lamiaceae	Latex
E.	**Rat bite**	*Senna occidentalis* (L.) Link	Fabaceae	Leaf extract
		Tecoma stans (L.) Juss. ex Kunth	Bignoniaceae	Leaf paste
F.	**Removal of caterpillar sting**	*Luffa acutangula* (L.) Roxb.	Cucurbitaceae	Fresh leaf juice
G.	**Scorpion bite**	*Citrus medica* L.	Rutaceae	Fruit juice
		Curculigo orchioides Gaertn.	Hypoxidaceae	Root paste
		Desmodium gangeticum (L.) DC.	Fabaceae	Root extract
		Tecoma stans (L.) Juss. ex Kunth	Bignoniaceae	Leaf, root paste
H.	**Snake bite**	*Acalypha indica* L.	Euphorbiaceae	Leaf juice
		Aegle marmelos (L.) Corrêa	Rutaceae	Crushed root bark
		Alstonia scholaris (L.) R. Br.	Apocynaceae	Bark paste
		Aristolochia indica L.	Aristolochiaceae	Leaf, root paste
		Artocarpus heterophyllus Lam.	Morceae	Leaf juice
		Cissampelos pareira L.	Menispermaceae	Root paste
		Commelina benghalensis L.	Commelinaceae	Plant juice
		Curcuma aromatica Salisb.	Zingiberaceae	Rhizome paste
		Desmodium gangeticum (L.) DC.	Fabaceae	Root extract
		Dicliptera paniculata (Forssk.) I.Darbysh.	Acanthaceae	Plant juice
		Euphorbia neriifolia L.	Euphorbiaceae	Latex
		Gloriosa superba L.	Colchicaceae	Crushed rhizome
		Gomphrena celosioides Mart.	Amaranthaceae	Plant juice
		Heliotropium indicum L.	Boraginaceae	Crushed rood
		Hemidesmus indicus (L.) R. Br. ex Schult.	Apocynaceae	Root paste
		Leucas lavandulifolia Sm.	Lamiaceae	Crushed leaf
		Nerium oleander L.	Apocynaceae	Root paste
		Plumbago zeylanica L.	Plumbaginaceae	Root bark paste
		Portulaca quadrifida L.	Portulacaceae	Warm plant juice

continued on following page

Table 3. Continued

Sl. No.	Disease	Plants	Family	Parts used
		Senna occidentalis (L.) Link	Fabaceae	Root extract
		Tecoma stans (L.) Juss. ex Kunth	Bignoniaceae	Leaf, root paste
		Typhonium trilobatum (L.) Schott	Araceae	Aerial plant part juice
I.	**Spider poisoning**	*Lablab purpureus* (L.) Sweet	Fabaceae	Leaf paste
9.	**Bleeding**	*Achyranthes aspera* L.	Amaranthaceae	Crushed leaves
		Alocasia macrorrhizos (L.) G.Don	Araceae	Leaf, petiole sap
		Amaranthus viridis L.	Amaranthaceae	Root juice
		Ayapana triplinervis (Vahl) R.M.King & H.Rob.	Compositae	Leaf juice
		Barleria cristata L.	Acanthaceae	Leaf juice
		Butea superba Roxb.	Fabaceae	Stem bark juice
		Cascabela thevetia (L.) Lippold	Apocynaceae	Latex
		Cayratia pedata (Lam.) Gagnep.	Vitaceae	Leaf paste
		Cayratia trifolia (L.) Domin	Vitaceae	Leaf juice
		Chromolaena odorata (L.) R.M.King & H.Rob.	Compositae	Leaf juice
		Colocasia esculenta (L.) Schott	Araceae	Petiole juice
		Croton bonplandianus Baill.	Euphorbiaceae	Latex
		Cucurbita pepo L.	Cucurbitaceae	Leaf juice
		Mikania micrantha Kunth	Compositae	Leaf juice
		Mimosa pudica L.	Fabaceae	Leaf juice
		Ocimum americanum L.	Lamiaceae	Seed smoke
		Tagetes erecta L.	Compositae	Leaf juice
		Tridax procumbens (L.) L.	Compositae	Crushed leaves
10.	**Boils/ Sores**	*Annona squamosa* L.	Annonaceae	Leaf paste, seed powder
		Azadirachta indica A.Juss.	Maliaceae	Leaf paste
		Cassia fistula L.	Fabaceae	Leaf and Stem bark paste
		Dendrocalamus strictus (Roxb.) Nees	Poaceae	Leaf paste
		Dentella repens (L.) J.R.Forst. & G.Forst.	Rubiaceae	Plant paste
		Desmodium velutinum (Willd.) DC.	Fabaceae	Stem bark paste
		Gardenia gummifera L.f.	Rubiaceae	Bud resin
		Ipomoea pes-caprae (L.) R. Br.	Convolvulaceae	Root paste
		Litsea glutinosa (Lour.) C.B.Rob.	Lauraceae	Bark paste
		Merremia umbellata (L.) Hallier f.	Convolvulaceae	Plant juice
		Mirabilis jalapa L.	Nyctaginaceae	Leaf decoction
		Murdannia nudiflora (L.) Brenan	Commelinaceae	Leaf paste
		Ocimum gratissimum L.	Lamiaceae	Seed paste
		Plumbago zeylanica L.	Plumbaginaceae	Leaf sap
		Pterocarpus marsupium Roxb.	Fabaceae	Leaf paste
		Sida cordifolia L.	Malvaceae	Smashed leaf
		Trichosanthes cucumerina L.	Cucurbitaceae	Plant extract
11.	**Burn**	*Amaranthus spinosus* L.	Amaranthaceae	Smashed plant
		Mangifera indica L.	Anacardiaceae	Leaf paste
		Tagetes erecta L.	Compositae	Leaf paste
12.	**Cardiac disorders**	*Catharanthus roseus* (L.) G.Don	Apocynaceae	Leaf juice
		Cocculus hirsutus (L.) W.Theob.	Menispermaceae	Crushed leaf
		Lygodium flexuosum (L.) Sw.	Lygodiaceae	Leaf paste
		Moringa oleifera Lam.	Moringaceae	Leaf juice
		Rauvolfia tetraphylla L.	Apocynaceae	Crushed root
		Terminalia arjuna (Roxb. ex DC.) Wight & Arn.	Combretaceae	Bark decoction
13.	**Development of skull bone**	*Celastrus paniculatus* Willd.	Celastraceae	Seed oil
		Gmelina arborea Roxb.	Lamiaceae	Fruit paste
14.	**Diabetes**	*Catharanthus roseus* (L.) G.Don	Apocynaceae	Plant juice
		Coccinia grandis (L.) Voigt	Cucurbitaceae	Root, leaf juice
		Eleusine indica (L.) Gaertn.	Poaceae	Plant extract
		Ficus hispida L.f.	Moraceae	Root exudates
		Ficus racemosa L.	Moraceae	Fruit juice
		Gymnema sylvestre (Retz.) R.Br. ex Sm.	Apocynaceae	Leaf decoction
		Pterocarpus marsupium Roxb.	Fabaceae	Bark decoction
		Saccharum spontaneum L.	Poaceae	Leaf juice
15.	**Ear-ache**	*Calotropis gigantea* (L.) Dryand.	Apocynaceae	Leaf juice
		Cleome viscosa L.	Cleomaceae	Leaf juice
		Crinum viviparum (Lam.) R.Ansari & V.J.Nair	Amaryllidaceae	Leaf juice

continued on following page

Table 3. Continued

Sl. No.	Disease	Plants	Family	Parts used
		Euphorbia tithymaloides L.	Euphorbiaceae	Leaf juice
		Gossypium barbadense L.	Malvaceae	Leaf juice
		Grangea maderaspatana (L.) Poir.	Compositae	Leaf juice
		Justicia gendarussa Burm.f.	Acanthaceae	Leaf juice
		Vitex negundo L.	Lamiaceae	Leaf juice
16.	**Excess saliva secretion**	*Aponogeton natans* (L.) Engl. & K.Krause	Aponogetonaceae	Leaf extract
	Eye diseases	*Ageratum conyzoides* (L.) L.	Compositae	Leaf juice
		Glinus oppositifolius (L.) Aug.DC.	Molluginaceae	Leaf juice
		Melochia corchorifolia L.	Malvaceae	Leaf juice
		Trapa natans var. *bispinosa* (Roxb.) Makino	Lythraceae	Petiole juice
17.	**Fever**	*Bauhinia acuminata* L.	Fabaceae	Leaf decoction
		Caesalpinia pulcherrima (L.) Sw.	Fabaceae	Flower extract
		Celastrus paniculatus Willd.	Celastraceae	Root paste
		Corchorus olitorius L.	Malvaceae	Leaf paste
		Desmodium gangeticum (L.) DC.	Fabaceae	Root extract
		Dillenia pentagyna Roxb.	Dilleniaceae	Stem bark paste
		Dioscorea pubera Blume	Dioscoreaceae	Tuber decoction
		Emilia sonchifolia (L.) DC. ex DC.	Compositae	Plant decoction
		Gmelina arborea Roxb.	Lamiaceae	Stem bark decoction
		Haldina cordifolia (Roxb.) Ridsdale	Rubiaceae	Stem bark
		Lagerstroemia speciosa (L.) Pers.	Lythraceae	Stem bark decoction
		Leonurus sibiricus L.	Lamiaceae	Leaf and root juice
		Limnophila heterophylla (Roxb.) Benth.	Plantaginaceae	Plant juice
		Limnophila sessiliflora (Vahl) Blume	Plantaginaceae	Plant juice
		Lygodium flexuosum (L.) Sw.	Lygodiaceae	Plant juice
		Ocimum americanum L.	Lamiaceae	Seed smoke
		Ocimum tenuiflorum L.	Lamiaceae	Leaf juice
		Phyllanthus debilis Klein ex Willd.	Phyllanthaceae	Root decoction
		Plumbago zeylanica L.	Plumbaginaceae	Root bark paste
		Premna mollissima Roth	Lamiaceae	Bark paste
		Pseudarthria viscida (L.) Wight & Arn.	Fabaceae	Plant juice
		Santalum album L.	Santalaceae	Wood paste
		Trichosanthes tricuspidata Lour.	Cucurbitaceae	Root paste
		Uraria lagopus DC.	Fabaceae	Plant juice
		Wrightia tinctoria R.Br.	Apocynaceae	Seed juice
		Ziziphus jujuba Mill.	Rhamnaceae	Plant decoction
18.	**Fissure of feet**	*Mangifera indica* L.	Anacardiaceae	Bark resin
19.	**Galactogogue**	*Alternanthera sessilis* (L.) R.Br. ex DC.	Amaranthaceae	Cooked plant juice
		Gnaphalium polycaulon Pers.	Compositae	Root extract
		Hemidesmus indicus (L.) R. Br. ex Schult.	Apocynaceae	Root decoction
20.	**Gastrointestinal disorders**			
A.	**Acidity and Stomach problems**	*Bambusa bambos* (L.) Voss	Poaceae	Bamboo crystals
		Cissampelos pareira L.	Menispermaceae	Root decoction
		Cymbopogon martini (Roxb.) W.Watson	Poaceae	Plant juice
		Cyperus iria L.	Cyperaceae	Root stock extract
		Diospyros exsculpta Buch.-Ham.	Ebenaceae	Fruit paste
		Leucas cephalotes (Roth) Spreng.	Lamiaceae	Plant juice
		Lygodium pinnatifidum Sw.	Sahizacaceae	Pinnae paste
		Persicaria orientalis (L.) Spach	Polygonaceae	Plant juice, seed paste
		Ricinus communis L.	Euphorbiaceae	Crushed seed
		Scoparia dulcis L.	Plantaginaceae	Leaf juice
		Strychnos nux-vomica L.	Longaniaceae	Seed juice
		Terminalia chebula Retz.	Combretaceae	Dry fruit pulp powder
		Trichosanthes cucumerina L.	Cucurbitaceae	Seed decoction
		Ziziphus jujuba Mill.	Rhamnaceae	Fruit juice
B.	**Diarrhea**	*Achyranthes aspera* L.	Amaranthaceae	Root paste
		Ailanthus excelsa Roxb.	Simaroubaceae	Bark extract
		Andrographis paniculata (Burm.f.) Nees	Acanthaceae	Leaf juice
		Desmodium gangeticum (L.) DC.	Fabaceae	Root extract
		Ipomoea nil (L.) Roth	Convolvulaceae	Seed extract

continued on following page

Table 3. Continued

Sl. No.	Disease	Plants	Family	Parts used
		Ludwigia adscendens (L.) H.Hara	Onagraceae	Leaf decoction
		Mallotus philippensis (Lam.) Müll.Arg.	Euphorbiaceae	Bark paste
		Mucuna pruriens (L.) DC.	Fabaceae	Leaf decoction
		Pterocarpus marsupium Roxb.	Fabaceae	Gum
		Shorea robusta Gaertn.	Dipterocarpaceae	Fruit juice, Resin
		Soymida febrifuga (Roxb.) A. Juss.	Meliaceae	Bark extract
		Trapa natans L.	Lythraceae	Fruit
		Wrightia tinctoria R.Br.	Apocynaceae	Seed juice
C.	**Dysentery**	*Acacia nilotica* (L.) Delile	Fabaceae	Vegetative bud
		Ailanthus excelsa Roxb.	Simaroubaceae	Bark extract
		Barleria cristata L.	Acanthaceae	Leaf juice
		Bauhinia acuminata L.	Fabaceae	Stem bark decoction
		Centella asiatica (L.) Urb.	Apiaceae	Leaf juice
		Chenopodium album L.	Amaranthaceae	Leaf juice
		Dalbergia sissoo DC.	Fabaceae	Leaf extract
		Diospyros exsculpta Buch.-Ham.	Ebenaceae	Bark decoction
		Enydra fluctuans DC.	Compositae	Leaf juice
		Erythrina variegata L.	Fabaceae	Stem decoction
		Evolvulus alsinoides (L.) L.	Convolvulaceae	Twig juice
		Evolvulus nummularius (L.) L.	Convolvulaceae	Twig juice
		Ficus racemosa L.	Moraceae	Fruit juice
		Ficus religiosa L.	Moraceae	Root decoction
		Hemigraphis hirta (Vahl) T.Anderson	Acanthaceae	Plant juice
		Holarrhena pubescens Wall. ex G.Don	Apocynaceae	Stem bark decoction
		Litsea glutinosa (Lour.) C.B.Rob.	Lauraceae	Leaf filtrate
		Luffa cylindrica (L.) M.Roem.	Cucurbitaceae	Seed kernel
		Paederia foetida L.	Rubiaceae	Leaf juice
		Psidium guajava L.	Myrtaceae	Leaf decoction
		Pterospermum acerifolium (L.) Willd.	Malvaceae	Flower paste
		Smilax zeylanica L	Smilacaceae	Root juice
		Streblus asper Lour.	Moraceae	Leaf decoction
		Syzygium jambos (L.) Alston	Myrtaceae	Bark decoction
		Trachelospermum lucidum (D.Don) K.Schum.	Apocynaceae	Plant juice
		Tylophora indica (Burm. f.) Merr.	Apocynaceae	Root bark extract
		Wrightia tinctoria R.Br.	Apocynaceae	Bark, seed juice
		Ziziphus jujuba Mill.	Rhamnaceae	Plant decoction
21.	**Haemorrhage**	*Hibiscus rosa-sinensis* L.	Malvaceae	Flower bud paste
22.	**Heat trouble**	*Portulaca oleracea* L.	Portulacaceae	Stem juice
23.	**Helminthiasis**	*Azadirachta indica* A.Juss.	Meliaceae	Twig paste
		Baccharoides anthelmintica (L.) Moench	Compositae	Seed soaked in water
		Blumea lacera (Burm.f.) DC.	Compositae	Leaf juice
		Butea monosperma (Lam.) Taub.	Fabaceae	Bark juice
		Carissa spinarum L.	Apocynaceae	Root bark powder
		Chenopodium album L.	Amaranthaceae	Leaf juice
		Clerodendrum indicum (L.) Kuntze	Lamiaceae	Root, leaf paste
		Clerodendrum infortunatum L.	Lamiaceae	Leaf juice
		Cyanthillium cinereum (L.) H.Rob.	Compositae	Plant juice
		Euphorbia hirta L.	Euphorbiaceae	Latex
		Holoptelea integrifolia Planch.	Ulmaceae	Leaf paste
		Ludwigia adscendens (L.) H.Hara	Onagraceae	Plant juice
		Martynia annua L.	Martyniaceae	Leaf juice
		Tabernaemontana divaricata (L.) R.Br. ex Roem. & Schult.	Apocynaceae	Crushed root
		Tagetes erecta L.	Compositae	Leaf paste
		Trichosanthes cucumerina L.	Cucurbitaceae	Plant extract
		Wrightia tinctoria R.Br.	Apocynaceae	Seed juice
24.	**Hepatic disorders**	*Ailanthus excelsa* Roxb.	Simaroubaceae	Leaf paste
		Andrographis paniculata (Burm.f.) Nees	Acanthaceae	Leaf juice
		Jatropha curcas L.	Euphorbiaceae	Latex
		Azadirachta indica A.Juss.	Meliaceae	Leaf paste
		Bauhinia variegata L.	Fabaceae	Stem bark paste
		Cajanus cajan (L.) Millsp.	Fabaceae	Leaf juice

continued on following page

Table 3. Continued

Sl. No.	Disease	Plants	Family	Parts used
		Cassia fistula L.	Fabaceae	Leaf paste
		Celastrus paniculatus Willd.	Celastraceae	Seed oil
		Centella asiatica (L.) Urb.	Apiaceae	Leaf juice
		Cissam pelospareira L.	Menispermaceae	Leaf paste
		Crotalaria retusa L.	Fabaceae	Plant juice
		Ecbolium viride (Forssk.) Alston	Acanthaceae	Crushed root
		Enydra fluctuans DC.	Compositae	Leaf juice
		Eriocaulon quinquangulare L.	Eriocaulaceae	Root paste
		Erythrina variegata L.	Fabaceae	Stem decoction
		Hyptis suaveolens (L.) Poit.	Lamiaceae	Calyx juice
		Ipomoea carnea Jacq.	Convolvulaceae	Leaf paste
		Madhuca longifolia (J.Koenig ex L.) J.F.Macbr.	Sapotaceae	Seed oil
		Martynia annua L.	Martyniaceae	Leaf juice
		Mussaenda roxburghii Hook.f.	Rubiaceae	Juice of colored leaf
		Plumbago zeylanica L.	Plumbaginaceae	Leaf paste
		Plumeria rubra L.	Apocynaceae	Latex
		Pogostemon benghalensis (Burm.f.) Kuntze	Lamiaceae	Leaf juice
		Pterocarpus marsupium Roxb.	Fabaceae	Leaf paste
		Ruellia suffruticosa Roxb.	Acanthaceae	Root extract
		Santalum album L.	Santalaceae	Wood paste
		Senna occidentalis (L.) Link	Fabaceae	Leaf, seed extract
		Sphaeranthus indicus L.	Compositae	Leaf paste
		Tagetes erecta L.	Compositae	Leaf paste
		Trema orientalis (L.) Blume	Cannabaceae	Leaf paste
		Trichosanthes cucumerina L.	Cucurbitaceae	Leaf, stem decoction
		Ventilago denticulata Willd.	Rhamnaceae	Seed extract
		Xyris pauciflora Willd.	Xyridaceae	Aerial part paste
25.	**Infection in nail**	*Argemone mexicana* L.	Papaveraceae	Plant juice
		Combretum decandrum Jacq.	Combretaceae	Twig juice
		Euphorbia hirta L.	Euphorbiaceae	Latex
		Gardenia jasminoides J.Ellis	Rubiaceae	Leaf juice
		Pergularia daemia (Forssk.) Chiov.	Apocynaceae	Milky latex
26.	**Inflammatory Disorders**			
A.	**Fracture**	*Kydia calycina* Roxb.	Malvaceae	Bark paste
B.	**Gout**	*Anisomeles indica* (L.) Kuntze	Lamiaceae	Leaf, root paste
		Byttneria herbacea Roxb.	Malvaceae	Pulverized root stock
		Celosia argentea L.	Amaranthaceae	Fresh root juice
		Cissampelos pareira L.	Menispermaceae	Root paste
		Euphorbia tithymaloides L.	Euphorbiaceae	Stem paste
		Helicteres isora L.	Malvaceae	Fruit extract
		Lawsonia inermis L.	Lythraceae	Leaf juice
		Limnophila indica (L.) Druce	Plantaginaceae	Plant juice
		Litsea glutinosa (Lour.) C.B. Rob.	Lauraceae	Bark paste
		Luffa cylindrica (L.) M.Roem.	Cucurbitaceae	Root paste
		Santalum album L.	Santalaceae	Wood paste
		Symplocos racemosa Roxb.	Symplocaceae	Bark paste
		Tectona grandis L.f.	Lamiaceae	Saw dust paste
		Urena lobata L.	Malvaceae	Leaf juice
		Vitex negundo L.	Lamiaceae	Leaf juice
C.	**Rheumatism**	*Ailanthus excelsa* Roxb.	Simaroubaceae	Leaf decoction
		Bauhinia vahlii Wight & Arn.	Fabaceae	Stem bark paste
		Chrysopogon zizanioides (L.) Roberty	Poaceae	Root decoction
		Crinum viviparum (Lam.) R.Ansari & V.J.Nair	Amaryllidaceae	Bulb decoction
		Dioscorea pentaphylla L.	Dioscoreaceae	Tuber paste
		Madhuca longifolia (J.Koenig ex L.) J.F.Macbr.	Sapotaceae	Seed oil
		Mallotus philippensis (Lam.) Müll.Arg.	Euphorbiaceae	Root bark paste
		Mallotus repandus (Willd.) Müll.Arg.	Euphorbiaceae	Root decoction
		Morinda citrifolia L.	Rubiaceae	Leaf juice
		Paederia foetida L.	Rubiaceae	Leaf juice
		Plumbago zeylanica L	Plumbaginaceae	Root bark paste
		Pongamia pinnata (L.) Pierre	Legumiaceae	Root paste

continued on following page

Table 3. Continued

Sl. No.	Disease	Plants	Family	Parts used
		Pothos scandens L.	Araceae	Bark, fruit paste
		Pseudarthria viscida (L.) Wight & Arn.	Fabaceae	Plant juice
		Pterospermum acerifolium (L.) Willd.	Malvaceae	Flower paste
		Scindapsus officinalis (Roxb.) Schott	Araceae	Fruit paste
		Syzygium cumini (L.) Skeels	Myrtaceae	fruit juice, bark paste
		Urena lobata L.	Malvaceae	Leaf juice
		Vanda tessellata (Roxb.) Hook. ex G.Don	Orchidaceae	Leaf, root decoction
27.	**Insanity**	*Aloe vera* (L.) Burm.f.	Xanthorrhoeaceae	Leaf flesh
28.	**Insecticide**	*Azadirachta indica* A.Juss.	Meliaceae	Leaves
29.	**Insomnia**	*Butea superba* Roxb.	Fabaceae	Stem bark juice
30.	**Killing hair lice**	*Pterospermum acerifolium* (L.) Willd.	Malvaceae	Dried flower powder
31.	**Loss of hair**	*Eclipta prostrata* (L.) L.	Compositae	Leaf juice
		Hibiscusrosa-sinensis L.	Malvaceae	Flower bud paste
		Sphagneticola calendulacea (L.) Pruski	Compositae	Leaf juice
32.	**Lung pain**	*Spinacia oleracea* L.	Amaranthaceae	Cooked juice
33.	**Mosquito repellant**	*Cymbopogon martini* (Roxb.) W.Watson	Poaceae	Seed oil
34.	**Nervous disability**	*Centipeda minima* (L.) A. Braun & Asch.	Compositae	Leaf and flowering top juice
		Barringtonia acutangula (L.) Gaertn.	Lecythidaceae	Fruit juice, seed Powder
		Cissus quinquangularis Chiov.	Vitaceae	Stem sap
		Ficus racemosa L.	Moraceae	Root decoction
		Ipomoea aquatica Forssk.	Convolvulaceae	Plant soup
		Justicia gendarussa Burm.f.	Acanthaceae	Leaf juice
		Lawsonia inermis L.	Lythraceae	Flower decoction
		Leucas aspera (Willd.) Link	Lamiaceae	Leaf juice
		Madhuca longifolia (J.Koenig ex L.) J.F.Macbr.	Sapotaceae	Seed oil
		Marsilea minuta L.	Marsileaceae	Leaf soup
		Mikania micrantha Kunth	Compositae	Leaf juice
		Momordica charantia L.	Cucurbitaceae	Leaf juice
		Ocimum gratissimum L.	Lamiaceae	Seed juice
		Santalum album L.	Santalaceae	Wood paste
		Tragia involucrata L.	Euphorbiaceae	Fruit paste
35.	**Nose congestion**	*Abelmoschus esculentus* (L.) Moench	Malvaceae	Fruit paste
		Annona squamosa L.	Annonaceae	Leaf juice
		Murraya paniculata (L.) Jack	Rutaceae	Leaf juice
36.	**Obesity**	*Bauhinia variegata* L.	Fabaceae	Root bark decoction
37.	**Piles**	*Barleria cristata* L.	Acanthaceae	Crushed root
		Cyanthillium cinereum (L.) H.Rob.	Compositae	Plant juice
		Mimosa pudica L.	Fabaceae	Root, leaf past
		Mirabilis jalapa L.	Nyctaginaceae	Cooked tuberous root
		Phyllodium pulchellum (L.) Desv.	Fabaceae	Bark decoction
		Plumbago zeylanica L.	Plumbaginaceae	Root bark paste
38.	**Pox**	*Azadirachta indica* A.Juss.	Meliaceae	Leaf decoction
		Blumea lacera (Burm.f.) DC.	Compositae	Leaf juice
		Jatropha curcas L.	Euphorbiaceae	Latex
		Moringa oleifera Lam.	Moringaceae	Cooked fruit
		Rungia pectinata (L.) Nees	Acanthaceae	Plant juice
		Terminalia bellirica (Gaertn.) Roxb.	Combretaceae	Fruit paste
39.	**Renal disorders**	*Ageratum conyzoides* (L.) L.	Compositae	Root juice
		Asparagus racemosus Willd.	Asparagaceae	Fresh root juice
		Cheilocostus speciosus (J.Koenig) C.D.Specht	Costaceae	Crushed rhizome
		Cissampelos pareira L.	Menispermaceae	Root decoction
		Desmodium triflorum (L.) DC.	Fabaceae	Crushed tap root
		Knoxia sumatrensis (Retz.) DC.	Rubiaceae	Plant juice
		Meyna spinosa Roxb. ex Link	Rubiaceae	Root paste
		Physalis minima L.	Solanaceae	Ripe fruit
		Saccharum spontaneum L.	Poaceae	Leaf juice
40.	**Respiratory disorders**			
A.	**Asthma**	*Ailanthus excelsa* Roxb.	Simaroubaceae	Bark extract
		Bambusa bambos (L.) Voss	Poaceae	Bamboo liquid
		Blumea membranacea DC.	Compositae	Root paste
		Cleome gynandra L.	Cleomaceae	Root paste

continued on following page

Table 3. Continued

Sl. No.	Disease	Plants	Family	Parts used
		Desmodium gangeticum (L.) DC.	Fabaceae	Root extract
		Ocimum gratissimum L.	Lamiaceae	Dried leaf powder
		Trichosanthes tricuspidata Lour.	Cucurbitaceae	Fruit smoke
B.	**Bronchitis**	*Barringtonia acutangula* (L.) Gaertn.	Lecythidaceae	Seed sap
		Datura metel L.	Solanaceae	Leaf smoke
		Lindenbergia indica Vatke	Plantaginaceae	Plant juice
C.	**Cough and Cold**	*Andrographis paniculata* (Burm.f.) Nees	Acanthaceae	Whole plant juice
		Chrozophora rottleri (Geiseler) A.Juss. ex Spreng.	Euphorbiaceae	Root ash
		Cissampelos pareira L.	Menispermaceae	Root decoction
		Gossypium barbadense L.	Malvaceae	Seed oil
		Lindenbergia muraria (Roxb. ex D. Don) Brühl	Plantaginaceae	Plant juice
		Oxalis corniculata L.	Oxalidaceae	Plant juice
		Piper nigrum L.	Piperaceae	Seed decoction
		Solanum torvum Sw.	Solanaceae	Fruit juice
		Tragia involucrata L.	Euphorbiaceae	Root powder
		Waltheria indica L.	Malvaceae	Plant juice
41.	**Sexual disorders**			
A.	**Abortion**	*Aloe vera* (L.) Burm.f.	Xanthorrhoeaceae	Resinous leaf juice
		Plumbago zeylanica L.	Plumbaginaceae	Root juice
B.	**Child birth**	*Ailanthus excelsa* Roxb.	Simaroubaceae	Bark and leaf decoction
		Dillenia pentagyna Roxb.	Dilleniaceae	Stem bark extract
		Kydia calycina Roxb.	Malvaceae	Stem mucilage
C.	**Contraceptive**	*Dioscorea bulbifera* L.	Dioscoreaceae	Tube juice
D.	**Gonorrhea**	*Azanza lampas* (Cav.) Alef.	Malvaceae	Fruit, root juice
		Byttneria herbacea Roxb.	Malvaceae	Seed paste
		Lindernia ciliata (Colsm.) Pennell	Linderniaceae	Plant juice
		Ocimum gratissimum L.	Lamiaceae	Nutlet decoction
		Plectranthus mollis (Aiton) Spreng.	Lamiaceae	Seed paste
		Pterospermum acerifolium (L.) Willd.	Malvaceae	Bark and petiole decoction
		Smilax zeylanica L.	Smilacaceae	Root juice
		Tinospora sinensis (Lour.) Merr.	Menispermaceae	Root decoction
		Typha elephantina Roxb.	Typhaceae	Fruit, inflorescence, rhizome juice
		Ziziphus jujuba Mill.	Rhamnaceae	Seed paste
E.	**Leucorrhoea**	*Sphaeranthus indicus* L.	Compositae	Root extract
F.	**Menstrual problem**	*Butea monosperma* (Lam.) Taub.	Fabaceae	Leaf juice
		Celosia argentea L.	Amaranthaceae	Flower paste
		Magnolia champaca (L.) Baill. ex Pierre	Magnoliaceae	Stem bark decoction
		Polycarpon prostratum (Forssk.) Asch. & Schweinf.	Caryophyllaceae	Cooked young twigs and leaves
		Sesamum indicum L.	Pedaliaceae	Seed decoction
		Solanum indicum L.	Solanaceae	Seed decoction
		Solanum sisymbriifolium Lam.	Solanaceae	Shoot, leaf juice
		Xanthium strumarium L.	Asteraceae	Twig sap
G.	**Syphilis**	*Acacia catechu* (L.f.) Willd.	Fabaceae	Crushed seed
		Areca catechu L.	Arecaceae	Unripe nut paste
		Azanza lampas (Cav.) Alef.	Malvaceae	Fruit, root juice
		Smilax zeylanica L.	Smilacaceae	Root juice
42.	**Skin diseases**	*Abutilon indicum* (L.) Sweet	Malvaceae	Crushed root and seed
43.	**Thirst**	*Plumeria rubra* L.	Apocynaceae	Sap wood
		Centipeda minima (L.) A. Braun & Asch.	Compositae	Root juice
44.	**Toothache**	*Acacia nilotica* (L.) Delile	Fabaceae	Bark, gum decoction
		Azadirachta indica A.Juss.	Meliaceae	Twig
		Calotropis gigantea (L.) Dryand.	Apocynaceae	Latex
		Eichhornia crassipes (Mart.) Solms	Pontederiaceae	Root powder
		Leucas aspera (Willd.) Link	Lamiaceae	Leaf filtrate
		Mimusops elengi L.	Sapotaceae	Bark decoction, seed powder
		Monochoria hastata (L.) Solms	Pontederiaceae	Root powder
		Persicaria hydropiper (L.) Delarbre	Polygonaceae	Dried root powder
		Persicaria orientalis (L.) Spach	Polygonaceae	Root paste
		Phyllanthus reticulatus Poir.	Phyllanthaceae	Stem, twig juice
		Portulaca oleracea L.	Portulacaceae	Fried seed
		Psidium guajava L.	Myrtaceae	Warm leaf decoction

continued on following page

Table 3. Continued

Sl. No.	Disease	Plants	Family	Parts used
		Pterocarpus marsupium Roxb.	Fabaceae	Gum
		Solanum torvum Sw.	Solanaceae	Fruit powder
		Spermacoce articularis L.f.	Rubiaceae	Root paste
		Tabernaemontana divaricata (L.) R.Br. ex Roem. & Schult.	Apocynaceae	Root paste
45.	**Tumours**	*Clerodendrum infortunatum* L.	Lamiaceae	Leaf juice
		Bryophyllum pinnatum (Lam.) Oken	Crassulaceae	Crushed leaf
46.	**Vomiting**	*Alangium salviifolium* (L.f.) Wangerin	Cornaceae	Leaf juice
		Citrus maxima (Burm.) Merr.	Rutaceae	Fruit
		Desmodium gangeticum (L.) DC.	Fabaceae	Root extract
47.	**Wounds**	*Ageratum conyzoides* (L.) L.	Compositae	Leaf juice
		Biophytum sensitivum (L.) DC.	Oxalidaceae	Seed powder
		Cleistanthus collinus (Roxb.) Benth. ex Hook.f.	Phyllanthaceae	Stem bark decoction
		Dendrophthoe falcata (L.f.) Ettingsh.	Loranthaceae	Leaf juice
		Dillenia indica L.	Dilleniaceae	Mucilaginous sap from calyx
		Ficus racemosa L.	Moraceae	Stem bark paste
		Gmelina arborea Roxb.	Lamiaceae	Root bark decoction

Table 4. Local use of polyherbal formulation to treat different disease.

	Disease	Name	Family	Plant parts and other materials used
1.	**Alcohol addiction**	*Amorphophallus paeoniifolius* (Dennst.) Nicolson	Araceae	Corm extract (75-100gm) + ethyl alcohol.
2.	**Anaemia**	*Olax scandens* Roxb.	Olacaceae	Bark decoction with *Hemidesmus indicus.*
		Portulaca oleracea L.	Portulacaceae	Leaf juice with NaCl.
3.	**Bleeding**	*Blumea lacera* (Burm.f.) DC.	Asteraceae	Crushed root.
4.	**Boil**	*Nymphoides cristata* (Roxb.) Kuntze	Menyanthaceae	Plant crushed with *Curcuma longa* rhizome.
		Senna occidentalis (L.) Link	Fabaceae	Crushed leaf mixed with butter.
5.	**Cardiac disorders**	*Anacardium occidentale* L.	Anacardiaceae	Crushed nut with sugar and water.
		Ocimum gratissimum L.	Lamiaceae	Twig sap with fermented decoction of boiled rice.
		Rauvolfia serpentina (L.) Benth. ex Kurz	Apocynaceae	Root powder with *Piper longum* seed.
		Sida cordifolia L.	Malvaceae	Root barks powder with cow milk.
6.	**Constitutional disorders**	*Trapa natans* var. *bispinosa* (Roxb.) Makino	Lythraceae	Dried fruit powder with *Piper longum* (3:2).
7.	**Contraceptive**	*Butea superba* Roxb.	Fabaceae	Flower paste with root paste of *Bombax ceiba* and *Cynodon dactylon* juice.
		Pergularia daemia (Forssk.) Chiov.	Apocynaceae	Root bark paste with *Piper longum* seed decoction with country liquor.
8.	**Diabetes**	*Coccinia grandis* (L.) Voigt	Cucurbitaceae	Root juice with *Swertia chirayita* decoction and *Syzygium cumini* seed.
		Cryptolepis dubia (Burm.f.) M.R.Almeida	Apocynaceae	Stem bark paste mixed with *Piper longum* seed powder.
9.	**Dog bite**	*Achyranthes aspera* L.	Amaranthaceae	Plant and root paste with *Piper longum* seed.
		Albizia amara (Roxb.) B. Boivin	Fabaceae	Seed powder with latex of *Euphorbia neriifolia.*
		Calotropis gigantea (L.) Dryand.	Apocynaceae	Flower paste with sugar.
		Vitex negundo L.	Lamiaceae	Unopened and open flower bud paste with sugar.
10.	**Ear ache**	*Cleome viscosa* L.	Cleomaceae	Seed boiled in mustard oil.
		Ocimum americanum L.	Lamiaceae	Plant juice with mustard oil (3:2).
		Pergularia daemia (Forssk.) Chiov.	Apocynaceae	Leaf extract with *Allium sativum.*
11.	**Eye disease**	*Achyranthes aspera* L.	Amaranthaceae	Root pasted with *Solanum surattense.*
		Rosa cymosa Tratt.	Rosaceae	Flower, leaf extract with dilute solution of potash alum.
12.	**Fever**	*Holarrhena pubescens* Wall. ex G.Don	Apocynaceae	Root with *Piper nigrum* seeds and ghee.
		Jasminum multiflorum (Burm.f.) Andrews	Oleaceae	Leaf juice with honey.
		Ludwigia perennis L.	Onagraceae	Plant boiled with oil.
		Nyctanthes arbor-tristis L.	Oleaceae	Leaf juice with honey.
		Polyalthia longifolia (Sonn.) Thwaites	Annonaceae	Root extracts with *Piper nigrum* seed.
		Solanum torvum Sw.	Solanaceae	Root decoction with root paste of *Achyranthes aspera* and *Desmodium gangeticum* (5:3:2).
		Stephania japonica (Thunb.) Miers	Menispermaceae	Leaf crushed with sugar and water.
13.	**Gastrointestinal disorders**			

continued on following page

Table 4. Continued

	Disease	Name	Family	Plant parts and other materials used
A.	***Acidity and stomach pain***	*Andrographis paniculata* (Burm.f.) Nees	Acanthaceae	Crushed root with *Aristolochia indica, Barleria cristata, Clerodendrum serratum, Rivea hypocrateriformis*, and dried *Chirata* and *Piper nigrum, Terminalia bellirica, Terminalia chebula, Zingiber officinale.*
		Citrus maxima (Burm.) Merr.	Rutaceae	Fruit juice with salt.
		Hemidesmus indicus (L.) R. Br. ex Schult.	Apocynaceae	Root decoction with honey.
		Jatropha curcas L.	Euphorbiaceae	Latex with molasses.
		Magnolia champaca (L.) Baill. ex Pierre	Magnoliaceae	Leaf juice with honey.
		Neolamarckia cadamba (Roxb.) Bosser	Rubiaceae	Leaf juice with molasses.
		Ocimum gratissimum L.	Lamiaceae	Leaf juice with NaCl.
		Phyllanthus emblica L.	Phyllanthaceae	Dried fruit with fruits of *Terminalia bellerica* and *Terminalia chebula* soaked in water.
		Punica granatum L.	Lythraceae	Young fruit decoction with root bark extract of *Holarrhena pubescens.*
		Rosa cymosa Tratt.	Rosaceae	Crushed bud with honey or sugar.
		Semecarpus anacardium L.f.	Anacardiaceae	Seed with *Curcuma longa*, mustard oil cake and NaCl.
B.	***Constipation***	*Aloe vera* (L.) Burm.f.	Xanthorrhoeaceae	Leaf juice with *Ricinus communis* seeds oil.
C.	***Diarrhea***	*Byttneria herbacea* Roxb.	Malvaceae	Root stock powder used with other drugs.
		Stephania japonica (Thunb.) Miers	Menispermaceae	Root crushed with *Piper nigrum* seed.
D.	***Dysentery***	*Annona squamosa* L.	Annonaceae	Root barks juice (20-25 drops)+7-8 spoonful milk.
		Centella asiatica (L.) Urb.	Apiaceae	Crushed root with *Dipteracanthus prostratus* root.
		Codariocalyx motorius (Houtt.) H.Ohashi	Fabaceae	Plant decoction with *Piper longum* seeds paste.
		Cynodon dactylon (L.) Pers.	Poaceae	Plant crushed with *Syzygium cumini* leaves.
		Drosera burmanni Vahl	Droseraceae	Plant paste with NaCl.
		Flacourtia indica (Burm.f.) Merr.	Salicaceae	Leaf decoction with NaCl.
		Grewia tiliifolia Vahl	Malvaceae	Bark decoction with paste of *Nymphaea* rhizomes (3:1).
		Jatropha curcas L.	Euphorbiaceae	Latex with sugar.
		Jatropha gossypiifolia L.	Euphorbiaceae	Latex with sugar.
		Ludwigia adscendens (L.) H.Hara	Onagraceae	Plant juice mixed with butter or milk.
		Mangifera indica L.	Anacardiaceae	Bark decoction with that of *Psidium guajava, Syzygium cumini* and *Terminalia arjuna.*
		Plumeria rubra L.	Apocynaceae	Bark juice with *Cocos nucifera*endosperm.
		Syzygium cumini (L.) Skeels	Myrtaceae	Leaf juice with goat milk, warm filtered leaf juice with that of *Cynodon dactylon* + milk.
		Tamarindus indica L.	Fabaceae	Leaf decoction with molasses.
		Tephrosia purpurea (L.) Pers.	Fabaceae	Root extract with sugar.
		Trigonella foenum-graecum L.	Fabaceae	Leaf juice with that of *Cynodon dactylon* with milk.
14.	**Goiter**	*Bauhinia purpurea* L.	Fabaceae	Root powder pasted with dry *Zingiber officinale.*
15.	**Haemorrhage**	*Echinochloa colona* (L.) Link	Poaceae	Leaf juice with turmeric powder.
16.	**Hair growth**	*Tridax procumbens* (L.) L.	Asteraceae	Leaf extract with *Cocos nucifera* seed oil.
17.	**Heat trouble**	*Sida acuta* Burm.f.	Malvaceae	Root extract mixed with ghee.
18.	**Helminthiasis**	*Chenopodium album* L.	Amaranthaceae	Leaf extract with leaf paste of *Punica granatum* fruit juice.
		Jasminum multiflorum (Burm.f.) Andrews	Oleaceae	Leaf juice with sugar.
		Nyctanthes arbor-tristis L.	Oleaceae	Leaf juice with sugar.
19.	**Hepatic disorders**	*Carica papaya* L.	Caricaceae	Latex with sugar.
		Dipteracanthus prostratus (Poir.) Nees	Acanthaceae	Root pasted with rhizome of *Centella asiatica.*
		Glycosmis pentaphylla (Retz.) DC.	Rutaceae	Root decoction with sugar and curd.
		Nymphoides indica (L.) Kuntze	Menyanthaceae	Root juice mixed with curd.
		Oldenlandia corymbosa L.	Rubiaceae	Plant juice with milk and sugar.
		Zingiber officinale Roscoe	Zingiberaceae	Dried rhizome crushed with seeds and roots of *Piper nigrum* and *Citrus* juice.
20.	**Inflammatory disorders**			
A.	***Fracture***	*Heteropogon contortus* (L.) P.Beauv. ex Roem. & Schult.	Poaceae	Root paste with *Litsea glutinosa* bark.
		Premna mollissima Roth	Lamiaceae	Stem bark with lime (3:1).
		Trapa natans L.	Lythraceae	Fruit paste crushed with lime.
		Uraria picta (Jacq.) DC.	Fabaceae	Stem bark crushed with lime.
B.	***Gout***	*Agave vivipara* L.	Asparagaceae	Leaf juice with *Piper nigrum* seed.
		Clerodendrum infortunatum L.	Lamiaceae	Root paste with that of the bulbs of *Allium sativum.*
		Dillenia pentagyna Roxb.	Dilleniaceae	Leaf decoction with NaCl.
		Holoptelea integrifolia Planch.	Ulmaceae	Stem bark paste with *Zingiber officinale* paste (3:2).

continued on following page

Table 4. Continued

	Disease	Name	Family	Plant parts and other materials used
		Plumbago zeylanica L.	Plumbaginaceae	Root boiled with oil and using as massage oil.
		Senna tora (L.) Roxb.	Fabaceae	Root bark paste with albumen of hen egg.
C.	***Rheumatism***	*Achyranthes aspera* L.	Amaranthaceae	Root paste with cumin paste and milk and root paste of *Solanum xanthocarpum*.
		Ageratum conyzoides (L.) L.	Asteraceae	Whole plant boiled in oil.
		Alangium salviifolium (L.f.) Wangerin	Cornaceae	Stem bark boiled in oil.
		Cotula anthemoides L.	Asteraceae	Plant heated with oil.
		Crateva nurvala Buch.-Ham.	Capparaceae	Leaf decoction with dried *Zingiber officinale*.
		Datura metel L.	Solanaceae	Leaves with Juice of *Calotropis procera*, stem juice of *Euphorbia tirucalli*, roots of*Premna mollissima*, seed oil of *Ricinus communis* and leaf juice of *Vitex negundo*.
		Ichnocarpus frutescens (L.) W.T.Aiton	Apocynaceae	Root crushed with *Justicia adhatoda* leaves and milk.
		Kydia calycina Roxb.	Malvaceae	Leaf paste with *Piper nigrum* seed.
		Premna mollissima Roth	Lamiaceae	Stem bark pasted with lime.
		Premna tomentosa Willd.	Lamiaceae	Root crushed with leaf juice of *Calotropis gigantea, Vitex negundo, Datura fastuosa* stem of *Euphorbia tirucalli, Ricinus communis* seed and ghee.
		Uraria lagopus DC.	Fabaceae	Stem bark crushed with lime.
		Vitex negundo L.	Lamiaceae	Shoot tip in mustard oil.
21.	**Insanity**	*Enydra fluctuans* Lour.	Asteraceae	Leaf juice with tuber extracts of *Asparagus sp., Musa sp.* and *Wedelia chinensis*.
22.	**Loss of memory**	*Bacopa monnieri* (L.) Wettst.	Plantaginaceae	Fresh leaves roasted with ghee.
23.	**Miscellenous snake bite**	*Clerodendrum japonicum* (Thunb.) Sweet	Lamiaceae	Stem bark decoction with *Piper nigrum* seed.
		Euphorbia antiquorum L.	Euphorbiaceae	Root paste mixed with *Piper nigrum* seed.
		Gardenia latifolia Aiton	Rubiaceae	Stem bark paste mixed with *Piper nigrum* seed.
		Melochia corchorifolia L.	Malvaceae	Stem and leaves boiled in oil.
		Opuntia dillenii (Ker Gawl.) Haw.	Cactaceae	Root paste mixed with *Piper nigrum* seed (2:5).
		Rauvolfia serpentina (L.) Benth. ex Kurz	Apocynaceae	Root powder with seeds of *Piper nigrum* seed.
		Senna tora (L.) Roxb.	Fabaceae	Root paste with *Piper nigrum* seed.
24.	**Nervous disability**	*Codariocalyx motorius* (Houtt.) H.Ohashi	Fabaceae	Root powder with NaCl with warm milk.
		Ficus racemosa L.	Moraceae	Stem bark paste with paste of *Piper longum* (5:2).
		Phoenix sylvestris (L.) Roxb.	Arecaceae	Seed paste mixed with root paste of *Aerva lanata* eaten with betel leaves.
		Trichosanthes tricuspidata Lour.	Cucurbitaceae	Root boiled in mustard oil. Fruit boiled in *Cocos nucifera* oil.
25.	**Piles**	*Bauhinia purpurea* L.	Fabaceae	Root powder with curd.
		Codariocalyx gyroides (Link) Hassk.	Fabaceae	Anthers pasted with butter.
		Curculigo orchioides Gaertn.	Hypoxidaceae	Root bark juice with milk and sugar.
		Ipomoea quamoclit L.	Convolvulaceae	Leaf powder with milk.
26.	**Pox**	*Bombax ceiba* L.	Malvaceae	Seed with ripe banana(*Musa paradisiaca*).
		Capparis zeylanica L.	Capparaceae	Root paste with camphor, *Piper nigrum* seed, *Curcuma longa* and Mustard oil.
		Curcuma longa L.	Zingiberaceae	Rhizome mixed with sugar.
		Solanum surattense Burm. f.	Solanaceae	Root chewed with *Piper nigrum* seed.
27.	**Renal**	*Bombax ceiba* L.	Malvaceae	Bark soaked in water, sugar disorders.
		Trema orientalis (L.) Blume	Cannabaceae	Stem bark decoction with seeds of *Piper longum*.
28.	**Respiratory disorders**	*Monochoria hastata* (L.) Solms	Pontederiaceae	Crushed rhizome with sugar.
A.	***Asthma and bronchitis***	*Justicia adhatoda* L.	Acanthaceae	Leaf juice with honey.
B.	***Cough and cold***	*Bacopa monnieri* (L.) Wettst.	Plantaginaceae	Leaf decoction with clove and sugar.
		Cinnamomum camphora (L.) J.Presl	Lauraceae	Leaf boiled with that of *Justicia adhatoda*.
		Eucalyptus tereticornis Sm.	Myrtaceae	Leaf juice with ghee.
		Justicia adhatoda L.	Acanthaceae	Leaf juice with honey.
		Justicia gendarussa Burm.f.	Acanthaceae	Leaf juice with honey.
		Ocimum tenuiflorum L.	Lamiaceae	Leaf juice with milk.
		Sesbania sesban (L.) Merr.	Fabaceae	Dry rhizome with NaCl.
		Zingiber officinale Roscoe	Zingiberaceae	Dry rhizome with NaCl.
C.	***Tuberculosis***	*Caesalpinia bonduc* (L.) Roxb.	Fabaceae	Root bark pasted with roots of *Asparagus racemosus, Azanza lampas, Flemingia strobilifera, Polygala arvensis, Rauvolfia serpentina, Smilax ovalifolia* and stem bark of *Schleichera oleosa* with goose egg shell, *Cinnamomum camphora, Ficus religiosa,Piper nigrum,Pterocarpus indicus* and *Zingiber officinale*.
29.	**Ricket**	*Cuscuta reflexa* Roxb.	Convolvulaceae	Plant pasted with *Curcuma aromatica*.
30.	**Sexual disorders**			

continued on following page

Table 4. Continued

	Disease	Name	Family	Plant parts and other materials used
A.	***Abnormality of semen***	*Aerva lanata* (L.) Juss.	Amaranthaceae	Crushed roots of *Abrus precatorius* with colyledons of *Jatropha curcas.*
		Alangium salviifolium (L.f.) Wangerin	Cornaceae	Root crushed with root that of *Butea monosperma, Chrozophora rottleri, Glossocardia bidens, Hemidesmus indicus, Phyllanthus emblica, Ziziphus jujuba.*
B.	***Abortion***	*Pandanus odorifer* (Forssk.) Kuntze	Pandanaceae	Stilt root extraction with milk.
		Symplocos racemosa Roxb.	Symplocaceae	Stem bark decoction with paste of *Piper longum*(3:1).
		Uraria lagopus DC.	Fabaceae	Plant juice with milk.
C.	***Child birth***	*Dillenia indica* L.	Dilleniaceae	Fruit extract with *Piper nigrum* seed paste (3:2).
D.	***Gonorrhea***	*Tribulus terrestris* L.	Zygophyllaceae	Leaf juice with milk and sugar.
E.	***Leucorrhoea***	*Amaranthus spinosus* L.	Amaranthaceae	2-3 inch. long root with 3 *Piper nigrum* seeds.
		Amaranthus viridis L.	Amaranthaceae	Root juice with decoction of unboiled rice.
		Clitoria ternatea L.	Fabaceae	Root paste with *Piper nigrum* seed.
		Pithecellobium dulce (Roxb.) Benth.	Fabaceae	Stem bark paste with seeds of *Piper nigrum.*
		Saraca asoca (Roxb.) Willd.	Fabaceae	Stem bark decoction with cow milk and water.
		Sida acuta Burm.f.	Malvaceae	Root paste with *Cuminum* powder and sugar.
F.	***Menstrual disorder***	*Abroma augusta* (L.) L.f.	Malvaceae	Root bark paste (2.5 gm) with *Piper nigrum* seed (2-3 nos.).
		Cardiospermum halicacabum L.	Sapindaceae	Leaf juice with *Piper nigrum* seed.
		Cassia fistula L.	Fabaceae	Stem bark powder and leaves with *Piper nigrum.*
		Nelumbo nucifera Gaertn.	Nelumbonaceae	Rhizome decoction with sugar.
		Rauvolfia tetraphylla L.	Apocynaceae	Bark crushed with stem bark of *Allium cepa, Allium sativum, Piper nigrum* and *Vitex negundo* .
G.	***Syphilis***	*Cannabis sativa* L.	Cannabaceae	Flower mixed with sulphur dust; *Euphorbia neriifolia* leaf extract heated with coconut oil.
		Euphorbia antiquorum L.	Euphorbiaceae	Latex with Brassica oil (2:1).
		Ficus benghalensis L.	Moraceae	Latex with honey.
		Litsea glutinosa (Lour.) C.B.Rob.	Lauraceae	Leaf extract with sugar.
		Sida rhombifolia L.	Malvaceae	Root extract mixed with that of *Sida cordifolia* root extract.
		Sphaeranthus indicus L.	Asteraceae	Plant decoction with NaCl (3:2).
		Spondias dulcis Parkinson	Anacardiaceae	Stem bark juice with sugar.
		Terminalia arjuna (Roxb. ex DC.) Wight & Arn.	Combretaceae	Bark decoction with *Curcuma longa.*
		Trema orientalis (L.) Blume	Cannabaceae	Root decoction with *Piper nigrum* seed paste (3:2).
31.	**Skin diseases**	*Baccharoides anthelmintica* (L.) Moench	Asteraceae	Seed decoction with *Piper nigrum* seed.
		Caesalpinia bonduc (L.) Roxb.	Fabaceae	Karnel powder with mustard oil.
		Clerodendrum infortunatum L.	Lamiaceae	Root bark boiled with *Allium sativum* and *Madhuca* seed oil.
		Curcuma longa L.	Zingiberaceae	Rhizome paste with water and mustard oil.
		Cyperus dubius Rottb.	Cyperaceae	Crushed root stock boiled in mustard oil.
		Datura metel L.	Solanaceae	Seed boiled/roasted in mustard oil.
		Elephantopus scaber L.	Asteraceae	Leaf juice boiled with coconut oil.
		Lindenbergia indica Vatke	Plantaginaceae	Plant juice with *Coriandrum sativum.*
		Merremia tridentata (L.) Hallier f.	Convolvulaceae	Leaf juice with latex of *Calotropis gigantea.*
		Opuntia dillenii (Ker Gawl.) Haw.	Cactaceae	Plant paste with paste of *Piper longum.*
		Rauvolfia tetraphylla L.	Apocynaceae	Leaf juice with castor oil.
		Senna occidentalis (L.) Link	Fabaceae	Seed powder with oil.
		Spondias pinnata (L. f.) Kurz	Anacardiaceae	Fruit endocarp with cow urine.
		Tiliacora racemosa Colebr.	Menispermaceae	Root juice with twigs of *Trichosanthes tricuspidata.*
		Trichosanthes dioica Roxb.	Cucurbitaceae	Leaf juice with roots of *Tinospora sinensis.*
32.	**Tetanus**	*Dalbergia volubilis* Roxb.	Fabaceae	Root paste with that of *Senna occidentalis.*
33.	**Toothache**	*Ocimum americanum* L.	Lamiaceae	Seed paste with NaCl (5:2).
34.	**Vomiting**	*Trema orientalis* (L.) Blume	Cannabaceae	Stem bark with *Citrus* juice (2:1).
35.	**Wounds**	*Morinda citrifolia* L.	Rubiaceae	Fruit paste with salt.
		Saccharum spontaneum L.	Poaceae	Ash obtained by burning inflorescence used with mustard oil (4:3).

cause a serious decline in plants population. It is thus recommended that cultivation techniques should be formulated, especially for the most important plants species that may be used widely and traded outside the region. Because the knowledge of the uses of the medicinal plants reported here belongs to the indigenous people of the study area, the benefits obtained from this knowledge should equally be shared with them.

ACKNOWLEDGEMENTS

We thank local individuals and medicine-man for their continued assistance.

References

Alcorn, J. B. (1989). Process as resource. *Advances in Economic Botany*, *7*, 63–77.

Amsalu, N., Bezie, Y., Fentahun, M., Alemayehu, A., & Amsalu, G. (2018). Use and conservation of medicinal plants by indigenous people of Gozamin Wereda, East Gojjam Zone of Amhara region, Ethiopia: An ethnobotanical approach. *Evidence-Based Complementary and Alternative Medicine*, *2018*, 1–23. doi:10.1155/2018/2973513 PMID:29743921

Andrewartha, H. G., & Birch, L. C. (1954). *The distribution and abundance of animals*. Chicago, IL: University of Chicago Press.

Balick, M. J., & Cox, P. A. R. (1996). *Plants, people, and culture: the science of ethnobotany* (p. 219). New York: Scientific American Library.

Berkes, F. (1993). Traditional ecological knowledge in perspective. In Traditional ecological knowledge: concepts and cases (Vol. 1). Canadian Museum of Nature/International Development Research Centre, Ottawa.

Berkes, F. (1999). *Sacred ecology: traditional ecological knowledge and resource management*. Philadelphia, PA: Taylor and Francis.

Berkes, F., Colding, J., & Folke, C. (2000). Rediscovery of traditional ecological knowledge as adaptive management. *Ecological Applications*, *10*(5), 1251–1262. doi:10.1890/1051-0761(2000)010[1251:ROTEKA]2.0.CO;2

Berkes, F., Folke, C., & Gadgil, M. (1994). Traditional ecological knowledge, biodiversity, resilience and sustainability. In *Biodiversity conservation* (pp. 269–287). Dordrecht, The Netherlands: Springer.

Braatz, S. M. (1992). *Conserving biological diversity: a strategy for protected areas in the Asia-Pacific region* (Vol. 23). Washington, DC: World Bank Publications.

Camara-Leret, R., Paniagua-Zambrana, N., Balslev, H., & Macia, M. J. (2014). Ethnobotanical knowledge in vastly under-documented in Northwestern South America. *PLoS One*, *9*(1). doi:10.1371/journal.pone.0085794 PMID:24416449

CBD. (1992). Convention on Biological Diversity, Rio de Janeiro, Brazil. Available from http://www.cbd.int/convention

Chepkosgey, M. P., & Jerotich, C. G. (2016). Trends for application of indigenous knowledge in natural resource management among Nandi people, Kenya. *Journal of Environmental Science. Computer Science and Engineering Technology.*, *5*(3), 513–527.

Clements, F. E., & Shelford, V. E. (1939). *Bio-ecology*. New York: John Wiley & Sons.

Colding, J. (1998). Analysis of hunting options by the use of general food taboos. *Ecological Modelling*, *110*(1), 5–17. doi:10.1016/S0304-3800(98)00038-6

Colding, J., & Folke, C. (2001). Social taboos: 'invisible' systems of local resource management and biological conservation. *Ecological Applications*, *11*, 584–600.

Colding, J., Folke, C., & Elmqvist, T. (2003). Social institutions in ecosystem management and biodiversity conservation. *Tropical Ecology*, *44*(1), 25–41.

Cunningham, A. B. (1993). *African Medicinal Plants setting priorities at the interface between conservation and primary health care*. Working Paper 1. Paris, France: UNESCO.

Danielsen, F., Burgess, N. D., Balmford, A., Donald, P. F., Funder, M., Jones, J. P., & Child, B. (2009). Local participation in natural resource monitoring: A characterization of approaches. *Conservation Biology*, *23*(1), 31–42. doi:10.1111/j.1523-1739.2008.01063.x PMID:18798859

Das, A. P., & Pandey, A. K. (Eds.). (2007). *Advances in ethnobotany*. Dehradun, India: Bishen Singh Mahendra Pal Singh.

Dobson, T. (1992). Loss of biodiversity: An international environmental policy perspective. *North Carolina Journal of International Law and Commercial Regulation*, *17*, 277.

Folke, C., Berkes, F., & Colding, J. (1998). Ecological practices and social mechanisms for building resilience and sustainability. In F. Berkes, & C. Folke (Eds.), *Linking social and ecological systems: management practices and social mechanisms for building resilience* (pp. 414–436). Cambridge, UK: Cambridge University Press.

Frankel, O. H., Brown, A. H., & Burdon, J. J. (1995). *The conservation of plant biodiversity*. Cambridge University Press.

Gadgil, M., Berkes, F., & Folke, C. (1993). Indigenous knowledge for biodiversity conservation. *Ambio*, *22*, 151–156.

Grenier, L. (1998). *Working with indigenous knowledge: a guide for researchers*. IDRC.

Hamilton, A. C. (2004). *Medicinal plants, conservation and livelihoods*. Kluwer Academic Publishers. Retrieved from http://www.cbd.int/doc/articles/2004/A

Hannan, M. T., & Freeman, J. (1989). *Organizational ecology*. Cambridge, MA: Harvard University Press.

Haunss, S. (2013). *Conflicts in the knowledge society: the contentious politics of intellectual property* (Vol. 20). Cambridge University Press; doi:10.1017/CBO9781139567633

Horton, D., & Mackay, R. (2003). Using evaluation to enhance institutional learning and change: Recent experiences with agricultural research and development. *Agricultural Systems*, *78*(2), 127–142. doi:10.1016/S0308-521X(03)00123-9

Jain, S. K., & Rao, R. R. (1977). *A Handbook of field and herbarium methods*. New Delhi, India: Today and Tomorrow's Printers and Publishers.

Johannes, R. E. (1998). The case for data-less marine resource management: Examples from tropical near-shore finfisheries. *Trends in Ecology & Evolution*, *13*(6), 243–246. doi:10.1016/S0169-5347(98)01384-6 PMID:21238285

Jones, G. E. (1987). *The conservation of ecosystems and species*. New York: Croom Helm.

Krebs, C. J. (1985). *Ecology: experimental analysis of distribution and abundance* (3rd ed.). New York: Harper & Row.

Little, K. (1965). *West African urbanization: a study of voluntary associations in social change*. CUP Archive.

Manicad, G. (1996). Biodiversity conservation and development: the collaboration of formal and non-formal institutions. *Biotechnology and development monitor, 26*, 15-17.

Marsh, D. M., & Trenham, P. C. (2008). Current trends in plant and animal population monitoring. *Conservation Biology*, *22*(3), 647–655. doi:10.1111/j.1523-1739.2008.00927.x PMID:18445076

Mavi, S., & Shava, S. (1997). *Traditional methods of conserving medicinal plants in Zimbabwe*. BGC-News 2. Journal archive.

McNaughton, S. J. (1989). Ecosystems and conservation in the twenty-first century. In D. Western, & M. C. Pearl (Eds.), Conservation for the twenty-first century (pp. 109-120). New York: Oxford University Press.

Monica, K. C., Mark, K., & Paul, O. (2016). Traditional controls of harvesting and conserving medicinal plants in Keiyo south sub-county, Kenya. *International Journal of Humanities and Social Science*, *6*(11), 54–63.

Naveh, Z., & Lieberman, A. S. (2013). *Landscape ecology: theory and application*. Springer Science & Business Media.

Njoroge, G. N., Kaibui, I. M., Njenga, P. K., & Odhiambo, P. O. (2010). Utilization of priority traditional medicinal plants and local people's knowledge on their conservation status in arid lands of Kenya (Mwingi district). *Journal of Ethnobiology and Ethnomedicine*, *6*(1), 1–22. doi:10.1186/1746-4269-6-22 PMID:20712897

Padulosi, S., Leaman, D., & Quek, P. (2002). Challenges and opportunities in enhancing the conservation and use of medicinal and aromatic plants. *Journal of Herbs, Spices, & Medicinal Plants*, *9*(4), 243–267. doi:10.1300/J044v09n04_01

Pringle, R. M. (2017). Upgrading protected areas to conserve wild biodiversity. *Nature*, *546*(7656), 91–99. doi:10.1038/nature22902 PMID:28569807

Robbins, P. (2011). *Political ecology: A critical introduction* (Vol. 16). John Wiley & Sons.

Sathe, V. (1985). *Culture and related corporate realities*. Homewood, IL: Irwin.

Schmeller, D. S. (2008). European species and habitat monitoring: Where are we now? *Biodiversity and Conservation*, *17*(14), 3321–3326. doi:10.100710531-008-9514-1

Schmink, M., Redford, K. H., & Padoch, C. (1992). Traditional peoples and the biosphere: framing the issues and defining the terms. In K. H. Redford, & C. Padoch (Eds.), *Conservation of neotropical forests: working from traditional resource use* (pp. 3–13). New York: Columbia University Press.

Sen, U. K. (2016). *Botanical and socio-cultural studies on some sacred groves of West Midnapore district, West Bengal.* (Ph.D. Thesis), Vidyasagar University, India.

Ugulu, I. (2011). Traditional ethnobotanical knowledge about medicinal plants used for external therapies in Alasehir, Turkey. *International Journal of Medicine and Aromatic Plants.*, *2*(1), 101–106.

UNESCO. (2002). Analytical report to governments and international partners on the follow-up to the World Conference on Science; June 26–July 1, 1999; Budapest, Hungary. Paris, France: UNESCO.

Warren, D. M. (1992). *Indigenous knowledge, biodiversity conservation and development.* Keynote address at the International Conference on Conservation of Biodiversity in Africa: Local Initiatives and Institutional Roles, August 30-September 3, 1992, Nairobi, Kenya.

Williams, N. M., & Baines, G. (Eds.). (1993). *Traditional ecological knowledge: wisdom for sustainable development. Centre for Resource and Environmental Studies.* Canberra, Australia: Australian National University.

Wilson, E. O. (1992). *The diversity of life.* Cambridge, MA: Harvard University Press.

World Resources Institute. (1992). *The World Conservation Union, and the United Nations Environment Programme, Global Biodiversity Strategy: Policy-makers' Guide.* Baltimore, MD: WRI Publications.

Section 2

Pharmacology of Ethnobotanical Plants

Chapter 5
Ethnic Use, Phytochemistry, and Pharmacology of *Cyperus rotundus*: A Medicinal Plant

Mohammed Rahmatullah
University of Development Alternative, Bangladesh

Khoshnur Jannat
University of Development Alternative, Bangladesh

Gerald R. Reeck
Kansas State University, USA

Rownak Jahan
University of Development Alternative, Bangladesh

Taufiq Rahman
University of Cambridge, UK

Nasrin A. Shova
University of Development Alternative, Bangladesh

Maidul Islam
University of Development Alternative, Bangladesh

ABSTRACT

Cyperus rotundus (nut grass in English) is a perennial erect sedge plant and is distributed in over 90 countries of the world, where it has been mostly classified as a highly invasive weed. Despite this classification, the plant has been considered from traditional times to be medicinally important. The traditional uses of the plant in various countries include uses against various gastrointestinal tract disorders, skin

DOI: 10.4018/978-1-7998-1320-0.ch005

diseases, leprosy, fever, and neurological disorders. Evaluation of the plant and especially its rhizomes in a scientific manner has revealed the presence of numerous phytochemicals and wide-ranging pharmacological activities, which include anti-microbial, gastrointestinal, wound healing, anti-diabetic, anti-cancer, anti-malarial, anti-obesity, hepatoprotective, and anti-pyretic activity. The scientific validation of a number of traditional uses strongly indicates that the plant may prove useful in the discovery of a number of lead compounds and novel drugs.

INTRODUCTION

Cyperus rotundus L. is a perennial herbaceous plant belonging to the Cyperaceae family. It has an extensive network of rhizomes, tubers, bulbs and roots through which it can propagate easily, which has made it a very invasive plant and as such been classified as an invasive weed in the more than 90 countries that the plant is now found. The plant is believed to have originated from India, where its Sanskrit name as in Ayurveda is 'nagarmotha' (Bajpay, Nainwal, Singh & Tewari, 2018). The plant is found at present in various countries of Asia, Africa, Europe, North America, South America, and the Pacific Islands. Being present in so many countries, the plant has numerous names in the local vernacular. Some of the vernacular names given in The Global Invasive Species Database (GISD) [http://www.iucngisd.org/gisd/species.php?sc=1448] are tiririca-vermelha (Portuguese, Brazil), oniani lau (Maori, Cook Islands), vuthesa (Fijian), souchet à tubercules (French), zigolo infestante (Italian), soro ni kabani (Fijian), pakopako (Tongan, Tonga), coco grass (English), mauku'oniani (Maori, Cook Islands), mothe (English, Nepal), chufa (Spanish), mumuta (Samoan), oniani tita (Maori, Cook Islands), ya khon mu (Thai), vucesa (Fijian), juncia (Spanish), tiririca (Portuguese, Brazil), sur-sur (Pampangan), capim-alho (Portuguese, Brazil), tuteoneon (Marshallese), mala-apulid (Pampangan), alho-bravo (Portuguese, Brazil), castanuela (Spanish), brown nut sedge (English), almendra de tierra (Spanish), Rundes Zypergras (German), xiang fu zi (Chinese), mutha (Tagalog), coquito (Spanish), coquillo purpura (Spanish), ya haeo mu (Thai, Central Thailand), coquillo (Spanish), tamanengi (Palauan), coco (Spanish), purple nut sedge (English), pakopako (Tagalog), herbe à oignons (French), and juncia real (Spanish). In Bangladesh, the plant is known as 'mutha' or 'mutha ghas'.

The plant also has a number of synonyms. Some of the synonyms according to the Plant List [http://www.theplantlist.org/tpl1.1/record/kew-238342] are *Chlorocyperus rotundus* (L.) Palla, *Chlorocyperus salaamensis* Palla, *Cyperus agrestis* Willd. ex Spreng. & Link, *Cyperus arabicus* Ehrenb. ex Boeckeler, *Cyperus bicolor* Vahl, *Cyperus bifax* C.B.Clarke, *Cyperus bulbosostoloniferus* Miq., and *Cyperus comosus* Sm. The taxonomic hierarchy of the plant is given below.

Kingdom: Plantae
SubKingdom: Viridiplantae
Division: Tracheophyta
Subdivision: Spermatophytina
Class: Magnoliopsida

Order: Poales
Family: Cyperaceae
Genus: *Cyperus*
Species: *Cyperus rotundus* L.

[https://www.itis.gov/servlet/SingleRpt/SingleRpt?search_topic=TSN&search_value=39900#null]

As reviewed by Bajpay et al. (2018), traditional medicinal uses in a number of countries of the plant include "indigestion, constipation, dysentery, abdominal distension, neurogenic gastralgia, chest pains, irregular as well as painful catatonia, skin diseases, furuncle infections, staphylococcal infections, leprosy, and sprains and bruises". The substantial number of ethnic or traditional medicinal uses combined with a plethora of pharmacological activities with the whole plant or plant parts, which include anti-microbial, gastrointestinal, wound healing, anti-diabetic, anti-cancer, anti-malarial, anti-obesity, hepatoprotective and anti-pyretic activity suggest that the plant can be a potential source of lead compounds and new drugs against possibly several diseases. The objective of the present review is to give an update on the ethnic uses, reported phytochemicals and pharmacological activities of the plant or plant parts along with a discussion on the possibilities of obtaining new drugs from the plant.

Ethnic Uses

In East Anatolia, Turkey, the root of the plant is used as a diuretic (Altundag & Ozturk, 2011). In Poncokusumo district, Malang, East Java Province, Indonesia, the rhizomes are used as stomachic and diuretic and also used for treatment of dysentery and blood disorders, headaches, indigestion, diarrhea and cholera (Batoro & Siswanto, 2017). Juice obtained from root is used to treat fever by the Newar community of Pharping village of Kathmandu district, Nepal (Balami, 2004). For treatment of malaria, the Bodo tribal people of Manas Biosphere Reserve in India take 200g of *Costus speciosus* rhizomes, 200g bark of *Cyperus rotundus*, and 200g bark of *Azadirachta indica*. The combination is boiled in one liter water till the volume is reduced to 250 ml. Two to four spoons of the decoction are taken orally after meal for 15 days and during this time the diet should be strictly vegetarian (Paul, Devi & Sarma, 2013). The rhizome is used for treatment of dengue by local traditional healers of Tiruchirappalli district in India (Selvam & Durai, 2018).

A number of traditional uses of the plant have been given by Singh, Singh and Singh (2018) without further elaboration as to the places or the people using the plant in the reported manner. Powder of *Cyperus rotundus*, *Holarrhena antidysenterica*, *Mesua ferrea*, *Glycyrrhiza glabra*, and *Emblica officinalis* (plant parts not mentioned) are taken in equal amounts and a uniform mixture prepared. The mixture is taken twice a day after meals as treatment for ulcerative colitis. For treatment of dental disorders, one part each of *Cyperus rotundus*, *Terminalia chebula*, and *Embelia ribes* is mixed with two part leaves of *Azadirachta indica* and homogenized in cow urine, which is then taken twice or thrice a day. A uniform mixture is prepared from equal amounts of powder of *Cyperus rotundus*, *Embelia ribes* and dried rhizomes of *Zingiber officinale*. The mixture is taken with honey once or twice as treatment for vomiting. Powder of *Cyperus rotundus* rhizome is to be taken for inducing uterine contractions and prevention of miscarriage. Pills prepared from *Cyperus rotundus*, *Glycyrrhiza glabra*, *Coriandrum sativum* and *Elettaria cardamomum* are taken to dispel bad breath. Plant extract (swaras or the most potent type of extracting plant according to Ayurveda) is boiled in cow ghee (clarified butter) and subjected to slow heat boiling for 30 minutes. The resultant decoction (lepa) is applied to wounds for healing. Pills prepared from 100g

Cyperus rotundus powder is mixed with 50g each of powder of *Cinnamomum camphora*, *Piper longum*, and *Ferula asafoetida* are taken to treat diarrhea, indigestion, cholera, and dyspepsia.

In five remote southern regions of Khyber Pakhtunkhwa province of Pakistan, it has been reported that the rhizomes are used for treatment of dyspepsia, diarrhea and vomiting (Tariq et al., 2015). Whole plant extract is mixed with sugar solution and taken orally for dysentery in Bannu district of the same area in Pakistan (Khan et al., 2013). Joshi (2005) has made a summary of the ethnic uses of the plant in Nepal. The tubers are considered astringent, carminative, sedative, stimulant and vermifuge. Pieces of tuber are chewed to treat fever due to cold, and coughs. Tuber is taken with pounded fruits of black pepper (*Piper nigrum*) for stomach ache. Pounded root is topically applied to cure boils and blisters. Root juice is taken orally for helmintic infections. The tribal people of Papikondalu Wildlife Sanctuary, Eastern Ghats of Andhra Pradesh, India make a paste of grounded roots of *Hemidesmus indicus* and *Aristolochia indica*, and tubers of *Cyperus rotundus* and apply the paste to affected areas in herpes infection; three or four tubers are crushed and the juice taken with a few drops of honey orally for about 3 days for diarrhea (Rao, Rao & Murthy, 2016).

The tribal peoples of Rajasthan, India, rub paste of rhizomes inwards on the tongue for tonsillitis; culm is chewed for snakebite (Sharma & Khandelwal, 2010). The Garasia tribe of Sirohi district, Rajasthan, India, chews 5-10 fresh stems and swallows the juice for water snakebite (Meena & Yadav, 2011). People of Villupuram district, Tamil Nadu, India, use whole plants for treatment of vomiting and wounds (Dhanam & Elayaraj, 2014). The villagers of Pauri (PAURI Garhwal), Uttarakhand, India, take the whole plant for abdominal problem and fever; tuber for its diuretic properties; and decoction of the plant for urinary trouble (Khajuria & Bisht, 2017). The Thengal Kacharies, a tribal people of Assam, India, use whole plant to treat fever (Dutta, 2017). The inhabitants of Hafizabad district, Punjab, Pakistan, use the plant as a diuretic, vermifuge, stomachic, galactagogue, and for treatment of dermatitis and hypersplenism (Umair, Altaf, & Abbasi, 2017).

The local people of Kusmi Forest, Uttar Pradesh, Gorakhpur, India, use decoction of the rhizome as treatment for malaria (Gautam, Rajkumar, Srivastava, & Singh, 2015). It has been reported that in Bojnord, North Khorasan Province, Iran, the people use decoction of the root to treat migraine, and as a sedative and to strengthen memory (Nadaf, Joharchi, & Amiri, 2019). Folk medicinal practitioners in Lalmonirhat district, Bangladesh, use whole plant, flowers and tubers to treat a number of ailments. Leaves of *Psidium guajava*, leaves of *Punica granatum* and whole plants of *Cyperus rotundus* are mixed, warmed and macerated to obtain juice. Half cup of the juice is orally taken with 10-15 drops of honey twice daily for 3 days as treatment of loss of libido in men and puerperal fever in women. One cup of juice obtained from macerated flowers is taken with 10-15 drops of honey as treatment for blood dysentery. As treatment for eczema and other skin diseases, leaves of *Lawsonia inermis* and *Azadirachta indica* are macerated with whole plants of *Cyperus rotundus* and topically applied to affected areas. For any type of skin infections, macerated tubers are topically applied to affected area(s). For constipation, half cup of juice obtained from macerated tubers is taken thrice daily (Jahan et al., 2011).

Phytochemical Constituents

It has been reported by Huang (1999) that the dried tubers of the plant contain sesquiterpenes (patchoulane, rotundane, eudesmane, guaiane, cadinane and caryophyllene types), flavonoids (visnagin, khellin, ammiol, isorhamnetin, and tricin), phenolic acids (salicylic acid, protocatechuic acid, caffeic acid and *p* coumaric acid), and steroids (steroidal glycoside, sitosteryl-(6'-hentriacontanoyl)-β-d-galactopyranoside) besides unidentified quinones, coumarins, alkaloids and saponins. Essential oil from the plant reportedly contain (+) oxo-α-ylangene, (+) α-cyperone, trans-pinocarveol, and cyperene. Essential oil from rhizome of the plant collected from Uttarakhand, India, showed the presence of α-pinene, 3-ethylidene-1-methyl cyclopentene, sabinene, β-pinene, *p*-cymene, 1-limonene, 8-cineole, *trans*-pinocarveol, terpinen-4-ol, citronellal, 4,4-dimethyl-tricyclo-(3,2,1) octan-6-on, *p*-cymen-8-ol, 1-α-terpineol, *cis*-dihydrocarvone, myrtenol, verbenone, 1-β-4,4-trimethyl-bicyclo (3,2) hept-6-en-2-ol, *trans*-carveol, carvone, carvenone, α-cubebene, dihydro-carvylacetate, α-copaene, isolongifoline, cyperene, *trans*-caryophyllene, dihydro-aromadendrene, aromadendrene-epoxide, naphthalene, α-silenene, *cis*-calamenene, *trans*-calamenene, elema-1,3,11(13)-trien-12-ol, caryophyllene oxide, *cis*-12-caryophyll-5-en-2-one, caryophylla-2(12),6(13) dien-5-one, cyclohexane, cyclohexenone, 10-epi-α-cyperone, (+) oxo-α-ylangene, α''cyperone, caryophyllenol, vulgarol A, vellerdiol, aristolone, vulgarol B, ledenoxide, dimethyl-7-isopropenyl-bicyclo-dec-1-en-3-one, longifolinaldehyde, and longipynocarvone (Bisht, Bisht, Singh, Gupta & Singh, 2011; El-Gohary, 2004). Essential oil obtained from two different locations (Empangeni-A and KwaDlangezwa-B, both in the Kwa-Zulu Natal Province of South Africa) yielded 41 and 43 different components, respectively. The major components from the second oil were β-pinene, α-pinene, α-cyperone, α-selinene and myrtenol; from the first oil, the main components were α-cyperone, myrtenol, caryophyllene oxide, and β-pinene (Lawal & Oyedeji, 2009).

Essential oil collected from tubers of the plant growing wild in Isfahan province, Iran showed sesquiterpenes as the major constituents of the oil. Cyperene, caryophyllene oxide, α-longipinane, and α-selinene were the major components; other compounds found were α-pinene, verbenene, β-pinene, *o*-cymene, limonene, 1,8-cineole, *p*-cymene, α-fenchol, *trans*-pinocarveol, *cis*-verbenol, pinocarvone, isopinocamphon, *para*-cymen-8-ol, α-terpineol, myrtenol, verbenone, *trans*-carveol, *trans*-myrtenyl acetate, cuminic aldehyde, carvone, cinnamaldehyde, *trans*-anethole, thymol, 2,4-decadienal, carvacrol, eugenol, α-ylangene, α-copaene, β-elemene, β-caryophyllene, β-gurjunene, α-guaiene, aromadendrene, isoaromadendrene, α-humulene, α-caryophyllene, rotundene, γ-gurjunene, γ-muurolene, n-dodecanol, n-dodecanol, β-selinene, α-farnesene, *cis*-γ-bisabolene, *trans*-calamenene, *trans*-γ-bisabolene, α-calacorene, β-calacorene, γ-elemene, spathulenol, humulene epoxide II, γ-gurjunene epoxide, aristolone, α-cyperone, *n*-hexadecanoic acid, phytol, and methyl linoleate (Ghannadi, Rabbani, Ghaemmaghami & Malekian, 2012).

Three novel sesquiterpene alkaloids, namely rotundines A, B and C have been reported from rhizomes (Jeong, Miyamoto, Inagaki, Kim, & Higuchi, 2000). Three new sesquiterpene hydrocarbons (-)-isorotundene, (-)-cypera-2,4(15)-diene, (-)-norrotundene and the ketone (+)-cyperadione were isolated as minor constituents of the essential oil of the plant (Sonwa & König, 2001). From aerial parts of the plant, a new steroid glycoside named sitosteryl (6'-hentriacontanoyl)-β-D-galactopyranoside in addition to three furochromones, khellin, visnagin and ammiol have been reported along with the compounds benzo-α-pyrone (coumarin), salicylic acid, caffeic acid, protocatechuic acid, *p*-coumaric acid, tricin and isorhamnetin (Sayed, Mohamed, Farag, Mohamed, & Proksch, 2007). Further studies of the chemical components of the aerial parts led to isolation of a fructose-amino acid conjugate, N-(1-deoxy-α-D-fructos-1-yl)-L-tryptophan and its tautomers, in addition to n-butyl-β-D-fructopyranoside, ethyl-α-D-glucopyranoside, adenosine, (-)-(E)-caffeoylmalic acid, vitexin, isovitexin, orientin, epiorientin, myricetin 3-O-β-D-galactopyranoside, luteolin 7-O-β-D-glucuronopyranoside-6''-methyl ester, chlorogenic acid, luteolin 4'-O-β-D-glucuronopyranoside, luteolin 7-O-β-D-glucuronopyranoside, uridine and ellagic acid (Sayedet al., 2008).

A novel norsesquiterpene, named norcyperone has been reported from rhizomes, the structure of which has been determined as 8,11,11-trimethylbicyclo[5.3.1]undecane-5α, 8α-epoxy-3-one. Along with this compound, other compounds reported in this study included (-)-clovane-2,9-diol, rosenonolactone, and 5α,8α-epidioxy-(20S,22E,24R)-ergosta-6,22-dien-3β-ol (Xu et al., 2008). Six sesquiterpenes have been isolated from rhizomes including two new, namely, epi-guaidiol A and sugebiol, the other four sesquiterpenes are guaidiol A, sugetriol triacetate, cyperenoic acid and cyperotundone (Xu, Zhang, Wan & Zou, 2009). From 70% ethanolic extract of rhizomes, the sesquiterpene derivatives (valencene, nootkatone, and caryophyllene α-oxide), monoterpenes (β-pinene, 1,8-cineole, and limonene) and 4-cymene have been isolated (Jin, Lee, Kim, & Kim, 2011).

From rhizomes of the plant the compounds salicylic acid, caffeic acid, protocatechuic acid, *p*-coumaric acid, pongamone and biochanin A have been reported besides two new phenolic compounds, methoxycyperotundol and cyperotundol (Zhou & Yin, 2012). Stilbene oligomers have been isolated from rhizomes; they are (+)- and (−)-(E)-cyperusphenol A, (E)-mesocyperusphenol A, cyperusphenols B, C and D, scirpusins A and B, and piceid and luteolin (Ito et al., 2012). Five iridoid glycosides have been isolated from rhizomes; they are rotunduside, 10-*O*-*p*-hydroxybenzoyltheviridoside, 10-*O*-vanilloyltheviridoside, 6^2-*O*-(*trans-p*-coumaroyl)-procumbide, and loganic acid (Zhou, Yin, Zhang, Feng, & Xia, 2013). Other iridoid glycosides reported from rhizomes are rotundusides G and H, negundoside, nishindaside, isooleuropein, and neonuezhenide (Zhou et al., 2016). Ethanol extract of rhizomes have been shown to contain patchoulane-type sesquiterpenoids like cyperene-3, 8-dione, 14-hydroxy cyperotundone, 14-acetoxy cyperotundone, 3β-hydroxycyperenoic acid, and sugetriol-3, 9-diacetate. Other sesquiterpenoids isolated with these compounds include cyperenol, cyperenoic acid, cyperotundone, sugeonol, scariodione, sugetriol triacetate, α-cyperone, [(4a*S*,7*S*),-7-hydroxy-1, 4a-dimethyl-7-(prop-1-en-2-yl),-4,4a,5,6,7,8-hexahydronaphthalen-2 (3*H*),-one], [(4a*S*,7*S*,8*R*),-8-hydroxy-1, 4a-dimethyl-7-(prop-1-en-2-yl),-4,4a,5,6,7,8-hexahydronaphtha-len-2 (3*H*)-one], cyperol, 1β-hydroxy-α-cyperone, 10-epieudesm-11-ene-3β,5α-diol, 3β-hydroxyilicic alcohol (11(13),-eudesmene-3,4,12-triol), cyperusol C, α-corymbolol, 3β,4α-dihydroxy-7-epi-eudesm-11(13),-ene, 2-oxo-α-cyperone, 7α(*H*),10β-eudesm-4-en-3-one-11,12-diol, rhombitriol, 7-epi-teucrenone, α-rotunol, nootkatone, 12-hydroxynootkatone, oxyphyllol C, 5-hydroxylucinone, oplopanone, 10-hydroxyamorph-4-en-3-one, cyperusol D, 2-hydroxy-14-calamenenone, 1-isopropyl-2,7 dimethylnaphthalene, argutosine D, and 4,5-seco-guaia-1(10),11-diene-4,5-dioxo (Xu et al. 2015).

From ethyl acetate extract of rhizomes, thirteen compounds have been identified; these compounds are cyperotundic acid, (4*S*, 5*E*, 10*R*)-7-oxo-trinoreudesm-5-en-4β-ol, 4-hydroxy-4, 7-dimethyl-1-tetralone, taraxerone, dammaradienyl acetate, zeorin, sarmentine, guineensine, pellitorine, caprolactam, liriodendrin, 3-hydroxy-1-(4-hydroxy-3,5-dimethoxyphenyl)-2-[4-(3-hydroxy-1-(E)-propenyl)-2,6-dimethoxyphenoxy]propyl-β-D-glucopyranoside, and 1-(3, 4-methylenedioxyphenyl)-1E-tetradecene (Xu et al., 2016). In another report, ethyl acetate fraction of rhizomes yielded nine compounds, namely scirpusin A, scirpusin B, luteolin, 6'-acetyl-3,6-diferuloylsucrose, 4',6' diacetyl-3,6-diferuloylsucrose, *p*-coumaric acid, ferulic acid, pinellic acid, and fulgidic acid (Sim et al., 2016). Two new cycloartane glycosides, cyprotusides A and B have been reported from rhizomes (Zhou, Lin & Yin, 2016). Twelve compounds isolated from rhizomes of *Cyperus rotundus* and *Cyperus scariosus* included stigmasterol, β-sitosterol, lupeol, gallic acid, quercetin, β-amyrin, oleanolic acid, β-amyrin acetate, 4- hydroxyl butyl cinnamate, 4- hydroxyl cinnamic acid, caffeic acid, and kaempferol (Kakarla et al. 2016). Besides rotundusolides A, B and C and secomacrogenin B, other compounds reported from dried rhizomes were 3,4-*O*-isopropylideneshikimic acid, dehydrocostuslactone, (+)-alismoxide, sugetriol triacetate,

2β-hydroxy-α-cyperone, eudesma-4(14),11(13)-diene-7α,8α,12-triol, and 3,4-*seco*-mansumbinoic acid (Yang & Shi, 2012).

Two new cycloartane glycosides, cyprotuosides C and D have been isolated from etrhanol extract of rhizomes. Their structures have been elucidated as, respectively, 24*R*-9,10-seco-cycloartan-1(10),9(11)-dien-3β,7β,24,25-tetraol 3-*O*-β-D-xylopyranosyl-(1®4)-[α-L-arabinopyranosyl-(1®6)]-β-D-glucopyranosyl-25-*O*- β-D-glucuronide and 9,10-seco-cycloartan-1(10),9(11),23(24)-trien-3β,7β,25-triol 3-*O*-β-D-xylopyranosyl-(1®4)-{α-L-arabinopyranosyl-(1®6)-[α-L-rhamnopyranosyl-(1®2)]}-β-D-glucopyranosyl-25-*O*-β-D-glucuronide (Lin, Zhou, & Li, 2018). A novel compound, (2)-acetyl-3 (5)-styryl-5 (3)-methylthiopyrazole has been detected in methanol extract of rhizomes (Kamala, Middha, Gopinath, Sindhura & Karigar, 2018). Two compounds with anti-inflammatory activity have been isolated from the plant, namely cyperalin A and sugetriol triacetate (Mohamed-Ibrahim, Mohamed, Abdullah Khayat, Zayed & Soliman El-Kholy, 2018).

Structures of some of the phytoconstituents present in the plant are shown in Figure 1.

Figure 1.Some phytochemicals isolated from various parts of Cyperus Rotundus

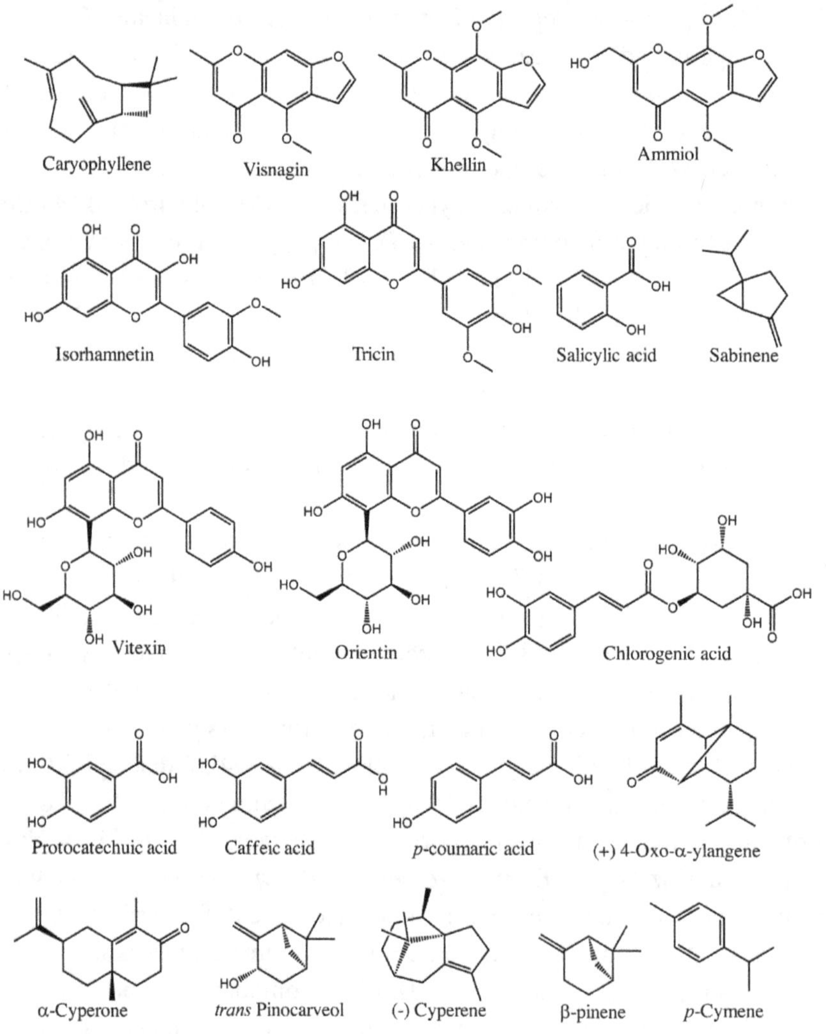

PHARMACOLOGICAL ACTIVITIES

Analgesic Activity

Essential oil from fresh rhizomes of the plant showed analgesic activity in acetic acid-induced writhing test in mice (Chen et al., 2011). Extracts of aerial parts of the plant reportedly showed analgesic activity as determined with acetic acid-induced writhing tests in mice (Soumaya et al., 2013). Hydromethanol extract of whole plant exhibited antinociceptive activity in thermal types (hot plate and tail immersion) and chemical-induced (formalin) nociception models in mice (Imam & Sumi, 2014).

Anti-Allergic Activity

From a 70% ethanolic extract of rhizomes of *C. rotundus*, a number of sesquiterpene derivatives, namely valencene, nootkatone, and caryophyllene α-oxide were isolated and which showed anti-allergic activity *in vitro* and *in vivo*. In rat basophilic leukemia (RBL)-1 cells, the compounds inhibited leukotrienes production by 5-lipoxygenase. The sesquiterpenes also inhibited β-hexosaminidase release by antigen-stimulated RBL-2H3 cells, valencene giving the strongest effect. Valencene was found to inhibit β-hexosaminidase degranulation by inhibiting the initial activation reaction, Lyn phosphorylation, in IgE-stimulated RBL-2H3 cells. When administered orally, valencene and nootkatone significantly inhibited the delayed-type hypersensitivity reaction in mice, clearly demonstrating that the sesquiterpenes possessed potent anti-allergic activities (Jin, Lee, Kim, & Kim, 2011).

Anti-Bacterial Activity

Aqueous extract of dried roots of the plant reportedly showed inhibitory activity against *Bacillus pumilis* (Sini & Malathy, 2005). Tuber extract has been found to reduce the growth and acid production of *Streptococcus mutans*, the causative bacteria in the formation of dental plaques and dental caries. The extract also markedly inhibited the adherence of the bacteria to saliva-coated hydroxyapatite beads (HAs). Around 50% repression of this adherence was noted with concentration of 0.5 mg/ml of the extract and 100% inhibition was observed with a concentration of 4 mg/ml extract (Yu, Lee, Seo & You, 2007). Inhibitory effect of the plant extract has also been reported against caries causing *S. mutans* and *Lactobacillus acidophilus* (Haghgoo, Mehran, Zadeh, Afshari, & Zadeh, 2017).

A marked inhibitory effect was observed against *Salmonella enteritidis*, *Staphylococcus aureus* and *Enterococcus faecalis* with total oligomers flavonoids (TOFs) and ethyl acetate extracts of tubers of *C. rotundus* (Kilani et al., 2008).

Anti-Cancer Activity

Visnagin, khellin and sitosteryl (6'-hentriacontanoyl)-β-D-galactopyranoside, isolated from aerial parts of the plant showed activity in brine shrimp lethality test and showed strong cytotoxic activity against L5178y mouse lymphoma cells (Sayed et al., 2007).

Total oligomer flavonoids and ethyl acetate extracts of *C. rotundus* tuber have been shown to suppress growth and proliferation of L1210 cells derived from murine lymphoblastic leukaemia; the cytotoxicity was found to be due to induction of apoptosis as shown in DNA fragmentation studies (Kilani, 2008).

Luteolin, a compound found in ethyl acetate extract was found to be the active component, which inhibited significantly the proliferation of K562 cells with an IC_{50}=25 microg/ml (Kilani-Jaziri et al., 2009).

Cyperusphenol D, a compound found in the rhizomes of the plant has been reported to suppress the growth of Jurkat cells (human T-cell leukemia cells) through induction of apoptosis (Ito et al., 2012). Orientin, isolated from butanol fraction of methanol extract of aerial parts of the plant exhibited DNA fragmentation (apoptotic) activity against K562 and L1210 cell lines (Soumaya et al., 2014).

The pro-apoptotic effect of ethanol extract of rhizomes has been investigated in a human breast carcinoma MDA-MB-231 cell model. Apoptosis, induced by the extract, was associated with upregulation of death receptor 4 (DR4), DR5 and pro-apoptotic Bax, as well as downregulation of anti-apoptotic surviving and Bcl-2. The extract also downregulated Bid expression and activated caspase-8 and -9, which respectively are the initiator caspases of the extrinsic and intrinsic apoptotic pathways (Park et al., 2014).

Ethanol extract of rhizomes have been reported to inhibit human ovarian cancer (A2780, SKOV3 and OVCAR3) and endometrial cancer (Hec1A and Ishikawa) cells. 6-Acetoxy cyperene, one of the constituents of the extract induced apoptosis in the ovarian cancer cell lines through activation of caspase-3, caspase-8 and caspase-9 and poly(ADP-ribose)polymerase (PARP) in a dose-dependent manner (Ahn et al., 2015).

Amentoflavone, isolated from the plant, has been shown to have a significant inhibitory effect on uterine tumors in rats. The study suggests that amentoflavone may be acting by elevating Bax protein expression, down-regulating Bcl-2 expression, forming homodimers Bax/Bax, and reducing plasma estradiol and progesterone to promote apoptosis of uterine fibroid cells (Ying & Bing, 2016).

The methanolic extract of rhizomes reportedly showed cytotoxic activity against different human cancer cell lines like breast (MCF-7), cervical (HeLa), liver (Hep G2), prostate (PC-3), and colorectal (HT-29) (Mannarreddy et al., 2017).

Anti-Diabetic Activity

Extract of the plant demonstrated both protein glycation inhibitory and anti-oxidant activity in a model of fructose-mediated protein glycoxidation. The extract also significantly reduced the formation of advanced glycation end products (AGE). Taken together, the findings suggest that the plant extract or its purified component(s) can play a significant role in ameliorating diabetes-induced complications (Ardestani & Yazdanparast, 2007).

A number of compounds isolated from methanol extract of rhizomes demonstrated α-glucosidase and α-amylase inhibitory activities as well as anti-oxidant activity. Compound 1, (2*RS*,3*SR*)-3,4',5,6,7,8-hexahydroxyflavane inhibited α-amylase; compound 2, cassigarol E inhibited both α-glucosidase and α-amylase activities; compounds 3 and 4, scirpusin A and B inhibited α-glucosidase. All four compounds showed significant 2,2-diphenyl-1-picrylhydrazyl (DPPH) scavenging activity, indicating that the four compounds were all good anti-oxidants (Tran et al., 2014). Oxidative stress during diabetes can lead to stroke, neuropathy, retinopathy and nephropathy (Asmat, Abad, & Ismail, 2016). From that view point, compounds that can act as both glucose lowering agent and antioxidant as in the present case has potential as effective drugs for treating diabetes mellitus.

Ethanol extract of rhizomes showed anti-diabetic activity in streptozotocin (STZ)-induced diabetic Swiss mice by lowering blood glucose, improving body weight, and reducing serum glutamic pyruvic transaminase (GPT), glutamic oxaloacetic transaminase (GOT), cholesterol, and triglyceride levels (Singh, Khosa, Mishra, & Jha, 2015).

Anti-Diarrheal Activity

The methanol extract of rhizome, when administered to mice in which diarrhea has been induced by castor oil, showed anti-diarrheal activity at doses of 250 and 500 mg per kg body weight (Uddin, Mondal, Shilpi, & Rahman, 2006).

Decoction of tubers showed anti-giardial activity, reduced adherence of enteropathogenic *Escherichia coli*, and reduced invasion of Hep-2 cells by enteroinvasive *E. coli* (EIEC) and *Shigella flexneri*. The anti-diarrheal activity did not seem to involve killing of enteric pathogens (Daswani, Brijesh, Tetali, & Birdi, 2011). Since diarrhea is a common problem, particularly in underdeveloped countries because of poor hygiene and absence of safe drinking water, the plant can prove beneficial as a handy remedy.

Anti-Fungal Activity

Essential oil obtained from roots of the plant grown in Brazil demonstrated inhibitory activity against *Candida albicans* (Duarte, Figueira, Sartoratto, Rehder, & Delarmelina, 2005). *Candida* infections can be difficult to cure, the most common infection being Candidiasis in vagina. Other *C. albicans* infections may range from superficial infections of the skin to life-threatening systemic infections (Mayer, Wilson, & Hube, 2013).

Anti-Inflammatory Activity

In one study, the methanol extract of rhizomes showed the inhibition of NO (nitric oxide) production by RAW 264.7 cells stimulated with interferon-γ with LPS (lipopolysaccharide). The inhibition of NO production was shown to occur due to the suppression of iNOS protein, as well as iNOS mRNA expression. The extract also suppressed the production of reactive oxygen species (O2-) in phorbol ester-stimulated RAW 264.7 cells. Overall, the results support the conclusion that the extract can produce an anti-inflammatory effect (Seo et al., 2001).

In two models of acute inflammation, that of carrageenan induced rat paw edema and acetic acid induced peritonitis in mice, the plant demonstrated anti-inflammatory activity (Dang, Parekar, Kamat, Scindia, & Rege, 2011). In another report, the anti-inflammatory activity of extract of rhizomes and extract constituents like nootkatone and valencene was attributed to induction of heme oxygenase-1 (HO-1). The increased expression of HO-1 was correlated with significant inhibition of iNOS/NO production in LPS-activated RAW264.7 cells. Additionally, the extract, nookatone and valencene showed marked inhibition of high mobility group box-1 (HMGB1) in LPS-activated macrophages and increased survival rates in cecal ligation and puncture (CLP)-induced sepsis in mice. The effects, when taken together, could be attributed to HO-1 induction (Tsoyi et al., 2011).

Extracts of aerial parts of the plant were found to decrease the mouse ear edema induced by xylene (Soumaya et al., 2013). The n-hexane fraction of the 80% EtOH extract from rhizomes was found to inhibit both nitric oxide (NO) and prostaglandin E2 (PGE2) production in RAW 264.7 cells. The active component was identified to be α-cyperone. Inhibition of PGE2 production by α-cyperone was found to be due to suppression of LPS-induced expression of inducible cyclooxygenase-2 (COX-2) at both mRNA and protein levels. Additionally, α-cyperone downregulated the production and mRNA expression of the inflammatory cytokine interleukin-6 (IL-6); moreover, α-cyperone suppressed the transcriptional activity of nuclear factor κB (NFκB) and the nuclear translocation of the p65 NFκB subunit in LPS-induced RAW 264.7 cells. The results are strongly indicative of α-cyperone having strong anti-inflammatory activity (Jung et al., 2013).

Atopic dermatitis (AD) is an inflammatory skin disease. It has been reported that a major constituent of the plant, (+)-nootkatone may suppress tumor necrosis factor α (TNF-α)/interferon γ (IFN-γ)-induced TARC/CCL17 and MDC/CCL22 expression in HaCaT cells by inhibiting of PKCζ and p38 MAPK signaling pathways that lead to activation of NF-κB, and as such, can be a therapeutic candidate for skin diseases like AD (Choi, Lee, & Jung, 2014).

Fulgidic acid, isolated from rhizomes, has been reported to reduce the production of nitric oxide (NO), prostaglandin E2 (PGE2), tumor necrosis factor-α (TNF-α), and interleukin-6 (IL-6) in lipopolysaccharide (LPS)-induced RAW264.7 macrophages. In addition, the compound suppressed the LPS-induced expression of inducible nitric oxide synthase (iNOS) and cyclooxygenase-2 (COX-2) at the protein level, as well as iNOS, COX-2, TNF-α, and IL-6 at mRNA levels. Suppression of LPS-induced transcriptional activity of activator protein-1 (AP-1) as well as phosphorylation of c-Fos and c-Jun was also observed with the compound. Together the results strongly indicate that the anti-inflammatory mechanism of fulgidic acid is associated with the suppression of iNOS, COX-2, TNF-α, and IL-6 expression through down-regulating AP-1 activation in LPS-induced RAW264.7 macrophages (Shin et al., 2015).

Isocyperol, isolated from the rhizomes was found to significantly inhibit lipopolysaccharide (LPS)-induced production of nitrite oxide (NO) and prostaglandin E2 (PGE2) and suppressed LPS-induced expression of inducible nitric oxide synthase (iNOS) and cyclooxygenase-2 (COX-2) at the mRNA and protein levels in RAW 264.7 macrophages. The compound furthermore downregulated the LPS-induced expression of several proinflammatory cytokines, such as interleukin-1β (IL-1β), IL-6, and monocyte chemotactic protein-1 (MCP-1) and suppressed the LPS-induced nuclear translocation and transcriptional activation of nuclear factor-κB (NF-κB) in macrophages. The activation of another pro-inflammatory signal STAT3 was also suppressed by the compound in LPS-stimulated RAW 264.7 cells. Pre-treatment with the compound induced heme oxygenase-1 (HO-1) expression and reduced LPS-stimulated reactive oxygen species (ROS) accumulation in macrophages. Furthermore, isocyperol significantly increased the survival rate and attenuated serum levels of NO, PGE2, and IL-6 in LPS-induced septic shock mouse model. The results, when taken cumulatively, indicate that isocyperol suppresses septic shock through negative regulation of pro-inflammatory factors through inhibition of the NF-κB and STAT3 pathways and ROS (Seo, Jeong, Lee, Jang, & Choi, 2016).

In experimental inflammatory bowel disease (IBD) induced in Sprague Dawley rats with dinitrobenzene sulfonic acid, chloroform extract of the plant was found to downregulate the expression of proinflammatory cytokine genes and so lower the severity of the disease (Johari, Joshi, & Gandhi, 2016).

A component of the plant, α-cyperone, has been shown to exert anti-inflammatory neuroprotective action in LPS-treated BV-2 cells (microglial cells used in model of neuroinflammation) by inhibiting the production of inflammatory cytokines through activating Akt/Nrf2/HO-1 and suppressing the NF-κB pathway (Huang et al., 2018). [Nrf2 is a basic leucine zipper (bZIP) protein that regulates the expression of antioxidant proteins that protect against oxidative damage triggered by injury and inflammation.]

Two terpenoids from rhizomes of the plant, namely cyperalin A and sugetriol triacetate have been reported from the plant having anti-inflammatory activity with inhibitory activity of PGE2, COX-2, and lipooxygenase-5 (LOX-5) (Mohamed-Ibrahim, Mohamed, Abdullah Khayat, Zayed, & Soliman El-Kholy, 2018).

Anti-Obesity Activity

Administration of hexane extract of tubers to obese Zucker rats showed that at doses of 45 or 220 mg/kg/day of the extract for 60 days induced a significant reduction in weight gain without affecting food consumption or inducing toxicity in the rats. Since *in vitro* experiments showed that the extract can stimulate lipolysis in 3T3-F442 adipocytes, it has been hypothesized that the observed anti-obesity effect may at least partially be mediated through activation of β3-adrenoreceptors (Lemaure et al., 2007).

Anti-Oxidant Activity

Hydroalcoholic extract of the plant has been found to exhibit high reduction capability and powerful free radical scavenging activity; the extract also inhibited lipid peroxidation in rat liver homogenate induced by Fe(2+)/ascorbate and prevented deoxyribose degradation in both non-site-specific and site-specific assays thus demonstrating the hydroxyl radical scavenging and metal chelating activity of the extract; the extract also demonstrated peroxidation inhibiting activity. Cumulatively, the results show that the plant has anti-oxidative potential (Yazdanparast & Ardestani, 2007).

Ethyl acetate extract of tubers showed significant anti-oxidant activity as demonstrated by its ability to inhibit nitroblue tetrazolium reduction by the superoxide radical in a non-enzymatic superoxide generating system (Kilani et al., 2008). Luteolin was found to be the significant compound in the ethyl acetate extract of tubers in reducing the production of thiobarbituric acid reactive substances TBARS) and protecting against H_2O_2/UV-photolysis induced DNA damage (Kilani-Jaziri et al., 2009).

(2*RS*,3*SR*)-3,4',5,6,7,8-hexahydroxyflavane, cassigarol E, and scirpusins A and B, isolated from methanol extract of rhizomes showed strong anti-oxidant activity as measured by DPPH radical scavenging assay (Tran et al., 2014).

Methanol and aqueous extracts of aerial parts showed anti-oxidant activity as demonstrated by inhibition of xanthine oxidase, lipid peroxidation, and OH. radical formation (Soumaya et al., 2014).

Anti-Malarial Activity

Extracts of tubers showed *in vitro* anti-malarial activity against the multidrug resistant K1 strain of *Plasmodium falciparum* (Weenen et al., 1990). Ethyl acetate extract of the plant collected from Eastern Ghats of South India was assayed for growth inhibition of asexual erythrocytic stages of chloroquine (CQ)-sensitive (3D7) and (CQ)-resistant (INDO) strains of *P. falciparum* in culture using the fluorescence-based SYBR Green I assay and showed good anti-plasmodial activities with IC_{50} values less than 10 microg/ml (Kaushik et al., 2013).

Anti-Platelet Activity

The anti-platelet activity of ethanol extract has been examined by seeing their effects on rat platelet aggregations *in vitro* and *ex vivo*, and on mice tail bleeding times. Concentration-dependent inhibitory effects were observed on collagen-, thrombin-, and arachidonic acid (AA)-induced platelet aggregation. Among the various constituents of the extract, (+)-nootkatone was the most potent; (+)-nootkatone treated mice exhibited significantly prolonged bleeding times and the compound had a significant inhibitory effect on rat platelet aggregation *ex vivo* (Seo et al., 2011).

Anti-Ulcer Activity

Ulcer was induced in rats with a single oral dose of aspirin (400 mg per kg body weight). Pre-treatment with 70% methanolic extract of the plant significantly inhibited aspirin-induced gastric ulceration in animals. Inhibition of ulceration was possibly due to the anti-oxidant effects exerted by the extract, which included significant increase in the activity of superoxide dismutase, cellular glutathione and glutathione peroxidase, and inhibition of lipid peroxidation in the gastric mucosa of ulcerated animals (Thomas et al., 2015). Anti-ulcer and anti-inflammatory activity has also been reported in another study with ulcer induced in rats through administration of aspirin (Ahmad et al., 2014).

Hepatoprotective Activity

It has been found that the hexane fraction of rhizomes of the plant reduced the elevated transcription levels of sterol regulatory element binding protein-1c (SREBP-1c) in primary hepatocytes following exposure to the liver X receptor α (LXRα) agonist. The extract was found to inhibit not only the LXRα-dependent activation of the synthetic LXR response element (LXRE) promoter, but also the activation of the natural SREBP-1c promoter. Additionally, the rhizome extract decreased the recruitment of RNA polymerase II to the LXRE of the SREBP-1c gene, the LXRα-dependent up-regulation of various lipogenic genes, and the LXRα-mediated accumulation of triglycerides in primary hepatocytes. From the results, it has been postulated that the extract may be a novel therapeutic remedy for fatty liver disease (Oh, Yoon, Lee, Kwak, & Kim, 2015).

Eudesmane-type sesquiterpenoids isolated from rhizomes have been found to be active against hepatitis B virus (HBV) and significantly inhibited HBV DNA replication (Xu et al., 2015). Rhizome fractions of the plant also showed very promising hepatoprotective and anti-HBV potential (Parvez, Al-Dosari, Arbab, & Niyazi, 2019). Hepatitis is becoming common in countries like Bangladesh, and so the plant can prove to be an effective means to combat this problem.

Effect on the Central Nervous System (CNS)

The effect of ethanol extract of roots and rhizomes on the central nervous system has been tested in mice. The extract was found to significantly potentiate the sleeping time of mice induced by standard hypnotics, demonstrated significant analgesic activity as determined with the acetic acid induced writhing test, potentiated analgesia induced by morphine and pethidine, and protected against strychnine and leptazol-induced convulsions. Overall, the extract was found to be CNS depressant (Pal, Dutta, & Sarkar, 2009).

Rotundusides G and H, isolated from rhizomes, in mice models of despair, showed significant anti-depressant activity (Zhou et al., 2016). The cycloartane glycosides, cyprotusides A and B, isolated from rhizomes also reportedly showed significant anti-depressant activity in despair mice models (Zhou, Lin, & Yin, 2016).

Neuroprotective Activity

In PC-12 cells, water extract of rhizomes protected cells against 6-hydroxydopamine (6-OHDA)-induced cell non-viability. The extract inhibited 6-OHDA-induced generation of reactive oxygen species and nitric oxide, reduction of mitochondrial membrane potential and caspase-3 activity. A significant protective effect was also seen with the extract against damage to dopaminergic neurons in primary mesencephalic culture suggesting that the cumulative protective effects of the extract against 6-OHDA-induced damages were due to its anti-oxidant and anti-apoptotic effects (Lee et al., 2010).

Total oligomeric flavonoids (TOFs), obtained from the plant, ameliorated neurological deficits, excitotoxicity and behavioral alterations induced by cerebral ischemic-reperfusion injury in rats. Oxidative stress was reduced as shown by significantly decreased malondialdehyde (MDA), and increased superoxide dismutase (SOD) and glutathione contents (Sunil et al., 2011).In another experiment behavioral analysis by Morris water maze experiment conducted with Sprague Dawley rats, it has been shown that TOF supplementation enhanced the spatial reference memory activity of the rats exposed to hypobaric hypoxia (Kandikattu et al., 2017).

Pre-treatment with rhizome extract has been shown to attenuate H_2O_2-induced cell damage in human neuroblastoma SH-SY5Y cells. The extract exhibited potent anti-oxidant activity and regulated the levels of a number of enzymes and proteins such as superoxide dismutase (SOD), catalase (CAT), glutathione peroxidase (GPx), glutathione reductase (GR), heat shock protein-70 (HSP-70), Caspase-3, and Bcl-2. Pre-treatment with the extract restored H_2O_2 (hydrogen peroxide)-induced cellular, nuclear, and mitochondrial morphologies and increased the expression of brain derived nerve growth factor (BDNF). The anti-oxidant and anti-apoptotic potential of the extract has been suggested can be a preventive factor in neurodegenerative disorders (Kumar & Khanum, 2013).

The protective action of rhizome extract has been also demonstrated against nitrosylation of tyrosine leading to formation of 3-nitro tyrosine (3-NT), a compound implicated in pathophysiology of various disorders particularly neurodegenerative conditions and aging. The extract also has been shown to protect against peroxynitrite or ONOO(-) induced apoptosis in human neuroblastoma SH-SY5Y cells (Hemanth Kumar, Tamatam, Pal, & Khanum, 2013).

Hypoxic injury (HI) can have long-term neurological complications with accompanying morbidity and mortality. It has been reported that ethanol extract of the plant gives a protective effect against sodium nitrite-induced hypoxia injury in rats. The extract was able to protect against the cognitive impairments, and the locomotor activity and muscular coordination defects, which are affected by sodium nitrite-induced hypoxic injury (Jebasingh, Devavaram, Venkataraman, Adeghate, & Starling, 2014).

The neuroprotective effect of ethanol extract of the plant has been studied in a model of global transient ischemia in rat, by evaluating the pathophysiology of the hippocampal tissue and spatial memory. It was observed that the extract was only able to prevent pyramidal cell loss in CA1 region of hippocampus after transient, global cerebral ischemia (Dabaghian et al., 2015).

Scirpusins A and B, isolated from rhizomes, have been shown to have neuroprotective activity and may serve as candidates for therapeutic agents against neurodegenerative disorders (Sim et al., 2016). Another compound, α-cyperone, has been shown to bind and interact with tubulin, which interaction can reduce inflammation and be beneficial in Alzheimer's disease (Azimi et al., 2016).

The activity of a combined extract of *C. rotundus* and *Zingiber officinale* (CPI) has been studied on the improvement of age-related dementia in male Wistar rats with ethylcholine mustard aziridinium ion or AF64A-induced memory deficits. CPI enhanced memory, increased neuronal density, decreased acetylcholinesterase (AChE) activity and decreased oxidative stress in the hippocampus of CP1-treated rats. CPI-induced improved memory function may be due to enhanced cholinergic function and decreased neurodegeneration and oxidative stress (Sutalangka & Wattanathorn, 2017).

Extract of the plant has been showed to improve learning impairment in male Wistar rats following amyloid β-peptide 1-40 (Aβ)-induced memory impairment; it has been suggested that the extract treatment can improve cognitive dysfunction as happens during Alzheimer's disease (AD) (Mehdizadeh et al., 2017).

Other Miscellaneous Pharmacological Activities

Valencene, isolated from rhizomes, inhibited skin photoaging-related ion channels and UV-induced melanogenesis in B16F10 melanoma cells (Nam, Nam & Lee, 2016). The epidermally expressed transient receptor potential vanilloid 1 channel is involved in heat- and UV-induced skin aging and also senses various noxious chemical and thermal stimuli. Oleanolic acid, isolated from rhizomes, partially inhibited transient receptor potential vanilloid 1 channel activity suggesting the therapeutic potential of the rhizomes in analgesia and UV-induced photoaging (Nam & Lee, 2015).

Oral extract of rhizomes, when administered to female rats at 300 and 600 mg stimulated milk production and increased the synthesis of prolactin significantly (Badgujar & Bandivdekar, 2015).

Ethanol extract of the plant showed cardioprotective action against isoprenaline-induced myocardial infarction (Khwaja, Mahmood, & Siddiqui, 2016). It has been reported in a study that the plant possesses spasmolytic, bronchodilator and vasodilator activities possibly through calcium channels blockade (Hussain et al., 2018).

CONCLUSION

The various phytochemicals and diverse pharmacological activities reported for the plant thus far strongly suggests that the plant may prove to be an excellent source for lead compounds and novel drugs against particularly difficult to cure diseases with current allopathic medicines. These diseases include inflammatory diseases, cancer, diabetes, hepatic diseases, and neurodegerative disorders. Cancer, diabetes and Alzheimers are fast reaching or have reached near endemic proportions in the world. The plant, as such, needs scientific interest and further research towards finding possible efficacious drugs against these diseases.

REFERENCES

Ahn, J. H., Lee, T. W., Kim, K. H., Byun, H., Ryu, B., Lee, K. T., ... Choi, J. H. (2015). 6-Acetoxy Cyperene, a Patchoulane-type Sesquiterpene Isolated from *Cyperus rotundus* Rhizomes Induces Caspase-dependent Apoptosis in Human Ovarian Cancer cells. *Phytotherapy Research*, *29*(9), 1330–1338. doi:10.1002/ptr.5385 PMID:26062076

Altundag, E., & Ozturk, M. (2011). Ethnomedicinal studies on the plant resources of east Anatolia, Turkey. *Procedia: Social and Behavioral Sciences*, *19*, 756–777. doi:10.1016/j.sbspro.2011.05.195

Ardestani, A., & Yazdanparast, R. (2007). *Cyperus rotundus* suppresses AGE formation and protein oxidation in a model of fructose-mediated protein glycoxidation. *International Journal of Biological Macromolecules*, *41*(5), 572–578. doi:10.1016/j.ijbiomac.2007.07.014 PMID:17765965

Asmat, U., Abad, K., & Ismail, K. (2016). Diabetes mellitus and oxidative stress – A concise review. *Saudi Pharmaceutical Journal*, *24*(5), 547–553. doi:10.1016/j.jsps.2015.03.013 PMID:27752226

Ahmad, M., Rookh, M., Rehman, A. B., Muhammad, N., Younus, M., & Wazir, A. (2014). Assessment of anti-inflammatory, anti-ulcer and neuro-pharmacological activities of *Cyperus rotundus* Linn. *Pakistan Journal of Pharmaceutical Sciences*, *27*(6). PMID:26045387

Azimi, A., Ghaffari, S. M., Riazi, G. H., Arab, S. S., Tavakol, M. M., & Pooyan, S. (2016). α-Cyperone of *Cyperus rotundus* is an effective candidate for reduction of inflammation by destabilization of microtubule fibers in brain. *Journal of Ethnopharmacology*, *194*, 219–227. doi:10.1016/j.jep.2016.06.058 PMID:27353867

Badgujar, S. B., & Bandivdekar, A. H. (2015). Evaluation of a lactogenic activity of an aqueous extract of *Cyperus rotundus* Linn. *Journal of Ethnopharmacology*, *163*, 39–42. doi:10.1016/j.jep.2015.01.019 PMID:25625349

Bajpay, A., Nainwal, R. C., Singh, D., & Tewari, S. K. (2018). Medicinal value of *Cyperus rotundus* Linn: An updated review. *Medicinal Plant*, *10*, 165–170.

Balami, N. P. (2004). Ethnomedicinal uses of plants among the Newar community of Pharping village of Kathmandu district, Nepal. *Tribhuvan University Journal*, *24*(1), 13–19. doi:10.3126/tuj.v24i1.251

Batoro, J., & Siswanto, D. (2017). Ethnomedicinal survey of plants used by local society in Poncokusumo district, Malang, East Java Province, Indonesia. *Asian Journal of Medical and Biological Research*, *3*(2), 158–167. doi:10.3329/ajmbr.v3i2.33563

Bisht, A., Bisht, G. R. S., Singh, M., Gupta, R., & Singh, V. (2011). Chemical composition and antimicrobial activity of essential oil of tubers of *Cyperus rotundus* Linn. collected from Dehradun (Uttarakhand). *International Journal of Research in Pharmaceutical and Biomedical Sciences*, *2*, 661–665.

Chen, Y., Zhao, Y. Y., Wang, X. Y., Liu, J. T., Huang, L. Q., & Peng, C. S. (2011). GC-MS analysis and analgesic activity of essential oil from fresh rhizoma of *Cyperus rotundus*. *Zhong Yao Cai*, *34*, 1225–1229. PMID:22233038

Choi, H. J., Lee, J. H., & Jung, Y. S. (2014). (+)-Nootkatone inhibits tumor necrosis factor α/interferon γ-induced production of chemokines in HaCaT cells. *Biochemical and Biophysical Research Communications*, *447*(2), 278–284. doi:10.1016/j.bbrc.2014.03.121 PMID:24704449

Dabaghian, F. H., Hashemi, M., Entezari, M., Movassaghi, S., Goushegir, S. A., Kalantari, S., ... Sharifi, Z. N. (2015). Effect of *Cyperus rotundus* on ischemia-induced brain damage and memory dysfunction in rats. *Iranian Journal of Basic Medical Sciences.*, *18*, 199–204. PMID:25825638

Dang, G. K., Parekar, R. R., Kamat, S. K., Scindia, A. M., & Rege, N. N. (2011). Antiinflammatory activity of *Phyllanthus emblica, Plumbago zeylanica* and *Cyperus rotundus* in acute models of inflammation. *Phytotherapy Research*, *25*(6), 904–908. doi:10.1002/ptr.3345 PMID:21132843

Daswani, P. G., Brijesh, S., Tetali, P., & Birdi, T. J. (2011). Studies on the activity of *Cyperus rotundus* Linn. tubers against infectious diarrhea. *Indian Journal of Pharmacology*, *43*(3), 340–344. doi:10.4103/0253-7613.81502 PMID:21713044

Dhanam, S., & Elayaraj, B. (2014). Ethnomedicinal aspects of some weeds from paddy fields of Villupuram district in Tamil Nadu, India. *International Letters of Natural Sciences*, *19*, 1–10. doi:10.18052/www.scipress.com/ILNS.19.1

Duarte, M. C., Figueira, G. M., Sartoratto, A., Rehder, V. L., & Delarmelina, C. (2005). Anti-*Candida* activity of Brazilian medicinal plants. *Journal of Ethnopharmacology*, *97*(2), 305–311. doi:10.1016/j.jep.2004.11.016 PMID:15707770

Dutta, M. L. (2017). Plants used as ethnomedicine by the Thengal Kacharies of Assam, India. *Asian Journal of Plant Science and Research*, *7*, 7–8.

El-Gohary, H. M. A. (2004). Study of essential oils of the tubers of *Cyperus rotundus* L. and *Cyperus alopecuroides* Rottb. *Bulletin of Faculty of Pharmacy, Cairo University*, *42*, 157–164.

Gautam, R. P., Rajkumar, S. D., Srivastava, S. K., & Singh, S. K. (2015). Folk medicinal uses of plants from Kusmi Forest, Uttar Pradesh, Gorakhpur, India. *International Journal of Current Microbiology and Applied Sciences*, *4*, 343–351.

Ghannadi, A., Rabbani, M., Ghaemmaghami, L., & Malekian, N. (2012). Phytochemical screening and essential oil analysis of one of the Persian sedges; *Cyperus rotundus* L. *International Journal of Pharmaceutical Sciences and Research*, *3*, 424–427.

Haghgoo, R., Mehran, M., Zadeh, H. F., Afshari, E., & Zadeh, N. F. (2017). Comparison Between Antibacterial Effect of Chlorhexidine 0.2% and Different Concentrations of *Cyperus rotundus* Extract: An *In vitro* Study. *Journal of International Society of Preventive & Community Dentistry*, *7*, 242–246. PMID:29026695

Hemanth Kumar, K., Tamatam, A., Pal, A., & Khanum, F. (2013). Neuroprotective effects of *Cyperus rotundus* on SIN-1 induced nitric oxide generation and protein nitration: Ameliorative effect against apoptosis mediated neuronal cell damage. *Neurotoxicology*, *34*, 150–159. doi:10.1016/j.neuro.2012.11.002 PMID:23174672

Huang, B., He, D., Chen, G., Ran, X., Guo, W., Kan, X., ... Liu, J. (2018). α-Cyperone inhibits LPS-induced inflammation in BV-2 cells through activation of Akt/Nrf2/HO-1 and suppression of the NF-κB pathway. *Food & Function*, *9*(5), 2735–2743. doi:10.1039/C8FO00057C PMID:29667667

Huang, K. C. (1998). *The dried tuber of Cyperus rotundus L. The pharmacology of Chinese herbs* (2nd ed., pp. 320–321). Boca Raton, FL: CRC Press.

Hussain, M., Waqas, H. M., Hussain, I., Majeed, A., Raza, S. M., & Janbaz, K. H. (2018). Pharmacological validation of the folkloric uses of *Cyperus rotundus* L. in different ailments: An *in vivo* and *in vitro* research. *Pakistan Journal of Pharmaceutical Sciences*, *31*, 95–102. PMID:29348090

Imam, M. Z., & Sumi, C. D. (2014). Evaluation of antinociceptive activity of hydromethanol extract of *Cyperus rotundus* in mice. *BMC Complementary and Alternative Medicine*, *14*(1), 83. doi:10.1186/1472-6882-14-83 PMID:24589067

Ito, T., Endo, H., Shinohara, H., Oyama, M., Akao, Y., & Iinuma, M. (2012). Occurrence of stilbene oligomers in *Cyperus* rhizomes. *Fitoterapia*, *83*(8), 1420–1429. doi:10.1016/j.fitote.2012.08.005 PMID:23339254

Jahan, F. I., Hasan, M. R. U., Jahan, R., Seraj, S., Chowdhury, A. R., Islam, M. T., ... Rahmatullah, M. (2011). A comparison of medicinal plant usage by folk medicinal practitioners of two adjoining villages in Lalmonirhat district, Bangladesh. *American-Eurasian Journal of Sustainable Agriculture*, *5*, 46–66.

Jebasingh, D., Devavaram, J. D., Venkataraman, S., Adeghate, E., & Starling, E. B. (2014). The protective effects of *Cyperus rotundus* on behavior and cognitive function in a rat model of hypoxia injury. *Pharmaceutical Biology*, *52*(12), 1558–1569. doi:10.3109/13880209.2014.908395 PMID:25026346

Jeong, S. J., Miyamoto, T., Inagaki, M., Kim, Y. C., & Higuchi, R. (2000). Rotundines A-C, three novel sesquiterpene alkaloids from *Cyperus rotundus. Journal of Natural Products*, *63*(5), 673–675. doi:10.1021/np990588r PMID:10843585

Jin, J. H., Lee, D. U., Kim, Y. S., & Kim, H. P. (2011). Anti-allergic activity of sesquiterpenes from the rhizomes of *Cyperus rotundus. Archives of Pharmacal Research*, *34*(2), 223–228. doi:10.100712272-011-0207-z PMID:21380805

Johari, S., Joshi, C., & Gandhi, T. (2016). Effect of *Cyperus Rotundus* on Cytokine Gene Expression in Experimental Inflammatory Bowel Disease. *Iranian Journal of Medical Sciences*, *41*, 391–398. PMID:27582588

Joshi, K. (2005). Ethnobotanical information on *Cyperus rotundus* L. in Nepal. *Bionotes*, *7*, 86–88.

Jung, S. H., Kim, S. J., Jun, B. G., Lee, K. T., Hong, S. P., Oh, M. S., ... Choi, J. H. (2013). α-Cyperone, isolated from the rhizomes of *Cyperus rotundus*, inhibits LPS-induced COX-2 expression and PGE2 production through the negative regulation of NFκB signalling in RAW 264.7 cells. *Journal of Ethnopharmacology*, *147*(1), 208–214. doi:10.1016/j.jep.2013.02.034 PMID:23500883

Kakarla, L., Katragadda, S. B., Tiwari, A. K., Kotamraju, K. S., Madhusudana, K., Kumar, D. A., & Botlagunta, M. (2016). Free radical scavenging, α-glucosidase inhibitory and anti-inflammatory constituents from Indian sedges, *Cyperus scariosus* R.Br and *Cyperus rotundus* L. *Pharmacognosy Magazine*, *12*(47), S488–S496. doi:10.4103/0973-1296.191467 PMID:27761080

Kamala, A., Middha, S. K., Gopinath, C., Sindhura, H. S., & Karigar, C. S. (2018). *In vitro* Antioxidant Potentials of *Cyperus rotundus* L. Rhizome Extracts and Their Phytochemical Analysis. *Pharmacognosy Magazine*, *14*(54), 261–267. doi:10.4103/pm.pm_228_17 PMID:29720842

Kandikattu, H. K., Deep, S. N., Razack, S., Amruta, N., Prasad, D., & Khanum, F. (2017). Hypoxia induced cognitive impairment modulating activity of *Cyperus rotundus. Physiology & Behavior*, *175*, 56–65. doi:10.1016/j.physbeh.2017.03.035 PMID:28351559

Kaushik, N. K., Bagavan, A., Rahuman, A. A., Mohanakrishnan, D., Kamaraj, C., Elango, G., ... Sahal, D. (2013). Antiplasmodial potential of selected medicinal plants from eastern Ghats of South India. *Experimental Parasitology*, *134*(1), 26–32. doi:10.1016/j.exppara.2013.01.021 PMID:23399920

Khajuria, A. K., & Bisht, N. S. (2017). Ethnomedicinal plants used to treat Nephrolithiasis: A case study Pauri (PAURI Garhwal), Uttarakhand. *International Journal of Herbal Medicine*, *5*, 10–13.

Khan, S. U., Khan, R. U., Mehmood, S., Khan, A., Ullah, I., & Bokhari, T. Z. (2013). Medicinal plants used to cure diarrhea and dysentery by the local inhabitants of district Bannu, Khyber Pakhtoon Khwa, Pakistan. *Advances in Pharmaceutical and Ethnomedicines*, *1*, 15–18.

Khwaja, S., Mahmood, T., & Siddiqui, H. H. (2016). Effect of ethanolic extract of *Cyperus rotundus* L. against isoprenaline induced cardiotoxicity. *Indian Journal of Experimental Biology*, *54*, 670–675. PMID:30084568

Kilani, S., Ben Sghaier, M., Limem, I., Bouhlel, I., Boubaker, J., Bhouri, W., ... Chekir-Ghedira, L. (2008). *In vitro* evaluation of antibacterial, antioxidant, cytotoxic and apoptotic activities of the tubers infusion and extracts of *Cyperus rotundus*. *Bioresource Technology*, *99*(18), 9004–9008. doi:10.1016/j.biortech.2008.04.066 PMID:18538563

Kilani-Jaziri, S., Neffati, A., Limem, I., Boubaker, J., Skandrani, I., Sghair, M. B., ... Chekir-Ghedira, L. (2009). Relationship correlation of antioxidant and antiproliferative capacity of *Cyperus rotundus* products towards K562 erythroleukemia cells. *Chemico-Biological Interactions*, *181*(1), 85–94. doi:10.1016/j.cbi.2009.04.014 PMID:19446539

Kumar, K. H., & Khanum, F. (2013). Hydroalcoholic extract of *Cyperus rotundus* ameliorates H_2O_2-induced human neuronal cell damage via its anti-oxidative and anti-apoptotic machinery. *Cellular and Molecular Neurobiology*, *33*(1), 5–17. doi:10.100710571-012-9865-8 PMID:22869350

Lawal, O. A., & Oyedeji, A. O. (2009). Chemical composition of the essential oils of *Cyperus rotundus* L. from South Africa. *Molecules (Basel, Switzerland)*, *14*(8), 2909–2917. doi:10.3390/molecules14082909 PMID:19701133

Lee, C. H., Hwang, D. S., Kim, H. G., Oh, H., Park, H., Cho, J. H., ... Oh, M. S. (2010). Protect.ive effect of Cyperi rhizoma against 6-hydroxydopamine-induced neuronal damage. *Journal of Medicinal Food*, *13*(3), 564–571. doi:10.1089/jmf.2009.1252 PMID:20521982

Lemaure, B., Touché, A., Zbinden, I., Moulin, J., Courtois, D., Macé, K., & Darimont, C. (2007). Administration of *Cyperus rotundus* tubers extract prevents weight gain in obese Zucker rats. *Phytotherapy Research*, *21*(8), 724–730. doi:10.1002/ptr.2147 PMID:17444573

Lin, S. Q., Zhou, Z. L., & Li, C. Y. (2018). Cyprotuoside C and Cyprotuoside D, Two New Cycloartane Glycosides from the Rhizomes of *Cyperus rotundus*. *Chemical & Pharmaceutical Bulletin*, *66*(1), 96–100. doi:10.1248/cpb.c17-00608 PMID:29118309

Mannarreddy, P., Denis, M., Munireddy, D., Pandurangan, R., Thangavelu, K. P., & Venkatesan, K. (2017). Cytotoxic effect of *Cyperus rotundus* rhizome extract on human cancer cell lines. *Biomedicine and Pharmacotherapy*, *95*, 1375–1387. doi:10.1016/j.biopha.2017.09.051 PMID:28946185

Mayer, F. L., Wilson, D., & Hube, B. (2013). *Candida albicans* pathogenicity mechanisms. *Virulence*, *4*(2), 119–128. doi:10.4161/viru.22913 PMID:23302789

Meena, K. L., & Yadav, B. L. (2011). Some ethnomedicinal plants used by the *Garasia* tribe of district Sirohi, Rajasthan. *Indian Journal of Traditional Knowledge*, *10*, 354–357.

Mehdizadeh, M., Hashem Dabaghian, F., Shojaee, A., Molavi, N., Taslimi, Z., Shabani, R., & Soleimani Asl, S. (2017). Protective Effects of *Cyperus Rotundus* Extract on Amyloid β-Peptide (1-40)-Induced Memory Impairment in Male Rats: A Behavioral Study. *Basic and Clinical Neuroscience*, *8*(3), 249–254. doi:10.18869/nirp.bcn.8.3.249 PMID:28781732

Mohamed-Ibrahim, S. R., Mohamed, G. A., Abdullah Khayat, M. T., Zayed, M. F., & Soliman El-Kholy, A. A. (2018). Anti-inflammatory terpenoids from *Cyperus rotundus* rhizomes. *Pakistan Journal of Pharmaceutical Sciences*, *31*, 1449–1456. PMID:30058534

Nadaf, M., Joharchi, M. R., & Amiri, M. S. (2019). Ethnomedicinal uses of plants for the treatment of nervous disorders at the herbal markets of Bojnord, North Khorasan Province, Iran. *Avicenna Journal of Phytomedicine*, *9*, 153–163. PMID:30984580

Nam, J. H., & Lee, D. U. (2015). Inhibitory effect of oleanolic acid from the rhizomes of *Cyperus rotundus* on transient receptor potential vanilloid 1 channel. *Planta Medica*, *81*, 20–25. PMID:25402944

Nam, J. H., Nam, D. Y., & Lee, D. U. (2016). Valencene from the Rhizomes of *Cyperus rotundus* Inhibits Skin Photoaging-Related Ion Channels and UV-Induced Melanogenesis in B16F10 Melanoma Cells. *Journal of Natural Products*, *79*(4), 1091–1096. doi:10.1021/acs.jnatprod.5b01127 PMID:26967731

Oh, G. S., Yoon, J., Lee, G. G., Kwak, J. H., & Kim, S. W. (2015). The Hexane Fraction of *Cyperus rotundus* Prevents Non-Alcoholic Fatty Liver Disease Through the Inhibition of Liver X Receptor α-Mediated Activation of Sterol Regulatory Element Binding Protein-1c. *The American Journal of Chinese Medicine*, *43*(03), 477–494. doi:10.1142/S0192415X15500305 PMID:25967664

Pal, D., Dutta, S., & Sarkar, A. (2009). Evaluation of CNS activities of ethanol extract of roots and rhizomes of *Cyperus rotundus* in mice. *Acta Poloniae Pharmaceutica*, *66*, 535–541. PMID:19894649

Park, S. E., Shin, W. T., Park, C., Hong, S. H., Kim, G. Y., Kim, S. O., ... Choi, Y. H. (2014). Induction of apoptosis in MDA-MB-231 human breast carcinoma cells with an ethanol extract of *Cyperus rotundus* L. by activating caspases. *Oncology Reports*, *32*(6), 2461–2470. doi:10.3892/or.2014.3507 PMID:25241797

Parvez, M. K., Al-Dosari, M. S., Arbab, A. H., & Niyazi, S. (2019). The *in vitro* and *in vivo* anti-hepatotoxic, anti-hepatitis B virus and hepatic CYP450 modulating potential of *Cyperus rotundus*. *Saudi Pharmaceutical Journal*, *27*(4), 558–564. doi:10.1016/j.jsps.2019.02.003 PMID:31061625

Paul, S., Devi, N., & Sarma, G. C. (2013). Ethnobotanical utilization of some medicinal plants by Bodo people of Manas Biosphere Reserve in the treatment of malaria. *International Research Journal of Pharmacy*, *4*(6), 102–105. doi:10.7897/2230-8407.04622

Rao, S., Rao, G. M. N., & Murthy, P. P. (2016). Diversity and indigenous uses of some ethno medicinal plants in Papikondalu wildlife Sanctuary, Eastern Ghats of Andhra Pradesh, India. *American Journal of Ethnomedicine*, *3*, 6–25.

Sayed, H. M., Mohamed, M. H., Farag, S. F., Mohamed, G. A., Omobuwajo, O. R., & Proksch, P. (2008). Fructose-amino acid conjugate and other constituents from *Cyperus rotundus* L. *Natural Product Research, 22*(17), 1487–1497. doi:10.1080/14786410802038556 PMID:19023813

Sayed, H. M., Mohamed, M. H., Farag, S. F., Mohamed, G. A., & Proksch, P. (2007). A new steroid glycoside and furochromones from *Cyperus rotundus* L. *Natural Product Research, 21*(4), 343–350. doi:10.1080/14786410701193056 PMID:17479423

Selvam, J. P., & Durai, M. (2018). An ethnobotanical study of medicinal plants used for the dengue in Tiruchirappalli district. *International Journal of Recent Scientific Research, 9*, 27108–27112.

Seo, E. J., Lee, D. U., Kwak, J. H., Lee, S. M., Kim, Y. S., & Jung, Y. S. (2011). Antiplatelet effects of *Cyperus rotundus* and its component (+)-nootkatone. *Journal of Ethnopharmacology, 135*(1), 48–54. doi:10.1016/j.jep.2011.02.025 PMID:21354294

Seo, W. G., Pae, H. O., Oh, G. S., Chai, K. Y., Kwon, T. O., Yun, Y. G., ... Chung, H. T. (2001). Inhibitory effects of methanol extract of *Cyperus rotundus* rhizomes on nitric oxide and superoxide productions by murine macrophage cell line, RAW 264.7 cells. *Journal of Ethnopharmacology, 76*(1), 59–64. doi:10.1016/S0378-8741(01)00221-5 PMID:11378282

Seo, Y. J., Jeong, M., Lee, K. T., Jang, D. S., & Choi, J. H. (2016). Isocyperol, isolated from the rhizomes of *Cyperus rotundus*, inhibits LPS-induced inflammatory responses via suppression of the NF-κB and STAT3 pathways and ROS stress in LPS-stimulated RAW 264.7 cells. *International Immunopharmacology, 38*, 61–69. doi:10.1016/j.intimp.2016.05.017 PMID:27240136

Sharma, L., & Khandelwal, S. (2010). Weeds of Rajasthan and their ethno-botanical importance. *Ethno Med, 4*(2), 75–79. doi:10.1080/09735070.2010.11886363

Shin, J. S., Hong, Y., Lee, H. H., Ryu, B., Cho, Y. W., Kim, N. J., ... Lee, K. T. (2015). Fulgidic Acid Isolated from the Rhizomes of *Cyperus rotundus* Suppresses LPS-Induced iNOS, COX-2, TNF-α, and IL-6 Expression by AP-1 Inactivation in RAW264.7 Macrophages. *Biological & Pharmaceutical Bulletin, 38*(7), 1081–1086. doi:10.1248/bpb.b15-00186 PMID:26133719

Sim, Y., Choi, J. G., Gu, P. S., Ryu, B., Kim, J. H., Kang, I., ... Oh, M. S. (2016). Identification of Neuroactive Constituents of the Ethyl Acetate Fraction from Cyperi Rhizoma Using Bioactivity-Guided Fractionation. *Biomolecules & Therapeutics, 24*(4), 438–445. doi:10.4062/biomolther.2016.091 PMID:27350341

Singh, L. R., Singh, K., & Singh, D. (2018). Wild underutilized medicinal plant; motha (*Cyperus rotundus* Linn.). *International Journal of Current Research, 10*, 64617–64619.

Singh, P., Khosa, R. L., Mishra, G., & Jha, K. K. (2015). Antidiabetic activity of ethanolic extract of *Cyperus rotundus* rhizomes in streptozotocin-induced diabetic mice. *Journal of Pharmacy & Bioallied Sciences, 7*(4), 289–292. doi:10.4103/0975-7406.168028 PMID:26681885

Sini, S., & Malathy, N. S. (2005). Antimicrobial properties of roots of medicinal plants. *Ancient Science of Life, 25*, 62–65. PMID:22557193

Sonwa, M. M., & König, W. A. (2001). Chemical study of the essential oil of *Cyperus rotundus*. *Phytochemistry, 58*(5), 799–810. doi:10.1016/S0031-9422(01)00301-6 PMID:11672746

Soumaya, K. J., Dhekra, M., Fadwa, C., Zied, G., Ilef, L., Kamel, G., & Leila, C. G. (2013). Pharmacological, antioxidant, genotoxic studies and modulation of rat splenocyte functions by *Cyperus rotundus* extracts. *BMC Complementary and Alternative Medicine*, *13*(1), 28. doi:10.1186/1472-6882-13-28 PMID:23388107

Soumaya, K. J., Zied, G., Nouha, N., Mounira, K., Kamel, G., Genvième, F. D., & Leila, G. C. (2014). Evaluation of *in vitro* antioxidant and apoptotic activities of *Cyperus rotundus. Asian Pacific Journal of Tropical Medicine*, *7*(2), 105–112. doi:10.1016/S1995-7645(14)60004-3 PMID:24461522

Sunil, A. G., Kesavanarayanan, K. S., Kalaivani, P., Sathiya, S., Ranju, V., Priya, R. J., ... Babu, C. S. (2011). Total oligomeric flavonoids of *Cyperus rotundus* ameliorates neurological deficits, excitotoxicity and behavioral alterations induced by cerebral ischemic-reperfusion injury in rats. *Brain Research Bulletin*, *84*(6), 394–405. doi:10.1016/j.brainresbull.2011.01.008 PMID:21272618

Sutalangka, C., & Wattanathorn, J. (2017). Neuroprotective and cognitive-enhancing effects of the combined extract of *Cyperus rotundus* and *Zingiber officinale. BMC Complementary and Alternative Medicine*, *17*(1), 135. doi:10.118612906-017-1632-4 PMID:28253877

Tariq, A., Mussarat, S., & Adnan, M., Abd_Allah, E. F., Hashem, A., Alqarawi, A. A., & Ullah, R. (2015). Ethnomedicinal evaluation of medicinal plants used against gastrointestinal complaints. *BioMed Research International*, 892947. PMID:26114117

Thomas, D., Govindhan, S., Baiju, E. C., Padmavathi, G., Kunnumakkara, A. B., & Padikkala, J. (2015). *Cyperus rotundus* L. prevents non-steroidal anti-inflammatory drug-induced gastric mucosal damage by inhibiting oxidative stress. *Journal of Basic and Clinical Physiology and Pharmacology*, *26*(5), 485–490. doi:10.1515/jbcpp-2014-0093 PMID:25915081

Tran, H. H., Nguyen, M. C., Le, H. T., Nguyen, T. L., Pham, T. B., Chau, V. M., ... Nguyen, T. D. (2014). Inhibitors of α-glucosidase and α-amylase from *Cyperus rotundus. Pharmaceutical Biology*, *52*(1), 74–77. doi:10.3109/13880209.2013.814692 PMID:24044731

Tsoyi, K., Jang, H. J., Lee, Y. S., Kim, Y. M., Kim, H. J., Seo, H. G., ... Chang, K. C. (2011). (+)-Nootkatone and (+)-valencene from rhizomes of *Cyperus rotundus* increase survival rates in septic mice due to heme oxygenase-1 induction. *Journal of Ethnopharmacology*, *137*(3), 1311–1317. doi:10.1016/j.jep.2011.07.062 PMID:21843620

Uddin, S. J., Mondal, K., Shilpi, J. A., & Rahman, M. T. (2006). Antidiarrhoeal activity of *Cyperus rotundus. Fitoterapia*, *77*(2), 134–136. doi:10.1016/j.fitote.2004.11.011 PMID:16376024

Umair, M., Altaf, M., & Abbasi, A. M. (2017). An ethnobotanical survey of indigenous medicinal plants in Hafizabad district, Punjab-Pakistan. *PLoS One*, *12*(6). doi:10.1371/journal.pone.0177912 PMID:28574986

Weenen, H., Nkunya, M. H., Bray, D. H., Mwasumbi, L. B., Kinabo, L. S., & Kilimali, V. A. (1990). Antimalarial activity of Tanzanian medicinal plants. *Planta Medica*, *56*(4), 368–370. doi:10.1055-2006-960984 PMID:2236289

Xu, H. B., Geng, C. A., Zhang, X. M., Ma, Y. B., Huang, X. Y., & Chen, J. J. (2016). Chemical structure of cyperotundic acid from rhizomes of *Cyperus rotundus. Zhongguo Zhongyao Zazhi*, *41*, 1066–1069. PMID:28875671

Xu, H. B., Ma, Y. B., Huang, X. Y., Geng, C. A., Wang, H., Zhao, Y., ... Chen, J. J. (2015). Bioactivity-guided isolation of anti-hepatitis B virus active sesquiterpenoids from the traditional Chinese medicine: Rhizomes of *Cyperus rotundus*. *Journal of Ethnopharmacology*, *171*, 131–140. doi:10.1016/j.jep.2015.05.040 PMID:26051832

Xu, Y., Zhang, H. W., Wan, X. C., & Zou, Z. M. (2009). Complete assignments of (1)H and (13)C NMR data for two new sesquiterpenes from *Cyperus rotundus* L. *Magnetic Resonance in Chemistry*, *47*(6), 527–531. doi:10.1002/mrc.2416 PMID:19288546

Xu, Y., Zhang, H. W., Yu, C. Y., Lu, Y., Chang, Y., & Zou, Z. M. (2008). Norcyperone, a novel skeleton norsesquiterpene from *Cyperus rotundus* L. *Molecules (Basel, Switzerland)*, *13*(10), 2474–2481. doi:10.3390/molecules13102474 PMID:18923338

Yang, J.-L., & Shi, Y.-P. (2012). Structurally diverse terpenoids from the rhizomes of *Cyperus rotundus* L. *Planta Medica*, *78*(01), 59–64. doi:10.1055-0031-1280216 PMID:21928169

Yazdanparast, R., & Ardestani, A. (2007). *In vitro* antioxidant and free radical scavenging activity of *Cyperus rotundus*. *Journal of Medicinal Food*, *10*(4), 667–674. doi:10.1089/jmf.2006.090 PMID:18158839

Ying, J., & Bing, X. (2016). Chemical constituents of *Cyperus rotundus* L. and their inhibitory effects on uterine fibroids. *African Health Sciences*, *16*, 1000–1006. PMID:28479892

Yu, H. H., Lee, D. H., Seo, S. J., & You, Y. O. (2007). Anticariogenic properties of the extract of *Cyperus rotundus*. *The American Journal of Chinese Medicine*, *35*(3), 497–505. doi:10.1142/S0192415X07005016 PMID:17597508

Zhou, Z., & Yin, W. (2012). Two novel phenolic compounds from the rhizomes of *Cyperus rotundus* L. *Molecules (Basel, Switzerland)*, *17*(11), 12636–12641. doi:10.3390/molecules171112636 PMID:23099619

Zhou, Z., Yin, W., Zhang, H., Feng, Z., & Xia, J. (2013). A new iridoid glycoside and potential MRB inhibitory activity of isolated compounds from the rhizomes of *Cyperus rotundus* L. *Natural Product Research*, *27*(19), 1732–1736. doi:10.1080/14786419.2012.750318 PMID:23356789

Zhou, Z. L., Lin, S. Q., & Yin, W. Q. (2016). New cycloartane glycosides from the rhizomes of *Cyperus rotundus* and their antidepressant activity. *Journal of Asian Natural Products Research*, *18*(7), 662–668. doi:10.1080/10286020.2016.1142976 PMID:26978535

Zhou, Z. L., Yin, W. Q., Yang, Y. M., He, C. H., Li, X. N., Zhou, C. P., & Guo, H. (2016). New Iridoid Glycosides with Antidepressant Activity Isolated from *Cyperus rotundus*. *Chemical & Pharmaceutical Bulletin*, *64*(1), 73–77. doi:10.1248/cpb.c15-00686 PMID:26726748

Chapter 6
The Therapeutic Potential of Ethnobotanical Plants in the Treatment of Different Diseases

Martha B. Ramírez-Rosas
https://orcid.org/0000-0003-4142-0493
UAM Reynosa-Aztlán, Universidad Autónoma de Tamaulipas, Mexico

Adriana L. Perales-Torres
UAM Reynosa-Aztlán, Universidad Autónoma de Tamaulipas, Mexico

Rubén Santiago-Adame
UAM Reynosa-Aztlán, Universidad Autónoma de Tamaulipas, Mexico

ABSTRACT

This is an overview of plant use for medicinal applications, a practice from old civilizations still used around the world. According to WHO, nearly 80% of people use herbal medicine plant extracts as their primary health solution. Ethnobotany emerges as a research field to document and understand the traditional knowledge about plants and their roles in society. Diseases like diabetes, hypertension, hypercholesterolemia, and cancer are the predominant challenges to global health, chronic diseases accounting for two-thirds of deaths worldwide. This document discusses ethnobotanical studies on many medicinal plants affecting these diseases.

INTRODUCTION

Several studies have shown that a large amount of medicinal knowledge has remained among tribes around the world, a common practice in old civilizations still used today. *Traditional medicine* is defined as indigenous medicine used to maintain health and to prevent, diagnose, and treat physical and mental illnesses. The World Health Organization (WHO) considers natural and traditional medicine safe and effective as a first-line treatment, and nearly 80% of people use herbal medicine plant extracts

DOI: 10.4018/978-1-7998-1320-0.ch006

as primary health care (WHO, 2013). This practice comes from ancient knowledge and still prevails in modern society, as a local culture or traditions this interaction between medicinal plants and humanity.

The study of how humans use endemic plants in a culture is called *ethnobotany* (De Santayana, Pieroni, & Kuri, 2010). Ethnobotany emerges as a research field for documenting and understanding the knowledge of the people about plants and their role in society. It is supported by popular knowledge coming from communities of rural remotes areas. This information is usually passed verbally from one generation to the next through family members and has not been fully documented yet (Faruque et al., 2018). While the uses of endemic plants vary from ornaments to agriculture, one of their principal uses is medicinal. A *medicinal plant* refers to the stem, leaves, roots bark, etc. It causes fewer or no side effects compared to regular medications when ingested. Interest in medicinal plants has increased as a source of new natural medicines. Medicinal plants play an important role in the development of new drugs.

Phytochemicals from medicinal plants are secondary metabolites, mainly small molecules (<2000 Da), absorbed and metabolized very quickly. This process gives health benefits due to their bioactive potential (Govea-Salas, Morlett-Chávez, Rodriguez-Herrera, & Ascacio-Valdés, 2017). Medicinal plants contain different phytochemicals and are potentially a principal source of bioactive phytocompounds. However, the majority remain unexplored with respect to pharmacological and therapeutic studies. Herbs and medicinal plants are rich in phytochemicals like polyphenols, alkaloids, terpenes, carotenoids, etc.

The variety of medicinal plants around the world provides the possibility of further investigation for researchers, and in some cases, these studies might lead to discovering new drugs or new compounds useful in medicine and pharmacology (Mohamadi, Sharififar, Koohpayeh, & Daneshpajouh, 2015). Medicinal plants could be efficient against the most important diseases in the modern world. Chronic diseases like diabetes, hypertension, hypercholesterolemia, and cancer account for two-thirds of deaths worldwide. A basic search of scientific literature shows a variety of medicinal plants and herbs that report alleviating chronic or metabolic diseases and their symptoms. Therefore, this chapter focuses on the results of scientific studies of medicinal plants used to treat various human diseases around the world, specifically chronic diseases such as diabetes, hypertension, hypercholesterolemia, obesity, cancers, among others.

BACKGROUND

Impact of Demographic, Socioeconomic, and Management of Leading Global Burden Diseases

Chronic diseases are the predominant challenge to global health, accounting for two-thirds of deaths worldwide. High-burden chronic conditions such as diabetes, hypertension, hypercholesterolemia, obesity, cancers, among others, are leading diseases with direct and indirect impacts on healthcare costs, productivity losses, and disability-adjusted life years (Ding et al., 2016). Chronic diseases are the main cause of poor health disability and death, being highly related to the increases in the health care cost. For these reasons, there is widespread concern over managing these illnesses since their incidence in the world (high-income to low and middle-income countries) is currently high, and it is projected to dramatically increase in the next decade (James et al., 2018).

Diabetes Mellitus Type 2

Diabetes is a chronic disease affecting not only high-income countries (HICs) but also in many low and middle-income countries (LMICs), becoming a major public health problem worldwide (Seuring, Archangelidi, & Suhrcke, 2015). Currently, over 451 million people suffer this disease, and by 2045 nearly 693 million will develop it, which means an increase of »54% in two decades, according to the International Diabetes Federation (Da Rocha Fernandes et al., 2016). Type 2 diabetes (T2D) is the most common with more than 90% of the cases and is growing rapidly worldwide (Federation, 2017). Accordingly, in the past decade the economic burden of type 2 diabetes has rapidly grown (Seuring et al., 2015), which is associated with a high risk for long-term complications, such as microvascular (involving small arteries and capillaries) and macrovascular (involving large arteries such as conduit vessels) diseases (Petrie, Guzik, & Touyz, 2018). In this sense, diabetes imposes a large economic burden on health care systems across the world. In 2014, it was estimated on average, health expenditures per person are \$1,583–2,842 in U.S. dollars, and annual global health expenditures are \$612–1,099 billion. These data show that diabetes needs to be considered a public health priority in order to reduce the financial burden (Da Rocha Fernandes et al., 2016). Therefore, new strategies for prevention and alternative treatments play an important role.

Hypertension

Hypertension is an important public health challenge, affecting approximately one-billion people worldwide, where primary or idiopathic hypertension is more common in patients (>90%). It is estimated that by 2025, about 1.56 million adults in economically developing countries will present disabilities related to hypertension (Kearney et al., 2005). Hypertension increases yearly, leading to increased morbidity and mortality, mainly in aging populations (McFarlane, Muniyappa, Shin, Bahtiyar, & Sowers, 2004).

Since life expectancy has increased in recent decades, we can predict that it will become an increasing worldwide burden (Varenna et al., 2013). Hypertension is considered one of the basic cost drivers in healthcare systems because several costs are related to the management of nonoptimal blood pressure and its resultant diseases, mainly related to stroke and myocardial infarction. Therefore, this disease generates high costs because of direct and indirect complications (Athanasakis, 2017). Accordingly, it was reported that expenditures for hypertension would reach 1 trillion U.S. dollars, approximately 10% of all expenditures in healthcare, globally (Gaziano, Bitton, Anand, & Weinstein, 2009).

Finally, it is important to highlight that hypertension and type 2 diabetes are common comorbidities, since several studies have shown that increases in blood pressure are commoner among people with diabetes than in the general population, and this is because both conditions share common risk factors related to pathophysiology (Petrie et al., 2018). Moreover, both diseases have a positive correlation with aging (Ferrannini & Cushman, 2012). Accordingly, patients with type 2 diabetes have shown high blood pressure (BP), which could be related to the damage to the renal system, as a consequence of dysregulation at micro and macrovascular levels (Ferrannini & Cushman, 2012). In addition, several pieces of evidence suggest that hypertensive patients have an alteration in carbohydrate metabolism (Rizos & Elisaf, 2014), suggesting an etiological relationship between diabetes and hypertension. Hence diabetic subjects exhibit poor control of hypertension; as a result, these diseases show a high correlation (Long & Dagogo-Jack, 2011).

Hypercholesterolemia

Hypercholesterolemia is associated with an enormous risk of heart disease. It is one of the major modifiable risk factors for atherosclerotic cardiovascular disease (CVD), and it is an important health problem (Soran, Adam, & Durrington, 2017). Cholesterol, an inorganic molecule from sterols functional group and lipids, which has a lot of bio-functions in animal systems, is present in a lot of foods from animal sources and is used in many human biological processes. A regular human male weighing 150 lb synthesizes around 1,000 mg of cholesterol daily, and his body has almost 35,000 mg, mostly in cell membranes. The Dietary Guidelines recommendation for cholesterol intake is 300 mg per day, maximum. The average female and male consume about 307 and 225 mg per day, respectively (Xu, McClure, & Appel, 2018). High concentration in the body is prejudicial to human health and is a precursor to symptoms related to cardiovascular diseases. Elevated serum lipid levels are strongly related to the risk of coronary events and atherosclerosis (Lachhiramka & Patil, 2016). Actual medicinal practices are related to consuming lipid-lowering precursors such as statins. However, it has secondary effects like myalgia, neuropathy, cognitive dysfunction, muscle weakness, and an elevated risk of diabetes (Ried, Toben, & Fakler, 2013).

Obesity

Obesity is considered a chronic, multifactorial disease, the result of the interaction between genetic load and environment that affects a large percentage of the population of all ages, sexes, and social conditions. It is defined as an abnormal or excessive accumulation of adipose tissue that can be harmful to health (>25% in men and >33% in women; Izaola, de Luis, Sajoux, Domingo, & Vidal, 2015) or as a body mass index (BMI) >30 kg/m^2. Obesity is considered a risk factor for presenting diabetes mellitus type 2. For example, compared to people with normal weights, those with a BMI of 25–30 have twice the risk of developing diabetes, those with BMI of 30–35 have more than triple the risk, and those with an index greater than 35 have a risk six times as high (Córdova Villalobos, 2016). The prevalence of obesity in the world has doubled in the last 30 years. It is a particularly serious problem in the Americas. Obesity has reached epidemic proportions among adults aged 18 years or older; 11% of men and 15% of women were obese in 2014, and more than 42 million children under the age of 5 years were overweight in 2013 (Arroyo-Johnson & Mincey, 2016). In 2014, the global economic impact of obesity was estimated at $2.0 billion in U.S. dollars, or 2.8% of the world's gross domestic product (GDP). In addition to the excess of expenses in medical care, obesity also imposes costs in the form of lost productivity and foregone economic growth as a result of lost workdays, lower productivity at work, mortality, and permanent disability (Tremmel, Gerdtham, Nilsson, & Saha, 2017). Several plants have been used to treat obesity, including those with effects on the central nervous system, used as antidepressants and anxiolytics because of their anxiolytic effects while on weight-loss diets (Alonso-Castro, Domínguez, Zapata-Morales, & Carranza-Álvarez, 2015).

Cancer

Cancer is characterized by cells in the human body continually multiplying, with the inability to be controlled or stopped, consequently forming tumors of malignant cells with the potential to be metastatic (Greenwell & Rahman, 2015). It is a public health problem with a total annual economic cost estimated at $1.16 trillion U.S. dollars in 2010, considering costs of prevention and treatment, as well as the annual economic cost of disability-adjusted life years. Based on another study, the total 2010 VSL (value of a statistical life) estimate of the global cost of cancer is $2.5 trillion, with approximately $1.7 trillion attributed to high-income countries and $800 billion attributed to low- and middle-income countries (Stewart & Wild, 2014). Around the world, Asia represents 48.4% of all cancer cases, Europe accounts for 23.4%, followed by 21% in the Americas, 5.8% in Africa, and 1.4% in Oceania. The 10 main types of cancer (without considering nonmelanoma skin cancers) registered in 2018 for both sexes were lung, breast, colorectum, prostate, stomach, liver, esophagus, cervix uteri, thyroid, and bladder. For both sexes combined, lung cancer is the most commonly diagnosed (11.6% of all cases), followed closely by female breast cancer (11.6%), colorectal cancer (10.2%), and prostate cancer (7.1%; Bray et al., 2018).

Risk factors for developing cancer have been classified as *intrinsic* and *non-intrinsic*, the latter is subdivided into *endogenous* (e.g., immune, metabolism, hormone levels) and *exogenous* (carcinogens, viruses, lifestyle factors). Intrinsic factors refer to inevitable spontaneous mutations that arise as a result of random errors in DNA replication. These processing errors of DNA replication occur at different speeds in different organisms (Wu, Zhu, Thompson, & Hannun, 2018). Some non-intrinsic factors are specifically related to a certain type of cancer, for example, prolonged exposure to high concentrations of estrogen (hormone levels) increases the risk of breast cancer; for example, this occurs in women with early menarche and late menopause who do not have children (Martín, Herrero, & Echavarría, 2015). The Epstein Barr virus is associated with Hodgkin's lymphoma and nasopharyngeal carcinoma, hepatitis B virus and hepatitis C virus with hepatocellular carcinoma, and human high-risk papillomaviruses with cervical, head and neck, and anogenital tract carcinomas (Morales-Sánchez & Fuentes-Pananá, 2014).

There are many types of cancer treatments, including surgery, radiotherapy, chemotherapy, immunotherapy, targeted therapy, hormone therapy, stem cell transplantation, and precision medicine (National Cancer Institute, 2019). In addition, several bioactive food components (BFCs) have been identified that have the potential to prevent cancer. These include methyl-group donors, polyphenols, selenium, retinoids, fatty acids, isothiocyanates, and allyl compounds. These BFCs affect the different stages of carcinogenesis through the modulation of biological processes such as carcinogen detoxification, DNA repair, oxidative stress, inflammation and cell proliferation, and death and differentiation (Ong, Moreno, & Ross, 2012). In fact, from the dawn of ancient medicine, chemical compounds derived from plants have been used to treat human diseases. Natural products have received increasing attention over the past 30 years for their potential as novel cancer preventive and therapeutic agents (Solowey et al., 2014).

For all the above data, there is an urgent need to research new strategies for complementary therapies to treat these chronic diseases, and at this point medical plants have proved to be an important source in the research for new drugs mainly because of the diversity of pharmacological properties and chemical structures (Pan et al., 2013), and it is well known that the medical effects of the plants are related to the presence of phytochemicals with bioactive potential (Sofowora, Ogunbodede, & Onayade, 2013). Useful wild plants are found in many parts of the world, and most of them contribute to herbal therapy being an important part of the culture and tradition of people to treat many diseases (Mathew & Negi, 2017).

Medicinal Plants and Traditional Medicine Bases of Ethnobotany

The use of plants for medicinal applications is an old practice in several civilizations, still used in the modern world (Bouasla & Bouasla, 2017). Traditional or native medicine is responsible for preserving this older knowledge through the years and diverse events of human history. For this reason, several ethnobotanical studies have been made over time, influenced by theories and methods of this science in different regions of the world. According to data from the WHO, nearly 80% of people use herbal medicinal plant extracts for primary health care worldwide; several reports show this mainly in populations of developing countries (Elujoba, Odeleye, & Ogunyemi, 2005; Esakkimuthu et al., 2016; World Health Organization, 1978). Accordingly, interest in natural products has significantly increased as a novel source of new medicines or natural treatments at industrial, commercial, and research levels (Govea-Salas et al., 2017).

At least 25% of medicines are derived from phytochemicals isolated from medicinal plants (Thatoi & Patra, 2011). Herbs contain hundreds of active ingredients, that could be used as potential therapeutic agents and those are related with at least 15 major phytochemical groups such as flavonoids, alkaloids, glucosides, volatile oils, resins, phytochromes, organic acids, amino acids, tannins, proteins, enzymes, trace elements, polysaccharides, mineral salts, among others (Pan et al., 2013). Thereby, *ethnobotany* is defined as a research field for documenting and understanding the knowledge of the people about plants, how plants form part of their society, as a result, the contribution of botanical field for medicine throughout evolution has been helpful, being an important alternative to use and manage of plants to treat diseases and injuries around the world (Casas, Blancas, & Lira, 2016).

Therapeutic Potential of Plants to Treat Diseases

For centuries, people have used plants for healing, whether as foods or botanical potions or powders. It is known that about 250,000 living plant species contain a much greater diversity of bioactive compounds that any report made by humans. It is widely accepted than more than 80% of drug substances are either directly derived from natural products or developed from natural compounds (Pan et al., 2013; Raskin et al., 2002) The following are examples of plants (mainly Mexican but also others) and their compounds used as complementary treatments with beneficial effects in leading diseases, with wide impacts on the worldwide health system.

Plants Used for Diabetes and Hypertension

Yarrow *(Achillea millefollium* L.) (*Asteraceae*) is a perennial herb used in Mexican folk medicine. It is used in different parts of the world to treat diabetes and some related complications. The hydroalcoholic extract of this herb has shown a potential antidiabetic effect (Ramírez, Zavala, Pérez, & Zamilpa, 2012), related to its capability to inhibit the α-glucosidase. This effect has also been observed in a NOD mice model (Petlevski et al., 2001). The hydroalcoholic extract showed a decrease of IL-1β and iNOS gene expression in STZ-induced diabetic rats. These findings could be related to PPARγ activation, and the effects would contribute to decreasing inflammatory cytokines. Thus, this extract has important effects, mainly related to antidiabetic, antihyperglycemic, hypoglycemic, and potential insulin sensitizer actions (Chávez-Silva et al., 2018). Those effects are related to phytochemicals compounds that include phenols, flavonoids, sesquiterpenoids, essential oils, hydrocarbon monoterpenes, oxygenated monoterpenes, sesquiterpene hydrocarbons, oxygenated sesquiterpenes, proazulenes (Ali, Gopalakrishnan, & Venkatesalu, 2017).

Mexican giant hyssop *(Agastache mexicana* (Kunth) Lint & Epling) (*Lamiaceae*) is a medicinal plant widely used in Mexican traditional medicine for the treatment of hypertension, diabetes, and other complications (Monroy-Ortíz & Castillo-España, 2007). Analysis shows an important relaxant effect related to its antihypertensive effect. The main phytochemicals involved in these responses are related to essential oils such as estragole, D-limonene, and linalyl anthranilate and compounds related to chemical structures of acacetin ursolic acid and oleanolic acid. All of these compounds are related to its strong action in modulating the myogenic tone. Therefore, its mechanism of action is related to blocking calcium entry. It also allows the opening of K+ channels in vascular smooth muscle cells and induces hyperpolarization of the membrane potential, which provides an important mechanism to dilate the arteries (Flores-Flores et al., 2016).

Cucumber tree *(Averrhoa bilimbi* Linn.) tree leaves show the presence of primary metabolites such as aldehydes, sugar, protein, and secondary metabolites like cardiac glycoside, flavonoid, alkaloid, phenol, tannin, and coumarin (Valsan & Regi Raphael, 2016). Among these metabolites, the phenolic compounds are the most ubiquitous groups, with such diverse properties as anti-atherosclerosis, cardiovascular protection, and improvement of endothelium function, among others. Therefore, this plant can act as an antidiabetic, antihypertensive, thrombolytic, antimicrobial, antioxidant, hepatoprotective, and hypolipidemic agent (Alhassan & Ahmed, 2016). Moreover, in the fruit of this tree volatile constituents have been identified, the main constituents were hexadecanoic acid (palmitic acid; 20.4%), 2-furaldehyde (19.1%), and (Z)-9-octadecenoic acid (10.2%; Wong & Wong, 1995). It has been observed that the leaves ethanol extracts (125 mg per kg of bodyweight twice daily, p.o.) significantly decreases blood glucose and triglyceride levels in comparison with its vehicle (Pushparaj, Tan, & Tan, 2000).

Basil (*Ocimum basilicum*) is among herbs and spices in the Mediterranean diet. Their leaves are commonly used in salads and most commonly in Italian *pesto*. The main constituents of their extract are rosmarinic and catechin (Shan, Cai, Sun, & Corke, 2005). Because of their phenols levels, a wide antidiabetic effect has been reported related to the inhibition of (a) α-glucosidase, (b) α-amylase, (c) aldose reductase, and (d) albumin glycation. Potent antioxidant and anti-inflammatory effects have also been reported (Bower, Marquez, & de Mejía, 2016).

Trumpet tree *(Cecropia obtusifolia* Bertol) (*Cecropiaceae*) is common throughout Mexico, typically used as a treatment for type 2 diabetes because of its hypoglycaemic effects, which are related to compounds such as β-sitosterol, stigmasterol, 4-ethyl-5-(n-3valeroil)-6-hexahydrocoumarin and 1-(2-methyl-1-nonen-8-il)-aziridine (Arqueta, 1994). Traditionally, its dry leaves (15 g) are boiled in water (500 mL) for regular consumption. Moreover, cecropia extract has two main bioactive compounds, chlorogenic acid, which could explain the hypoglycaemic effect, and isoorientin, which is related to potent antioxidant activity. The median lethal dose was determinate to be 673 g for a 60 kg person, and the common dose used is 15 g per person per day. Therefore the commonly used extract used has low toxicity (Andrade-Cetto & Heinrich, 2005).

Chaya *(Cnidoscolus chayamansa* Mc Vaugh) (*Euphorbiaceae*), commonly known as *chaya,* is used as a traditional remedy for treatment of type 2 diabetes because of the compounds: coumarins, glycosidic flavonoids, cyanogenic glycosides, flavonoids dihydromyricetin, astragalin, and kaempferol-3-rutinoside, which are commonly used for their anti-inflammatory and cardioprotective effects (García-Rodríguez et al., 2014).

Chilacayote *(Cucurbita ficifolia* Bouché) (*Cucurbitaceae*) is better known as *chilacayote*. Its fruit is used to treat diabetes, usually ingested by maceration, with water. The main constituents of the plant are lectins present in the stems and roots, but there is no report about the constituents of the fruit extract. Its effect is mainly related to the hypoglycaemic effect (García et al., 2017).

Mexican giant horsetail, *(Equisetum myriochaetum* Schlecht and Cham) (*Equisetaceae*) is used to treat diabetes. Its main constituents are kaempferol-3-O-sophoroside, kaempferol-3,7-di-O-β-glucoside, caffeoyl-methylate-4-β-glucopyranoside, and kaempferol-3-O-sophoroside-4-O-β-glucoside. Pinocembrin, chrysin, β-sitosterol, β-D-glycosyl-sitosterol, β-D-glucose, and fatty acids are also constituents of this plant (Camacho et al., 1991). It is prepared as a traditional tea to drink during the day. Its therapeutic effects have been associated with flavonoids, mainly quercetin, kaempferol, and myricetin (Knekt et al., 2002).

Drumstick tree ***(Moringa oleifera)*** is an Indian tree cultivated in different regions of Mexico, belonging to the family *Moringaceae* (Villarruel-López et al., 2018). It is important because its flowers, pods, and leaves have medicinal uses. Between them, a capacity to decrease blood glucose has been reported, because they contain polyphenols such as quercetin-3-glycoside, rutin, kaempferol, and glycosides (Jaiswal et al., 2009). Accordingly, the leaves of the tree have been used as antidiabetic drugs. The great activity of the Moringa is related to its phytochemical compounds, where the majority are related with polyphenols, which have antioxidant activities associated with preventing degenerative diseases. Doses of 50–100 mg per kg of bodyweight have been connected with beneficial effects in diabetes mellitus cases (Olayaki et al., 2015).

Nopal (*Opuntia ficus-indica*) is used in traditional Mexican medicine to treat diabetes. It is also considered a functional food because it is a proven source of dietary fiber and bioactive compounds with antioxidant activity, such as flavonoids flavonols, carotenes, and ascorbic acid, and its calorie content is low. In diabetes, regulation of postprandial glucose is important, and between the compounds involved in their regulation, incretins are closely involved (Kim & Egan, 2008). In this sense, the nopal has low glycemic, insulinemic, and GIP indexes and could be recommended for patients with type 2 diabetes, because it has antihyperglycemic and antihyperinsulinemic effects, it prevents postprandial blood glucose peaks, and its consumption increases the antioxidant activity in both healthy and type 2 diabetic people (López-Romero et al., 2014).

Sage (*Salvia officinalis*) is native to the Mediterranean and has exceptional pharmacologic properties. Its extract may reduce hyperglycemia because of high levels of polyphenols, mainly the rosmarinic acid, able to display several properties such as anti-apoptotic, neuroprotective, and antioxidant effects (Hasanein, Felehgari, & Emamjomeh, 2016). Moreover, an infusion of 4–6 g of dry *Salvia officinalis* leaves is used as an anti-hyperglycemic agent. Studies have reported its pharmacologic effects in several experimental conditions related to agonistic activity on the PPARγ, pancreatic lipase inhibitor, lipid absorption inhibitory, and anti-hypertriglyceridemic effects, and in rats, an antioxidant and lipid peroxidation inhibitory effects have been observed. In clinical studies, taking *Salvia officinalis* tea twice a day (4 g dry leaves on 300 mL of boiling water), improved lipid profiles, since decreased plasma LDL-C, total cholesterol, and LDL-C/HDL-C levels were observed. Plasma HDL-C levels saw increases without hepatotoxic or other adverse effects (Kianbakht, Khalighi-Sigaroodi, & Dabaghian, 2013).

Plants used to treat hypertension have presented vasodilatory activity; the results from the pharmacological evaluation of the selected compounds indicated that 3-α-hydroxymasticadienonic acid (*Amphypterygium adstringens*), corosolic acid (*Crataegus gracilior*), galphimidin (*Galphimia glauca*), meso-dihydroguaiaretic acid and 5,8,4¢-trihydroxy-3,7-dimethoxyfla (*Larrea tridentata*), and pinostrobin (*Teloxys graveolens*) elicited significant vasodilation (Luna-Vázquez et al., 2018). Moreover, several traditional therapeutic plants have been used in different cultures worldwide to treat hypertension, mostly from the *Asteraceae*, *Polygonaceae*, *Rosaceae*, and *Oleaceae* families, because they contain flavonoids and terpenes components exerting antioxidant and anti-inflammatory properties. Examples of these plants are *Allium sativum* L., *Berberis vulgaris* L., *Ribes orientale*, *Rheum ribes* L, among others (Baharvand-Ahmadi & Asadi-Samani, 2016).

Plants Used for Hypercholesterolemia

Activity contributes to the protective effects of a diet. Accordingly, several scientific studies have analyzed multiple herbs and medicinal plants because of their therapeutic properties against hypercholesterolemia. Some of them are listed below.

Caraway *(Carum carvi)* fruit effect on lipid metabolism in normal and streptozocin-induced diabetics rats (STZ) was reported in 2006 by Lemhadri et al., observing potent lipid-lowering activity in normal and severe hyperglycemic rats six hours after oral administration.

Curry (*Murraya koenigii*) leaf extract was examined in 2006 by Xie et al., shown to reduce blood cholesterol levels diabetic mice. This study starts from reports of the hypolipidemic effect of curry leaf on high-fat cholesterol rats after three months of consumption. The plasma lipoprotein profile showed remarkable effects in cholesterol, triglycerides, phospholipids, and low-density lipoproteins reduction (Khan et al., 1996; Khan et al.,1998).

Garlic *(Allium sativum)* preparations have related to positive cardiovascular effects (Banerjee & Maulik, 2002; Rahman & Lowe, 2006). Several studies have investigated the effect of garlic on cholesterol and reported a reduction of 13% ($P < 0.001$), from 269.30 mg/dL to 233.93 mg/dL and 10% ($P < 0.001$) from 260.30 to 233.90 mg/dL at 90 days of treatment in male and female patients, respectively (Lachhiramka & Patil, 2016). These studies concluded garlic has a role in cholesterol management as helper therapy in several cases of cholesterol diseases and related diseases.

Roselle *(Guazuma ulmifolia* Lam, *Gynura procumbens* (Lour.) Merr., *Muraya paniculata* and *Hibiscus sabdariffa* L.) (roselle) have been used in Indonesia as traditional cholesterol reducers. These herbs have in common their phytochemical compositions including epicatechin, procyanidin oligomers, flavonoids, sterols, steroids, and phenolic acids. This study showed the water and ethanolic extracts can reduce dietary fat absorption by blocking lipase activity (Sari, I. P., Nurrochmad, a. N. and Setiawan, 2013)

Roselle *(Hibiscus sabdariffa)* is a plant used for a long time as a soft drink and medicinal herb; however, some studies have been found that can reduce blood lipids in animals (Chen et al., 2003; Wang et al., 2000). *Hibiscus sabdariffa* is a traditional Chinese rose tea used in folk medicines for several diseases (Chewonarin et al., 1999; Dafallah & Al-Mustafa, 2004; Haji Faraji & Haji Tarkhani, 1999). Lin et al., 2007 explain the health effect is related to its chemical components, including anthocyanins, flavonoids, and polyphenols. They report a clinical study conducted using an oral preparation of *Hibiscus sabdariffa* extract (HSE) in 42 human subjects for a period of 4 weeks. After HSE had been administered for 15 days, serum cholesterol levels from all groups were reduced by 7.8–8.2%, and similarly, after four weeks of taking supplements, levels were reduced by 8.3–14.4%. This study concludes that a dosage of 2 HSE extract concentration for 30 days can lower serum cholesterol level significantly and could be effective in hypercholesterolemic patients.

Mustard tree *(Salvadora persica)* stems were evaluated in diet-induced rat hypercholesterolemia for periods of 15 and 30 days. Significantly, the *S. persica* decoction lowered cholesterol and LDL plasma levels, but the HDL and triglycerides did not change (Galati et al., 1999).

Plants Used for Obesity

Amaranth *(Amaranthus hypochondriacus* L.) (amaranth) is considered a pseudo-cereal whose cultivation can survive in areas with extreme climatic conditions, from dry to humid environments, and from high regions at sea level (Martínez Salvador, 2016). *Amaranth* comes from Greek and means "one that does not fade." It is a cultivated plant, domesticated and used in Mexico for more than 4,000 years. Its production is concentrated in the central zone of Mexico, highlighting the states of Puebla, State of Mexico, Morelos, Tlaxcala and the Federal District (Ayala Garay et al., 2014). Amaranth contains minerals such as magnesium (848 μg/g), phosphorus (330 μg/g), iron (65 μg/g) and calcium (519.3 μg/g). Calcium helps improve bone health as well as reducing obesity (Bhat, Satpathy, & Gupta, 2015). In a study by Mithila & Khanum (2015), three different diets were administered to 24 rats. Group 1 (the control group) was given a diet with casein, group 2 was fed a diet with quinoa, and group 3, amaranth. There was a significant decrease in food intake in the quinoa and amaranth groups compared to the control, and the average body weight increase was significantly reduced in the amaranth group ($p < 0.05$) compared to the other two groups (Mithila & Khanum, 2015).

Turmeric ***(Curcuma longa)*** contains curcumin. In primary human adipocytes and murine 3T3-L1 adipocytes, curcumin treatment suppressed the expression of adipogenic genes peroxisome proliferator-activated receptor γ (PPARγ), CCAAT/enhancer-binding protein α (C/EBP α), downstream genes, sterol regulatory element-binding protein (SREBP)-1, and fatty acid synthase (FAS). In addition to the antiadipogenic effects, curcumin also suppresses the differentiation of preadipocytes to mature adipocytes and induced apoptosis (Aggarwal, 2010). The maximum recommended daily dose of curcumin is 1 mg per kg of bodyweight, according to WHO and FAO. However, hepatic toxicity has been reported with the chronic use of curcumin, and high doses can induce gastrointestinal upset, and inflamed skin (Zhao et al., 2017).

Artichoke ***(Cynara scolymus*** **L.)** is commonly consumed as a vegetable. Its leaves are frequently used in folk medicine for the treatment of hepatitis, hyperlipidemia, dyspeptic disorders, and obesity (Miraj & Kiani, 2016). Its slimming action is based on its depurative effect and the presence of fructose polysaccharides that satiate (Espinosa, 2016). The effect of artichoke on the BMI in 20 adult patients (20 to 52 years old) with obesity and overweight without other pathologies was evaluated. The average BMI of the first consultation was 33.07 ± 5.97 kg/m^2, in the second consultation 32.86 ± 5.98 kg/m^2, in the third consultation 32.98 ± 6.22 kg/m^2, in the fourth consultation 32.94 ± 6.30 kg/m^2, in the fifth consultation 32.91 ± 6.24 kg/m^2, and in the final consultation 33.06 ± 6.25 kg/m^2. They concluded that homeopathic doses of Cynarascolymus do not impact on the body mass index (Misael, De Guadalupe, Paola, & Carlos, 2014).

Tamarindo malabar ***(Garcinia cambogia)*** is a plant belonging to the family of *Guttifereae*. The orange-yellow fruit has the shape of a pumpkin with thin skin and vertical sections that form deep lobes. The shell contains hydroxycitric acid that plays an important role in the management of body weight and appetite as it inhibits the production of hepatic enzyme ATP citrate-lyase that transforms carbohydrates does not consume in fat limiting the formation of fat from storage (Ferrara, 2014). However, hepatotoxicity due to *Garcinia cambogia* extract has been reported. The intake of "Garcinia Cambogia Extract 5:1" at a dose of two 80 mg capsules, three times a day before meals, for five months resulted in fulminant hepatic failure in a 34-year-old male patient who required a liver transplant (Lunsford, Bodzin, Reino, Wang, & Busuttil, 2016).

Guarana ***(Paullinia cupana)*** contains relatively large amounts of caffeine and has been reported to increase the rate of gastric emptying (Najafian, Abdar-Esfahani, Arab-Momeni, & Akhavan-Tabib, 2014). At the molecular level, guarana can negatively regulate the expression of pro-adipogenic genes such as Pparγ and Creb1 and to regulate the expression of anti-adipogenic genes such as FoxO1, Gata3, and Dlk1. In addition, it induces the expression of three important regulators of the Wnt pathway (Wnt1, Wnt3a, and Wnt10b), the activation of Wnt10b stabilizes β-catenin in the cytoplasm favoring its translocation to the nucleus, contributing to the inhibition of adipogenesis (Da Silva Lima et al., 2017). However, due to the high caffeine content of their seeds, excessive consumption of guarana can cause symptoms of caffeine poisoning (nausea, palpitations, anxiety, excitement, insomnia, and irritability). The database of adverse drug reactions of the World Health Organization (WHO-ADR) reported that guarana is associated with cases of personality disorders, manic reaction, insomnia, asthenia, fatigue, anxiety, and impaired concentration (Bersani et al., 2015).

Plants Used for Cancer

There are plants that have been reported widely useful to treat cancer. Some examples used in Mexico for this purpose are listed below.

Achiote *(Bixa Orellana)* grows in various regions of South and Central America, including Mexico, where it is known as achiote and was used as a coloring agent in pre-Hispanic Mayan religious ceremonies. Today, it is still used to color and flavor traditional dishes (Rivera-Madrid, Aguilar-Espinosa, Cárdenas-Conejo, & Garza-Caligaris, 2016). The commercial importance of the plant lies in the seed, since the pericarp contains the pigments. About 80% of this pigment is the carotenoid known as bixin, which is an antioxidant (Vilar et al., 2014). Bixin concentrated in seed extracts correlates with a high capacity to eliminate scavenge reactive oxygen species (ROS). Additionally, achiote seeds contain geranylgeraniol, squalene, and beta-sitosterol, which also have anticancer effects (Rivera-Madrid et al., 2016).

Choisy (bugambilia) *(Bougainvillea glabra)*, also called the paper flower, is a climbing evergreen woody ornamental shrub in warmer climates. It is native to Brazil and is now also seen in the Middle East, Bangladesh, India, Pakistan, and North America (Islam, Hossain, Hossen, Akter, & Mokammel, 2016). Methanolic extracts of *B. glabra* leaves contain a variety of phytochemical compounds, which can effectively protect the body from oxidative damage by free radicals (DNA damage can cause mutagenesis in the cells and initiate the development of cancer which is mediated by free radical; Bhaskara Rao et al., 2015). BD extracts (*B. glabra* aerial dichloromethane extract) has been reported as active against MCF-7 (breast cancer) and CaSki (cervical cancer), while BM extract (*B. glabra* aerial methanol extract) has shown high anti-proliferative activity for SW-480 (colon cancer) and MCF-7 cell lines (Saleem et al., 2019).

Chaya *(Cnidoscolus chayamansa* McVaugh) leaves contain flavonoids such as amentoflavone (diapigenin), astragalin (kaempferol-3-O-glucoside), kaempferol-3-O-rutinoside, dihydromyricetin, and cyanogenic glycosides (Jiménez-Arellanes, García-Martínez, & Rojas-Tomé, 2014). Additionally, a novel simple fatty acid with in-vitro antineoplastic activity has been identified: 8-methyl-6-nonanoic acid methyl ester (Fort et al., 2018).

Coffee fruit *(Coffea arabica)*: Chlorogenic acid is found in foods and herbs such as coffee and has an anti-inflammatory and anti-cancer effect (Santana-Gálvez, Cisneros-Zevallos, & Jacobo-Velázquez, 2017). Air-dried whole coffee fruits from China, India, and Mexico have been analyzed by HPLC and Orbitrap exact mass spectrometry. The content of chlorogenic acids was similar for Mexican and Indian coffee fruits but was much lower in samples from China. The highest levels of flavan-3-ols were found in samples from India, while Mexican samples had more flavanols (Varzakas, Zakynthinos, & Verpoort, 2016). Likewise, the total antioxidant capacity of coffee fruit and its extracts has been measured by ORAC FN analysis (oxygen radical absorbance capacity; coffee fruits extract-1: multistep ethanol extraction; coffee fruits extract-2: single-step extraction; coffee fruits powder-1: freeze-dried; coffee fruits powder-2: air-dried), observing that the total content of chlorogenic acid correlated with the total antioxidant capacity. The extracts presented a higher chlorogenic acid content (76.5% extract-1 and 45% extract-2) compared to powders (8.79% powder-1 and 4.53% powder-2) and therefore, higher ORAC activity (64.354 ± 2584 extract-1 and 28, 237 ± 782 extract-2) unlike dust-1 (4,768 ± 285) and dust-2 (3,439 ± 134; results expressed in Trolox Equivalent μmole (TE)/g ± SD; Yashin, Yashin, Wang, & Nemzer, 2013).

Saffron *(Crocus sativus* L.) has been used as an herbal remedy for various ailments including cancer. In animal models, the administration of aqueous extracts has inhibited the progression of gastric tissue cancer, a result attributed to the apoptosis induced by crocin. In prostate cancer, both saffron extracts and crocin decrease cell proliferation of all malignant cell lines in a time-and concentration-dependent manner (Bhandari, 2015).The dimethyl-crocetin, crocetin, and crocin have an inhibitory effect on the growth of human chronic myelogenous leukemia K562 and promyelocytic leukemia HL-60 cells at concentrations of 0.8, 2, and 2 mM, respectively (50% inhibition reached). In addition, it has been observed that extracts of saffron and its major constituent (crocin) inhibit the growth of colorectal cancer cells without affecting normal cells (Samarghandian & Borji, 2014).

Mohuite *(Justicia spicigera* Schltdl.) is an evergreen perennial shrub, native to Mexico and South America. It is well known in Mexican traditional medicine for healing different diseases (Sengupta & Paul, 2016). Extracts have been evaluated in human breast T47D cancer cells and HeLa cervical cancer cells. The ethanol extract of *J. spicigera* showed cytotoxic activity in T47D cells. In the HeLa cell line of uterine cancer, increased cytotoxic activity was observed with colchicine that is used as an antitumor drug compared with the ethanol extract of *J. spicigera* (Baqueiro-Peña & Guerrero-Beltrán, 2014).

Creosote bush *(Larrea tridentate)*, also known as *chaparral* or *greasewood*, grows in deserts and serves as an herbal remedy among Native Americans in the Southwestern United States and Northern Mexico. This plant is well known for its antioxidant properties and for inducing apoptosis in tumor cells (Rodriguez-Fragoso, Reyes-Esparza, Burchiel, Herrera-Ruiz, & Torres, 2008). Lignans are one of the most important classes of natural products in *Larrea* species, the molecular mechanisms underlying the anticancer activities of creosote bush lignans have been shown to involve a dozen cancer pathways (Gnabre, Bates, & Huang, 2015). In addition, concentrations of 200, 212, and 115 mg GAE/L have been observed in ethanol extract, ethanol-water mixture, and water, respectively (Skouta, Morán-Santibañez, Valenzuela, Vasquez, & Fenelon, 2018). On the other hand, creosote bush has been linked to extensive liver damage, cystic renal disease, and adenocarcinoma of the kidney (Mazzio & Soliman, 2009).

Oregano *(Origanum vulgare* L.) belonging to the *Lamiaceae* (lipped) family, is used for medicinal purposes for its antioxidant properties, in particular the methanolic extract and essential oils (phenolic monoterpenes, carvacrol, and thymol). These substances are beneficial at therapeutic doses during a small period of time but in excess can be toxic to the liver, kidneys, and nervous system (Martins et al., 2014). It has been observed that flavonoids and phenolic acids from hydroalcoholic extracts of *Origanum vulgare* L. ssp. *Viridulum* inhibits cell proliferation against human breast cancer cells by 44% (MFC-7), proliferation against liver cancer cells (HepG2) is inhibited by approximately 64%, and for cells of colorectal cancer (LoVo), the inhibition corresponds to 40%. Extracts of *Origanum vulgare* appear to exhibit a selective antiproliferative activity against HepG2 (Gutiérrez-Grijalva et al., 2018).

Anise *(Pimpinella anisum)* belongs to the *Umbelliferae* family. Aqueous and ethanolic extracts of anise have been shown to have strong antioxidant activity, reducing power, DPPH radical and superoxide anion scavenging, hydrogen peroxide scavenging, and metal chelating activities compared to BHA (butylated hydroxyanisole), BHT (butylated hydroxytoluene), and α-tocopherol, mainly the aqueous extract (Shojaii & Abdollahi Fard, 2012). Alsalhi et al., developed silver nanoparticles (AgNPs) by using an aqueous extract of Pimpinella anisum and evaluated their cytotoxicity on colon cancer cells. The overall results suggest that concentrations of AgNPs >10 μg/mL are toxic to the cells, causing either apoptosis or necrosis. At concentrations >10 μg/mL, necrotic cells increased in volume, and dead orange-red fluorescence cells were observed. Toxicity was higher in HT115 cancer cells (colon cancer cells) compared to other cell types (Alsalhi et al., 2016).

Romero *(Rosmarinus officinalis)* is a medicinal plant that belongs to the *Lamiaceae* family and is commonly known as *rosemary*. Its antioxidant activity protects against carcinogenic and toxic agents. It contains carnosic acid and carnosol, which represent approximately 5% of the weight of the dried leaves of *R. officinalis*. Several in vitro studies have been carried out to evaluate the cytotoxicity of carnosol and carnosic acid, observing decrease in cell viability in a dependent manner of the dose in human cancer cells (HepG2, COLO 205 and HL-60), breast cancer cells and colon cancer cells (Andrade et al., 2018) Diterpenoids and flavonoids (flavones, flavanones, and flavonols) have been identified in rosemary extracts, as well as some phenolic and lignan acids, and a triterpenoid (Mena et al., 2016).

Ajedrea *(Satureja hortensis)* has important applications in the treatment of diseases such as cancer due to the type of biomolecules that are present in its extracts and essential oils. The volatile oil isolated from *S. hortenisis* contains thymol (0.3–28.2%), γ-terpinene (15.30–39%), carvacrol (11–67%), and *p*-cymene (3.5–19.6%). The methanolic extract obtained from aerial parts contains rosmarinic acid (24.9 mg/g), caffeic acid (1.3 mg/g), naringenin (1.1 mg/g), isoferulic acid (220 μg/g) and apigenin (165 μg/g; Fierascu et al., 2018). Feyzioglu and Tornuk in 2016, produced chitosan nanoparticles (NPs) loaded with *S. hortensis* essential oil (EO) and observed that the sample of NP prepared without EO did not show antioxidant activity, while the NP with EO showed AOA levels that ranged between 43.66% and 56.99% (Feyzioglu & Tornuk, 2016).

Cempasuchil *(Tagetes erecta* L.) is commonly known as *Aztec marigold*. It is a flowered annual herbaceous plant native from Mexico used for ornamental and medicinal purposes (Moliner et al., 2018). It is considered a rich source of lutein, since dried marigold flowers have 0.1–0.2% carotenoid dry matter, of which 80% are lutein diesters. Lutein is a yellow pigment that belongs to the family of carotenoids and is used for the preparation of (Šivel, Kráčmar, Fišera, Klejdus, & Kubáň, 2014). Likewise, nine phenolic compounds have been identified: digallic acid, laricitrin-di-hexoside, myricetin-hexoside, laricitrin-di-hexoside, laricitrin-galloyl-hexoside, laricitrin-hexoside, laricitrin-hexoside, laricitrin-hexoside, and laricitrin (Moliner et al., 2018). *T. erecta* showed that its compounds have anticancer activity (Salehi et al., 2018).

Damiana *(Turnera diffusa)* has been shown in phytochemical studies to have demonstrated the presence of tetraphyllin B, gonzalitozin I, damianin, tricosan-2-one, hexaconazole, squalene, ρ-arbutin, α-pinene, β-pinene, ρ-cymene, 1,8-cineole, β-sitosterol, acacetin, pinocembrin, teuhetenone A, eremophyllane, echinaticin, apigenin, and simple sugars (Avelino-Flores, Cruz-López, Jiménez-Montejo, & Reyes-Leyva, 2015). Several studies with acacetin have shown that it can modulate various signaling pathways to inhibit cancer growth (Kandhari, Mishra, & Singh, 2019).

DISCUSSION

Medicinal plants are widely used by people of different cultures and regions in the world (Baharvand-Ahmadi & Asadi-Samani, 2016). The relationship between society and nature is very important since traditional medicine has become in the dominant health care system in several countries, mainly in low- and middle-income countries. This is due to several reasons including affordability, accessibility, and low cost. Therefore, 80% of the world's population use traditional medicine practitioners to meet their primary healthcare needs and also to treat global burden diseases (Oyebode, Kandala, Chilton, & Lilford, 2016).

The use of plants to cure various types of diseases is related to their effect on health and is closely related to the parts of plants such as leaves, stem, bark, root, among others. Most plants are utilized alone or in mixtures with other plants with related therapeutic effects. Most medicinal plants are consumed alone or eventually in mixtures with other plants with therapeutic effects related and generally consumed without the support or recommendation of the traditional doctors (Panthong, Kanjanapothi, Taesotikul, & Taylor, 1991). For these reasons, an important link must be established between traditional and complementary medicine systems.

This chapter focuses on the use of traditional medicines to prevent or treat the main diseases of global burden, such as type 2 diabetes, hypertension, hypercholesterolemia, obesity, and cancer. These diseases are responsible for an important impact on health services as well as the socio-economic sector around the world. For each disease, we established the relationship between the medicinal plants and the disease, pointed out the novel phytochemicals involved and their therapeutic effects. Therefore, the phytochemical expressed in each medicinal plant has effects such as antioxidant, anti-inflammatory, anticarcinogenic, antidiabetic, antihypertensive, antithrombolytic, antimicrobial, hepatoprotective, hypolipidemic, hypoglycaemic, antihyperglycemic, and antihyperinsulinemic activity, to serve as preventative agents or definitive treatments. However, ingesting medicinal plants, fractions, or their extracts must be under the supervision and control of specialized people because, like conventional medicinal drugs, they could produce side effects, and there are not enough studies on this yet.

FUTURE RESEARCH DIRECTIONS

Medicinal plants are a wide field of study since many people use them as a first option to prevent or treat some symptoms or diseases. Some plants have yet unknown pharmacological and therapeutic properties, although they have been used as a traditionally for ages, recommended by word to mouth between families and elders. For these reasons, it is extremely important to perform studies that help identify the chemical compounds (phytochemicals) in each part of medicinal plants, the changes in concentration that occur through their life cycle, and safe doses and concentrations for use against specific diseases and to investigate the therapeutic and side effects of bioactive phytochemicals in medicinal plants and their mixtures. Accordingly, the classification of different extracts involved to obtain secondary metabolites is an important issue in medicinal plant research. For these reasons, pharmacological studies are necessary to properly understand these natural products.

CONCLUSION

This chapter documents uses of traditional medicine in the modern world and the scientific interest in analyzing it. The use of medicinal plants proves humanity's confidence in resources provided by nature and ancient knowledge. The broad variety of medicinal plants discussed here gives a great number of bioactive phytochemicals differentiated only by their chemical functional groups, physical state, and chemical structure. Even though scientific studies report medicinal plants are useful to prevent or treat the main diseases modern world such as type 2, diabetes, hypertension, hypercholesterolemia, obesity, and cancer, many traditional formulations as well scientific studies have a significant gray area regarding safe doses, effectiveness, toxicity, interactions human-medicinal plant, side effects, and other issues. Expanded scientific studies are needed on medicinal plants and their effects against different diseases. Additionally, the inclusion of medicinal plants into the diet could improve the health of the consumers; this might be associated with phytochemicals diverse, which have been and will be an interesting human health research area across the time.

REFERENCES

Aggarwal, B. B. (2010). Targeting inflammation-induced obesity and metabolic diseases by curcumin and other nutraceuticals. *Annual Review of Nutrition*, *30*(1), 173–199. doi:10.1146/annurev.nutr.012809.104755 PMID:20420526

Alhassan, A. M., & Ahmed, Q. U. (2016). Averrhoa bilimbi Linn.: A review of its ethnomedicinal uses, phytochemistry, and pharmacology. *Journal of Pharmacy & Bioallied Sciences*, *8*(4), 265–271. doi:10.4103/0975-7406.199342 PMID:28216948

Ali, S. I., Gopalakrishnan, B., & Venkatesalu, V. (2017). Pharmacognosy, phytochemistry and pharmacological properties of *Achillea millefolium* L.: A review. *Phytotherapy Research*, *31*(8), 1140–1161. doi:10.1002/ptr.5840 PMID:28618131

Alonso-Castro, A. J., Domínguez, F., Zapata-Morales, J. R., & Carranza-Álvarez, C. (2015). Plants used in the traditional medicine of Mesoamerica (Mexico and Central America) and the Caribbean for the treatment of obesity. *Journal of Ethnopharmacology, 175*(June 2016), 335–345. . 2015.09.029 doi:10.1016/j.jep

Alsalhi, M. S., Devanesan, S., Alfuraydi, A. A., Vishnubalaji, R., Munusamy, M. A., Murugan, K., ... Benelli, G. (2016). Green synthesis of silver nanoparticles using *Pimpinella anisum* seeds: Antimicrobial activity and cytotoxicity on human neonatal skin stromal cells and colon cancer cells. *International Journal of Nanomedicine*, *11*, 4439–4449. doi:10.2147/IJN.S113193 PMID:27660438

Andrade, J. M., Faustino, C., García, C., Ladeiras, D., Reis, C. P., & Rijo, P. (2018). *Rosmarinus officinalis* L.: An update review of its phytochemistry and biological activity. *Future Science OA, 4*(4).

Andrade-Cetto, A., & Heinrich, M. (2005). Mexican plants with hypoglycaemic effect used in the treatment of diabetes. *Journal of Ethnopharmacology, 99*(3), 325–348. . 2005.04.019 doi:10.1016/j.jep

Arqueta, A. (1994). *Atlas de las plantas de la medicina tradicional mexicana* (Vol. 3). Universidad Autónoma de Chiapas, Consorcio de Ciencias Agropecuarias, Instituto de Estudios Indígenas.

Arroyo-Johnson, C., & Mincey, K. D. (2016). Obesity epidemiology trends by race/ethnicity, gender, and education. *Gastroenterology Clinics of North America*, *45*(4), 571–579. doi:10.1016/j.gtc.2016.07.012 PMID:27837773

Athanasakis, K. (2017). The Socioeconomic effects of uncontrolled hypertension. *Current Vascular Pharmacology, 16*(1), 5–9. doi:10.2174/1570161115666170413145125 PMID:28412912

Avelino-Flores, M. D. C., Cruz-López, M. D. C., Jiménez-Montejo, F. E., & Reyes-Leyva, J. (2015). Cytotoxic activity of the methanolic extract of *Turnera diffusa* Willd. on breast cancer cells. *Journal of Medicinal Food, 18*(3), 299–305. doi:10.1089/jmf.2013.0055 PMID:25299247

Ayala Garay, A., Rivas Valencia, P., Cortes-Espinoza, L., de la Olán, M., Escobedo-López, D., & Espitia Rangel, E. (2014). La rentabilidad del cultivo de amaranto (*Amaranthus* spp.) en la región centro de México. *CIENCIA ergo-sum, 21*(1), 47–54.

Baharvand-Ahmadi, B., & Asadi-Samani, M. (2016). A mini-review on the most important effective medicinal plants to treat hypertension in ethnobotanical evidence of Iran. *Journal of Nephropharmacology, 6*(1), 3–8. PMID:28197520

Banerjee, S. K., & Maulik, S. K. (2002). Effect of garlic on cardiovascular disorders: A review. *Nutrition Journal, 1*(1), 4. doi:10.1186/1475-2891-1-4 PMID:12537594

Baqueiro-Peña, I., & Guerrero-Beltrán, J. A. (2014). Uses of *Justicia spicigera* in medicine and as a source of pigments. *Functional Foods in Health and Disease, 4*(9), 401–414. doi:10.31989/ffhd.v4i9.150

Bersani, F. S., Coviello, M., Imperatori, C., Francesconi, M., Hough, C. M., Valeriani, G., . . . Corazza, O. (2015). Adverse psychiatric effects associated with herbal weight-loss products. *BioMed Research International, 2015*, 10 pages. doi:10.1155/2015/120679

Bhandari, P. R. (2015). *Crocus sativus* L. (saffron) for cancer chemoprevention: A mini review. *Journal of Traditional and Complementary Medicine, 5*(2), 81–87. doi:10.1016/j.jtcme.2014.10.009 PMID:26151016

Bhaskara Rao, K. V., Nidhi, H., Dipankar, D., Garima, D., Kumar, G., & Karthik, L. (2015). Phytochemical profile, in vitro antioxidant property and HPTLC analysis of methanol extract of *Bougainvillea glabra* (*Nyctaginaceae*). *International Journal of Pharmaceutical Sciences Review and Research, 31*(2), 235–241.

Bhat, A., Satpathy, G., & Gupta, R. K. (2015). Evaluation of Nutraceutical properties of *Amaranthus hypochondriacus* L. grains and formulation of value-added cookies. *Journal of Pharmacognosy and Phytochemistry, 3*(5), 51–54.

Bouasla, A., & Bouasla, I. (2017). Ethnobotanical survey of medicinal plants in northeastern of Algeria. *Phytomedicine, 36*(September), 68–81. doi:10.1016/j.phymed.2017.09.007 PMID:29157830

Bower, A., Marquez, S., & de Mejía, E. G. (2016). The health benefits of selected culinary herbs and spices found in the traditional Mediterranean diet. *Critical Reviews in Food Science and Nutrition, 56*(16), 2728–2746. doi:10.1080/10408398.2013.805713 PMID:25749238

Bray, F., Ferlay, J., Soerjomataram, I., Siegel, R. L., Torre, L. A., & Jemal, A. (2018). Global cancer statistics 2018: GLOBOCAN estimates of incidence and mortality worldwide for 36 cancers in 185 countries. *CA: a Cancer Journal for Clinicians, 68*(6), 394–424. doi:10.3322/caac.21492 PMID:30207593

Camacho, M. R., Chávez, D., Mata, R., & Palacios-Ríos, M. (1991). Chemical studies on Mexican plants used in traditional medicine XXII: Constituents of *Equisetum myriochaetum* Schlecht. et Cham. *Fitoterapia1, 63*(5), 471–472.

Casas, A., Blancas, J., & Lira, R. (2016). Ethnobotany of Mexico, (April). doi:10.1007/978-1-4614-6669-7

Chávez-Silva, F., Cerón-Romero, L., Arias-Durán, L., Navarrete-Vázquez, G., Almanza-Pérez, J., Román-Ramos, R., . . . Estrada-Soto, S. (2018). Antidiabetic effect of *Achillea millefollium* through multitarget interactions: α-glucosidases inhibition, insulin sensitization and insulin secretagogue activities. *Journal of Ethnopharmacology, 212*, 1–7. . 2017.10.005 doi:10.1016/j.jep

Chen, C. C., Hsu, J. D., Wang, S. F., Chiang, H. C., Yang, M. Y., Kao, E. S., ... Wang, C. J. (2003). *Hibiscus sabdariffa* extract inhibits the development of atherosclerosis in cholesterol-fed rabbits. *Journal of Agricultural and Food Chemistry*, *51*(18), 5472–5477. doi:10.1021/jf030065w PMID:12926900

Chewonarin, T., Kinouchi, T., Kataoka, K., Arimochi, H., Kuwahara, T., Vinitketkumnuen, U., & Ohnishi, Y. (1999). Effects of roselle (*Hibiscus sabdariffa* Linn.), a Thai medicinal plant, on the mutagenicity of various known mutagens in *Salmonella typhimurium* and on formation of aberrant crypt foci induced by the colon carcinogens Azoxymethane and 2-Amino-1-methyl-6-ph. *Food and Chemical Toxicology*, *37*(6), 591–601. doi:10.1016/S0278-6915(99)00041-1 PMID:10478827

Córdova Villalobos, J. Á. (2016). La obesidad: La verdadera pandemia del siglo XXI. *Cirugia y Cirujanos*, *84*(5), 351–355. doi:10.1016/j.circir.2016.08.001 PMID:27601182

Da Rocha Fernandes, J., Ogurtsova, K., Linnenkamp, U., Guariguata, L., Seuring, T., Zhang, P., ... Makaroff, L. E. (2016). IDF Diabetes Atlas estimates of 2014 global health expenditures on diabetes. *Diabetes Research and Clinical Practice*, *117*, 48–54. doi:10.1016/j.diabres.2016.04.016 PMID:27329022

Da Silva Lima, N., De Paula Numata, E., De Souza Mesquita, L. M., Dias, P. H., Vilegas, W., Gambero, A., & Ribeiro, M. L. (2017). Modulatory effects of guarana (*Paullinia cupana*) on adipogenesis. *Nutrients*, *9*(6), 635. doi:10.3390/nu9060635 PMID:28632199

Dafallah, A. A., & Al-Mustafa, Z. (2004). Investigation of the anti-inflammatory activity of *Acacia nilotica* and *Hibiscus sabdariffa. The American Journal of Chinese Medicine, 24*(03n04), 263–269. doi:10.11420192415x96000323

De Santayana, M. P., Pieroni, A., & Kuri, R. K. (2010). *Ethnobotany in the new Europe: People, health, and wild plant resources* (Vol. 14; Studies In). Berghahn Books.

Ding, D., Lawson, K. D., Kolbe-Alexander, T. L., Finkelstein, E. A., Katzmarzyk, P. T., van Mechelen, W., & Pratt, M. (2016). The economic burden of physical inactivity: A global analysis of major non-communicable diseases. *Lancet*, *388*(10051), 1311–1324. doi:10.1016/S0140-6736(16)30383-X PMID:27475266

Elujoba, A., Odeleye, O., & Ogunyemi, C. (2005). Traditional medicine development for medical and dental primary health care delivery system in Africa. *African Journal of Traditional, Complementary, and Alternative Medicines*, *2*(1), 46–61. doi:10.4314/ajtcam.v2i1.31103

Esakkimuthu, S., Mutheeswaran, S., Arvinth, S., Paulraj, M. G., Pandikumar, P., & Ignacimuthu, S. (2016). Quantitative ethnomedicinal survey of medicinal plants given for cardiometabolic diseases by the non-institutionally trained siddha practitioners of Tiruvallur district, Tamil Nadu, India. *Journal of Ethnopharmacology*. . 2016.04.017 doi:10.1016/j.jep

Espinosa, L. N. (2016). Tratamiento de la obesidad con productos naturales. *Enfermería Investiga, Investigación, Vinculación. Docencia y Gestión, 1*(4), 164–168.

Faruque, M. O., Uddin, S. B., Barlow, J. W., Hu, S., Dong, S., Cai, Q., ... Hu, X. (2018). Quantitative ethnobotany of medicinal plants used by indigenous communities in the Bandarban district of Bangladesh. *Frontiers in Pharmacology*, *9*(40), 1–12. doi:10.3389/fphar.2018.00040 PMID:29467652

Federation, I. D. (2017). *IDF Diabetes Atlas* (8th ed.). Brussels, Belgium: International Diabetes Federation; doi:10.1016/S0140-6736(16)31679-8

Ferrannini, E., & Cushman, W. C. (2012). Diabetes and hypertension: The bad companions. *Lancet*, *380*(9841), 601–610. doi:10.1016/S0140-6736(12)60987-8 PMID:22883509

Ferrara, L. (2014). The *Garcinia cambogia* in phytotreatment of obesity: Activities of the hydroxycitric acid. *European Scientific Journal*, *10*(21), 291–301.

Feyzioglu, G. C., & Tornuk, F. (2016). Development of chitosan nanoparticles loaded with summer savory (*Satureja hortensis* L.) essential oil for antimicrobial and antioxidant delivery applications. *Lebensmittel-Wissenschaft + Technologie*, *70*, 104–110. doi:10.1016/j.lwt.2016.02.037

Fierascu, I., Dinu-Pirvu, C. E., Fierascu, R. C., Velescu, B. S., Anuta, V., Ortan, A., & Jinga, V. (2018). Phytochemical profile and biological activities of *Satureja hortensis* L.: A review of the last decade. *Molecules (Basel, Switzerland)*, *57*(24), 11557–11563. doi:10.3390/molecules23102458 PMID:30257512

Flores-Flores, A., Hernández-Abreu, O., Rios, M. Y., León-Rivera, I., Aguilar-Guadarrama, B., Castillo-España, P., ... Estrada-Soto, S. (2016). Vasorelaxant mode of action of dichloromethane-soluble extract from *Agastache mexicana* and its main bioactive compounds. *Pharmaceutical Biology*, *54*(12), 2807–2813. doi:10.1080/13880209.2016.1184690 PMID:27252080

Fort, R. S., Trinidad Barnech, J. M., Dourron, J., Colazzo, M., Aguirre-Crespo, F. J., Duhagon, M. A., & Álvarez, G. (2018). Isolation and structural characterization of bioactive molecules on prostate cancer from Mayan traditional medicinal plants. *Pharmaceuticals*, *11*(3), 78. doi:10.3390/ph11030078 PMID:30110911

Galati, E. M., Monforte, M. T., Forestieri, A. M., Miceli, N., Bade, A., & Trovato, A. (1999). *Salvadora persica* L.: Hypolipidemic activity on experimental hypercholesterolemia in rat. *Phytomedicine*, *6*(3), 181–185. doi:10.1016/S0944-7113(99)80007-5 PMID:10439483

García, J. G., García, M. L., Zamilpa, A., Almanza, J. C. P., Jasso, I. E., Ramos, R. R., & Alarcón-Aguilar, F. J. (2017). Chemical characterization of a hypoglycemic extract from *Cucurbita ficifolia* Bouche that induces liver glycogen accumulation in diabetic mice. *African Journal of Traditional, Complementary, and Alternative Medicines*, *14*(3), 218–230. doi:10.21010/ajtcam.v14i3.24 PMID:28480434

García-Rodríguez, R. V., Gutiérrez-Rebolledo, G. A., Méndez-Bolaina, E., Sánchez-Medina, A., Maldonado-Saavedra, O., Domínguez-Ortiz, M. Á., . . . Cruz-Sánchez, J. S. (2014). *Cnidoscolus chayamansa* Mc Vaugh, an important antioxidant, anti-inflammatory and cardioprotective plant used in Mexico. *Journal of Ethnopharmacology, 151*(2), 937–943. . 2013.12.004 doi:10.1016/j.jep

Gaziano, T. A., Bitton, A., Anand, S., & Weinstein, M. C. (2009). The global cost of nonoptimal blood pressure. *Journal of Hypertension, 27*(7), 1472–1477. doi:10.1097/HJH.0b013e32832a9ba3 PMID:19474763

Gnabre, J., Bates, R., & Huang, R. C. (2015). Creosote bush lignans for human disease treatment and prevention: Perspectives on combination therapy. *Journal of Traditional and Complementary Medicine, 5*(3), 119–126. doi:10.1016/j.jtcme.2014.11.024 PMID:26151022

Govea-Salas, M., Morlett-Chávez, J., Rodriguez-Herrera, R., & Ascacio-Valdés, J. (2017). Some Mexican plants used in traditional medicine. In *Aromatic and Medicinal Plants: Back to Nature* (pp. 191–200). BoD: Books on Demand.

Greenwell, M., & Rahman, P. K. S. M. (2015). Medicinal plants: Their use in anticancer treatment. *International Journal of Pharmaceutical Sciences and Research, 6*(10), 4103–4112. doi:10.13040/IJPSR.0975-8232.6(10).4103-12 PMID:26594645

Gutiérrez-Grijalva, E. P., Picos-Salas, M. A., Leyva-López, N., Criollo-Mendoza, M. S., Vazquez-Olivo, G., & Heredia, J. B. (2018). Flavonoids and phenolic acids from Oregano: Occurrence, biological activity and health benefits. *Plants, 7*(1), 2. doi:10.3390/plants7010002 PMID:29278371

Haji Faraji, M., & Haji Tarkhani, A. H. (1999). The effect of sour tea (*Hibiscus sabdariffa*) on essential hypertension. *Journal of Ethnopharmacology, 65*(3), 231–236. doi:10.1016/S0378-8741(98)00157-3 PMID:10404421

Hasanein, P., Felehgari, Z., & Emamjomeh, A. (2016). Preventive effects of *Salvia officinalis* L. against learning and memory deficit induced by diabetes in rats: Possible hypoglycaemic and antioxidant mechanisms. *Neuroscience Letters, 622*, 72–77. doi:10.1016/j.neulet.2016.04.045 PMID:27113201

Islam, M. Z., Hossain, M. T., Hossen, F., Akter, M. S., & Mokammel, M. A. (2016). In-vitro antioxidant and antimicrobial activity of *Bougainvillea glabra* flower. *Research Journal of Medicinal Plant, 10*(3), 228–236. . 2016.228.236 doi:10.3923/rjmp

Izaola, O., de Luis, D., Sajoux, I., Domingo, J. C., & Vidal, M. (2015). Inflamación y obesidad (lipoinflamación). *Nutrición Hospitalaria, 31*(6), 2352–2358. PMID:26040339

Jaiswal, D., Rai, P. K., Kumar, A., Mehta, S., & Watal, G. (2009). Effect of *Moringa oleifera* Lam. leaves aqueous extract therapy on hyperglycemic rats. *Journal of Ethnopharmacology, 123*(3), 392–396. . 2009.03.036 doi:10.1016/j.jep

James, S. L., Abate, D., Abate, K. H., Abay, S. M., Abbafati, C., Abbasi, N., ... Murray, C. J. L. (2018). Global, regional, and national incidence, prevalence, and years lived with disability for 354 diseases and injuries for 195 countries and territories, 1990–2017: A systematic analysis for the Global Burden of Disease Study 2017. *Lancet, 392*(10159), 1798–1858. doi:10.1016/S0140-6736(18)32279-7 PMID:30496104

Jiménez-Arellanes, M. A., García-Martínez, I., & Rojas-Tomé, S. (2014). Biological potential of medicinal species of *Cnidoscolus* genus. *Revista Mexicana de Ciencias Farmacéuticas, 45*(4), 1–6.

Kandhari, K., Mishra, J. P. N., & Singh, R. P. (2019). Acacetin inhibits cell proliferation, survival, and migration in human breast cancer cells. *International Journal of Pharmacy and Biological Sciences-IJPBS, 9*(1), 443–452.

Kearney, P. M., Whelton, M., Reynolds, K., Muntner, P., Whelton, P. K., & He, J. (2005). Global burden of hypertension: Analysis of worldwide data. *Lancet*, *365*(9455), 217–223. doi:10.1016/S0140-6736(05)17741-1 PMID:15652604

Khan, B. A., Abraham, A., & Leelamma, S. (1996). Biochemical response in rats to the addition of curry leaf (*Murraya koenigii*) and mustard seeds (*Brassica juncea*) to the diet. *Plant Foods for Human Nutrition (Dordrecht, The Netherlands)*, *49*(4), 295–299. doi:10.1007/BF01091978 PMID:8983055

Khan, B. A., Abraham, A., & Leelamma, S. (1998). Influence of spices: *Murraya koenigii* and *Brassica juncea* on rats fed atherogenic diet. *Journal of Food Science and Technology (Mysore), 35(1), 66-68.*

Kianbakht, S., Khalighi-Sigaroodi, F., & Dabaghian, F. H. (2013). Improved glycemic control in patients with advanced type 2 diabetes mellitus taking *Urtica dioica* leaf extract: A randomized double-blind placebo-controlled clinical trial. *Clinical Laboratory*, *59*(9–10), 1071–1076. PMID:24273930

Kim, W., & Egan, J. M. (2008). The role of incretins in glucose homeostasis and diabetes treatment. *Pharmacological Reviews*, *60*(4), 470–512. doi:10.1124/pr.108.000604 PMID:19074620

Knekt, P., Kumpulainen, J., Järvinen, R., Rissanen, H., Heliövaara, M., Reunanen, A., ... Aromaa, A. (2002). Flavonoid intake and risk of chronic diseases. *The American Journal of Clinical Nutrition*, *76*(3), 560–568. doi:10.1093/ajcn/76.3.560 PMID:12198000

Lachhiramka, P., & Patil, S. (2016). Cholesterol lowering property of garlic (*Allium sativum*) on patients with hypercholesterolemia. *International Journal of Medical Science and Public Health*, *5*(11), 2249. doi:10.5455/ijmsph.2016.28032016449

Lemhadri, A., Hajji, L., Michel, J. B., & Eddouks, M. (2006). Cholesterol and triglycerides lowering activities of caraway fruits in normal and streptozotocin diabetic rats. *Journal of Ethnopharmacology, 106*(3), 321–326. . 2006.01.033 doi:10.1016/j.jep

Lin, T. L., Lin, H. H., Chen, C. C., Lin, M. C., Chou, M. C., & Wang, C. J. (2007). *Hibiscus sabdariffa* extract reduces serum cholesterol in men and women. *Nutrition Research (New York, N.Y.)*, *27*(3), 140–145. doi:10.1016/j.nutres.2007.01.007

Long, A. N., & Dagogo-Jack, S. (2011). Comorbidities of diabetes and hypertension: Mechanisms and approach to target organ protection. *Journal of Clinical Hypertension*, *13*(4), 244–251. doi:10.1111/j.1751-7176.2011.00434.x PMID:21466619

López-Romero, P., Pichardo-Ontiveros, E., Avila-Nava, A., Vázquez-Manjarrez, N., Tovar, A. R., Pedraza-Chaverri, J., & Torres, N. (2014). The effect of nopal (*Opuntia Ficus Indica*) on postprandial blood glucose, incretins, and antioxidant activity in Mexican patients with type 2 diabetes after consumption of two different composition breakfasts. *Journal of the Academy of Nutrition and Dietetics*, *114*(11), 1811–1818. doi:10.1016/j.jand.2014.06.352 PMID:25132122

Luna-Vázquez, F. J., Ibarra-Alvarado, C., Del Rayo Camacho-Corona, M., Rojas-Molina, A., Rojas-Molina, J. I., García, A., & Bah, M. (2018). Vasodilator activity of compounds isolated from plants used in Mexican traditional medicine. *Molecules (Basel, Switzerland)*, *23*(6), 1474. doi:10.3390/molecules23061474 PMID:29912156

Lunsford, K. E., Bodzin, A. S., Reino, D. C., Wang, H. L., & Busuttil, R. W. (2016). Dangerous dietary supplements: *Garcinia cambogia*-associated hepatic failure requiring transplantation. *World Journal of Gastroenterology*, *22*(45), 10071–10076. doi:10.3748/wjg.v22.i45.10071 PMID:28018115

Martín, M., Herrero, A., & Echavarría, I. (2015). El cáncer de mama. *Arbor*, *191*(773), a234. doi:10.3989/arbor.2015.773n3004

Martínez Salvador, L. (2016). Seguridad alimentaria, autosuficiencia y disponibilidad del amaranto en México. *Problemas del Desarrollo*, *47*(186), 107–132. doi:10.1016/j.rpd.2016.08.004

Martins, N., Barros, L., Santos-Buelga, C., Henriques, M., Silva, S., & Ferreira, I. C. F. R. (2014). Decoction, infusion and hydroalcoholic extract of *Origanum vulgare* L.: Different performances regarding bioactivity and phenolic compounds. *Food Chemistry*, *158*, 73–80. doi:10.1016/j.foodchem.2014.02.099 PMID:24731316

Mathew, N. S., & Negi, P. S. (2017). Traditional uses, phytochemistry and pharmacology of wild banana (*Musa acuminata* Colla): A review. *Journal of Ethnopharmacology, 196*(20), 124–140. 2016.12.009

Mazzio, E. A., & Soliman, K. F. A. (2009). In Vitro screening for the tumoricidal properties of international medicinal herbs. *Phytotherapy Research*, *23*(3), 385–398. doi:10.1002/ptr.2636 PMID:18844256

McFarlane, S. I., Muniyappa, R., Shin, J. J., Bahtiyar, G., & Sowers, J. R. (2004). Osteoporosis and cardiovascular disease: Brittle bones and boned arteries, is there a link? *Endocrine*, *23*(1), 1–10. doi:10.1385/ENDO:23:1:01 PMID:15034190

Mena, P., Cirlini, M., Tassotti, M., Herrlinger, K. A., Dall'Asta, C., & Del Rio, D. (2016). Phytochemical profiling of flavonoids, phenolic acids, terpenoids, and volatile fraction of a rosemary (*Rosmarinus officinalis* L.) extract. *Molecules (Basel, Switzerland)*, *21*(11), 1576. doi:10.3390/molecules21111576 PMID:27869784

Miraj, S., & Kiani, S. (2016). Study of therapeutic effects of *Cynara scolymus* L.: A review. *Der Pharmacia Lettre*, *8*(9), 168–173.

Misael, P. C. E., De Guadalupe, T. M. B. M., Paola, D. C. S. G. M., & Carlos, R. L. J. (2014). Effect of *Cynara scolymus* (artichoke) in homeopathic doses on body mass index in obese and overweight patients. *Biomedical & Pharmacology Journal*, *7*(2), 525–533. doi:10.13005/bpj/520

Mithila, M. V., & Khanum, F. (2015). Effectual comparison of quinoa and amaranth supplemented diets in controlling appetite; a biochemical study in rats. *Journal of Food Science and Technology*, *52*(10), 6735–6741. doi:10.100713197-014-1691-1 PMID:26396423

Mohamadi, N., Sharififar, F., Koohpayeh, A., & Daneshpajouh, M. (2015). Traditional and Ethnobotanical uses of medicinal plants by ancient populations in Khabr and Rouchon of Iran. *Journal of Applied Pharmaceutical Science*, *5*(11), 101–107. doi:10.7324/JAPS.2015.501117

Moliner, C., Barros, L., Dias, M. I., López, V., Langa, E., Ferreira, I. C. F. R., & Gómez-Rincón, C. (2018). Edible flowers of *Tagetes erecta* L. as functional ingredients: Phenolic composition, antioxidant and protective effects on *Caenorhabditis elegans. Nutrients*, *10*(12), 2002. doi:10.3390/nu10122002 PMID:30567311

Monroy-Ortíz, C., & Castillo-España, P. (2007). *Medicinal plants used in the State of Morelos*. Morelos, México: Centro de Investigaciones Biológicas, Universidad Autónoma del Estado de Morelos.

Morales-Sánchez, A., & Fuentes-Panañá, E. M. (2014). Human viruses and cancer. *Viruses*, *6*(10), 4047–4079. doi:10.3390/v6104047 PMID:25341666

Najafian, J., Abdar-Esfahani, M., Arab-Momeni, M., & Akhavan-Tabib, A. (2014). Safety of herbal medicine in treatment of weight loss. *ARYA Atherosclerosis*, *10*(1), 55–58. PMID:24963315

National Cancer Institute. (2019). Types of cancer treatment. Retrieved August 8, 2019, from Methanolic extract of *Moringa oleifera* leaves improves glucose tolerance, glycogen synthesis and lipid metabolism in alloxan-induced diabetic rats. *Journal of Basic and Clinical Physiology and Pharmacology, 26*(6), 585–593. doi:10.1515/jbcpp-2014-0129

Ong, T. P., Moreno, F. S., & Ross, S. A. (2012). Targeting the epigenome with bioactive food components for cancer prevention. *Journal of Nutrigenetics and Nutrigenomics*, *4*(5), 275–292. doi:10.1159/000334585 PMID:22353664

Oyebode, O., Kandala, N. B., Chilton, P. J., & Lilford, R. J. (2016). Use of traditional medicine in middle-income countries: A WHO–SAGE study. *Health Policy and Planning*, *31*(8), 984–991. doi:10.1093/heapol/czw022 PMID:27033366

Pan, S. Y., Zhou, S. F., Gao, S. H., Yu, Z. L., Zhang, S. F., Tang, M. K., ... Ko, K. M. (2013). New perspectives on how to discover drugs from herbal medicines: CAM'S outstanding contribution to modern therapeutics. *Evidence-Based Complementary and Alternative Medicine*, *2013*, 1–25. doi:10.1155/2013/627375 PMID:23634172

Panthong, A., Kanjanapothi, D., Taesotikul, T., & Taylor, W. C. (1991). Ethnobotanical review of medicinal plants from Thai traditional books, part II: Plants with antidiarrheal, laxative and carminative properties. *Journal of Ethnopharmacology*, *31*(2), 121–156. doi:10.1016/0378-8741(91)90001-T PMID:2023424

Petlevski, R., Hadžija, M., Slijepčevič, M., & Juretič, D. (2001). Effect of "antidiabetis" herbal preparation on serum glucose and fructosamine in NOD mice. *Journal of Ethnopharmacology*, *75*(2), 181–184. doi:10.1016/S0378-8741(01)00177-5 PMID:11297848

Petrie, J. R., Guzik, T. J., & Touyz, R. M. (2018). Diabetes, hypertension, and cardiovascular disease: Clinical insights and vascular mechanisms. *The Canadian Journal of Cardiology*, *34*(5), 575–584. doi:10.1016/j.cjca.2017.12.005 PMID:29459239

Pushparaj, P., Tan, C. H., & Tan, B. K. H. (2000). Effects of *Averrhoa bilimbi* leaf extract on blood glucose and lipids in streptozotocin-diabetic rats. *Journal of Ethnopharmacology*, *72*(1–2), 69–76. doi:10.1016/S0378-8741(00)00200-2 PMID:10967456

Rahman, K., & Lowe, G. M. (2006). Garlic and cardiovascular disease: A critical review. *The Journal of Nutrition*, *136*(3), 736S–740S. doi:10.1093/jn/136.3.736S PMID:16484553

Ramírez, G., Zavala, M., Pérez, J., & Zamilpa, A. (2012). In vitro screening of medicinal plants used in Mexico as antidiabetics with glucosidase and lipase inhibitory activities. *Evidence-Based Complementary and Alternative Medicine*, *2012*, 1–6. doi:10.1155/2012/701261 PMID:23082084

Raskin, I., Ribnicky, D. M., Komarnytsky, S., Ilic, N., Poulev, A., Borisjuk, N., ... Fridlender, B. (2002). Plants and human health in the twenty-first century. *Trends in Biotechnology*, *20*(12), 522–531. doi:10.1016/S0167-7799(02)02080-2 PMID:12443874

Ried, K., Toben, C., & Fakler, P. (2013). Effect of garlic on serum lipids: An updated meta-analysis. *Nutrition Reviews*, *71*(5), 282–299. doi:10.1111/nure.12012 PMID:23590705

Rivera-Madrid, R., Aguilar-Espinosa, M., Cárdenas-Conejo, Y., & Garza-Caligaris, L. E. (2016). Carotenoid derivates in Achiote (*Bixa orellana*) seeds: Synthesis and health promoting properties. *Frontiers in Plant Science*, *7*, 1406. doi:10.3389/fpls.2016.01406 PMID:27708658

Rizos, C. V., & Elisaf, M. S. (2014). Antihypertensive drugs and glucose metabolism. *World Journal of Cardiology*, *6*(7), 517–530. doi:10.4330/wjc.v6.i7.517 PMID:25068013

Rodriguez-Fragoso, L., Reyes-Esparza, J., Burchiel, S. W., Herrera-Ruiz, D., & Torres, E. (2008). Risks and benefits of commonly used herbal medicines in Mexico. *Toxicology and Applied Pharmacology*, *227*(1), 125–135. . 2007.10.005 doi:10.1016/j.taap

Saleem, H., Zengin, G., Ahmad, I., Lee, J. T. B., Htar, T. T., Mahomoodally, F. M., ... Ahemad, N. (2019). Multidirectional insights into the biochemical and toxicological properties of *Bougainvillea glabra* (Choisy.) aerial parts: A functional approach for bioactive compounds. *Journal of Pharmaceutical and Biomedical Analysis*, *170*(5), 132–138. doi:10.1016/j.jpba.2019.03.027 PMID:30921647

Salehi, B., Valussi, M., Flaviana Bezerra Morais-Braga, M., Nalyda Pereira Carneiro, J., Linkoln Alves Borges Leal, A., Douglas Melo Coutinho, H., ... Sharifi-Rad, J. (2018). *Tagetes* spp. Essential oils and other extracts: Chemical characterization and biological activity. *Molecules (Basel, Switzerland)*, *23*(11), 2847. doi:10.3390/molecules23112847 PMID:30388858

Samarghandian, S., & Borji, A. (2014). Anticarcinogenic effect of saffron (*Crocus sativus* L.) and its ingredients. *Pharmacognosy Research*, *6*(2), 99–107. doi:10.4103/0974-8490.128963 PMID:24761112

Santana-Gálvez, J., Cisneros-Zevallos, L., & Jacobo-Velázquez, D. A. (2017). Chlorogenic acid: Recent advances on its dual role as a food additive and a nutraceutical against metabolic syndrome. *Molecules (Basel, Switzerland)*, *22*(3), 358. doi:10.3390/molecules22030358 PMID:28245635

Sari, I. P., Nurrochmad, N., & Setiawan, I. M. (2013). Indonesian herbals reduce cholesterol levels in diet-induced hypercholesterolemia through lipase inhibition. *Malaysian Journal of Pharmaceutical Sciences*, *11*(1), 13–20.

Sengupta, S. D., & Paul, R. (2016). Exploration of knowledge on new ethno-botanical value of *Justicia spicigera* SCHLTDL. *Global Journal of Research on Medicinal Plants & Indigenous*, *5*(10), 261–266.

Seuring, T., Archangelidi, O., & Suhrcke, M. (2015). The economic costs of type 2 diabetes: A global systematic review. *PharmacoEconomics*, *33*(8), 811–831. doi:10.100740273-015-0268-9 PMID:25787932

Shan, B., Cai, Y. Z., Sun, M., & Corke, H. (2005). Antioxidant capacity of 26 spice extracts and characterization of their phenolic constituents. *Journal of Agricultural and Food Chemistry*, *53*(20), 7749–7759. doi:10.1021/jf051513y PMID:16190627

Shojaii, A., & Abdollahi Fard, M. (2012). Review of pharmacological properties and chemical constituents of *Pimpinella anisum*. *ISRN Pharmaceutics*, *2012*, 1–8. doi:10.5402/2012/510795 PMID:22848853

Šivel, M., Kráčmar, S., Fišera, M., Klejdus, B., & Kubáň, V. (2014). Lutein content in marigold flower (*Tagetes erecta* L.) concentrates used for production of food supplements. *Czech Journal of Food Sciences*, *32*(6), 521–525. doi:10.17221/104/2014-CJFS

Skouta, R., Morán-Santibañez, K., Valenzuela, C. A., Vasquez, A. H., & Fenelon, K. (2018). Assessing the antioxidant properties of *Larrea tridentata* extract as a potential molecular therapy against oxidative stress. *Molecules (Basel, Switzerland)*, *23*(7), 1826. doi:10.3390/molecules23071826 PMID:30041415

Sofowora, A., Ogunbodede, E., & Onayade, A. (2013). The role and place of medicinal plants in the strategies for disease prevention. *African Journal of Traditional, Complementary, and Alternative Medicines: AJTCAM/African Networks on Ethnomedicines*, *10*(5), 210–229.

Solowey, E., Lichtenstein, M., Sallon, S., Paavilainen, H., Solowey, E., & Lorberboum-Galski, H. (2014). Evaluating medicinal plants for anticancer activity. *The Scientific World Journal*, *2014*, 1–12. doi:10.1155/2014/721402 PMID:25478599

Soran, H., Adam, S., & Durrington, P. N. (2017). Are recent statin recommendations to employ fixed doses and abandon targets effective for treatment of hypercholesterolaemia? Investigation based on number needed to treat. *European Journal of Preventive Cardiology*, *24*(1), 76–83. doi:10.1177/2047487316667785 PMID:27609614

Stewart, B. W., & Wild, C. P. (2014). World cancer report 2014. *World Health Organization*.

Thatoi, H., & Patra, J. K. (2011). Biotechnology and pharmacological evaluation of medicinal plants: An overview. *Journal of Herbs, Spices, & Medicinal Plants*, *17*(3), 214–248. doi:10.1080/10496475.2011.602471

Tremmel, M., Gerdtham, U. G., Nilsson, P. M., & Saha, S. (2017). Economic burden of obesity: A systematic literature review. *International Journal of Environmental Research and Public Health*, *14*(4), 1–18. doi:10.3390/ijerph14040435 PMID:28422077

Valsan, A., & Regi Raphael, K. (2016). Pharmacognostic Profile of *Averrhoa bilimbi* Linn. leaves. *South Indian Journal of Biological Sciences*, *2*(1), 75–80. doi:10.22205ijbs/2016/v2/i1/100347

Varenna, M., Manara, M., Galli, L., Binelli, L., Zucchi, F., & Sinigaglia, L. (2013). The association between osteoporosis and hypertension: The role of a low dairy intake. *Calcified Tissue International*, *93*(1), 86–92. doi:10.100700223-013-9731-9 PMID:23652773

Varzakas, T., Zakynthinos, G., & Verpoort, F. (2016). Plant food residues as a source of nutraceuticals and functional foods. *Foods*, *5*(4), 88. doi:10.3390/foods5040088 PMID:28231183

Vilar, D. D. A., Vilar, M. S. D. A., Moura, T. F. A. D. L. E., Raffin, F. N., De Oliveira, M. R., Franco, C. F. D. O., ... Barbosa-Filho, J. M. (2014). Traditional uses, chemical constituents, and biological activities of *Bixa orellana* L.: A review. *The Scientific World Journal*, 2014. PMID:25050404

Villarruel-López, A., López-de la Mora, D. A., Vázquez-Paulino, O. D., Puebla-Mora, A. G., Torres-Vitela, M. R., Guerrero-Quiroz, L. A., & Nuño, K. (2018). Effect of Moringa oleifera consumption on diabetic rats. *BMC Complementary and Alternative Medicine*, *18*(1), 127. doi:10.118612906-018-2180-2 PMID:29636032

Wang, C. J., Wang, J. M., Lin, W. L., Chu, C. Y., Chou, F. P., & Tseng, T. H. (2000). Protective effect of *Hibiscus anthocyanins* against tert-butyl hydroperoxide-induced hepatic toxicity in rats. *Food and Chemical Toxicology*, *38*(5), 411–416. doi:10.1016/S0278-6915(00)00011-9 PMID:10762726

Wong, K. C., & Wong, S. N. (1995). Volatile constituents of *Averrhoa bilimbi* L. fruit. *The Journal of Essential Oil Research*, *7*(6), 691–693. doi:10.1080/10412905.1995.9700533

World Health Organization. (1978). *Alma-Ata 1978: Primary health care, health for all series (No. 1)* (p. 47). Geneva, Switzerland: WHO.

World Health Organization. (2013). *WHO Traditional Medicine Strategy 2014-2023. World Health Organization (WHO).*

Wu, S., Zhu, W., Thompson, P., & Hannun, Y. A. (2018). Evaluating intrinsic and non-intrinsic cancer risk factors. *Nature Communications*. doi:10.103841467-018-05467-z PMID:30154431

Xie, J.-T., Chang, W.-T., Wang, C.-Z., Mehendale, S. R., Li, J., Ambihaipahar, R., ... Yuan, C.-S. (2006). Curry Leaf (*Murraya koenigii* Spreng.) reduces blood cholesterol and glucose levels in ob/ob mice. *The American Journal of Chinese Medicine*, *34*(02), 279–284. doi:10.1142/S0192415X06003825 PMID:16552838

Xu, Z., McClure, S. T., & Appel, L. J. (2018). Dietary cholesterol intake and sources among U.S. adults: Results from national health and nutrition examination surveys (NHANES), 2001–2014. *Nutrients*, *10*(6), 771. doi:10.3390/nu10060771 PMID:29903993

Yashin, A., Yashin, Y., Wang, J. Y., & Nemzer, B. (2013). Antioxidant and antiradical activity of coffee. *Antioxidants*, *2*(4), 230–245. doi:10.3390/antiox2040230 PMID:26784461

Zhao, Y., Chen, B., Shen, J., Wan, L., Zhu, Y., Yi, T., & Xiao, Z. (2017). The beneficial effects of quercetin, curcumin, and resveratrol in obesity. *Oxidative Medicine and Cellular Longevity, 2017*, 8 pages. doi:10.1155/2017/1459497

ADDITIONAL READING

Akhtar, M. S., & Swamy, M. K. (Eds.). (2018). *Anticancer plants: Natural products and biotechnological implements*. Springer Singapore. doi:10.1007/978-981-10-8064-7

Cotton, C. M., & Wilkie, P. (1996). *Ethnobotany: Principles and application*. Chichester: John Wiley & Sons.

Cseke, L. J., Kirakosyan, A., Kaufman, P. B., Warber, S., Duke, J. A., & Brielmann, H. L. (2016). *Natural Products From Plants*. CRC Press. doi:10.1201/9781420004472

Dahanukar, S. A., Kulkarni, R. A., & Rege, N. N. (2000). Pharmacology of medicinal plants and natural products. *Indian Journal of Pharmacology, 32*(4), S81–S118.

Kuete, V. (Ed.). (2013). *Medicinal plant research in Africa: Pharmacology and chemistry*. Newnes.

Muñoz, O. (2001). *Plantas medicinales de uso en Chile: Química y farmacología*. Editorial Universitaria.

Rai, M. K., Cordell, G. A., Martinez, J. L., Marinoff, M., & Rastrelli, L. (Eds.). (2012). *Medicinal plants: Biodiversity and drugs*. CRC Press. doi:10.1201/b12527

Varela, A., & Ibañez, J. (2010). *Medicinal plants: Classification, biosynthesis and pharmacology*. Nova Science Publishers, Incorporated.

Wiart, C. (2007). *Ethnopharmacology of medicinal plants: Asia and the Pacific*. Springer Science & Business Media.

Wiart, C. (2017). *Medicinal plants in Asia for metabolic syndrome: Natural products and molecular basis*. CRC Press. doi:10.1201/9781315177755

KEY TERMS AND DEFINITIONS

Alkaloids: A class of naturally occurring organic compounds that mostly contain basic nitrogen atoms. This group also includes some related compounds with neutral and even weakly acidic properties.

Atherosclerosis: A disease in which the inside of an artery narrows due to the buildup of plaque.

Diabetes: A group of metabolic disorders characterized by high blood sugar levels over a prolonged period.

Ethnobotany: The study of a region's plants and their practical uses through the traditional knowledge of the local culture and people.

Hypercholesterolemia: The presence of high levels of cholesterol in the blood. It is a form of hyperlipidemia, high blood lipids, and hyperlipoproteinemia (elevated levels of lipoproteins in the blood).

Hypertension: A long-term medical condition in which the blood pressure in the arteries is persistently elevated.

Hypoglycemia: Low blood sugar, when blood sugar decreases to below normal levels.

Polyphenols: A structural class of mainly natural, but also some synthetic or semisynthetic, organic chemicals characterized by the presence of large multiples of phenol structural units.

Terpenoids: A large and diverse class of naturally occurring organic chemicals derived from terpenes. Most are multicyclic structures with oxygen-containing functional groups.

Chapter 7
Phytochemistry and Antigenotoxic Properties of Six Ethnobotanically Important Members From the Family *Zingiberaceae*

Anish Nag
Christ University (Deemed), Bangalore, India

ABSTRACT

Genotoxicity is considered as a potential cause of various diseases including cancer. During the last decade, herbal extracts attained a great deal of attention due to its safe and effective applications against various DNA damaging agents. However, the mechanism of DNA strand breaks by various mutagens and genotoxins is often correlated with the generation of Reactive Oxygen Species (ROS). Herbal extracts constitute a number of phytochemicals and those are reported to have considerable antioxidant properties, which are in turn capable of neutralizing ROS mediated DNA damage. The botanical family Zingiberaceae is reported to have significant antioxidant and antigenotoxic potential by various researchers. Among a number of species belonging to this family, six species, namely Alpinia galanga, A. zerumbet, Curcuma amada, C. caesia, Zingiber officinale, and Z. zerumbet, attract notable attention due to their remarkable ethnobotanical and medicinal importance. This chapter deals with phytochemical composition, antioxidant, and antigenotoxic properties of these six Zingiberaceous plant extracts.

INTRODUCTION

Considerable attention has been focused on dietary and medicinal phytochemicals that inhibit, reverse or retard diseases caused by oxidative and inflammatory processes. Among the various classes of phytochemicals, focus has been given on the anti-inflammatory and antioxidant properties of polyphenols found in various botanical agents. The family Zingiberaceae comprises of 53 genera and over 1,200 species. It is

DOI: 10.4018/978-1-7998-1320-0.ch007

considered to be the powerhouse of active phytochemicals. The rhizomes are used in traditional medicine and are consumed as condiment and spices. In this review, the phytochemical composition, antioxidant as well as cytotoxic and genotoxic properties of six important plants namely *Alpina galanga*, *A. zerumbet*, *Curcuma amada*, *C. caesia*, *Zingiber officinale,* and *Z. Zerumbet* of Zingiberaceae are discussed.

ETHNOBOTANICAL IMPORTANCE

Ethnobotanical importance of Zingiberaceae plants is acknowledged worldwide. In Japan, *A. zerumbet* is used as a staple diet along with rice. Rhizomes are used as to make seasonings and beverages (Victório 2011). On the other hand, *A. galanga* or galangal is traditionally used as a flavoring agent, anti-ringworm, tonic and vegetables (Daimei and Kumar, 2014). Ginger (*Z. officinale*) is utilized as traditional medicines against various ailments such as nausea, vomiting, asthma, cough, palpitation, inflammation, dyspepsia, loss of appetite, constipation, indigestion, pain, acute and chronic cough, common cold, fever, allergic rhinitis, sinusitis, bronchitis, respiratory troubles and headache (Grzanna, Lindmark & Frondoza, 2005; Khare 2007). *Z. zerumbet* (shampoo ginger) is commonly used as traditional medicine as cough, tonsil, constipation in the North-Eastern part of India (Devi, Das & Singh, 2017). *Curcuma* is another important member of Zingiberaceae. *C. amada* is also known as mango ginger due to its raw mango-like aroma. Mango ginger is extensively utilized as a stomachic, as to improve blood quality, carminative and topically for contusions and sprains (Jatoi, Kikuchi, Gilani & Watanabe, 2007). Similarly, *C. caesia* or black turmeric is used extensively as to treat smooth muscle relaxant activity, Hemorrhoids, Leprosy, Asthma, Cancer, Epilepsy, Fever, Wound, Vomiting, Menstrual disorder, Antihelmentic, Aphrodisiac, Inflammation, Gonorrhoeal discharges, etc. (Sasikumar 2005; Karmakar, Dolai, Saha, Sarkar, Bala & Haldar, 2011).

PHYTOCHEMICAL COMPOSITION

Phytochemical compounds are ubiquitous in the plant kingdom and studies on such compounds indicate their inherent pharmaceutical and nutraceutical values. The botanical family Zingiberaceae is reported to be the powerhouse of such chemicals. Various classes of Phyto-compounds such as phenolics, flavonoids, terpenoids, and curcuminoids are reported from the important members of this family such as *A. galanga*, *A. zerumbet*, *C. amada*, *C. caesia*, *Z. officinale,* and *Z. zerumbet.*

These phytochemicals especially the content of phenolics vary broadly depending on the plant sources, climate, geography, their parts or the procedures of extraction, polar solvents such as ethanol, methanol, water and non-polar solvents such as hexane, dichloromethane are the popular choices for the extraction of bioactive compounds from the plants. Polar solvents are reported to yield a comparatively higher amount of phenolic substances. For example, ethanolic extract of *A. galanga* rhizome contained the highest concentration of total phenolic compounds as 31.49 mg gallic acid equivalent (GAE)/g) and flavonoids as 13.78 mg catechin equivalent (CE)/g. The water extract contained the total phenolic content of 8.25 mg GAE/g and the total flavonoid content of 1.48 mg CE/g. Further, the essential oil had a total phenolic as 5.01 mg GAE/g and total flavonoid as 0.20 mg CE/g (Mahae and Chaiseri, 2009). To add on, methanolic extracts of *C. amada* and *C. caesia* rhizomes are reported to have total phenolic content as 37.64 and 44.33 mg tannic acid equivalents (TAE)/g dry materials, respectively (Krishnaraj, Manibhushanrao & Mathivanan, 2010). Similarly, *Z. officinale* rhizome extract contained 95.2 mg/g of dry extract of total phenols and the hexane extract of fresh ginger had 87.5 mg/g dry extract (El-Ghorab, Nauman, Anjum, Hussain & Nadeem, 2010).

The mechanisms of bioactivity of plant extracts are often complicated due to their complex mixture of compounds. To address their proper composition and bio-function principles, various authors isolated and identified several classes ranging from phenolics to terpenoids from various plant sources.

Phenolics

Phenolics, especially phenolic acids are sourced from different parts of plants. The hexane extract of fresh rhizomes of *A. zerumbet* and ethyl acetate extract of leaves are rich in phenolics like ferulic and p-hydroxybenzoic acid. Further, the HPLC analysis showed that p-hydroxybenzoic, ferulic and syringic acids were the major phenolics in the ethyl acetate extract of flowers, whilst p-hydroxybenzoic, syringic acids were the predominant phenolics in the seeds of *A. zerumbet* (Elzaawely, Xuan & Tawata 2007). The free phenolics of the mango ginger (*C. amada*) contained caffeic (26%), gentisic (24%), ferulic (20%), gallic (10%), cinnamic (7%) and protocatechuic (7%) acids. Bound phenolics of *C. amada* whereas, contained a significant amount of ferulic (47%), cinnamic (29%), p-coumaric (11%), and syringic (5%) acids as major phenolic acids (Siddaraju and Dharmesh, 2007). It was also found that syringic (38%), gallic (18%) and cinnamic (14%) acids were the major ones in the free phenolic fractions of *Z. officinale* rhizomes. Cinnamic (48%), p-coumaric (34%) and caffeic (6%) acids whereas, were identified in the hydrolyzed phenolics.

Kaempferol, an important flavonoid is reported to have significant antioxidant and biological properties. Chemically, this compound is characterized as a flavonol under the broad group of flavonoids. Various authors isolated and identified kaempferol and its glycosides as major flavonoids mostly from leaves and rhizomes of Zingiberaceous plants. Kaempferol-3-O-rutinoside and kaempferol-3-O-glucuronide were isolated from the aqueous extract of *A. zerumbet* leaves (Mpalantinos, Soares de Moura, Parente & Kuster 1998). Further, kaempferol-3-O-glucuronide was also detected in the distilled water and 70% ethanol extracts of the leaves of *A. zerumbet*. The relative concentration of kaempferol-3-O-glucuronide was dependent on the method of extraction. The content of kaempferol-3-O-glucuronide was higher when extracted in the microwave (6.64 mg/ g dried leaves) than extracted by ultrasonic (5.62 mg /g dried leaves) assisted method (Victorio, Leitão & Lage, 2009).

Z. zerumbet is a rich source of kaempferol derivatives. Kaempferol-3, 4¢-O-dimethyl ether,kaempferol-3-O-methyl ether,kaempferol-3-O-(2,4-O-diacetyl-α-l-rhamnopyranosid),kaempferol-3-O-(2,4-O-diacetyl-α-l-rhamnopyranoside) and kaempferol-3-O-(3,4-O-diacetyl-α-l-rhamnopyranoside) were isolated from the chloroform-soluble part of the extract of the fresh rhizomes of *Z. zerumbet* (Nakatani, Jitoe, Masuda & Yonemori, 1991).

Chloroform fraction of *Z. zerumbet* rhizomes also gave other six kaempferol derivatives namely kaempferol-3,4¢,7-O-trimethyl ether, kaempferol-3-O-methyl ether, kaempferol-3,4¢-O-dimethyl ether, kaempferol-3-O-(4-O-acetyl-α-L-rhamnopyranoside), kaempferol-3-O-(2,4-O-diacetyl-α-L-rhamnopyranoside) and kaempferol-3-O-(3,4-O-diacetyl-α-L-rhamnopyranoside) (Jang, Han, Jhon & Seo, 2004). Alternatively, other extraction solvents also yielded a good number of kaempferol glycosides. The antioxidant active fraction from the aqueous extract *Z. zerumbet* rhizomes led to isolation of five components, including kaempferol 3-O-rhamnoside, compound 5 [kaempferol 3-O-(2^2-O-acetyl) rhamnoside or kaempferol 3-O-(3^2-O-acetyl) rhamnoside (5b)], kaempferol 3-O-(4^2-O-acetyl) rhamnoside, kaempferol 3-O-(3^2,4^2-O-diacetyl) rhamnoside and kaempferol 3-O-(2^2,4^2-O-diacetyl) rhamnoside (Ruslay et al., 2007).

Subsequently, 3-O-methyl kaempferol, kaempferol-3-O-(2,4-di-O-acetyl-α-l-rhamnopyranoside) and kaempferol-3-O-(3,4-di-O-acetyl-α-l-rhamnopyranoside) were isolated from the ethyl acetate fraction of methanolic extract *Z. zerumbet* rhizomes (Chien, Chen, Lee & Wang, 2008). Apart from kaempferol, few other phenolics were also identified from the various plant extracts, such as gallic acid glycoside (Galangogalloside) from methanolic (Jaju, Indurwade, Sakarkar, Fuloria & Ali, 2009), myricetin from the aqueous and catechin from the ethanolic extracts of the *A. galanga* rhizome (Mahae and Chaiseri, 2009), rutin from 70% ethanolic extract (Victório, Alviano Alviano & Lage, 2009) and catechin and (−)-epicatechin from the aqueous extract (Mpalantinos, Soares de Moura, Parente & Kuster, 1998) of *A. zerumbet* leaves and vanillin from chloroform fraction of *Z. zerumbet* rhizomes.

Although, phenolics are one of the most known groups of phytochemicals, however, various other important groups such as phenylpropanoids, gingerol related compounds, curcuminoids, etc are also present in the plant system.

Phenylpropanoids

Numerous phenylpropanoids and their derivatives were isolated and identified from *A. galanga*. 1'-acetoxychavicol acetate, acetoxyeugenol acetate were isolated from the hexane, chloroform, and ethyl acetate extracts of *A. galanga* rhizome (Kaur, Singh, Dey, Sharma, Bhutani, & Singh, 2010). N-pentane/diethyl ether extract of dried rhizomes contained 1'-Acetoxychavicol acetate, 1'-acetoxyeugenol acetate and 1'-hydroxychavicol acetate (Janssen and Scheffer, 1985). Seeds of *A. galanga* were also found to be rich in 1'-acetoxychavicol acetate and 1'-acetoxyeugenol acetate (Mitsui, Kobayashi, Nagahori & Ogiso, 1976). In agreement, many researchers such as Yang and Eilerman (1999), Matsuda, Pongpiriyadacha, Morikawa, Ochi, & Yoshikawa (2003) and Xiao-liu, Ming-Hua, Jian-Guang, Huang & Ling-Yi (2009) demonstrated *A. galanga* as a rich source of phenylpropanoids.

Curcuminoids (Diarylheptanoids)

Curcuminoids are the diarylheptanoid class of compounds predominantly found in *Curcuma* sp, although not limited to. Curcuminoids comprising curcumin, demethoxy curcumin, and bis-demethoxy curcumin were identified in the acetone extract of *C. amada* rhizomes (Gupta, Gupta, & Kumar, 1999). Further, all three types of curcuminoids were also detected in the methanolic extract of *C. caesia* rhizomes. However, the methanolic extract of leaves contained only curcumin, not the other analogs (Behar et al., 2013). In contrast, the absence of curcumin in similar extraction procedure was reported elsewhere (Lawand & Gandhi, 2013), indicating the changes in chemical constituents due to the geographical and seasonal variations between similar species. Apart from *Curcuma* sp, the presence of curcuminoids was also recorded from the other species of Zingiberaceae. Curcumin was isolated from the chloroform extract of *Z. zerumbet* rhizomes (Nakatani, Jitoe, Masuda & Yonemori, 1991).

Ethanolic extract of *Z. zerumbet* rhizome spent was found to be rich in demethoxy curcumin, and bis-demethoxy curcumin (Sreevani, Puranaik, & Naidu, 2013). A trace amount of curcumin was also detected in the acetone extract of *A. galanga* rhizomes (Jitoe et al., 1992). Other diaryl compounds such as diarylheptenones like gingerenone A, gingerenone B, isogingerenone B, and gingerenone C had been reported from *Z. officinale* (Endo, Kanno, & Oshima, 1990).

Gingerol Related Compounds

Oleoresin of *Z. officinale* was reported to be rich in gingerol related compounds. Oleoresin of ginger rhizome contained several such compounds such as gingerols, methylgingerols, gingerol acetates, shogaols, paradols, gingerdiols, mono- and diacetyl gingerdiols, and dehydrogingerdiones (Jiang, Sólyom, Timmermann & Gang, 2005). Dichloromethane extract of *Z. officinale* rhizomes was reported to contain a similar mixture of gingerol related compounds (Jolad, Lantz, Chen, Bates & Timmermann, 2005). Various observations often indicated 6 gingerol as one of the major components of *Z. officinale* rhizomes (Agarwal, Walia, Dhingra, & Khambay, 2001; Zancan, Marques, Petenate, & Meireles, 2002; Rai et al., 2006).

Volatiles (Terpenes and Terpenoids, Diarylheptanoids)

Essential oils are the volatile aromatic constituents of the plants. These oils find its uses from antioxidants to insect repellants. Typically, essential oils consist of terpenoids (as well as terpenes), an isoprenoid class of compounds. The plants from the family Zingiberaceae harbor a vast range of terpenoids such as monoterpenoids, sesquiterpenoids, etc. in their essential oil composition along with some other compounds such as diarylheptanoids.

A. galanga is extensively studied for its essential oil as well as aromatic constituents. Terpinen-4-ol was found to be the major component of the fresh and dried essential oils of *A. galanga* rhizomes (Janssen and Scheffer, 1985). Cineole is an aromatic monoterpenoid and abundantly available among the essential oils. Four isomers of acetoxy cineoles, (trans and cis)-2- and 3-acetoxy-1, 8-cineole were isolated from the rhizomes of *A. galanga*. The concentration of trans-3-acetoxy-1,8-cineole was found to be highest among the isomers (Kubota, Nakamura, Kobayashi, & Amaike, 1998). Possible precursors of acetoxy-1,8-cineoles namely hydroxy-1,8-cineole glucopyranosides, trans-2-hydroxy-1,8-cineole β-D-glucopyranosides, and trans-3-hydroxy-1,8-cineole β-D-glucopyranoside were isolated from the essential oils of *A. galanga* (Someya, Kobayashi, & Kubota, 2001). Various authors reported diverse terpenoid entities from its essential oils. limonene, 1,8-cineole, camphor, α-terpineol, α-fenchyl and (E)-methyl cinnamate (Mallavarapu et al., 2002), sesquiterpenoids, β-bisabolene (8.4%), (Z,E)-farnesol (3.8%), β-caryophyllene (3.6%) and (E)- β-farnesene (3.2%) (Jantan, Ahmad, & Ahmad, 2004) and sixteen compounds including zerumbone (44.8%) (Arambewela, Arawwawala, Owen & Jarvis, 2007) were detected as the major ones from the essential oils of *A. galanga* rhizome.

Further, two skeleton diterpenes (galangal A and B) and a new labdane type diterpene (Morita and Itokawa, 1986), sesquiterpenoids with the major ones being β-bisabolene (37.6%), (E)- β-farnesene (22.7%), (E,E)-farnesyl acetate (7.9%), (Z,E)-farnesol (3.9%) and β-caryophyllene (3.0%), Matsuda, Pongpiriyadacha, Morikawa, Ochi, & Yoshikawa, 2003) from seed oils and pinene (6.6%), camphene (5.0%), β-pinene (21.5%), 1,8-cineole (34.4%) and camphor (7.8%) from leaf oils (Mallavarapu et al., 2002) of *A. galanga* were also identified.

Different parts of *A. zerumbet* presented a large variety of terpenoid content. Seed oils of this plant constituted a new labdane-type diterpene, named zerumin (Xu, Hui & Sim,1995), two new labdane-type diterpenes, zerumin A and zerumin B, together with two known compounds, (E)-15,16-bisnorlabda-8(17),11-diene-13-one and coronarin E (Xu, Dong, & Sim, 1996), cadinol, T-muurolol, α-terpineol, δ-cadinene and terpinene-4-ol (Elzaawely, Xuan & Tawata 2007). Monoterpene constituent represented 95% of the essential oil of *A. zerumbet* leaves. p-Cymene was the most abundant compound, followed by 1,8-cineole, terpinen-4-ol, alpha-pinene, beta-pinene, and limonene. The amount of sesquiterpene content of the essential oil of *A. zerumbet* was small, mostly represented by beta-caryophyllene and alpha-caryophyllene (Murakami, Li, Matsuura, Satou, Hayashi & Koike, 2009). Further, the leaf oils of *A. zerumbet* revealed a high content of oxygenated monoterpenes (52.5%), terpinen-4-ol, 1,8 cineole, γ-terpinene (Victório, Alviano Alviano & Lage, 2009) and also alpha-pinene, p-cymene, 1,8-cineole, and limonene (Satou, Murakami, Matsuura, Hayashi & Koike, 2010) as the major constituents. Seasonal variation was also noted in the leaf oil of *A. zerumbet*. In April, the main constituents were found as terpinen-7-al (40.5%) and sabinene hydrate (15.4%) while in August, the major components were terpinen-4-ol (29.4%) and 1,8-cineole (23.1%) (Victório, Leitão, & Lage, 2010). Besides, few terpenoids like 1,8-Cineol, camphor, methyl cinnamate, borneol, and α-cadinol were also identified in the essential oil of *A. zerumbet* flowers.

The genus *Curcuma* contains a significant amount of volatiles. The steam-distilled oil obtained from the fresh rhizomes of *C. amada* was reported to contain 26 components, of which 24, comprising 96.1% of the oil, were completely identified. The oil contained 10 monoterpenes and 14 sesquiterpenes accounting for 12.2% and 83.9%, respectively. The predominant monoterpene identified was camphor (5.5%). The sesquiterpene fraction consisted mainly of hydrocarbons (ca 83.6%) and the prominent hydrocarbons were (Z)-β-farnesene (21.9%), guaia-6, 9-diene (19.8%), α-longipinene (14.8%) and α-guaiene (14.5%). The aromatic constituent contributing to the odor of the oil was thymol (4.9%) (Mustafa and Khan, 2005). In another study, essential oil of *C. amada* rhizomes was reported to contain ar-curcumene (28.1%), β-curcumene (11.2%), camphor (11.2%), curzerenone (7.1%) and 1,8-cineole (6.0%) (Srivastava, Srivastava, & Shah, 2001). Volatile rhizomes oil of *C. caesia* was found to contain 30 components, representing 97.48% of the oil, with camphor (28.3%), ar-turmerone (12.3%), (Z)-β-ocimene (8.2%), ar-curcumene (6.8%), 1,8-cineole (5.3%), β-elemene (4.8%), borneol (4.4%), bornyl acetate (3.3%) and γ-curcumene (2.82%) as the major constituents (Pandey and Chowdhury, 2003).

Leaf oils of *Curcuma* sp are also comprised of certain variations of terpenoids. For example, the leaf essential oil of *C. amada* was mainly composed of furano-sesquiterpenoids, namely epi-curzerenone (10.76%), curzerenone (9.53%), curzerene (3.95%) and furanogermenone (1.77%). Camphor (17.90%), isoborneol (7.30%), camphene (3.57%), borneol (1.87%) and camphene hydrate (1.25%) were the major constituents of 36.79% of the monoterpenoid fraction of leaf essential oil composition (Sheeja and Nair, 2012). Parallel to that, leaf oil of *C. caesia* was dominated by the 1, 8-cineole (27%) and camphor (16.8%) (Behura and Srivastava, 2004) along with the compound zederone (Asem & Laitonjam, 2012).

Z. officinale, also known ginger is extensively used as a spice in Indian as well as South Asian cuisines. Geranial (25.9%) was the major component of the essential oil of *Z. officinale* rhizome (Singh et al., 2008). In another study, the analysis of volatile oils of fresh and dried ginger rhizomes showed camphene, p-cineole, α-terpineol, zingiberene, and pentadecanoic acid as major constituents (El-Ghorab, Nauman, Anjum, Hussain, S., & Nadeem, 2010). Interestingly, non-terpenoid curcumene was reported as one of the major constituents in the ginger rhizome oil (Agarwal, Walia, Dhingra & Khambay, 2001).

In continuation with these efforts, in characterizing the composition of the essential oils from various plants of Zingiberaceae, several scientific attempts were also taken for *Z. zerumbet*. The oil of *Z. zerumbet* rhizome was found to be rich in oxygenated derivatives of α-humulene, in particular, zerumbone (65.3%) (Lechat-Vahirua, François, Menut, Lamaty, & Bessiere, 1993). The study revealed that essential oil *Z. zerumbet* rhizomes contained 36 compounds including curzerenone (14.4%), zerumbone (12.6%), camphor (12.8%), isoborneol (8.9%) and 1,8-cineole (7.1%), were found as major compounds (Srivastava, Srivastava, & Shah, 2001). Various authors indicated that zerumbone was the major compound in the essential oil of the *Z. zerumbet* rhizomes; however, variations in terpenoid classes were also notes. The rhizome oil obtained by Chane-Ming, Vera & Chalchat (2003) was rich in zerumbone (37%), α-humulene (14.4%) and camphene (13.8%). A new humulene sesquiterpene, 5-hydroxyzerumbone, and a known compound, zerumbone oxide were also isolated from the rhizome of *Z. zerumbet* rhizome oil (Jang and Seo, 2005). The south Indian species of *Z. zerumbet* reported containing 76.3–84.8% zerumbone content in their rhizome oils. The Malaysian accession recorded comparatively lower content of zerumbone (68.9%), whereas, high content of the bioactive compound zerumbone was detected from *Z. zerumbet* rhizome oil from Kerala (Baby et al., 2009). Thirty components were identified from *Z. zerumbet* rhizome oil of Bangladesh with the main components being in zerumbone (46.83%); a-caryophyllene (19.00%) and 1,5,5,8-tetramethyl-12-oxabicyclo[9.1.0]dodeca-3,7-dieneÂ (4.28%) (Bhuiyan, Chowdhury & Begum, 2008).

Similarly, in another study, thirty-three compounds, accounting for 93.6% of the oil including twenty compounds in minor (0.1-0.6%) and eight in trace (<0.05%) were identified from the essential oil of *Z. zerumbet* rhizomes. Consistently, the oil was rich in zerumbone (69.9%) along with alpha-humulene (12.9%), humulene epoxide II (2.5%), caryophyllene oxide (1.1%) and camphene (1.9%) (Rana, Verdeguer & Blazquez, 2012).

However, researchers have also given much attention to elucidating the composition of essential oils from other parts of *Z. zerumbet,* besides the rhizomes. Stem and leaf oils of *Z. zerumbet* contained 40 compounds, accounting for more than 84% and 82% of the oils, respectively. The major components of these oils appeared to be (Z)-nerolidol (16.8% and 22.3%, respectively), β-caryophyllene (10.4%, 11.2%), zerumbone (21.3%, 2.4%), and trans-phytol (7.0%, 12.6%). Predominant minor constituents included β-pinene (5.4%, 5.2%), α-humulene (2.5%, 2.9%), caryophyllene oxide (1.1%, 5.5%) and linalool (1.1%, 2.4%). The volatile flower oil was found to contain more than 60 components, of which 45 compounds, making up 85% of the oil. (Z)-nerolidol (36.3%) and β-caryophyllene (13.2%) were the major constituents (Dũng, Chính & Leclercq, 1995). The oils from leaves and flowers differed appreciably that from rhizomes by the presence of large amounts of (E)-nerolidol (21.4% and 34.9%), β-caryophyllene (6.9% and 10.2%), and linalool (7.7% and 17.1%), respectively (Chane-Ming, Vera & Chalchat, 2003). The root oil constituents of *Z. zerumbet* were reported to be (Z)-citral (26.1%), camphene (16.3%), sabinene (14.6%), zingiberene (7.2%) and lavandulyl acetate (6.7%). Unlike rhizomes, roots contained a low amount of zerumbone (1.2%) (Dai, Thang, Chau & Ogunwande, 2013).

Literature showed that the plants *A. galanga*, *A. zerumbet*, *C. amada*, *C. caesia*, *Z. officinale*, and *Z. zerumbet* are rich in diverse phytochemicals.

ANTIOXIDANT PROPERTIES

Free radicals are capable of damaging various biomolecules such as lipid, protein, DNA which in turn led into various diseases. Antioxidants play a crucial role by scavenging these free radicals through different mechanisms to prevent such diseases. Phytochemicals such as phenolics, terpenoids, etc are rich in antioxidant properties due to their typical chemical structures. The family Zingiberaceae especially the species, *A. galanga, A. zerumbet, C. amada, C. caesia, Z. officinale,* and *Z. zerumbet* harbor a wide range of antioxidant molecules and attain the key attention of the researchers.

The plant *Alpinia galanga* was extensively studied for its antioxidant properties. Various antioxidant assays involving an array of reaction principles such as free radical scavenging activity, inhibition of lipid peroxidation, reducing capacity, etc were vastly studied to understand the inherent mechanism. The rhizome of this plant, in particular, received much attention. Ethanolic extract of *A. galanga* rhizomes exhibited strong superoxide anion scavenging activity, Fe^{2+} chelating activity and reducing power in a concentration-dependent manner. Antioxidant activity of the extract correlated well with reducing power (Juntachote and Berghofer, 2005). Further, the antioxidant activity of ethanolic extract of *A. galanga* rhizomes was established by the β-carotene bleaching method as 70.3% (Mayachiew and Devahastin, 2008). Galangal extract using 50% ethanol in water was studied for its antioxidant activity and composition in comparison with two other samples based on a water extract and the essential oil. The antioxidant activities were determined using the 2, 2-diphenyl-1-picrylhydrazyl (DPPH) and oxygen radical absorbance capacity (ORAC) methods. The ethanolic extract showed the highest DPPH free radical scavenging ability as well as the highest ORAC value when compared to the water extract or the essential oil. The IC50 values of the galangal ethanolic extract (10.66 mg/ml), water extract (55.48mg/ml) and essential oil (455.43 mg/ml) were higher than those of α-tocopherol (1.45 mg/ml) and butylated hydroxyanisole (BHA; 0.41 mg/ml) (Mahae and Chaiseri, 2009). The essential oil of *A. galanga* rhizome was also noted to have significant DPPH and ABTS+ radical scavenging activities (Tachakittirungrod and Chowwanapoonpohn, 2007).

In a comparative study, among methanolic extract, non-polymeric phenolic fractions, and polymeric tannin fractions of leaves and rhizomes of *A. galanga*, polymeric tannin fraction of *A. galanga* rhizomes showed highest DPPH activity (Chan, Tan & Low, 2011). Water extracts from fresh and dried rhizomes and ethanolic extract from dried rhizomes of Malaysian variety of *A. galanga* exhibited higher DPPH scavenging activity than the standard antioxidant butylated hydroxytoluene (BHT) (Pillai et al., 2012). Within the concentration range of 0.05 to 0.5 mg/ml, the DPPH, superoxide anion and hydroxyl free radical scavenging activities of the rhizome extract of *A. galanga* were found to be 94.4, 80.7, and 16.7% respectively (Wang, Niu & Duan, 2011). DPPH and 2,2'-azinobis-(3-ethylbenzothiazoline-6-sulfonic acid) (ABTS+) radical scavenging properties of the water extract of *A. galanga* rhizome with the IC50 values of 21.26 and 25.54 µg/ml was also confirmed (Nampoothiri, Esakkidurai & Pitchumani, 2015).

Literature study showed that the extracts from *A. galanga* possessed significant lipid peroxidation inhibition activities. Lipids are the key ingredients of the cell membrane and vulnerable to the free radicals attack. Oxidation of lipids may thus result in cellular death. Malondialdehyde (MDA) is the end-product of lipid peroxidation and the production of this aldehyde is used as a biomarker to measure the level of oxidative stress in an organism. MDA reacts with TBA (Thiobarbituric acid) to form thiobarbituric acid reactive substances (TBARS) and enables the measurement of the MDA content in an oxidized lipid system. The effectiveness of dried *A. galanga* rhizome powder and its ethanolic extract to inhibit lipid oxidation of cooked ground pork during storage at 5°C for 14 days were studied by using parameters like thiobarbituric acid-reactive substances (TBARS) value, peroxide value (POV), conjugated diene and hexanal content. Dried powder of concentrations 0.05%, 0.10% and 0.15% w/w and its ethanolic extract with 0.17%, 0.43% and 0.51% w/w significantly inhibited the formation of TBARS value, POV, conjugated diene and hexanal content in a dose-dependent manner. Dried powder, however at a concentration of 0.51% was the most effective. At the end of storage, TBARS values, POV, conjugated diene and hexanal content were 70.5%, 66.8%, 22.7%, and 89.7%, respectively, less than the untreated sample. Addition of dried galangal powder to cooked ground pork was more effective than its ethanolic counterpart in controlling lipid peroxidation (Juntachote and Berghofer, 2005). Furthermore, the thiobarbituric acid (TBA) values of raw and cooked minced beef samples containing galangal extract at 5 and 10% (w/w) were significantly lower than those of the controls. In raw beef, galangal extract at 10% (w/w) was as effective as α-tocopherol at 0.10% (w/w) and butylated hydroxytoluene (BHT) at 0.02% (w/w) in inhibiting lipid oxidation (Cheah and Hasim, 2000). In a similar report, it was found that galangal powder could improve the oxidative stability of minced beef as measured by monitoring malonaldehyde formation, using the thiobarbituric acid reactive substances method. Galangal extract at higher concentrations of 0.05% and 0.10% (w/w) were also found to extend the shelf-life of minced beef (Cheah and Gan, 2000). However, the ethanolic extract of the *A. galanga* rhizomes showed comparatively weak inhibition of lipid oxidation activity than Holi basil extract when tested in β-carotene–linoleic acid emulsion system (Juntachote and Berghofer, 2005).

Various parts of *A. zerumbet* were found to have significant antioxidant properties. Ethyl acetate extract of leaves of *A. zerumbet* showed higher DPPH radical scavenging activity than that from rhizomes. Ethyl acetate extract from wastewater of leaves possessed the strongest inhibition to β-carotene oxidation. Furthermore, the ethyl acetate extract of flowers and seeds possessed a high antiradical activity and prevented the bleaching of β-carotene (Elzaawely, Xuan & Tawata, 2007). Methanol extracts of *A. zerumbet* and *A. zerumbet variegata* leaves were evaluated for antioxidant activities. High DPPH and reducing activities were found in the leaves of *A. zerumbet* and *A. zerumbet variegata* (Wong, Lim & Omar, 2009). Ethanol, water and hexane extracts of *A. zerumbet* rhizomes showed comparable antioxidant activities (DPPH, ABTS+, and β-carotene/linoleic bleaching test) to that of the synthetic antioxidants TBHQ (Abo El-Maati, Labib, Al-Gaby & Ramadan, 2012).

Curcuma is an important genus from the family Zingiberaceae. Besides *Curcuma longa* (turmeric); *C. amada* (mango ginger) and *C. caesia* (black turmeric) are the two prominent species of *Curcuma* that are reported to possess strong radical scavenging properties. The antioxidant activity of ethanol and aqueous extracts of *C. amada* rhizomes was studied using various antioxidant assays such as DPPH, nitric Oxide and superoxide free radical scavenging assays (Sahu, Singh, & Roy, 2009). In DPPH assay the IC50 value of ethanolic extract of *C. amada* was 265.33 μg/ml and that of the aqueous extract was 217.90 μg/ml. Whereas, in superoxide scavenging assay the IC50 values were 526.87 μg/ml and 440.89 μg/ml for ethanolic and aqueous extracts. The extracts could inhibit NO moderately.

The antioxidant activities of the ethanolic extract of *C. amada* rhizomes were confirmed by phosphomolybdenam, ABTS+ and DPPH methods (Prema, Kamaraj, Achiraman, & Udayakumar, 2014). Oleoresins from mango ginger exhibited high DPPH radical scavenging activity and ferric reducing power (Rajamma, Bai, & Nambisan 2012). Mango ginger free phenolics and mango ginger bound phenolics had free radical scavenging with IC50 of 2.2± 0.17 and 4.2 ± 0.36 μg/ml, reducing the power of 193 and 104 units/g and inhibition of lipid peroxidation with IC50 of 10.3±0.91 and 15.6±1.6 μg/ml respectively (Siddaraju and Dharmesh, 2007). The aqueous extract of the *C. amada* rhizomes possessed a strong antioxidant activity with an IC50 value of 297.3 μg/ml on iron chelating capacity and IC50 value of 323.8 μg/ml on hydroxyl radical scavenging activity (Vishnupriya, Nishaa, Sasikumar, & Gopalakrishnan, 2012).

In a comparative study, it was found that the reducing power, superoxide, ABTS+ and DPPH radical scavenging activities of *C. caesia* were higher than *C. amada* (Krishnaraj, Manibhushanrao & Mathivanan, 2010). The methanolic extract of *C. caesia* rhizomes showed significant dose-dependent antioxidant activities such as scavenging of DPPH free radicals with IC50 value 94.03 ± 0.67 μg/ml, nitric oxide with IC50 value 155.59 ± 3.03 μg/ml, superoxide with IC50 value 68.10 ± 1.24 μg/ml, hydroxyl with IC50 value 21.07 ± 1.78 μg/ml, peroxynitrite with IC50 value 260.56 ± 12.65 μg/ml and hypochlorous acid with IC50 value 33.33 ± 0.52 μg/ml (Karmakar, Dolai, Saha, Sarkar, Bala & Haldar, 2011). Further, the ethanolic fraction of *C. caesia* had the highest antioxidant activity as revealed by DPPH assay with an IC50 value of 418 µg/ml for ethanolic extract when compared to the methanolic (441.90 µg/ml) and ethyl acetate extract (561 µg/ml) (Devi, Mazumder, & Devi, 2015).

Zingiber officinale (ginger) is widely used as a major spice across the South Asian realm. The antioxidant properties of the methanol extract from the leaves, stems, and rhizomes of two *Z. officinale* varieties (Halia Bentong and Halia Bara) were determined using DPPH radical scavenging and the ferric reducing/antioxidant potential (FRAP) assays. The antioxidant activity of the leaves as determined by the DPPH assay was higher than those of the rhizomes and stems. On the other hand, the FRAP activity of the rhizomes was higher than that of the leaves. At low concentration, the values of the leaves' inhibition activity in both varieties were significantly higher than or comparable to those of the young rhizomes. Halia Bara had higher antioxidant activities as in comparison with Halia Bentong (Ghasemzadeh, Jaafar & Rahmat, 2010). The concentration of Carbon dioxide (CO_2) was reported as a major contributor to influence the antioxidant capacities of ginger. As CO_2 concentration was increased from 400 to 800 µmol/mol, free radical scavenging power (DPPH) increased about 30% in Halia Bentong and 21.4% in Halia Bara in leaves; and the rhizomes exhibited more enhanced free radical scavenging power, with 44.9% in Halia Bentong and 46.2% in Halia Bara (Ghasemzadeh, Jaafar & Rahmat, 2010). The aqueous extracts from two varieties of ginger [red ginger (*Z. officinale* var. Rubra) and white ginger (*Z. officinale*) caused significant inhibition of Fe2+ induced lipid peroxidation in rat brain *in vitro* in a dose-dependent manner. However, the aqueous extract of red ginger had a significantly higher inhibitory effect than that of white ginger. This higher inhibitory effect of red ginger was attributed to its significantly higher phytochemical content, Fe^{2+} chelating ability, OH radical scavenging ability and reducing power (Oboh, Akinyemi & Ademiluyi, 2012).

Alcoholic extract of dried ginger reached 90.1% of DPPH radical scavenging activity and exceeded that of the standard BHT with the IC50 0.64 µg/ml. Temperature plays an important role in the antioxidant activities of the ginger extract. In a comparison between reaction temperatures (80°C and 37 °C), it was found that inhibition of lipid oxidation activity in a linoleic acid/water emulsion system, determined by means of thiobarbituric acid reactive substances (TBARS) was highest at 37°C – 73.2%, and 71.6% when the formation of conjugated dienes was inhibited i.e. at 80°C. At 80°C, the antioxidant activity at the highest concentration of a ginger extract was less efficient: 65.7% for conjugated dienes formation and 68.2% for TBARS. The ginger extract inhibited the hydroxyl radicals 79.6% at 37°C and 74.8% at 80 °C, which showed higher antioxidant activity than quercetin. The IC50 for inhibiting OH radicals at 37°C was slower than that at 80 °C – 1.90 and 2.78 µg/ml, respectively. The ginger extract also chelated Fe3+ in the solution (Stoilova, Krastanov, Stoyanova, Denev, & Gargova, 2007).

The dried ginger essential oil and fresh ginger essential oil had the DPPH activities of 83.87 ± 0.50 and 83.03 ± 0.54% respectively. The FRAP of essential oils showed almost comparative results with DPPH (El-Ghorab, Nauman, Anjum, Hussain & Nadeem, 2010). Further, the essential oil and oleoresin extracts (ethanol, methanol, CCl4, and isooctane) of *Z. officinale* also revealed significant antioxidant activity in mustard oil by peroxide, anisidine, thiobarbituric acid (TBA), ferric thiocyanate (FTC) and DPPH radical scavenging methods (Singh et al., 2008).

The DPPH scavenging activity of callus culture of *Z. zerumbet* rhizomes was concentration-dependent with an EC50 value of 254.55 µg (Stanly, Bhatt, Ali, Keng, & Lim, 2011). Similarly, the IC50 values for scavenging of DPPH radical and nitric oxide by the aqueous extract of *Z. zerumbet* rhizomes were 114.25 and 112.45 µg/ml respectively (Rout, Acharya & Mishra, 2011). The methanolic extract of *Z. zerumbet* rhizomes also possessed DPPH scavenging activity with IC50 value as 1659.70 µg/ml (Suksamerkun et al., 2013).

Taking into consideration the various antioxidant assays it was observed that the plants *A. galanga, A. zerumbet, C. amada, C. caesia, Z. officinale,* and *Z. zerumbet* were a potential source of natural antioxidants.

CYTOTOXICITY AND GENOTOXICITY/ANTIGENOTOXICITY

Cytotoxicity

Man since time immemorial has been using herbs or plant products as medicine for combating ailments like cold, joint pains, fever, etc. and building up immunity or resistance. The use of herbal products in amelioration of such common diseases is credited to the traditional systems of medicines. Plants belonging to the family Zingiberaceae are used in South East Asia for their medicinal properties as well as an herb in cooking. A majority of them has been exploited for their beneficial properties and therefore testing for their cytotoxicity is an important facet.

Crude aqueous extract of *Alpinia galanga* rhizome was tested for cytotoxicity, apoptosis and DNA damage in six different human cell lines (Muangnoi et al., 2007). The cell lines displayed a broad range of cytotoxicity. Exposure to as little as 100 µg/ml galangal extract generated a significant level of DNA strand breaks. The three major compounds in the aqueous extract identified were 1¢-acetoxychavicol acetate and its deacetylated derivatives. Aqueous extract of these compounds was non-cytotoxic to A549 human lung adenocarcinoma cells. Lee and Houghton (2005) reported a weak activity of galangal in human lung cancer cell line (COR L23) and breast cancer cell line (MC17). Methanolic extract of *A. galanga* seeds was found to be cytotoxic on lung cancer (H522), leukemia (K562), breast cancer (MCF-7/ADR) and prostate cancer (DU145) cell lines. A new compound, a diterpenoid was responsible for the cytotoxic activity. Yang Y.P. (2012) prepared nine extracts (Z1~Z9) from *Zingiber zerumbet* by refluxing with acetone, methanol or hot water on human oral carcinoma Cal27 cells. Results also showed that Z1, Z2, Z4, Z5, Z7, and Z8 induced apoptosis and inhibited the growth of Cal27 cells. Furthermore, Z1, Z2, Z4 and Z5 induced cell cycle arrest at the S phase, and Z7 induced cell cycle arrest at G2/M phase.

A. galanga was cytotoxic to human leukemic HL-60 and U937 cells due to the presence of 4'-hydroxycinnamaldehyde (4'-HCA) in a dose-dependent manner ($p<0.05$). The apoptotic death involved cytochrome c release, increase in Bax level and concomitant decreases in levels of Bcl-2 and Bcl-xL (Banjerdpongchai, Punyati, Nakrob, Pompimon, & Kongtawelert, 2011). *In vitro* monolayer of rat hepatocytes revealed that the LC50 of the aqueous extract of dried rhizome powder of *A. zerumbet* was more than 1000 µg/ml and that of the total methanolic extract of dried rhizome was 500 – 750 µg/ml in concentration compared to the control cells. The *in vitro* cytotoxicity study showed that the aqueous extract was toxic to U937 cells at a concentration of 100 µg/ml while it was safe on PBMC cells while the methanolic extract, was safe on both U937 and PBMC (Peripheral Mononuclear Blood Cells) cells at concentration 100 µg/ml and above (El-Harwary et al., 2013). The ethanolic extracts of *Alpinia galanga* exhibited excellent (100%) phytotoxic activity against *Lemna minor* and show antifungal activities against *Trichophyton longifusus* (60%). In the brine shrimp lethality bioassay, it was found to be toxic with an LD50 of 109 µg/ml (Khattak, Shah, Ahmad & Ahmad, 2005).

Acetone, methanol, ethanol and aqueous extracts of *C. amada* were assessed for anti-proliferative activity in BHK-21 (Baby Hamster Kidney 21) cells. Methanolic and ethanolic extracts showed better response than that of its aqueous and acetone extracts. The concentration of 10 mg/ml of ethanolic extract inhibited the cancerous cell growth. In BHK-21 cells, cell growth was inhibited within 24 h and the cytopathological changes involved clumping of cells, detachment of cells, flagging of cells and apoptosis (Joshi and Chauhan, 2013). Cytotoxicity properties of *C. amada* were confirmed in HeLa (Henrietta Lacks), DLA (Dalton Lymphoma Ascites) cells (Prema, Kamaraj, Achiraman & Udayakumar, 2014).

Methanol extracts, water extracts and volatile oils of the fresh rhizomes of *A. galanga* and *Zingiber officinale* have been assessed for cytotoxic activity against MCF7 (breast adenocarcinoma) and LS174T (colon adenocarcinoma) cell lines. Cytotoxicity of *A. galanga* varied in the two cell lines (Zaeoung, Plubrukarn & Keawpradub, 2005). Five compounds were isolated from the chloroform-soluble fraction of the methanolic extract of the dried rhizomes of *Zingiber officinale* (Zingiberaceae). Among them, 6-shogaol exhibited the most potent cytotoxicity against human A549, SK-OV-3, SK-MEL-2, and HCT15 tumor cells (Kim et al., 2008).

Twenty-nine phenolic compounds were isolated from the root bark of fresh (Yunnan) ginger and their structures were fully characterized by Peng et al. (2012). Three compounds, 6-shogaol, 10-gingerol and an enone-diarylheptanoid analog of curcumin (compound 6) were identified to be cytotoxic in nine cell lines tested, with KB and HL60 cells most susceptible to 6-shogaol and the curcumin analog with IC50 < 10 μM. QSAR analysis revealed cytotoxicity was related to compound lipophilicity and chemical reactivity. In a study conducted by Abdullah et al. (2010) ginger extract makes use of its antitumor effects on colon cancer cells by suppressing its growth, arresting the G0/G1 -phase, reducing DNA synthesis and inducing apoptosis. [6]-Gingerol treatment exhibited considerable cytotoxicity as indicated by growth inhibition of A431 cells mediated via generation of reactive oxygen species (ROS) (Nigam, Bhui, Prasad, George & Shukla, 2009).

The rhizome of *Z. zerumbet* was cytotoxic to *Artemia salina* (Hossain, Kader, Nikkon & Yeasmin, 2012). It was found to have larvicidal activity on *Aedes aegypti* and *Anopheles nuneztovari* mosquitoes (Bucker, Falcao-Bucker, Nunez, Pinheiro & Tadei, 2013).

In general, the rhizomes of the plants *A. galanga*, *A. zerumbet*, *C. amada*, *C. caesia*, *Z. officinale*, and *Z. zerumbet* exhibited cytotoxicity to different cells.

Genotoxicity/ Anti Genotoxicity

Genotoxicity describes the property of chemical agents that damages the genetic information within a cell causing mutations, which may lead to cancer. DNA damage induces mutation, inhibits replication that eventually leads to the lethal collapse of cellular machinery as well as various inheritable genetic disorders. Plant extracts, rich in antioxidants are capable of inhibiting such genetic damages and thus demand elaborate research. The genotoxic and anti-genotoxic properties of plants belonging to the family Zingiberaceae are scarce.

Crude aqueous extract of *A. galanga* rhizome was found to induce DNA damage in six different human cell lines including normal and p53-inactive fibroblasts, normal epithelial and tumor mammary cells and a lung adenocarcinoma cell line (Muangnoi et al., 2007). In a clinical trial, healthy volunteers were administered with *A. zerumbet* tea for 28 consecutive days and DNA damage was evaluated by comet assay in human peripheral lymphocyte cells. There was no evidence of DNA damage in the volunteers (Santana, 2009).

Cavalcanti et al., (2012) carried extensive study on the effects of the essential oil (EO) of *A. zerumbet* leaves on human leukocytes. There was the absence of DNA damage at concentrations (50–300 μg/ml), and at the concentration 500 μg/ml, increased DNA damage was observed. *In vivo* experiments in mice, EO (400 mg/kg) failed to show genotoxicity in the bone marrow and peripheral blood cells. Co-administration of the essential oil (EO) of *A. zerumbet* leaves with hydrogen peroxide and post-treatment of EO decreased the frequency of H2O2- induced DNA strand breaks. However, a synergistic effect was observed *in vitro* experiments. The result suggested that EO at concentrations up to 300 µg/ml or doses up to 400 mg/kg were not genotoxic to leukocytes and bone marrow cells of mice, but could ameliorate the DNA damage induced by H_2O_2 (Cavalcanti et al., 2012).

Antimutagenic properties of *Curcuma caesia* rhizomes as ethanol, methanol and ethyl acetate extracts were tested against indirect-acting mutagen cyclophosphamide (CP) using *Salmonella typhimurium* strains TA98 and TA 100. All the extracts showed the antimutagenic effect that was dose-dependent (Devi, Mazumder & Devi, 2015). Male Swiss albino mice were treated with crude aqueous extract and essential oil of *Z. officinale* rhizome (ginger) at doses of 0.5, 1, 2, 5, 10 g/kg body weight and 0.625, 1.25 and 2.50 ml/kg body weight respectively. Ginger oil in comparison with the aqueous extract gave a higher frequency of chromosomal aberrations in bone marrow cells of the mice (Mukhopadhyay and Mukherjee, 2000). In another study, it was found that ginger was non-genotoxic, and could not alter the DNA damage levels induced by the N-butyl-N-(4-hydroxibutyl) nitrosamine (BBN)/N-methyl-N-nitrosourea (MNU) in Swiss albino mice (Bidinotto, Spinardi-Barbisan, Rocha, Salvadori & Barbisan, 2006). Ethanol extract of *Z. officinale* rhizome could partially alleviate H_2O_2-induced changes in pBR-322 plasmid and Fenton reaction-induced DNA fragmentation in rat liver (Ajith, 2010). The antigenotoxic effect of ginger was reported on paracetamol-induced genotoxicity in the liver and renal tissues of male rats (Salah, Abdouh & Booles, 2012); against anti-inflammatory drug (tenoxicam®) in *Saccharomyces cerevisia* strain D7 (Mutawakil, 2012). Ethanol extract of *Z. zerumbet* rhizome (150–5000 µg per plate) did not show any increase in the number of revertant colonies, regardless of the metabolic activation system (S-9 mix) used in the histidine-dependent auxotrophic mutants of *Salmonella typhimurium* (strains TA97, TA98, TA100, TA102, and TA1535). At concentrations - 150–5000 µg/m1 the number of structural aberrations in Chinese hamster lung cells did not increase in presence or absence of S-9 mix. An oral administration of the extract to ICR mice, with doses of up to 2000 mg/kg, caused no significant increases in the number of micronucleated polychromatic erythrocytes (MNPCEs) and mean ratio of polychromatic erythrocytes to total erythrocytes (Chang, Tzeng, Liou, Chang & Liu, 2012).

FUTURE RESEARCH DIRECTIONS

The future scope of research for Zingiberaceae is extensive and diverse. Metabolic profiling and phytochemical fingerprinting for all medicinal plants including Zingiberaceae is one of the emerging areas and expected to be highly correlated with plants' pharmacological properties. Tissue culture-based study is another thrust area which needs to be more exploited in case of Zingiberaceae, few members of which are still facing difficulties in growing at *in vitro* condition due to its rich polyphenolic content.

CONCLUSION

The members of the botanical family Zingiberaceae have significant antioxidant and antigenotoxic properties as well as rich in bioactive molecules. The presence of such molecules can be well correlated with their pharmacological properties as discussed by various scientists across the globe. The plant members, discussed here are of remarkable ethnobotanical importance and vastly used by various ethnic populations especially in the southern part of Asia. Further research can, therefore, explore the expanded horizon of medicinal values of these plants.

REFERENCES

Abdullah, S., Abidin, S. A. Z., Murad, N. A., Makpol, S., & Ngah, W. Z. W., & Yusof, Y. A. M. (2010). Ginger extract (*Zingiber officinale*) triggers apoptosis and G0/G1 cells arrest in HCT 116 and HT 29 colon cancer cell lines. *African Journal of Biochemistry Research*, *4*(4), 134–142.

Abo El-Maati, M. F., Labib, S. M., Al-Gaby, A. M. A., & Ramadan, M. F. (2012). Antioxidant properties of different extracts from five medicinal plants. *Zagazig Journal of Agricultural Research*, *39*(4), 1–13.

Agarwal, M., Walia, S., Dhingra, S., & Khambay, B. P. S. (2001). Insect growth inhibition, antifeedant and antifungal activity of compounds isolated/derived from *Zingiber officinale* Roscoe (ginger) rhizomes. *Pest Management Science*, *57*(3), 289–300. doi:10.1002/ps.263 PMID:11455660

Ajith, T. A. (2010). Ameliorating reactive oxygen species-induced in vitro lipid peroxidation in brain, liver, mitochondria and DNA damage by *Zingiber officinale* Roscoe. *Indian Journal of Clinical Biochemistry*, *25*(1), 67–73. doi:10.100712291-010-0014-1 PMID:23105887

Arambewela, L. S., Arawwawala, M., Owen, N. L., & Jarvis, B. (2007). Volatile oil of *Alpinia galanga* willd. of Sri Lanka. *The Journal of Essential Oil Research*, *19*(5), 455–456. doi:10.1080/10412905.2007.9699950

Asem, S. D., & Laitonjam, W. S. (2012). Investigation of the structure-nonlinearity relationship of zederone from the rhizomes of *Curcuma caesia* Roxb. *Indian Journal of Chemistry-Part B Organic Including Medicinal, 51*(12), 1738-1742.

Baby, S., Dan, M., Thaha, A. R., Johnson, A. J., Kurup, R., Balakrishnapillai, P., & Lim, C. K. (2009). High content of zerumbone in volatile oils of *Zingiber zerumbet* from southern India and Malaysia. *Flavour and Fragrance Journal*, *24*(6), 301–308. doi:10.1002/ffj.1940

Banjerdpongchai, R., Punyati, P., Nakrob, A., Pompimon, W., & Kongtawelert, P. (2011). 4¢-Hydroxycinnamaldehyde from *Alpinia galanga* (Linn.) induces human leukemic cell apoptosis via mitochondrial and endoplasmic reticulum stress pathways. *Asian Pacific Journal of Cancer Prevention*, *12*(3), 593–598. PMID:21627350

Behar, N., Tiwari, K. L., Jadhav, S. K., Vairappan, C. S., Elias, U. M., Ramachandram, T. R., ... Haldar, P. K. (2013). Comparative phytochemical screening of bioactive compounds in *Curcuma caesia* Roxb. and *Curcuma longa*. *Research Journal of Medicinal Plant*, *7*(2), 113–118. doi:10.3923/rjmp.2013.113.118

Behura, S., & Srivastava, V. K. (2004). Essential oils of leaves of *Curcuma* species. *The Journal of Essential Oil Research*, *16*(2), 109–110. doi:10.1080/10412905.2004.9698664

Bhuiyan, M. N. I., Chowdhury, J. U., & Begum, J. (2008). Chemical investigation of the leaf and rhizome essential oils of *Zingiber zerumbet* (L.) Smith from Bangladesh. *Bangladesh Journal of Pharmacology*, *4*(1), 9–12. doi:10.3329/bjp.v4i1.845

Bidinotto, L. T., Spinardi-Barbisan, A. L. T., Rocha, N. S., Salvadori, D. M. F., & Barbisan, L. F. (2006). Effects of ginger (*Zingiber officinale* Roscoe) on DNA damage and development of urothelial tumors in a mouse bladder carcinogenesis model. *Environmental and Molecular Mutagenesis*, *47*(8), 624–630. doi:10.1002/em.20248 PMID:16878317

Bucker, A., Falcao-Bucker, N. C., Nunez, C. V., Pinheiro, C. C. D. S., & Tadei, W. P. (2013). Evaluation of larvicidal activity and brine shrimp toxicity of rhizome extracts of *Zingiber zerumbet* (L.) Smith. *Revista da Sociedade Brasileira de Medicina Tropical*, *46*(3), 377–380. doi:10.1590/0037-8682-1309-2013 PMID:23856868

Cavalcanti, B. C., Ferreira, J. R., Cabral, I. O., Magalhães, H. I., de Oliveira, C. C., Rodrigues, F. A., ... Moraes, M. O. (2012). Genetic toxicology evaluation of essential oil of *Alpinia zerumbet* and its chemoprotective effects against H2O2-induced DNA damage in cultured human leukocytes. *Food and Chemical Toxicology*, *50*(11), 4051–4061. doi:10.1016/j.fct.2012.03.038 PMID:22446814

Chan, E. W., Ng, V. P., Tan, V. V., & Low, Y. Y. (2011). Antioxidant and antibacterial properties of *Alpinia galanga, Curcuma longa*, and *Etlingera elatior* (Zingiberaceae). *Pharmacognosy Journal*, *3*(22), 54–61. doi:10.5530/pj.2011.22.11

Chane-Ming, J., Vera, R., & Chalchat, J. C. (2003). Chemical composition of the essential oil from rhizomes, leaves and flowers of *Zingiber zerumbet* Smith from Reunion Island. *The Journal of Essential Oil Research*, *15*(3), 202–205. doi:10.1080/10412905.2003.9712114

Chang, C. J., Tzeng, T. F., Liou, S. S., Chang, Y. S., & Liu, I. M. (2012). Absence of genotoxic and mutagenic effects of *Zingiber zerumbet* (L.) Smith (Zingiberaceae) extract. *Evidence-Based Complementary and Alternative Medicine*, *2012*, 1–7. PMID:22844331

Cheah, P. B., & Abu Hasim, N. H. (2000). Natural antioxidant extract from galangal (*Alpinia galanga*) for minced beef. *Journal of the Science of Food and Agriculture*, *80*(10), 1565–1571. doi:10.1002/1097-0010(200008)80:10<1565::AID-JSFA677>3.0.CO;2-7

Cheah, P. B., & Gan, S. P. (2000). Antioxidative/antimicrobial effects of galangal and α-tocopherol in minced beef. *Journal of Food Protection*, *63*(3), 404–407. doi:10.4315/0362-028X-63.3.404 PMID:10716573

Chien, T. Y., Chen, L. G., Lee, C. J., Lee, F. Y., & Wang, C. C. (2008). Anti-inflammatory constituents of *Zingiber zerumbet*. *Food Chemistry*, *110*(3), 584–589. doi:10.1016/j.foodchem.2008.02.038

Dai, D. N., Thang, T. D., Chau, L. T., & Ogunwande, I. A. (2013). Chemical Constituents of the Root Essential Oils of *Zingiber rubens* Roxb., and *Zingiber zerumbet* (L.) Smith. *American Journal of Plant Sciences*, *4*(1), 7–10. doi:10.4236/ajps.2013.41002

Daimei, P., & Kumar, Y. (2014). Ethnobotanical uses of gingers in Tamenglong district, Manipur, Northeast India. *Genetic Resources and Crop Evolution*, *61*(1), 273–285. doi:10.100710722-013-0054-5

Devi, H. P., Mazumder, P. B., & Devi, L. P. (2015). Antioxidant and Antimutagenic activity of *Curcuma caesia* Roxb. rhizome extracts. *Toxicology Reports*, *2*, 423–428. doi:10.1016/j.toxrep.2014.12.018 PMID:28962377

Devi, N. B., Das, A. K., & Singh, P. K. (2017). Ethnobotanical importance of *Zingiber* in the Imphal Valley District, Manipur with *Zingiber chrysanthum* Roscoe as a new record for Manipur. *Journal of Pharmaceutical and Biological Sciences*, *12*(3), 29–32.

Dũng, N. X., Chính, T. D., & Leclercq, P. A. (1995). Chemical investigation of the aerial parts of *Zingiber zerumbet* (L.) Sm. from Vietnam. *The Journal of Essential Oil Research*, *7*(2), 153–157. doi:10.1080/10412905.1995.9698490

El-Ghorab, A. H., Nauman, M., Anjum, F. M., Hussain, S., & Nadeem, M. (2010). A comparative study on chemical composition and antioxidant activity of ginger (*Zingiber officinale*) and cumin (*Cuminum cyminum*). *Journal of Agricultural and Food Chemistry*, *58*(14), 8231–8237. doi:10.1021/jf101202x PMID:20590154

El-Hawary, S. S., Kassem, H. A., Motaal, A. A., Tawfik, W. A., Hassanein, H. M., & El-Shamy, S. S. (2013). GC-MS Analysis of the essential oil of *Alpinia zerumbet* (Pers.) BL and in vitro hepatoprotection and cytotoxicity study. MPC-4. *Planta Medica*, *79*(13), PI106. doi:10.1055-0033-1352195

Elzaawely, A. A., Xuan, T. D., & Tawata, S. (2007). Essential oils, kava pyrones and phenolic compounds from leaves and rhizomes of *Alpinia zerumbet* (Pers.) BL Burtt. & RM Sm. and their antioxidant activity. *Food Chemistry*, *103*(2), 486–494. doi:10.1016/j.foodchem.2006.08.025

Endo, K., Kanno, E., & Oshima, Y. (1990). Structures of antifungal diarylheptenones, gingerenones A, B, C and isogingerenone B, isolated from the rhizomes of *Zingiber officinale*. *Phytochemistry*, *29*(3), 797–799. doi:10.1016/0031-9422(90)80021-8

Ghasemzadeh, A., Jaafar, H. Z., & Rahmat, A. (2010). Antioxidant activities, total phenolics and flavonoids content in two varieties of Malaysia young ginger (*Zingiber officinale* Roscoe). *Molecules (Basel, Switzerland)*, *15*(6), 4324–4333. doi:10.3390/molecules15064324 PMID:20657444

Grzanna, R., Lindmark, L., & Frondoza, C. (2005). Ginger - A herbal medicinal product with broad anti-inflammatory actions. *Journal of Medicinal Food*, *8*(2), 125–132. doi:10.1089/jmf.2005.8.125 PMID:16117603

Gupta, A. P., Gupta, M. M., & Kumar, S. (1999). Simultaneous determination of curcuminoids in *Curcuma* samples using high performance thin layer chromatography. *Journal of Liquid Chromatography & Related Technologies*, *22*(10), 1561–1569. doi:10.1081/JLC-100101751

Hossain, S., Kader, G., Nikkon, F., & Yeasmin, T. (2012). Cytotoxicity of the rhizome of medicinal plants. *Asian Pacific Journal of Tropical Biomedicine*, *2*(2), 125–127. doi:10.1016/S2221-1691(11)60205-0 PMID:23569881

Jaju, S., Indurwade, N., Sakarkar, D., Fuloria, N., & Ali, M. (2009). Isolation of galangogalloside from rhizomes of *Alpinia galanga*. *International Journal of Green Pharmacy*, *3*(2), 144. doi:10.4103/0973-8258.54907

Jang, D. S., Han, A. R., Park, G., Jhon, G. J., & Seo, E. K. (2004). Flavonoids and aromatic compounds from the rhizomes of *Zingiber zerumbet*. *Archives of Pharmacal Research*, *27*(4), 386–389. doi:10.1007/BF02980078 PMID:15180302

Jang, D. S., & Seo, E. K. (2005). Potentially bioactive two new natural sesquiterpenoids from the rhizomes of *Zingiber zerumbet*. *Archives of Pharmacal Research*, *28*(3), 294–296. doi:10.1007/BF02977795 PMID:15832816

Janssen, A. M., & Scheffer, J. J. (1985). Acetoxychavicol Acetate, an Antifungal Component of *Alpinia galanga*1. *Planta Medica*, *51*(6), 507–511. doi:10.1055-2007-969577 PMID:17345272

Jantan, I. B., Ahmad, F. B., & Ahmad, A. S. (2004). Constituents of the rhizome and seed oils of greater galangal *Alpinia galangal* (L.) Willd. from Malaysia. *The Journal of Essential Oil Research*, *16*(3), 174–176. doi:10.1080/10412905.2004.9698687

Jatoi, S. A., Kikuchi, A., Gilani, S. A., & Watanabe, K. N. (2007). Phytochemical, pharmacological and ethnobotanical studies in mango ginger (*Curcuma amada* Roxb.; Zingiberaceae). Phytotherapy Research. *An International Journal Devoted to Pharmacological and Toxicological Evaluation of Natural Product Derivatives*, *21*(6), 507–516. PMID:17397131

Jiang, H., Sólyom, A. M., Timmermann, B. N., & Gang, D. R. (2005). Characterization of gingerol-related compounds in ginger rhizome (*Zingiber officinale* Rosc.) by high-performance liquid chromatography/electrospray ionization mass spectrometry. *Rapid Communications in Mass Spectrometry*, *19*(20), 2957–2964. doi:10.1002/rcm.2140 PMID:16189817

Jitoe, A., Masuda, T., Tengah, I. G. P., Suprapta, D. N., Gara, I. W., & Nakatani, N. (1992). Antioxidant activity of tropical ginger extracts and analysis of the contained curcuminoids. *Journal of Agricultural and Food Chemistry*, *40*(8), 1337–1340. doi:10.1021/jf00020a008

Jolad, S. D., Lantz, R. C., Chen, G. J., Bates, R. B., & Timmermann, B. N. (2005). Commercially processed dry ginger (*Zingiber officinale*): Composition and effects on LPS-stimulated PGE 2 production. *Phytochemistry*, *66*(13), 1614–1635. doi:10.1016/j.phytochem.2005.05.007 PMID:15996695

Joshi, A., & Chauhan, R. S. (2013). Phytochemical Analysis and Cytotoxicity Studies of *Curcuma amada* rhizomes in BHK-21 Cells. *International Journal of Scientific Research in Environmental Sciences*, *1*(12), 365–371. doi:10.12983/ijsres-2013-p365-371

Juntachote, T., & Berghofer, E. (2005). Antioxidative properties and stability of ethanolic extracts of Holy basil and Galangal. *Food Chemistry*, *92*(2), 193–202. doi:10.1016/j.foodchem.2004.04.044

Karmakar, I., Dolai, N., Saha, P., Sarkar, N., Bala, A., & Haldar, P. K. (2011). Scavenging activity of *Curcuma caesia* rhizome against reactive oxygen and nitrogen species. *Oriental Pharmacy and Experimental Medicine*, *11*(4), 221–228. doi:10.100713596-011-0030-6

Kaur, A., Singh, R., Dey, C. S., Sharma, S. S., Bhutani, K. K., & Singh, I. P. (2010). Antileishmanial phenylpropanoids from *Alpinia galanga* (Linn.) Willd. *Indian Journal of Experimental Biology*, *48*, 314–317. PMID:21046987

Khare, C. P. (2007). Indian Medicinal Plants - An Illustrated Dictionary, Springer Science+Business Media, 733-734.

Khattak, S., Shah, H. U., Ahmad, W., & Ahmad, M. (2005). Biological effects of indigenous medicinal plants *Curcuma longa* and *Alpinia galanga*. *Fitoterapia*, *76*(2), 254–257. doi:10.1016/j.fitote.2004.12.012 PMID:15810156

Kim, J. S., Im Lee, S., Park, H. W., Yang, J. H., Shin, T. Y., Kim, Y. C., ... Kim, D. K. (2008). Cytotoxic components from the dried rhizomes of *Zingiber officinale* Roscoe. *Archives of Pharmacal Research*, *31*(4), 415–418. doi:10.100712272-001-1172-y PMID:18449496

Krishnaraj, M., Manibhushanrao, K., & Mathivanan, N. (2010). A comparative study of phenol content and antioxidant activity between non-conventional *Curcuma caesia* Roxb. and *Curcuma amada* Roxb. *International Journal of Plant Production*, *4*(3), 169–174.

Kubota, K., Nakamura (Murayam), K., Kobayashi, A., & Amaike, M. (1998). Acetoxy-1, 8-cineoles as Aroma Constituents of *Alpinia galanga* Willd. *Journal of Agricultural and Food Chemistry*, *46*(12), 5244–5247. doi:10.1021/jf9804239

Lawand, R. V., & Gandhi, S. V. (2013). Comparison of *Curcuma caesia* Roxb. with other Commonly Used *Curcuma* Species by HPTLC. *Journal of Pharmacognosy and Phytochemistry*, *2*(4), 126–131.

Lechat-Vahirua, I., François, P., Menut, C., Lamaty, G., & Bessiere, J. M. (1993). Aromatic plants of French Polynesia. I. Constituents of the essential oils of rhizomes of three Zingiberaceae: *Zingiber zerumbet* Smith, *Hedychium coronarium* Koenig and *Etlingera cevuga* Smith. *The Journal of Essential Oil Research*, *5*(1), 55–59. doi:10.1080/10412905.1993.9698170

Lee, C. C., & Houghton, P. (2005). Cytotoxicity of plants from Malaysia and Thailand used traditionally to treat cancer. *Journal of Ethnopharmacology*, *100*(3), 237–243. doi:10.1016/j.jep.2005.01.064 PMID:15888378

Mahae, N., & Chaiseri, S. (2009). Antioxidant activities and antioxidative components in extracts of *Alpinia galanga* (L.) Sw. *Witthayasan Kasetsat Witthayasat*, *43*(2), 358–369.

Mallavarapu, G. R., Rao, L., Ramesh, S., Dimri, B. P., Rajeswara Rao, B. R., Kaul, P. N., & Bhattacharya, A. K. (2002). Composition of the volatile oils of *Alpinia galanga* rhizomes and leaves from India. *The Journal of Essential Oil Research*, *14*(6), 397–399. doi:10.1080/10412905.2002.9699900

Matsuda, H., Pongpiriyadacha, Y., Morikawa, T., Ochi, M., & Yoshikawa, M. (2003). Gastroprotective effects of phenylpropanoids from the rhizomes of *Alpinia galanga* in rats: Structural requirements and mode of action. *European Journal of Pharmacology*, *471*(1), 59–67. doi:10.1016/S0014-2999(03)01785-0 PMID:12809953

Mayachiew, P., & Devahastin, S. (2008). Antimicrobial and antioxidant activities of Indian gooseberry and galangal extracts. *Lebensmittel-Wissenschaft + Technologie*, *41*(7), 1153–1159. doi:10.1016/j.lwt.2007.07.019

Mitsui, S., Kobayashi, S., Nagahori, H., & Ogiso, A. (1976). Constituents from seeds of *Alpinia galanga* Wild. and their anti-ulcer activities. *Chemical & Pharmaceutical Bulletin*, *24*(10), 2377–2382. doi:10.1248/cpb.24.2377 PMID:1017082

Morita, H., & Itokawa, H. (1986). New diterpenes from *Alpinia galanga* Wild. *Chemistry Letters*, *15*(7), 1205–1208. doi:10.1246/cl.1986.1205

Mpalantinos, M. A., Soares de Moura, R., Parente, J. P., & Kuster, R. M. (1998). Biologically active flavonoids and kava pyrones from the aqueous extract of *Alpinia zerumbet*. *Phytotherapy Research*, *12*(6), 442–444. doi:10.1002/(SICI)1099-1573(199809)12:6<442::AID-PTR320>3.0.CO;2-Y

Muangnoi, P., Lu, M., Lee, J., Thepouyporn, A., Mirzayans, R., Le, X. C., ... Changbumrung, S. (2007). Cytotoxicity, apoptosis and DNA damage induced by *Alpinia galanga* rhizome extract. *Planta Medica-Natural Products and Medicinal Plant Research*, *73*(8), 748–754. PMID:17611930

Mukhopadhyay, M. J., & Mukherjee, A. (2000). Clastogenic effect of ginger rhizome in mice. *Phytotherapy Research*, *14*(7), 555–557. doi:10.1002/1099-1573(200011)14:7<555::AID-PTR659>3.0.CO;2-J PMID:11054851

Murakami, S., Li, W., Matsuura, M., Satou, T., Hayashi, S., & Koike, K. (2009). Composition and seasonal variation of essential oil in *Alpinia zerumbet* from Okinawa Island. *Journal of Natural Medicines*, *63*(2), 204–208. doi:10.100711418-008-0306-4 PMID:19067113

Mustafa, A., Ali, M., & Khan, N. Z. (2005). Volatile Oil Constituents of the Fresh Rhizomes of *Curcuma amada* Roxb. *The Journal of Essential Oil Research*, *17*(5), 490–491. doi:10.1080/10412905.2005.9698974

Mutawakil, M. H. (2012). Antigenotoxic Effects of *Zingiber officinale* Extract on *Saccharomyces cerevisiae* D7. *Researcher*, *4*(2), 51–55.

Nakatani, N., Jitoe, A., Masuda, T., & Yonemori, S. (1991). Flavonoid constituents of *Zingiber zerumbet* Smith. *Agricultural and Biological Chemistry*, *55*(2), 445–460. doi:10.1080/00021369.1991.10870577

Nampoothiri, S. V., Esakkidurai, T., & Pitchumani, K. (2015). Identification and Quantification of Phenolic Compounds in *Alpinia galanga* and *Alpinia calcarata* and its Relation to Free Radical Quenching Properties: A Comparative Study. *Journal of Herbs, Spices, & Medicinal Plants*, *21*(2), 140–147. doi :10.1080/10496475.2014.923358

Nigam, N., Bhui, K., Prasad, S., George, J., & Shukla, Y. (2009). [6]-Gingerol induces reactive oxygen species regulated mitochondrial cell death pathway in human epidermoid carcinoma A431 cells. *Chemico-Biological Interactions*, *181*(1), 77–84. doi:10.1016/j.cbi.2009.05.012 PMID:19481070

Oboh, G., Akinyemi, A. J., & Ademiluyi, A. O. (2012). Antioxidant and inhibitory effect of red ginger (*Zingiber officinale* var. Rubra) and white ginger (*Zingiber officinale* Roscoe) on Fe2+ induced lipid peroxidation in rat brain *in vitro*. *Experimental and Toxicologic Pathology*, *64*(1), 31–36. doi:10.1016/j.etp.2010.06.002 PMID:20598871

Pandey, A. K., & Chowdhury, A. R. (2003). Volatile constituents of the rhizome oil of *Curcuma caesia* Roxb. from central India. *Flavour and Fragrance Journal*, *18*(5), 463–465. doi:10.1002/ffj.1255

Peng, F., Tao, Q., Wu, X., Dou, H., Spencer, S., Mang, C., ... Hao, X. (2012). Cytotoxic, cytoprotective and antioxidant effects of isolated phenolic compounds from fresh ginger. *Fitoterapia*, *83*(3), 568–585. doi:10.1016/j.fitote.2011.12.028 PMID:22248534

Pillai, M. K., Asmawi, M. Z., Choon, T. S., Sasidharan, S., Lachumy, S. J., & Ismail, R. (2012). Evaluation of radical scavenging activity of rhizome extracts of *Alpinia galanga* and *Zingiber officinale. Journal of Medicinal Plants Research, 6*(41), 5406–5411. doi:10.5897/JMPR11.084

Prema, D., Kamaraj, M., Achiraman, S., & Udayakumar, R. (2014). *In vitro* antioxidant and cytotoxicity studies of *Curcuma amada* Roxb. (Mango ginger). *International Journal of Scientific and Research Publications, 4*(4), 1–6.

Rai, S., Mukherjee, K., Mal, M., Wahile, A., Saha, B. P., & Mukherjee, P. K. (2006). Determination of 6-gingerol in ginger (*Zingiber officinale*) using high-performance thin-layer chromatography. *Journal of Separation Science, 29*(15), 2292–2295. doi:10.1002/jssc.200600117 PMID:17120812

Rajamma, A. G., Bai, V., & Nambisan, B. (2012). Antioxidant and antibacterial activities of oleoresins isolated from nine *Curcuma* species. *Phytopharmacology, 2*, 312–317.

Rana, V. S., Verdeguer, M., & Blazquez, M. A. (2012). Chemical composition of the essential oil of *Zingiber zerumbet* var. darcyi. *Natural roduct. Communications, 7*(10), 1369–1370. PMID:23157013

Rout, O. P., Acharya, R., & Mishra, S. K. (2011). In-Vitro Antioxidant potentials in leaves of Coleus aromaticus Benth and rhizomes of *Zingiber zerumbet* (L.) SM. *Journal of Applied Pharmaceutical Science, 1*(8), 194–198.

Ruslay, S., Abas, F., Shaari, K., Zainal, Z., Sirat, H., Israf, D. A., & Lajis, N. H. (2007). Characterization of the components present in the active fractions of health gingers (*Curcuma xanthorrhiza* and *Zingiber zerumbet*) by HPLC–DAD–ESIMS. *Food Chemistry, 104*(3), 1183–1191. doi:10.1016/j.foodchem.2007.01.067

Sahu, R. K., Singh, H., & Roy, A. (2009). Antioxidative characteristics of ethanol and aqueous extracts of *Curcuma amada* rhizomes. *Research Journal of Pharmacognosy and Phytochemistry, 1*(1), 41–43.

Salah, S. H., Abdouh, S., & Booles, H. (2012). Effect of *Zingiber Officinale* on paracetamol-induced genotoxicity in male rats. *Journal of Medicinal Plants Research, 6*(41), 5425–5434. doi:10.5897/JMPR12.218

Santana, A. P. M. (2009) Safety and genotoxicity evaluation of the *Alpinia zerumbet* (pers.) Burtt & Smith tea on healthy volunteers. *Cadernos de Agroecologia, 8*(2).

Sasikumar, B. (2005). Genetic resources of Curcuma: Diversity, characterization and utilization. *Plant Genetic Resources, 3*(2), 230–251. doi:10.1079/PGR200574

Satou, T., Murakami, S., Matsuura, M., Hayashi, S., & Koike, K. (2010). Anxiolytic effect and tissue distribution of inhaled *Alpinia zerumbet* essential oil in mice. *Natural Product Communications, 5*(1), 143–146. doi:10.1177/1934578X1000500133 PMID:20184040

Sheeja, D. A., & Nair, M. S. (2012). Phytochemical constituents of *Curcuma amada. Biochemical Systematics and Ecology, 44*, 264–266. doi:10.1016/j.bse.2012.06.008

Siddaraju, M. N., & Dharmesh, S. M. (2007). Inhibition of gastric H+, K+-ATPase and Helicobacter pylori growth by phenolic antioxidants of *Curcuma amada. Journal of Agricultural and Food Chemistry, 55*(18), 7377–7386. doi:10.1021/jf070719r PMID:17725316

Singh, G., Kapoor, I. P. S., Singh, P., de Heluani, C. S., de Lampasona, M. P., & Catalan, C. A. (2008). Chemistry, antioxidant and antimicrobial investigations on essential oil and oleoresins of *Zingiber officinale*. *Food and Chemical Toxicology*, *46*(10), 3295–3302. doi:10.1016/j.fct.2008.07.017 PMID:18706468

Someya, Y., Kobayashi, A., & Kubota, K. (2001). Isolation and identification of trans-2-and trans-3-hydroxy-1, 8-cineole glucosides from *Alpinia galanga*. *Bioscience, Biotechnology, and Biochemistry*, *65*(4), 950–953. doi:10.1271/bbb.65.950 PMID:11388479

Sreevani, K. H., Puranaik, G. S. J., & Naidu, M. M. (2013). Studies on antioxidant activity of *Zingiber zerumbet* spent and its constituents through *in vitro* models. *Wudpecker Journal of Food Technology*, *1*(13), 48–55.

Srivastava, A. K., Srivastava, S. K., & Shah, N. C. (2001). Constituents of the rhizome essential oil of *Curcuma amada* Roxb. from India. *The Journal of Essential Oil Research*, *13*(1), 63–64. doi:10.1080/10412905.2001.9699608

Stanly, C., Bhatt, A., Ali, H. M. D., Keng, C. L., & Lim, B. P. (2011). Evaluation of free radical scavenging activity and total phenolic content in the petiole-derived callus cultures of *Zingiber zerumbet* Smith. *Journal of Medicinal Plants Research*, *5*(11), 2210–2217.

Stoilova, I., Krastanov, A., Stoyanova, A., Denev, P., & Gargova, S. (2007). Antioxidant activity of a ginger extract (*Zingiber officinale*). *Food Chemistry*, *102*(3), 764–770. doi:10.1016/j.foodchem.2006.06.023

Suksamerkun, W., Thongsomchitt, S., Wongkrajang, Y., Temsiririrkkul, R., Kitphati, W., & Thongpraditchote, S. (2013). Screening of antioxidant activity of vegetables in Thailand. *Journal of Asian Association of Schools of Pharmacy*, *2*, 254–261.

Tachakittirungrod, S., & Chowwanapoonpohn, S. (2007). Comparison of antioxidant and antimicrobial activities of essential oils from *Hyptis suaveolens* and *Alpinia galanga* growing in Northern Thailand. *CMU Journal of Natural Science*, *6*(1), 31–41.

Victório, C. P. (2011). Therapeutic value of the genus *Alpinia*, Zingiberaceae. *Revista Brasileira de Farmacognosia*, *21*(1), 194–201. doi:10.1590/S0102-695X2011005000025

Victório, C. P., Alviano, D. S., Alviano, C. S., & Lage, C. L. (2009). Chemical composition of the fractions of leaf oil of *Alpinia zerumbet* (Pers.) BL Burtt & RM Sm. and antimicrobial activity. *Revista Brasileira de Farmacognosia*, *19*(3), 697–701. doi:10.1590/S0102-695X2009000500008

Victório, C. P., Leitão, S. G., & Lage, C. L. (2010). Chemical composition of the leaf oils of *Alpinia zerumbet* (Pers.) Burtt et Smith and *A. purpurata* (Vieill) K. Schum. from Rio de Janeiro, Brazil. *The Journal of Essential Oil Research*, *22*(1), 52–54. doi:10.1080/10412905.2010.9700264

Vishnupriya, M., Nishaa, S., Sasikumar, J. M., & Gopalakrishnan, V. K. (2012). Antioxidant Activity And Hydroxyl Radical Induced DNA Damage Protection Effect of Aqueous Extract of *Curcuma amada* Roxb. *Research Journal of Pharmaceutical, Biological and Chemical Sciences*, *3*(4), 89.

Wang, B. B., Niu, F. G., & Duan, Y. F. (2011). Comparison of Antioxidant Activity of Total Flavonoids from *Alpinia officinarum* Hance and *Alpinia galangal* Willd [J]. *Food Science*, *7*, 028.

Wong, L. F., Lim, Y. Y., & Omar, M. (2009). Antioxidant and antimicrobial activities of some *Alpina* species. *Journal of Food Biochemistry*, *33*(6), 835–851. doi:10.1111/j.1745-4514.2009.00258.x

Xiao-Lu, Z. H. U., Ming-Hua, Y. A. N. G., Jian-Guang, L. U. O., Huang, X. F., & Ling-Yi, K. O. N. G. (2009). A new phenylpropanoid from *Alpinia galanga. Chinese Journal of Natural Medicines*, *7*(1), 19–20.

Xu, H. X., Dong, H., & Sim, K. Y. (1996). Labdane diterpenes from *Alpinia zerumbet. Phytochemistry*, *42*(1), 149–151. doi:10.1016/S0031-9422(97)00900-X

Xu, H. X., Hui, D., & Sim, K. Y. (1995). The isolation of a new labdane diterpene from the seeds of *Alpinia zerumbet. Natural Product Letters*, *7*(1), 29–34. doi:10.1080/10575639508043183

Yang, X., & Eilerman, R. G. (1999). Pungent principal of *Alpinia galangal* (L.) Swartz and its applications. *Journal of Agricultural and Food Chemistry*, *47*(4), 1657–1662. doi:10.1021/jf9808224 PMID:10564034

Yang, Y. P. (2012). Anti-inflammatory activity and oral cancer cells cytotoxicity of extracts from *Alpinia galangal* and *Zingiber zerumbet*. (Master's Thesis).

Zaeoung, S., Plubrukarn, A., & Keawpradub, N. (2005). Cytotoxic and free radical scavenging activities of Zingiberaceous rhizomes. *Songklanakarin Journal of Science and Technology*, *27*(4), 799–812.

Zancan, K. C., Marques, M. O., Petenate, A. J., & Meireles, M. A. A. (2002). Extraction of ginger (*Zingiber officinale* Roscoe) oleoresin with CO_2 and co-solvents: A study of the antioxidant action of the extracts. *The Journal of Supercritical Fluids*, *24*(1), 57–76. doi:10.1016/S0896-8446(02)00013-X

ADDITIONAL READING

Bhandary, M. J., Chandrashekar, K. R., & Kaveriappa, K. M. (1995). Medical ethnobotany of the siddis of Uttara Kannada district, Karnataka, India. *Journal of Ethnopharmacology*, *47*(3), 149–158. doi:10.1016/0378-8741(95)01274-H PMID:8569239

de Cortes Sánchez-Mata, M., & Tardío, J. (Eds.). (2016). *Mediterranean wild edible plants: ethnobotany and food composition tables*. Springer. doi:10.1007/978-1-4939-3329-7

Hulley, I. M., & Van Wyk, B. E. (2019). Quantitative medicinal ethnobotany of Kannaland (western Little Karoo, South Africa): Non-homogeneity amongst villages. *South African Journal of Botany*, *122*, 225–265. doi:10.1016/j.sajb.2018.03.014

Kress, W. J., Prince, L. M., & Williams, K. J. (2002). The phylogeny and a new classification of the gingers (Zingiberaceae): Evidence from molecular data. *American Journal of Botany*, *89*(10), 1682–1696. doi:10.3732/ajb.89.10.1682 PMID:21665595

Listiani, L., & Abrori, F. M. (2019). Ethnobotanical Study on Tidung Tribe in Using Plants for Medicine, Spice, and Ceremony. *IPTEK The Journal for Technology and Science*, *29*(1), 18–24. doi:10.12962/j20882033.v29i1.3057

Narchi, N. E. (2019). Kumeyaay Ethnobotany: Shared Heritage of the Californias. By Michael Wilken-Robertson. 2018. Sunbelt Publications, San Diego. 281 pp. *Ethnobiology Letters*, *10*(1), 16–17. doi:10.14237/ebl.10.1.2019.1521

Razak, N. I. A., Othman, R., & Pahang, J. T. (2018). Ethnobotanical Study on Plant Materials Used in Malay Traditional Post-partum Bath (Mandi Serom) Among Malay Midwives in Kedah. In *Proceedings of the Second International Conference on the Future of ASEAN (ICoFA) 2017–Volume 2* (pp. 891-897). Springer, Singapore. 10.1007/978-981-10-8471-3_88

Schmidt, B. M., & Cheng, D. M. K. (Eds.). (2017). *Ethnobotany: A Phytochemical Perspective*. John Wiley & Sons. doi:10.1002/9781118961933

Whereat-Phillips, P., & Turner, N. J. (2016). *Ethnobotany of the Coos, Lower Umpqua, and Siuslaw Indians*. Oregon State University Press.

KEY TERMS AND DEFINITIONS

Antioxidant: A compound which prevents the oxidation of a substance by reducing it.

IC50: Inhibition Concentration 50, the particular concentration of a substance which is capable of inhibiting 50% of the corresponding biological or biochemical function.

Nutraceuticals: These are the substances which can provide nutrition which is otherwise not available and able to provide certain physiological benefits.

Phenolics: Phenolics are the antioxidant molecules; chemically they are aromatic benzene rings with one or more hydroxyl groups.

QSAR: Quantitative structure activity relationship, which is based on the hypothesis if the structure of any molecule changes its corresponding activity, shall also change simultaneously.

Zingiberaceae: Zingiberaceae is the botanical family consists of around 50 genera and more than 1600 species.

Chapter 8
Ethnobotanical, Phytochemical, and Pharmacological Aspects of *Hemidesmus indicus*: A Herbal Bliss for Mankind

Minakshi Rajput
Gurukula Kangri University, India

Navneet
Gurukula Kangri University, India

Akash
Gurukula Kangri University, India

ABSTRACT

Hemidesmus indicus (L.) R. Br. is commonly known as Indian Sarsaparilla or Anantmula. Traditionally, it has been utilised as a vital herb for the treatment of several disorders. Indian Sarsaparilla is rich in a wide range of phytoconstituents such as pregnane glycosides, steroids, terpenoids, aromatic aldehydes, lignans, saponins, flavonoids and aliphatic acids which may further contribute to its pharmacological properties. This chapter gathers and compiles the traditional ethnobotanical and ayurvedic aspects of H. indicus and recently updated knowledge regarding the pharmacology, phytochemistry, adulteration, and current trends of this medicinally important herb in the field of modern phytomedicine. It also presents the ayurvedic pharmacology of this herb and summarizes the biomedical researches in as much as it helps glean a better understanding of H. indicus safety and effectiveness in humans, and describes the various natural products and polyherbal medicines containing H. indicus.

DOI: 10.4018/978-1-7998-1320-0.ch008

INTRODUCTION

The ethnobotanical evaluation of medicinal herbs can bring out many efficient plant species for the treatment of various human disorders. The herb *Hemidesmus indicus* (L.) R. Br. has been widely utilized for its therapeutic properties in Ayurveda and also an official drug in Indian and British pharmacopeia for nearly a thousand years (Swathi, Amareshwari & Venkatesh, 2019). Formerly, it was placed under the family Asclepiadaceae but on the basis of pollinial characters, it was transferred to family Periplocaceae (George, Tushar, Unnikrishnan, Hasim & Balachandran, 2008). It is commonly known as Indian Sarsaparilla or Sariba or "Anantamula" is a Sanskrit word means "endless root" (Gupta, 2006). Roots are sweet in taste, woody and aromatic due to the presence of essential oil. Over the past few decades, several researchers highlighted the potential pharmacological properties of *H. indicus* with numerous evidences from *in vitro* and *in vivo* experimentations.

Taxonomy

Kingdom Plantae
Phylum Tracheophyta
Class Magnoliopsida
Order Gentianales
Family Periplocaceae
Sub-family Asclepiadiaceae
Genus *Hemidesmus*
Species *indicus*

Distribution

H. indicus (Figure 1 and Figure 2) is found in tropical and subtropical parts of India, from the upper Gangetic plain eastwards to Assam and in some places of western, central and southern India, Sri Lanka, Bangladesh, Pakistan, Iran, and Moluccas. It grows as a deciduous bush up to an altitude of 600 m under mesophytic to semi-dry conditions in plains and quite common in uncultivated soil, open scrub jungles and hedges (Kawlni, Bora, Upadhyay, Mukherjee & Hazra, 2017).

Climate and Soil

H. indicus is profusely found in loamy to silt-clay loamy soils with an appropriate amount of humus. The pH of the soil should be slightly alkaline with 7.5-8.5.

Morphological Characteristics

Form: Perennial, slender, twining/prostrate, laticiferous, wiry, semi-erect shrub.
Leaf: Simple, short, petioled, entire, opposite, exstipulate, very variable, elliptic-oblong to linear-lanceolate, polymorphous, dark green variegated with white above and silvery white and pubescent beneath, vary in shape and size according to their age. Basically, leaves are around 5-15 cm long and about 1.25 cm broad.

Figure 1. Hemidesmus indicus

Stem: Stem and branches are numerous, twine anticlockwise, profusely laticiferous, narrow, elongate, terete, wiry, deep purple to purplish brown in color and surface slightly ridged and thickened at the nodes.

Flower: Crowded in sub-sessile axillary fascicle, dull yellow-light purplish inside, greenish yellow-greenish purple outside, deeply five-lobed calyx, gamopetalous corolla, five stamens, inserted near the base of corolla with a thick coronal scale, bicarpellary pistil, free ovaries, many ovules with distinct styles.

Fruit: Narrowly cylindrical, widely divergent follicle 10-20 cm long and 0.5-0.6 cm thick.

Seed: Many, flat, oblong having white silky hairs.

Root: Woody, cylindrical but irregularly bent, curved/slightly twisted, rigid, very strong lateral roots, rootlets are prominent and root length is around 3.5 m or more, hence named as 'Anantamula' and 0.5-1 cm thick in diameter. Root flesh is purplish brown in color and masked with irregular patches of dull green color. The surface of the older roots is transversely cracked and longitudinally fissured. The root bark is easily detachable from the hard inner core of the root and this is the aromatic part. The aroma is characteristic and the taste is sweetish to slightly acrid. The surface of the root

Figure 2. Roots of Hemidesmus indicus

is smooth in young age but thicker in older roots and appears rough because of the formation of lenticels (Warrier, Nambiar & Ganapathy, 2000).

AYURVEDIC ASPECT

According to the shloka (Figure 3) Anantamula is sweet in taste, unctuous, heavy, peacify vata, pitta and used to improve the quality of sperms. It is used to treat several disorders as, asthma, skin problems, cough, indigestion, dysentery, fever, poisoning, menorrhagia and urinary disorders.

Charka Samhita: In Charaka Smahita, Anantmula is described in following ganas (groups):

Jwarahara (Antipyretic)
Varnya (Complexation enhancing drugs)
Dhahaprashaman (Relieve in burning sensation)
Kandhya (Beneficial for throat and voice)
Purishasangrahaniya (Rendering the faeces consistent)
Stanyashodhana (Detoxify breast milk)
Madhuraskandha (Sweet in taste)
Sushruta: In Sushruta Samhita, Anantmula is described in following ganas (groups):
Sarivadi (Alleviating thrist, removing burning sensation)
Vidarigandhadi (Subduing the action of deranged)
Vallipancamula (Medicinal creepers)
Ayurvedic Pharmacodynamic Properties
Rasa (Taste): Madhur (Sweet), Tikta (Bitter)
Guna (Qualities/Physical property): Guru (Heavy), Snigdha (Oily)
Veerya (Potency): Shita (Cold)
Vipaka (Metabolic property after digestion): Madhur (Sweet)
Doshakarma: Tridoshashamak (Alleviate all three doshas i.e vata, pitta, and kapha) (Sastri, 2013; Sharma, 2015)

Figure 3. Ancient verses or shlokas about H. indicus (Sharma, 1998)

× × ×

'सारिवा दाडिमीपत्रा श्वेतरेखांकितच्छदा ।
दुग्धगर्भा शिंबिफला तूलिनी कृष्णवल्लरी ॥' (शि.)
'सारिवा जम्बूपत्रा दुग्धगर्भा वल्ली स्वनामप्रसिद्धा कृष्णसारिवा ''' चन्दनगंधा ।'(डल्हण)
सारिवायुगलं स्वादु स्निग्धं शुक्रकरं गुरु । अग्निमांद्यारुचिश्वासकासामविषनाशनम् ॥
दोषत्रयास्रप्रदरज्वरातीसारनाशनम् । (भा. प्र.)
'सारिवे द्वे तु मधुरे पित्तवातास्रनाशने । कण्डूकुष्ठज्वरहरे मेहदुर्गन्धनाशने ॥' (ध.नि.)

W. I., V, 33–34.

Other medicinal properties of *H. indicus*

Shukrala (aphrodisiac), Kanduhara (destroys itching), Raktapittahara (useful in bleeding disorders such as nasal bleeding), Kaphavatasranashanam (alleviates kapha, and vata), Durgandhanashana (destroys foul smell), Kushtahara (useful in skin diseases), Mehanashana (useful in urinary tract disorders and diabetes), Atisarahara (useful in diarrhea and dysentery), Agnisada (useful in low digestion strength), Vishapaha (antitoxic), Pradaranut (useful in menorrhagia), Kasahara/Shvasa (useful in respiratory disorders such as cold, cough, and asthma) (Pandey, 2013; Sharma & Sharma, 1998; Tripathi, 2003; https://easyayurveda.com).

ETHNOBOTANICAL RELEVANCE AND TRADITIONAL KNOWLEDGE

Traditional medicine practitioners and tribals have used Anantmula as a healing as well as a magical-spiritual dream herb from hundreds of years. There is a wide variety of ethnomedicinal uses of *H. indicus,* the most important of which are the treatment of dysentery and diarrhea, skin disease, post-partum recovery, menorrhagia, stomach ulcer, gastric ailments, fever, headache, pain, sore mouth, inflammation, epileptic fits, poisoning, rat bites, eye disease, wounds, venereal disorders such as gonorrhea and syphilis, impotence, etc. It is also used as a blood purifier, appetite stimulant, cooling tonic, alterative, demulcent, diaphoretic, carminative, and to promote health and vitality, neutralize scorpion sting and snake bite (Das, Dash, & Padhy, 2003a; Austin, 2008; Siddique et al., 2004). According to Ayurveda, it is said to be good for the treatment of gout, catarrhal problems as well as for relieving flatulence, ringworms, and other skin problems. It is also used in herpes, insanity, abdominal distention, intestinal gas, debility, chronic nervous diseases, and turbid urine. It purifies urinogenital tract, blood and cleanses the mind from negative emotions, therefore it is very effective in many nervous disorders and promotes health and vigor. The root decoction is given in the cases of elephantiasis, loss of sensation, hemiplegia, cancer, for kidney and urinary disorders, etc. (Tabassum, Das, & Mazumder, 2015; Turrini et al., 2015). According to Charaka, the leaf decoction of *H. indicus* is prescribed for the sallow complexion, loss of voice, cough, menstrual disorders, and dysentery whereas entire plant decoction is prescribed for the treatment of asthma, abdominal swelling, cough, and aching limbs. As indicated by Sushruta, Anantmula in given for the treatment of respiratory infection and wasting diseases (Khare, 2004). Traditionally medicated ghee prepared through *H. indicus* along with few other herbs is given in chronic fever, cough, asthma, headache, burning of body, hiccup, and vitiation of digestive fire. Syrup from the roots of Anantmula was included as official medicine in British Pharmacopoeia (BP) by 1864 and is also in Indian Pharmacopoeia (IP). In Ayurveda, the syrup is prescribed for dyspepsia, leucorrhoea, loss of appetite, skin diseases, fever, chronic rheumatism, and ulceration due to syphilis. Infusion of root powder possesses sudorific properties and a very good blood purifier and also given to children along with milk and sugar as a tonic in cases of diarrhea and chronic cough (Gupta, 2006). According to Unani system of medicine, stem, and root of Sariva is used as a diaphoretic, laxative, diuretic and utilized in the treatment of syphilis, biliousness, and leukoderma. The powdered root is used in pre as well as postnatal care. The roots are also used by tribal India to cure leucoderma, gonorrhea, bleeding piles, dysentery, and jaundice. The paste of roots is used in the scorpion sting by the tribal of Rajasthan. The tribals of Asad, Mainpat, Dindo, Kusmi, Sonhat, Janakpur, and Rampur of Ambikapur district (Chhattisgarh) used *H. indicus* for stomach ulcer. The powder is prepared through fresh roots of Anantmula by crushing them with

8-10 black pepper seeds (*Piper nigrum*) and about 2 g of powder is given orally on an empty stomach for 7 days for the treatment of stomach ulcer. Root syrup is also used in flavoring medicinal mixtures and called 'Sugandha' due to the wonderful fragrance (Panchal, Panchal & Patel, 2009; Fimognari et al., 2011). The root is also used with cow milk for the treatment of renal calculi and it is also used to promote a healthy pregnancy and to reduce the chances of a miscarriage (Ravishankara, Shrivastava, Padh, & Rajani, 2002; Fitwe, Rathod, Sarnaik & Wakade 2013). Another study reported that Kani tribes in Thoduhills of Kerala, South India and their neighboring indigenous communities in India used *H. indicus* for blood purification, as liver tonic, diuretic, and the treatment of digestive problem, eczema, diabetes, leucorrhea, stomach ulcer, bronchitis, abdominal pain and for many other medicinal purposes (Xavier, Kannan, Lija, Auxillia & Rose, 2014). Paliyar/Paliyan aboriginal community of Western Ghats in the hilly regions of district Virudhunagar of Tamil Nadu used roots of *H. indicus* for the preparation of decoction with water to reduce body heat and to manage face scorching sun light and also used to prepare Nannari syrup (Bose, Aron & Mehalingam, 2014).

ADULTERATION

Since the demand for roots of *H. indicus* is very high due to the enormous medicinal value, there is a tendency to substitute or adulterate the roots by unscrupulous suppliers. The plant species which commonly employed for this purpose is *Ichnocarpus frutescens*. Some important characteristics through which the adulterant can be differentiated are as follows (Warrier, Nambiar & Ganapathy, 2000).

1. The root of *H. indicus* is slender, woody, long, profusely curved or bent, purplish brown in color with very strong lateral roots and rootlets whereas the root of *I. frutescens* is long, cylindrical having shallow transverse constrictions, rusty or pinkish brown in color and lateral roots are few or absent.
2. The surface of the *Hemidesmus* root is slightly exfoliating and often cracked transversely and longitudinally both whereas in *Ichnocarpus* root the surface is fairly soft, smooth, and non-exfoliating. The surface in dry roots appears firmly longitudinally wrinkled.
3. The bark in *H. indicus* is hard crustaceous, thick, and easily detachable from the inner part of the root and composed of several thick-walled rows of cells whereas the surface skin of *I. frutescens* is very soft, thin, composed of a small number of thin-walled cells and not easily peelable.
4. The root of *H. indicus* has a characteristic odor and sweetish aromatic taste whereas in case of *I. frutescens* the bark is slightly astringent and there is no aromatic taste in roots.
5. The cork tissue of the *Hemidesmus* consists of many rows of narrow rectangular empty cells and cortex is composed of several rows of thin-walled cells with richly loaded starch grains whereas in case of *Ichnocarpus,* cortical cells are scantily loaded with starch but several latex tubes, secretory cells, and traces of mechanical elements are present in it.
6. In the bark of *Hemidesmus,* the mechanical elements are completely absent whereas mechanical elements are present in *Ichnocarpus*.
7. The phloem of *Hemidesmus* is composed of narrow medullary rays and narrow zone of thin-walled elements whereas the phloem of *Ichnocarpus* is clearly differentiated as an annular zone which is composed mostly of thin-walled elements.

8. The wood of *H. indicus* is composed of varying size vessels and thick-walled parenchyma which is alternate with narrow uniseriate medullary rays whereas the wood of *I. frutescens* forms bulk of the root which is dull white to light yellow in color and appears diffusely porous in transverse section.
9. Pith is absent in the case of *H. indicus* whereas a pith like central part made up of thin-walled cells is present in older roots of *I. frutescens.*

Except for *I. frutescens*, there are few more herb species which are used as an adulterant or substitute for *H. indicus* root are *Cryptolepis buchanani* Roemer and Schultes, *Decalepis hamiltonii* Wight and Arn. and *Utleria salicifolia* Bedd. ex Hook. fil. (George, Tushar, Unnikrishnan, Hasim & Balachandran, 2008; Warrier, Nambiar & Ganapathy, 2000).

HERBAL MEDICINES/NATURAL PRODUCTS CONTAINING *H. INDICUS*

80% of world inhabitants are using herbal medicines for their primary health requirements. In traditional medicinal systems, the medicinal herbs are primarily used as medicines in the form of fresh plant parts, juice or crude powder and also dispensed as water or ethanolic extract. To be an herbal medicine, medicinal plant parts should be free from harmful agents like heavy metals, pesticides, microbial or radioactive contaminants and it is also very important that plant parts to be utilized as medicine are authenticated (Agarwal, Fatima & Singh, 2012). Among a huge variety of medicinal plants, *H. indicus* is one of the potential and widely utilized medicinal herb. The dried roots of *H. indicus* are most commonly used in several medicinal preparations and polyherbal formulations. Many herbal medicines and natural products contain *H. indicus* as a key ingredient along with many other medicinal plants or herbs as, Mathala Rasayanam (Acharya Pharma) is used for the treatment of cold, cough, asthma, bleeding disorders, abscess, ulcers in intestines, vomiting, chest injury, cardiac pain, liver disorders, splenomegaly, Maha Vishagarbha Taila (Tattva's Herbs) is indicated for the diseases due to vata dosha, sciatica, tetanus, stiffness and tightness in back and limbs, tinnitus and lack of tactile sensation, Sarivadyasavam (Bhardwaj Pharmaceutical Works) is used for the treatment of gout, diabetes, skin infections, blood detoxification, rheumatism, excess sweating, burning sensation, hyperhidrosis, cracked heels, psoriasis, eczema, tinea infection, pruritus, acne vulgaris, boils, pustules, blisters, prostatitis, excess burning urination, pemphigus vulgaris, Mahamanjistadi Kashayam (Ashoka pharmaceuticals and Everest Ayurvedic Medicines) is indicated for skin diseases, gout, syphilis, non-healing wounds, obesity, hypercholesterolemia, eye diseases, Manasamitra Vatakam (Kottakkal and Amruta Drugs) is utilized for the treatment of insomnia, sleep disturbances, adult autism, schizophrenia, paranoia, panic attack, panic disorders, phobia, manic depressive psychosis, weak nervous system, nervous shivering, nervous speech problems, Blood Cleaner (Herbsforever) is mainly indicated for skin problems, ringworm, blemishes, irritation/itching sensation, bleeding disorders, spider veins, scabies, depigmentation, sepsis, acute infections, and blood purification, Mahanarayan Oil (Banyan Botanicals) supports the comfortable joint movement, promotes healthy muscles and tendons, soothes pain, stiffness, and inflammation, pacifies excess vata, EveCare (Himalaya Drug Company) regularizes menstrual cycle, repairs endometrium, stimulates the immune system, increase the level of hemoglobin in the blood, improve fertility by normalizing the hormone levels, used in premenstrual syndrome, dysmenorrhea (painful periods), menstrual irregularities as menorrhagia (heavy/prolonged menstruation), oligomenorrhoea (infrequent/scanty menstruation), metrorrhagia (heavy uterine bleeding), dysfunctional uterine bleeding, Kapha oil (Vadik Herbs) restores balance to kapha constitutions, useful in treatment blackheads, excessive oiliness, Acne, Styplon (Himalaya Drug Company) is indicated for epistaxis, gynecological bleeding conditions which include abnormal uterine bleeding, spotting, post intrauterine contraceptive device bleeding, bleeding hemorrhoids, hemorrhage, open wound, bleeding disorders, bleeding gums, hematuria, and as an adjuvant in hemoptysis etc. Details regarding some other herbal products containing *H. indicus* as a key component are given in Table 1. Most listings show polyherbal formulations as oral tablets, syrups, topical oils or creams. The inclusion criteria of natural products were solely based on the presence of *H. indicus* in the ingredients of the product and appearance in search listings. The authors have no competing financial interests.

This table is for informational purposes only; formulations are not standardized to any particular constituent and the FDA has not approved any health claims for Anantmula's use.

SAFETY

In the available literature, there are a very few reported adverse effects of using *H. indicus* as a therapeutic agent. The acute and sub-acute toxicity of *H. indicus* was described by Austin & Jegadeesan (2002). The suspension of root powder has been tested for acute toxicity in Swiss albino mice. Different doses of this suspension were administered once (p.o.) to different groups of mice and kept under observation. There is no conspicuous alteration in behavior and lethality effect was noticed in the herbal drug administered mice up to the concentration of 5 g/kg (Kotnis, Patel, Menon & Sane, 2004) and 12 g/kg. Diarrhea or dysentery is one of the main indications for the herbal medicine prepared from *H. indicus,* but the method of preparation of medicine from root is crucial if it is to be used for this purpose, as ethanolic extract of root inhibits electrolyte and water absorption from the intestine, which would be harmful to diarrhea or dysentery patients, whereas the water extract has the opposite effect which can stimulate water and electrolyte absorption hence beneficial (Evans, Rajasekharan & Subramoniam, 2004). Excessive use of *H. indicus* may cause intestinal discomfort due to which it is not recommended for long-term use. Various manufacturer websites having knowledge regarding natural products, do not recommend taking Anantamul during pregnancy, but this disagrees with some other anecdotal reports that claim "fetus stabilizing" properties of *H. indicus* (Weissner, 2014). The clinically prescribed doses are considered safe and beneficial, but an overdose can be toxic (Fabricant & Farnsworth, 2001). These all reports indicate that the plant contains one or more phyto-components which may toxic in certain conditions, and these should be identified and removed for the safer use of the herbal medicines prepared from this dream herb.

BIOACTIVE PRINCIPLES OF *H. INDICUS*

Secondary metabolites are the phytochemicals compounds found in plants which are responsible for color, organoleptic profile and the protection against plant diseases, insect attacks and a number of other protective functions (Kennedy & Wightman, 2011). Indian Sarsaparilla has two variants which are, var. *indicus* and var. *pubescens*. The phytoconstituents present in them are almost similar, although var. *pubescens* has been found to have a higher amount of tannins and β-sitosterol whereas var. *indicus* has a higher amount of free amino acids and phenols (Austin, 2008; Das & Bisht, 2012). The steam distil-

Table 1. Information about some natural products containing H. indicus as a key ingredient

NATURAL PRODUCT	MANUFACTURER	FORMULATION	INGREDIENTS	DIRECTIONS FOR USE	INDICATIONS
Radiant Skin-Detox Management	Vpk Maharishi Ayurveda	Polyherbal standardized tablets	*Picrorhiza kurrooa, Tribulus terrestris,* ***Hemidesmus indicus****, Tinospora cordifolia, Glycyrrhiza glabra, Cyperus rotundus, Pterocarpus santalinus, Berberis aristata, Rubia cordifolia, Cassia fistula, Azadirachta indica, Swertia chirayita, Withania somnifera, Santalum album,* Coral, Iron oxide, Gum *Acacia*, Hydroxypropyl methylcellulose, Rice hulls, Rice extract	Take 1 tablet twice daily for the first two weeks; Thereafter, take 2 tablets twice daily; Continue for at least 3 months after the desired result is achieved	To maintain liver function, skin temperature, glow, moisture and the suppleness of skin, helps in digestive fire, reduces internal toxins
Anantmool Powder	Planet ayurveda	**Dried root powder**	***Hemidesmus indicus***	1/2 to 1 teaspoonful twice daily with plain water after the meals	Skin disorders, Maintain oral health and balance vata, pitta dosha
Hite-g Height Increasing Formula Oil	Nidanam	Polyherbal topical oil	*Withinia somnifera, Flacourtia cataphracta, Acorus calmus,* ***Hemidesmus indicus,*** *Aspargus racemosus, Terminalis chebula, Cedrus deodara, Boerhavia diffusa, Prunus amygdalus, Olea europea,* Sesame oil	Apply Hite-g oil on ankles and knees and gently massage in the circular motion for 10 to 20 minutes 2 times a day for 3 months	Helpful in increasing height, beneficial in nourishing joints and its condition, helpful in reducing tensile stress on joints making them relaxed and stress-free, improve joints health and may restrict arthritic conditions in future
DermaCare Skin Restoration & Repair	Himalaya Herbal Healthcare	Polyherbal standardized capsues	*Senegalia catechu, Mallotus philippensis, Sphaeranthus indicus, Tinospora cordifolia, Azadirachta indica,* ***Hemidesmus indicus,*** *Cyperus rotundus, Wrightia tinctoria, Plumbago zeylanica, Psoralea corylifolia, Embelia ribes, Curcuma longa, Operculina turpethum, Commiphora wightii,* Triphala (*Embelica officinalis, Terminalia chebula, Terminalia bellerica*)	Take 2 capsules twice daily with meals	Support normal dermal health, body's normal cleansing process, the immune system, normal skin metabolism
Rehmannia Complex	Professional Formulas	Polyherbal 30% alcoholic extract	*Bupleurum chinenese,* ***Hemidesmus indicus,*** *Rehmannia glutinosa, Tanacetum parthenium,* Ethanol, Purified water	1 tablet 2-4 times daily or as directed by your health practitioner	Support a healthy immune system, adrenal gland, and liver health, support effective toxin removal, facilitate the normal response to occasional stress in the body, promote a healthy response to environmental stress
HealThy Skin	Banyan Botanicals	Polyherbal tablets	***Hemidesmus indicus,*** *Rubia cordifola, Azadirachta indica, Cucuma longa, Centella asiatica, Tinospora cordifolia, Phyllanthus amarus, Emblica officinalis, Glycyrrhiza glabra*	1-2 tablets, once or twice daily or as directed by your health practitioner	Supports healthy, radiant skin, Promotes a clear complexion and a smooth appearance of the skin, Soothes excess heat and pitta in the body, supports the body's natural ability to cleanse and remove toxins
Anti-Dandruff Shampoo	Himalaya Drug Company	Polyherbal topical	***Hemidesmus indicus****, Aloe vera, Cicer arietinum,* Tea tree oil, Grape seed	Gently massage the shampoo into wet hair and scalp and rinse thoroughly	Removes dandruff, nourishes and strengthens hair roots
Renalka	Himalaya Drug Company	Polyherbal syrup	*Tribulus terrestris,* ***Hemidesmus indicus****, Cyperus rotundus, Asparagus racemosus, Crataeva magna, Crataeva magna, Elettaria cardamomum,* Kshara parpati, Trikatu (*Zingiber officinale, Piper longum, Piper nigrum*)	The dosage that best suits the condition, as directed by a health practitioner	**Relieves in** dysuria, **burning urination, normalizes pH balance of urine,** used in urethritis and cystitis (urinary bladder inflammation), recurrent UTIs, hematuria
DeTox	Yogi	Polyherbal tea blend	***Hemidesmus indicus,*** *Cinnamomum verum, Piper nigrum, P. longum, Zingiber officinalis, Glycyrrhiza glabra, Arctium lappa, Taraxacum officinale, Elettaria cardamomum, Syzygium aromaticum, Lonicera japonica, Forsythia suspensa, Gardenia jasminoides, Coptis chinensis, Rheum rhabarbarum, Benincasa hispida, Juniperus communis, Phellodendron amurense*	Bring water to boiling and steep 7 minutes; For stronger tea, use 2 tea bags	Supports healthy liver and kidney function; a gentle way to help the body cleanse itself; supports blood flow, lose weight
Fair Skin Cream	Dhathri	Polyherbal topical cream	*Curcuma longa, Nelumbium speciosum, Jasminum grandiflorum, Aloe vera, Ocimum tenuiflorum,* ***Hemidesmus indicus,*** *Pterocarpus santalinus*	Apply sufficient amount (5-7 g) of cream onto clean face and neck and massage gently; Use twice daily	Helps enhance fairness and complexion of skin, gives skin glossy nature, removes black scars, spots of skin and dark shade below the eyelid

continued on following page

Table 1. Continued

NATURAL PRODUCT	MANUFACTURER	FORMULATION	INGREDIENTS	DIRECTIONS FOR USE	INDICATIONS
Soothing Skin Balm	Banyan Botanicals	Polyherbal topical skin balm	*Rubia cordifolia* ***Hemidesmus indicus***, *Tinospora cordifolia, Azadirachta indica, Glycyrrhiza glabra,* Essential oil of *Chrysopogon zizanioides,* Castor oil, Sunflower oil, Beeswax	For external use only; Apply to the skin 3-4 times daily or as prescribed anywhere that needs soothing	Supports the natural healing process of irritated, red skin and scaly, dry rashes, soothes and hydrates the skin, removes natural toxins and excess heat
Uriflow	BioNeutrix Healthcare	*Polyherbal capsules*	*Boerhaavia diffusa, Crataeva nurvala, Tribulus terrestris, Lawsonia inermis, Ficus racemosa, Bergenia ligulata, Didymocarpus pedicellata, Achyranthes aspera,* ***Hemidesmus indicus***, *Phyllanthus niruri, Raphanus sativus, Asphaltum,* Mineral pitch, Chanca piedra	Take 2 capsules daily, 1 with breakfast and one with dinner	Help disintegrate kidney stones, flush out stone gravel, prevent urinary infections from stone movement, prevent stones from forming again, Stop kidney stone pain
Elim-Tox	Vpk Maharishi Ayurveda	Polyherbal StandardizedTablets	*Tribulus terrestris,* ***Hemidesmus indicus,*** *Tinospora* cordifolia, *Glycyrrhiza glabra, Rosa centifolia, Andrographis paniculata, Berberis aristata, Cyperus rotundus, Ipomoea purge, Sphaeranthus indicus, Embelia ribes, Curcuma longa, Rubia cordifolia, Santalum album, Cassia fistula, Swertia chirayita, Withania somnifera, Azadirachta indica,* Iron, Coral, Gum *Acacia,* Talcum, Silicon dioxide	Take 1 tablet twice daily with water for the first 15 days; Then increase to 2 tablets twice daily; For cumulative benefit, take for 3 months after the desired result is achieved	Supports detoxification of the gastrointestinal tract, cleanses the blood, promotes healthy digestion and absorption of nutrients, help regulate elimination, address digestive impurities, help cleanse the sweat glands, aid the natural production of pure nutritional fluid, blood tissue and fat tissue, gently cleanse the liver and colon of impurities, balance vata, pitta, kapha
Kidney Formula	Banyan Botanicals	Polyherbal tablets	*Tribulus terrestris, Boerhavia diffusa, Tinospora cordifolia, Rubia cordifolia, Cyperus rotundus,* ***Hemidesmus indicus***, *Passiflora incarnata, Emblica officinalis, Terminalia bellirica, T. chebula, Coriandrum sativum, Foeniculum vulgare,* Organic gum *Acacia,* Organic rice flour	1-2 tablets, once or twice daily, or as directed by your health practitioner	Strengthens and rejuvenates the kidneys and adrenals, supports healthy urinary composition and flow, soothes and tones the urinary tract
Ayurvedic Digestive Formula	Terry Naturally	Polyherbal capsules	*Curcuma longa, Alpinia galangal, Andrographis paniculata, Boerhavia diffusa, Cassia fistula, Centella asiatica, Cyperus rotundus, Glycyrrhiza glabra,* ***Hemidesmus indicus,*** *Piper longum, Terminalia chebula, Tinospora cordifolia, Zingiber officinale, Murraya koenigii*	2 capsules taken with 8 oz. of water before bedtime	Remove toxins from the body, reduce occasional bloating, constipation, promotes beneficial gut microflora, supports healthy digestion and regularity, maintain bowel regularity, intestinal flora, and gentle detoxification
Kava Stress Relief Tea	Yogi	Polyherbal tea blend	*Ceratonia siliqua,* ***Hemidesmus indicus***, *Cinnamomum verum, Zingiber officinalis,* Organic Barley Malt (*Hordeum vulgare*), *Piper methysticum, Elettaria cardamomum, Stevia rebaudiana,* Natural hazelnut flavor, Natural licorice flavor, Natural cinnamon bark oil	Bring water to boiling and steep 5 to 10 min; For a stronger effect, use 2 tea bags; On especially challenging days, drink 3 to 4 cups	Used to calm the body, mind and encourage good sleep and to help melt away the stress
Skinelle Cream	Charak	Polyherbal topical cream	*Randia dumetorum, Bombax ceiba, Hedychium spicatum, Glycyrrhiza glabra, Acacia catechu, Rubia cordifolia,* ***Hemidesmus indicus,*** *Zingiber officinale, Acorus Calamus, Berberis aristata, Symplocos racemosa, Coriandrum sativum, Curcuma longa, C. amada, C. zedoaria, Brassica nigra, Pterocarpus santalinus, Saussurea obvallata,* Rock salt, Rose, Varunamarubak, Kupilushuddha	Wash the affected area with soap and warm water dry with a clean towel apply Skinelle cream regularly, twice a day or as directed by the physician	Effective and safe natural treatment for acne and pimples
Anantamul	Best Nutrition Products	Standardized herbal extract	***Hemidesmus indicus***	Take 3 capsules daily; Preferable 1 capsule, 30 minutes to 1 h before each meal with a glass of water	Supports proper function of the kidneys and urinary tract, helps stimulate the flow of bile, helps maintain healthy skin and a clear complexion, remove toxins from the body, used as a tonic, diuretic, demulcent, diaphoretic and blood purifier

continued on following page

Table 1. Continued

NATURAL PRODUCT	MANUFACTURER	FORMULATION	INGREDIENTS	DIRECTIONS FOR USE	INDICATIONS
OGN-KB	Beyond Balance Inc.	Polyherbal capsules	*Vaccinium uliginosum, Barosma betulina, Prunus avium, Apium graveolens,* ***Hemidesmus indicus****, Hyssopus officinalis, Fucus vesiculosus, Arctostaphylos uva-ursi, Stigma maydis*	Consult your physician to prescribe the dosage that best suits your condition	Support kidney/bladder and urinary tract health
Hers	Veria Id	Polyherbal tablets	*Asparagus officinalis, Withania somnifera, Centella asiatica, Vitex, Cissus quadrangularis,* ***Hemidesmus indicus***	Consult your physician to prescribe the dosage that best suits your condition	Supports normal hormonal balance and vitality

lation extract of roots of *H. indicus* (yield 0.25%) contains 2-hydroxy 4-methoxy benzoic acid (HMBA, 91%) as the major component, with 40 other minor constituents (Nagarajan, Rao & Gurudutt, 2001). In a quantitative estimation Ananthi, Chandra & Santhiya (2010a) evaluated the phytochemical constituents of crude aqueous extract of *H. indicus* root which yielded the following: saponins 12.55%, tannins 3.06%, flavonoids 1.12%, terpenoids 0.79%, coumarins 0.91%, alkaloids 1.23%, and phenols 1.1%. In another study Darekar, Khetre, Sinha, Jeswani & Damle (2008) reported that lupeol octacosanoate could be used as a marker compound for the quality control of *H. indicus* and the average amount of lupeol octacosanoate was 36.5 mg/g in the root powder. Roots of *H. indicus* contain a series of novel coumarino-lignans known as hemidesmins (Mandal, Das, Joshi, Das & Chatterjee, 1991; Das et al., 1992), and steroidal glycosides called hemidesmosides A-C and these phytoconstituents are thought to contribute the therapeutic potential of this herb (Zhao et al., 2013). Subramaniam & Nair (1968) recognized the flavonoid glycosides viz. hyperoside, isoquercitin, and rutin in the flowers, whereas only hyperoside and rutin were isolated from the leaves of *H. indicus*. 2.5% tannins were found present in the leaves and roots which contain sitosterol. Chatterjee, Chakravarty & Gomes (2006) identified a new ester, lupeol octacosanoate, and also reported some known constituents as lupeol (α-amyrin), lupeol acetate (α-amyrin acetate), and hexatriacontane. A stearopten smilasperic acid was also obtained through distillation with water. The major phytoconstituents of *H. indicus* are summarized in Table 2 together with their biological activities.

PHARMACOLOGICAL ASPECTS OF *H. INDICUS*

A number of *in vivo* and *in vitro* studies have been undertaken to evaluate the therapeutic potential of *H. indicus* and some of these are summarized below and in Table 3.

1. Anti-Inflammatory Activity

It has been demonstrated that anti-inflammatory activity of *H. indicus* in laboratory experimentation mediated by the inhibition of the NF-kB, transcription factor, reduction of proinflammatory IL-8 expression (Guerrini et al., 2014) and alterations in other mediators of inflammation, as reactive oxygen species (ROS) and pro-inflammatory cytokines (Jain & Basal, 2003). The ethanolic root extract of *H. indicus* exhibited significant anti-inflammatory activity at a dose of 350 mg/kg p.o. as compared to control (Dutta, Sen & Sikdar, 1982). A saponin found in Anantmula was known to have anti-inflammatory activity against formalin-induced edema (Alam, Auddy & Gomes, 1994). Ethyl acetate extract of the root

Table 2. Some major phytoconstituents of H. indicus and their respective biological activities.

CLASS	PHYTOCONSTITUENTS	BIOLOGICAL ACTIVITIES	REFERENCES
Aromatic aldehydes	2-hydroxy 4-methoxy benzoic acid (HMBA), 2-Hydroxy 4-methoxy benzaldehyde, 4-hydroxy 3-methoxy benzaldehyde (Vanillin), 3-hydroxy 4-methoxy benzaldehyde, Salicylaldehyde, Methyl salicylate	Antivenom, Antidiabetic, Antioxidant, Anti-inflammatory, Antipyretic	Nagarajan, Rao & Gurudutt, 2001; Mahalingam & Kannabiran, 2008; Alam & Gomes, 1998a; 1998b; Saravanan, Rajasankar & Nalini, 2007
Steroids	Hemidosterol, Hemidesmol, β-sitosterol, β-sitosterylglucuronate	Immunomodulatory, Antimicrobial, Hormone regulatory, Lipid regulatory	Das & Devaraj, 2007; Das & Devaraj, 2008; Dutta, Ghosh, S., & Chopra, 1938; Roy, Ali, Sharma & Ramachandram, 2002; Chatterjee & Bhattacharya, 1955; Gupta, Verma & Misra, 1992
Terpenoids	Ledol, Nerolidol, Lupeol, Lupeoloctacosanoate, Lupeol acetate, Cis-caryophyllene, Isocaryophyllene, Camphor, Tetracyclic triterpene alcohols, β-selinene, Borneol, Dehydrolupanyl-3 acetate, Dehydrolupeol acetate1,8cineol, Pinocarveol, β-pinenoxide, Pinocarvone, Borneol, Terpenenol, Bornyl acetate, Myrtenal, α-terpineol, Verbenone, Myrtenol, Linalyl acetate, Isobornyl acetate, Isobornyl acetate, Dihydrocarvyl acetate, α-terpinyl acetate, β-elemene, Ciscaryophyllene, Isocaryophyllene, β-selinene, α-amyrin, β-amyrin, β-amyrin acetate, Hexatriacontane, Terpenoid lactone (3-Oxo-12-lupen-28, 21olide), Terpenoid ketone, Lupanone, Δ12-dehydrolupanyl-3β-acetate, Δ12-dehydrolupeol acetate, Olean-12-en-21 β-yl acetate, Olean-12-en-3α-yl acetate, Ursenes (16(17)-seco-urs12,20(30)-dien-18α-H-3β-ylactetate; Urs-20(30)-en-18β-H-3β-yl acetate; 16(17)-secours-12,20(30)dien-18-α-H-3β-ol; 2,6,10,14,18,22-hexamethyltetracos-1-oic acid; n-octyl-2,6,10,14-tetramethyl hexadec-7-ol-10-en-13-on-1-octanoate; n-non-2'-en-1'-yl-13(15,19,19-trimethyl-cyclohex -14, 16-dienyl)-2, 6, 10-trimethyl-tetradec-6-ol-13-on-1-oate	Antivenom, Antioxidant Antimicrobial, Anti-inflammatory	Nagarajan, Rao & Gurudutt, 2001; Chatterjee, Chakravarty & Gomes, 2006; Padhy, 1973; Roy, Ali, Sharma & Ramachandram, 2001; Sharma, Yelne, Dennis, Joshi & Billore, 2000; Gupta et al., 1992; Chatterjee & Bhattacharya, 1955; Murti & Seshadri, 1941; Gupta, 1981
Pregnane Glycosides	Hemidesmosides A-C, Indicusin is 11α,12β-di-O-acetyl-orgogenin-3-O-β-Dcymaropyranosyl(1®4)-O-β-D-cymaropyranosyl(1®4)-O-β-D-cymaropyranoside, Medidesmine is sarcostin-3-O-α-D-glucopyranosyl(1®4)-O-β-D-digitoxopyranosyl (1®4)-O-β-D-oleandropyranoside, Hemisine is calogenin-3-O-β-D-cymaropyranosyl (1®4)-O[3-O-methyl]β-D-glucopyranosyl(1®4)-O-β-D-glucopyranosyl(1®4)-O-β-D-cyma ropyranoside, Desmisine is calogenin-3-O-β-D-xylopyranosyl(1®4)-O-β-Digitoxopyranosyl(1®4)-O-β-D-xylopyranosyl(1®4)-O-β-D-digitoxopyranoside, Desinine is β-3-O-β-D-oleandropyranosyl-(1®4)-βD-oleandropyranoside, Denicunine is calogenin-3-O-3-O-methyl-β-D-fucopyranosyl-(1®4)-O-β-D-oleandropyranoside, Heminine is calogenin-3-O-β-D-cymaropyranosyl-(1®4)-O-β-D-digitoxopyranoside, Hemindicusin is calogenin-3-O-methyl-α-Lrhamnopyranoside, Hindicusine is 20-O-benzoyl calogenin-3-O-β -D-digitoxopyranoside, Indicine is calogenin-3-O-β-D-digitoxopyranoside, Hemidine is calogenin-3-O-β-Dboivinopyranoside, Hemidescine is 20-O-acetyl calogenin 3-O-β-Digitoxopyranosyl(1®4)-O-β-D-oleandropyranoside, Emidine is calogenin-3-O-β-Ddigitoxopyranosyl(1®4)-O-β-D-digitoxopyranosyl (1®4)-O-β-D-digitox opyranoside, Plocoside A	Antioxidant, Antiadhesive, Antidyslipidemic	Sethi et al., 2010; Sethi, Srivastav & Srivastav, 2006; Das & Devaraj, 2006; Oberai, Khare & Khare, 1985; Zhao et al., 2013; Prakash, Sethi, Deepak, Khare & Khare, 1991; Sigler et al., 2006; Deepak, Srivastav & Khare, 1995a; Deepak, Srivastav & Khare, 1995b; Deepak, Srivastav & Khare, 1997; Chandra, Deepak & Khare, 1994
Lignans	Hemidesmin-1, Hemidesmin-2, Hemidesminin	Cardioprotective, Antioxidant, Anti-inflammatory, Cytotoxic	Das et al., 1992; Mandal, Das, Joshi, Das & Chatterjee, 1991; Mary, Achuthan, Babu & Padikkala, 2003; Kostova, 2005
Saponins	Sarsasaponin, Smilasaponin, Sarsaparilloside, Sitosterolglucoside	Antioxidant, Cytotoxic, Anti-inflammatory, Cardioprotective	Mary, Achuthan, Babu & Padikkala, 2003; Kostova, 2005; Dandekar, Bajpai & Sakharwade, 2018
Flavonoids	Rutin, Hyperoside, Quercetin, Iso-quercitrin	Antiedema, Antioxidant, Antiplatelet aggregation activity, Anti-inflammatory, Vasodilatory	Verma, Joharapurkar, Chatpalliwar & Asnani, 2005; Mary, Achuthan, Babu & Padikkala, 2003; Subramanian & Nair, 1968; Khandelwal, Balaraman, Ondrejcakova, Pancza & Ravingerova, 2010
Aliphatic acids	Hexadecanoic acid, Dodecanoic acid, Linalyl acetate, Hexatriacontane, Dihydrocarvyl acetate	-	Sharma, Yelne, Dennis, Joshi & Billore, 2000

shows potential anti-inflammatory activity in acute and subacute cases of inflammation. Oral administration of root extract blocked both inflammatory and neurogenic pains. The aqueous root extract showed sufficient anti-inflammatory activity when compared to diclofenac sodium gel (Dutta, Sen & Sikdar,

1982). Shaikh (2011), evaluated the prominent dose-dependent inhibition of ethanol root extract (100, 200 mg/kg, p.o) in the experimental rat and mice models. Treatment with hydroalcoholic root extract of *H. indicus* at different concentrations (100, 200 and 300 mg/kg b.w., p.o.) significantly prevented an increase in the volume of paw edema and formation of granulation tissue in dose-dependent manner. The maximum effect was observed at the dose of 300 mg/kg b.w which was comparable to phenylbutazone 100 mg/kg b.w., i.p. (Farook, Atlee, Kannan, Kumar & Davey, 2011). In carrageenan-induced paw edema, the methanolic root extract showed a significant reduction in volume only in 2-4 h after treatment (Lakshman, Shivaprasad, Jaiprakash & Mohan, 2006).

2. Antioxidant Activity

Antioxidant potential of *H. indicus* has been investigated in a number of laboratory and animal studies and researchers reported that it might be the phenolics as an active constituent (Jayaram & Dharmesh, 2011). In animal models with ethanol-induced liver injury, extract of *H. indicus* roots protected against free radical-mediated oxidative stress in the erythrocytes, plasma, and liver (Saravanan & Nalini, 2007a). In another study, the protective effect of *H. indicus* against bromobenzene-induced oxidative stress has been observed in rat liver mitochondria (Gopi & Setty, 2010). Methanolic root bark extracts of Anantmula exhibited free radical scavenging properties in several *in vitro* and *ex vivo* models, as well as *in vitro* inhibition of lipid peroxidation, scavenge hydroxyl and superoxide radicals *in vitro* (Ravishankara, Shrivastava, Padh, & Rajani, 2002; Mary, Achuthan, Babu & Padikkala, 2003). The root extract of *H. indicus* possesses antioxidant properties which can significantly reduce the oxidative stress and thereby toxicity induced by Doxorubicin. Methanolic extract (70%) contains large amounts of flavonoids and phenolic compounds, which exhibited high antioxidant and free radical scavenging activities and it may also chelate iron and has reducing power. The methanolic extract exhibited a dose-dependent inhibition of superoxide radicals, 1, 1-diphenyl-2-picrylhydrazyl (DPPH) radical, and moderate nitric oxide scavenging activity because of the presence of polar phytochemicals (Ravishankara, Shrivastava, Padh, & Rajani, 2002). Ascorbate in rat liver homogenate was effectively inhibited and hemolysis of erythrocytes by phenylhydrazine was also inhibited. Similar effects were enlisted by Rao, Venkateswararao, Rawat, Pushpangadan & Shirwaikar (2005) by using 50% aqueous ethanolic extract of Anantmula along with hepatoprotective effect. Saravanan & Nalini (2007b), postulated that ethanolic extract of *H. indicus* exhibited potential antioxidant efficacy and provided protection against free radical mediated oxidative stress in kidney of rats with ethanol-induced nephrotoxicity. The terpenoidal fraction obtained through successive extraction of *H. indicus* roots possesses potent free radical scavenging activity (Kumar et al., 2008). Mahalingam & Kannabiran (2008) reported that the administration of aqueous extract of *H. indicus* roots (500mg/kg/day) in streptozotocin-induced diabetic rats, for the period of 12 weeks decreased the lipid peroxidation index which was attributed to its antioxidant action.

3. Anti-Arthritic Activity

Roots of *H. indicus* possess extensively high protective activity against arthritis, probably due to the presence of sterols, terpenes, and phenolic compounds in hydroalcoholic and ethyl acetate fractions. These fractions exhibited higher anti-arthritic efficacy than chloroform and other residual fraction (Mehta, Sethiya, Mehta & Shah, 2012). Desai, Babaria, Nakarani, Shah & Paranjape (2017) evaluated the effect of ethanolic root extract on osteoporosis in ovariectomized rats and revealed that it prevents bone loss in dorsal ovariectomy-induced osteoporosis without estrogen-like adverse effects. In an *in vitro* experimentation, the inhibition of protein denaturation emphasizes the anti-arthritic potency of *H. indicus* root extract comparable of the standard drug diclofenac sodium (Abiraamasri & Lakshmi, 2016).

4. Anti-Cancerous/Chemopreventive Activity

Several animal studies indicated that *H. indicus* possess potential anti-cancer efficacy and it may induce apoptosis as well as differentiation. Cytostatic, cytotoxic, and cytodifferentiative potency of root decoction of *H. indicus* was reported in the human promyelocytic leukemia cell line (HL-60) (Ferruzzi et al., 2013). Fimognari et al. (2011) demonstrated the antileukemic potential of *H. indicus* in human cell lines. Traditional medical practitioners of Sri Lanka recommended a decoction made of seeds of *Nigella sativa*, root bark of *H. indicus*, and rhizome of *Smilax glabra* to cancer patients (Iddamaldeniya, Thabrew, Wickramasinghe, Ratnatunge & Thammitiyagodage, 2006; Thabrew, Mitry, Morsy & Hughes, 2005; Iddamaldeniya, Wickramasinghe, Thabrew, Ratnatunge & Thammitiyagodage, 2003). Several laboratory experiments based on animal studies revealed that, this polyherbal formulation modulated apoptosis in human hepatocellular carcinoma (HepG2 cells) (Samarakoon, Thabrew, Galhena & Tennekoon, 2012), anticarcinogenic activity (Iddamaldeniya, Wickramasinghe, Thabrew, Ratnatunge & Thammitiyagodage, 2003), as well as exerted a strong dose-dependent *in vitro* cytotoxicity (Thabrew, Mitry, Morsy & Hughes, 2005; Samarakoon, Thabrew, Galhena, De Silva & Tennekoon, 2010), inflammatory activity, (Galhena et al., 2012), and long-term (about16 months) effectiveness in protecting rat livers against diethylnitrosamine-mediated carcinogenic changes which may induce tumor development (Iddamaldeniya, Thabrew, Wickramasinghe, Ratnatunge & Thammitiyagodage, 2006). *H. indicus* was an effective chemopreventive agent against skin carcinogens in mice (Sultana, Alam, Khan & Sharma, 2003a), and capable of ameliorating hydroperoxide-induced cutaneous oxidative stress and tumor induction (Sultana, Khan, Sharma & Alam, 2003b). Sterols and fatty acids found in *H. indicus* root extract were capable of taming *Salmonella enterica Typhimurium* induced cytotoxicity in a human intestinal epithelial cell line by suppressing its cytotoxic activity (Das & Devaraj, 2007). The chloroform fraction of *H. indicus* roots blocked the cytotoxic efficacy of *S. enterica Typhimurium*-mediated apoptosis in a murine macrophage cell line (Das & Devaraj, 2008). In another laboratory study Aqil, Zahin & Ahmad (2008) evaluated that the methanolic stem extracts of *H. indicus* exhibited antimutagenic potency. In a recent piece of work Turrini et al. (2018) evaluated the chemopreventive activity of soxhlet extract of *H. indicus* on acute lymphoblastic leukemia cell line (CCRF-CEM). They revealed the cytotoxic effect of the extract via. extrinsic as well as intrinsic apoptotic pathways and showed that the extract also hindered the cell cycle in S phase and alleviated DNA damage. The methanol root extract of *H. indicus* possesses remarkable anticancerous activity against MCF7 breast cancer cell line, Ehrlich Ascites Tumor and cytotoxic effect against HT29 colon cancer cell line. Moreover, it was also found that it could significantly enhance the antitumor activity of 3 commonly used chemotherapeutic drugs- 6-thioguanine, methotrexate, cytarabine (Zarei & Javarappa, 2012).

5. Antivenom Activity

Ethnobotanical survey of folk plants revealed that *H. indicus* was a potent herb in neutralizing snakebite venom (Samy, Thwin, Gopalakrishnakone & Ignacimuthu, 2008). Compounds such as 2-hydroxy-4-methoxy benzoic acid (Alam & Gomes, 1998a) and lupeol acetate have neutralized the snake venom in laboratory testing (Chatterjee, Chakravarty & Gomes, 2006). It was reported that viper venom-induced lethal, hemorrhagic, coagulant, anticoagulant and free radical formation caused by venom-induced inflammation were inhibited or treated by *H. indicus* root extracts (Alam, Auddy & Gomes, 1994; Alam & Gomes, 1998a). Inflammation induced by viper venom was treated by root extract of Anantmula which possibly reduce the reactive oxygen species (ROS) and inflammatory cytokines. It also nullifies *Naja kaouthia* venom-induced neurotoxicity, cardiotoxicity, and respiratory changes (Chatterjee, Chakravarty & Gomes, 2006).

CONCLUSION

Nowadays, the demand for safe, novel and efficient therapies is on the rise. Plant secondary metabolites or phytochemicals of medicinal value are the potential sources of the new medicines and one such source is a wonder herb *H. indicus*. *H. indicus* has a long history as a vital herb with diverse therapeutic properties. It has been traditionally claimed for a wide list of ayurvedic uses and pharmacological activities. Several *in vivo* and *in vitro* experimental studies conducted on *H. indicus* have verified many of the ayurvedic or traditional uses of this herb and revealed many new properties of this plant. Recently, the cosmetic industries have shown their interest in *H. indicus* for utilizing its skin diseases protection, healing, and sunscreen properties. Phytochemistry and pharmacology of *H. indicus* have been widely investigated but there is a very few clinical data or literature is available to prove the efficacy of *H. indicus* root extracts in humans for any of the indications specified and studies on safety and toxicology of the extracts in different solvents are very few. The ultimate conclusion of this study supports that the phytoconstituents of *H. indicus* contains a vast series of biologically active compounds such as pregnane glycosides, steroids, terpenoids, aromatic aldehydes, lignans, saponins, flavonoids and aliphatic acids which may further contribute to its pharmacological properties as hepatoprotection, hypolipidaemia,

Table 3. Summary of studies conducted on H. indicus extracts evaluating the different pharmacological activities of the herb

Biological Activity	Report/Experiment	Reference
Antidiabetic Activity	Administration of the alcoholic extract (400 mg/kg) in rat for 4 weeks decreased triglycerides, serum cholesterol, phospholipids, free fatty acids, and sugar levels. The aqueous extract (500 mg/kg) and 2-hydroxy 4-methoxy benzoic acid (HMBA) (500 mg/kg), exerted hypocholesterolemic and hypoglycemic effects in streptozotocin-induced diabetic rats. It restored liver functions and also normalize the levels of total cholesterol, glycosylated hemoglobin, triglycerides, and low-density lipoprotein cholesterol, and decreased the liver and kidney lipid peroxidation.	Sowmia & Kokilavani, 2007; Murshed, Begum & Nahar, 2005 Mahalingam & Kannabiran, 2008; 2009a; 2009b
Hepatoprotective Activity	In male Wistar rat the oral treatment of ethanol extract (100 mg/kg) for 15 days, significantly prevented hepatotoxicity induced by rifampicin and isoniazid. The administration of the methanolic extract (100, 250, 500 mg/kg) in rat for 7 days showed antihepatotoxic effects in CCl_4 and paracetamol models of liver damage. 2-hydroxy 4-methoxy benzoic acid extracted from *H. indicus* (200 mg/kg) for 30 days significantly inhibited liver injury caused by ethanol administration as well as reduced the severity of liver damage in the forms of body weight, oxidative stress, hepatic marker enzymes, antioxidant status, and histological changes. The administration of the aqueous extract (100 mg/kg) prior to bromobenzene, exhibited beneficial effects which include prevention of the rise in lipid peroxides and protein carbonyls, respiratory stimulation, and also enhance sulphydryl group levels back to control levels.	Prabakan, Anandan & Devaki, 2000 Baheti, Goyal & Shah, 2006 Saravanan, Rajasankar & Nalini, 2007 Gopi & Setty, 2011
Antihyperlipidaemic Activity	200 mg/kg dose of 2-hydroxy 4-methoxy benzoic acid showed antihyperlipidaemic potential in ethanol-induced hyperlipidaemia and significantly decreased the level of plasma and hepatic lipids as well as increased the level of plasma Low-density lipoprotein concentrations.	Bopanna, Bhagyalakshmi, Rathod, Balaraman & Kannan, 1997; Saravanan & Nalini, 2007b
Antimicrobial Activity	In an *in vitro* analysis, 500-1000 mg/ml concentration of methanolic extract showed antimicrobial potential against *Escherichia coli, Salmonella typhimurium,* and *Shigella flexneri.* The extract of *H. indicus* (5 mg/disc) was found more active than standard antibiotics against *Helicobacter pylori.* The ethanolic extract (100 mg/ml) exerted strong inhibitory effects on causative agents of acne vulgaris, *Propionibacterium acnes,* and *Staphylococcus epidermidis.* 10 mg/ml concentration of saponin fraction exhibited potent antibacterial activity against *Salmonella typhi, Staphylococcus aureus, Klebsiella pneumoniae,* and antifungal efficacy against *Aspergillus flavus, A. fumigatus,* and *A. niger.*	Das, Prakash & Devaraj, 2003b Austin, Jagadeesan & Gowrishankar, 2003 Kumar et al., 2007 Khanna & Kannabiran, 2008
Otoprotective Activity	The extract of *H. indicus* (25, 50 mg/ml) significantly counteracted the toxic effect of gentamicin on hair cells and inhibited gentamicin-induced apoptosis.	Previati et al., 2007
Genotoxic and Antigenotoxic Activity	The single dose of aqueous extract of *H. indicus* (50, 100, 200 mg/kg) and split dose (10, 20, 40 mg/kg) were investigated *in vivo* for their effects against cisplatin-induced cytogenetic damage. Results revealed that extract protected bone marrow cells from cisplatin-induced genotoxicity but was cytotoxic. In case of split dose regime, it conferred a higher level of genoprotection and also was not cytotoxic at lower doses. There was a reduction in cisplatin-induced sister chromatid exchanges and micronucleated or binucleate cells was observed at the lower concentrations (4 and 8 mg/ml) of ethanol extract whereas an increase in chromosome aberrations was found at higher concentrations (2-32 mg/ml).	Ananthi, Chandra & Santhiya, 2010a Ananthi, Chandra, Santhiya & Ramesh, 2010b
Antileprotic Activity	The extract of *H. indicus* when administrated in infected mice, delayed the multiplication of *Mycobacterium leprae.*	Gupta, 1981
Neuroprotective Activity/ Nootropic Effect	The ethanol extract of *H. indicus* (30, 100, 300 mg/kg for 7 days) increased the discrimination index in the object recognition test and reduced the reaction time in the hot plate test. In both acute and chronic studies it was found that extract potentiated the haloperidol-induced catalepsy and delayed the onset of death in sodium nitrite-induced respiratory arrest. The n-butanol fraction of ethanol extract (3, 10, 30 mg/kg for 7 days) significantly improved the learning and memory power in a mouse model. The administration of methanol extract (200, 400 mg/kg) in rat was investigated on cerebral infarct by the four vessel occlusion method and it was found that it improved the vestibulomotor and neuromuscular functions, increased brain antioxidant enzymes, serotonin, and dopamine, and decreased the level of glutamate, acetylcholine esterase, and monoamine oxidase.	Shete & Bodhankar, 2009 Shete & Bodhankar, 2010 Sivaraman et al., 2012
Immunomodulatory Activity	1 mg/ml concentration of *H. indicus* extract stimulated the proliferation and viability of peripheral blood lymphocytes, increased the production of IgG, enhance adenosine deaminase activity, and suppressed both humoral and cell-mediated components of the immune system.	Kainthla et al., 2006
Antidiarrheal Activity	The methanol extract (500-1500 mg/kg) inhibited the castor oil-induced diarrhea in the rat by reducing the intestinal motility and amount of wet feces. The aqueous extract (5-10 mg/ml), when injected into the ligated jejunal sac in rat, increased the absorption of water, Na+, K+, but not glucose, from the sac.	Das, Prakash & Devaraj, 2003b Evans, Rajasekharan & Subramoniam, 2004
Antiulcerogenic Effects	The ethanol extract (150, 300, 450 mg/kg) when administrated, showed antiulcerogenic effects in the cysteamine-induced, modified pyloric ligated, and aspirin-induced ulcer models. These properties were thought to be mainly due to the mucoprotective activity and a selective increase in prostaglandin content. The ethanol extract in the concentration of 200, 400 mg/kg when injected in rat, showed ulcer protective activity comparable of omeprazole in the indomethacin (20 mg/kg) induced ulcer model, and this was possibly due to the cytoprotective action or strengthening of the gastric mucosa.	Austin & Jegadeesan, 2003 Vishali, Kavitha, Rajesh & Perumal, 2011
Mosquito larvicidal Activity	The aqueous extract (1, 2, 3, 4 and 5% for 3 days) produced larval mortality of 100% in *Culex quinquefasciatus* at a concentration of 5% after 2 days. The chloroform extract (20 mg/ml) showed a larvicidal efficacy of 100% mortality (20 mg/ml) to *Aedes aegypti* larvae.	Sujatha & Anusha, 2010 Mallikarjun, Mesta, Sudharshan & Vinayaka, 2010

continued on following page

Table 3. Continued

Biological Activity	Report/Experiment	Reference
Diuretic Effect	The methanol extract showed antiurolithiatic activity in rat against ethylene glycol (0.75%) induced urolithiasis and hyperoxaluria, as well as increased renal excretion of Ca and PO_4^{3-}. It also reduced elevated urinary oxalate and showing a regulatory action on endogenous oxalate synthesis. It was found that the administration of the aqueous extract (400 mg/kg) in rat caused an increment in urinary Na+ and K+ levels over 5 h, and comparable to that of hydrochlorothiazide and frusemide.	Patel, Kanzariya, Patel, Mehta & Modh, 2010 Gadge & Jalalpure, 2011
Cardioprotective Activity	The cardioprotective potential of *H. indicus* was evaluated in salt water-induced ventricular hypertrophy in the rat. The aqueous and methanol extracts in the concentration of 100, 200 and 400 mg/kg exhibited inhibitory effects on microalbuminuria, calcium, serum urea, myocyte diameter, creatinine levels, and retention of Na+ and water. The extract in the concentration of 90, 180, 360 mg/kg for 15 min, reduced the number of ectopic beats, infarction size, and the duration of ventricular tachycardia in mouse heart and resulting in positive inotropic effects, vasodilation, and cardioprotection.	Chidrawar et al., 2009 Khandelwal, Balaraman, Pancza & Ravingerova, 2011
Analgesic/Anti-inflammatory/ Antinociceptive /Antipyretic Effects	Oral administration of the ethanol extract (25, 50, 100 mg/kg) in mice, prior to pain induction, exerted dose-dependent antinociceptive effects and blocked the neurogenic and inflammatory pain in the formalin (paw licking), acetic acid (writhing), and hot plate tests. In rats, the methanol extract (100, 200, 400 mg/kg) exhibited antipyretic and anti-inflammatory properties in the carrageenan-induced rat paw edema model and Brewer's yeast-induced pyrexia. Ethanol extracts (100, 200 mg/kg) showed dose-dependent inhibition in various chronic and subchronic models of inflammation, and delayed-type hypersensitivity using egg white lysozyme as an antigen.	Verma, Joharapurkar, Chatpalliwar & Asnani, 2005 Lakshman, Jayaprakash & Joshi, 2005; Lakshman, Shivaprasad, Jaiprakash & Mohan, 2006 Shaikh, 2011
Renoprotective Activity	The root powder of *H. indicus* showed efficacy in managing the gentamicin-induced nephrotoxicity in albino rats. Efficacy of *H. indicus* root extract evaluated against gentamycin induced hepatotoxicity in Wister albino rats 5 g/kg single dose, p.o. 90, last 6 days of treatment reduced renal impairment. The extracts of *H. indicus* (250, 500 mg/kg) protected renal tissue in the rat from cisplatin-induced oxidative damage by reducing the levels of lipid peroxidation. The ethanol extract (250, 500 mg/kg) produced the nephroprotective activity by a dose-dependent reduction in the elevated blood urea and serum creatinine, and inhibition of lipid peroxidation, in cisplatin-induced renal injury in the rat.	Kotnis, Patel, Menon & Sane, 2004 Verma, Joharapurkar, Chatpalliwar & Asnani, 2005 Sandeep & Nair, 2010 Kaur, Singh, Shirwaikar & Setty, 2011
Antithrombotic Activity/Antioxidant/ Radioprotective Effects	The pregnane glycoside hindicusine isolated from *H. indicus* (100 mg/kg) exhibited anti-dyslipidemic and anti-oxidant activities. The intravenous administration of a methanolic extract (5 mg/kg) in rabbit delayed the plasma recalcification time and enhanced the release of lipoprotein lipase enzymes. In an *in vitro* experiment, it was found that the methanol extract (50-250 mg/ml) inhibited the ADP-induced platelet aggregation, comparable to commercial heparin. *H. indicus* extract (10-100 mg/ml) was demonstrated for its antioxidant and radioprotective potential by using inhibition of lipid peroxidation in liver microsomes and plasmid DNA, where it protected from radiation-induced strand breaks. In an *in vitro* thrombolytic model (Whole blood), the extract of *H. indicus* (10 mg/ml) exerted a weak effect on clot lysis. The phenolics of *H. indicus* (2-25 mg/ml) inhibited the lipid peroxidation, supplemented DNA and cytoprotective activities.	Sethi et al., 2010 Mary, Achuthan, Babu & Padikkala, 2003 Mary, Achuthan, Babu & Padikkala, 2003 Shetty, Satav & Nair, 2005 Prasad et al., 2007 Jayaram & Dharmesh, 2011
Cytotoxic Effect	The aqueous and ethanol extracts *H. indicus* (300-4800 mg/ml) showed cytotoxicity in HepG2 cells using 3-(4, 5-dimethylthiazol-2-yl)-2,5-diphenyl tetrazolium bromide (MTT) and sulphorhodamine B assays and the aqueous extract was more potential than the ethanol extract. The methanol extract (12.5-1000 mg/ml) was shown cytotoxicity in the MCF7 breast cancer cell line, with an IC_{50} value of 48±0.02 mg/ml. *H. indicus* extract (1-100 mg/ml) modulated the components of intracellular signaling pathways involved in cell proliferation and cell viability, altering protein expression and leading to the tumor cell death. It was also found that it enhanced the antitumor property of 6-thioguanine, methotrexate, and cytarabine.	Thabrew, 2005; Samarakoon, Thabrew, Galhena, De Silva & Tennekoon, 2010 Mazumder, Das, Das & Das 2010 Fimognari et al., 2011
Wound Healing Activity	The alcoholic root extract of *H. indicus*, formulated as 5% and 10% ointment increased the rate of wound contraction and period of epithelization.	Moideen, Varghese, Kumar & Dhanapal, 2011
Anti-psychotic Activity	The antipsychotic activity of the extract was screened by apomorphine-induced stereotyped behavior and haloperidol-induced catalepsy models. Results revealed that *H. indicus* extract significantly inhibited the stereotyped behavior in rats induced by apomorphine and also potentiate the catalepsy induced by haloperidol.	Madhu, Gupta, Arali, Chellappan & Dua, 2017
Anticonvulsant Activity	The ethanol root extract of *H. indicus* (100, 200 mg/kg) significantly reduced the duration of tonic extensor phase, postictal depression in Maximal Electro Shock method (MES) and also the duration of clonus in pentylenetetrazol method in albino rats by using the standard drug as phenobarbitone. Hence, the ethanol extract was known to have antiepileptic activity. The aqueous root extract of *H. indicus* when administrated in rats in the concentrations of 100, 300, 500 mg/kg b.w., significantly reduced the time spent in hind limb extensor phase (Maximal Electro Shock method) and the onset of convulsions (INH).	Malathi & Maharani, 2011 Madhu, Keerthi, Singh & Shivalinge, 2009
Antiangiogenic Activity	*H. indicus* showed significant inhibition of cell proliferation and tube formation in both the conditions normoxia and hypoxia. It differently regulated multiple molecular targets according to oxygen availability related to angiogenesis. The inhibition of endothelial cell migration and invasion by *H. indicus* was more pronounced in normoxia than in hypoxia, possibly because of the physiological enhanced induction of invasion characteristic of hypoxia.	Turrini et al., 2015
Anti-cataractous Activity	The anti-cataractous activity of methanol root extract of *H. indicus* was evaluated by streptozotocin-induced diabetic cataract in a rodent model. The extract inhibited the aldose reductase activity, delayed progression of cataract, lowered blood glucose, decreased osmotic stress and prevented loss of antioxidants.	Tirumani et al., 2018

cardioprotection, neuroprotection, anti-inflammatory, antidiabetic, antipyretic, antiarthritic, antioxidant, antipsychotic, anti-diarrheal, antigenotoxic, antiangiogenic, wound healing, antiulcer, larvicidal, antivenom, antimicrobial activities etc. In future, the *in vivo* and *in vitro* studies need to be focused to decipher the mechanisms of action of the bioactive compounds isolated from *H. indicus*. Pure compounds isolated from Anantmula can be utilized for preclinical studies and consequent translation to clinical application by drug designing and development. Further, the bio-availability and pharmacokinetics studies of *H. indicus* are very much necessary to fully understand the mode of action of the potential bioactive compounds for the development of new drugs in the future.

REFERENCES

Abiraamasri, B. L., & Lakshmi, T. (2016). In vitro anti-arthritic activity of *Hemidesmus indicus* root extract. *International Journal of Pharmaceutical Sciences Review and Research*, *41*(2), 15–17.

Agarwal, P., Fatima, A., & Singh, P. P. (2012). Herbal medicine scenario in India and European countries. *Journal of Pharmacognosy & Phytochemistry*, *1*(4), 88–93.

Alam, M. I., Auddy, B., & Gomes, A. (1994). Isolation, purification and partial characterization of viper venom inhibiting factor from the root extract of the Indian medicinal plant sarsaparilla (*Hemidesmus indicus* R. Br.). *Toxicon*, *32*(12), 1551–1557. doi:10.1016/0041-0101(94)90314-X PMID:7725324

Alam, M. I., & Gomes, A. (1998a). Viper venom-induced inflammation and inhibition of free radical formation by pure compound (2-hydroxy-4-methoxy benzoic acid) isolated and purified from anantamul (*Hemidesmus indicus* R. BR) root extract. *Toxicon*, *36*(1), 207–215. doi:10.1016/S0041-0101(97)00070-6 PMID:9604294

Alam, M. I., & Gomes, A. (1998b). Adjuvant effects and antiserum action potentiation by a (herbal) compound 2-hydroxy-4-methoxy benzoic acid isolated from the root extract of the Indian medicinal plant 'sarsaparilla' (*Hemidesmus indicus* R. Br.). *Toxicon*, *36*(10), 1423–1431. doi:10.1016/S0041-0101(98)00076-2 PMID:9723840

Ananthi, R., Chandra, N., & Santhiya, S. T. (2010a). Protective effect of *Hemidesmus indicus* R. Br. root extract against cisplatin-induced cytogenetic damage in mouse bone marrow cells. *Genetics and Molecular Biology*, *33*(1), 182–185. doi:10.1590/S1415-47572010005000011 PMID:21637624

Ananthi, R., Chandra, N., Santhiya, S. T., & Ramesh, A. (2010b). Genotoxic and antigenotoxic effects of *Hemidesmus indicus* R. Br. root extract in cultured lymphocytes. *Journal of Ethnopharmacology*, *127*(2), 558–560. doi:10.1016/j.jep.2009.10.034 PMID:19896526

Aqil, F., Zahin, M., & Ahmad, I. (2008). Antimutagenic activity of methanolic extracts of four ayurvedic medicinal plants. *Indian Journal of Experimental Biology*, *46*(9), 668–672. PMID:18949897

Austin, A. (2008). A review on Indian sarsaparilla, *Hemidesmus indicus* (L.) R. Br. *The Journal of Biological Sciences*, *8*(1), 1–12. doi:10.3923/jbs.2008.1.12

Austin, A., Jagadeesan, M., & Gowrishankar, R. (2003). Antimicrobial activity of *Hemidesmus indicus* var. *indicus* R. Br. against human isolates of *Helicobacter pylori*. *Natural Product Sciences*, *9*(1), 1–3.

Austin, A., & Jegadeesan, M. (2002). Acute and sub-acute toxicity studies of *Hemidesmus indicus* var. pubescens R. Br. *Ancient Science of Life*, *22*(2), 34–41. PMID:22557084

Austin, A., & Jegadeesan, M. (2003). Biochemical studies on the anti-ulcerogenic potential of *Hemidesmus indicus* R. Br. var. *indicus*. *Journal of Ethnopharmacology*, *84*(2-3), 149–156. doi:10.1016/S0378-8741(02)00291-X PMID:12648808

Baheti, J. R., Goyal, R. K., & Shah, G. B. (2006). Hepatoprotective activity of *Hemidesmus indicus* R. Br. in rats. *Indian Journal of Experimental Biology*, *44*, 399–402. PMID:16708894

Bopanna, K. N., Bhagyalakshmi, N., Rathod, S. P., Balaraman, R., & Kannan, J. (1997). Cell culture derived *Hemidesmus indicus* in the prevention of hypercholesterolemia in normal and hyperlipidemic rats. *Indian Journal of Pharmacology*, *29*(2), 105–109.

Bose, M. F. J. N., Aron, S., & Mehalingam, P. (2014). An ethnobotanical study of medicinal plants used by the Paliyars aboriginal community in Virudhunagar district, Tamil Nadu, India. *Indian Journal of Traditional Knowledge*, *13*(3), 613–618.

Chandra, R., Deepak, D., & Khare, A. (1994). Pregnane glycosides from *Hemidesmus indicus*. *Phytochemistry*, *35*(6), 1545–1548. doi:10.1016/S0031-9422(00)86891-0

Chatterjee, I., Chakravarty, A. K., & Gomes, A. (2006). *Daboia russellii* and *Naja kaouthia* venom neutralization by lupeol acetate isolated from the root extract of Indian sarsaparilla *Hemidesmus indicus* R. Br. *Journal of Ethnopharmacology*, *106*(1), 38–43. doi:10.1016/j.jep.2005.11.031 PMID:16426782

Chatterjee, R. C., & Bhattacharya, B. K. (1955). A note on the isolation of ß-sitosterol from *Hemidesmus indicus*. *Journal of the Indian Chemical Society*, *32*, 485–486.

Chidrawar, V. R., Ushir, Y. V., Sudarshan, S., Patel, K. N., Patel, N. J., & Vadalia, K. R. (2009). Possible role of natural nephroprotective; *Hemidesmus indicus* in congestive heart failure. *Pharmacognosy Research*, *1*(6), 367–374.

Dandekar, S. P., Bajpai, N. D., & Sakharwade, S. N. (2018). A multifunctional *Hemidesmus indicus* as cosmetic agent: A review. *International Journal of Scientific Development and Research*, *3*(10), 62–69.

Darekar, R. S., Khetre, A. B., Sinha, P. K., Jeswani, R. M., & Damle, M. C. (2008). Quantitation of lupeol octacosanoate in *Hemidesmus indicus* R. Br. root powder by HPTLC. *Journal of Planar Chromatography - Modern TLC*, *22*(6), 453–456. doi:10.1556/JPC.22.2009.6.13

Das, P. C., Joshi, P. C., & Mandal, S., Das, A., Chatterjee, A., & Banerji, A. (1992). New Coumarino-Lignoids from *Hemidesmus indicus* R. Br. *ChemInform*, *23*(37), 342–345.

Das, S., Dash, S. K., & Padhy, S. N. (2003a). Ethno-medicinal informations from Orissa State, India, A review. *Journal of Human Ecology (Delhi, India)*, *14*(3), 165–227. doi:10.1080/09709274.2003.11905616

Das, S., & Devaraj, S. N. (2006). Glycosides derived from *Hemidesmus indicus* R. Br. root inhibit adherence of *Salmonella typhimurium* to host cells: Receptor mimicry. *Phytotherapy Research*, *20*(9), 784–793. doi:10.1002/ptr.1963 PMID:16807882

Das, S., & Devaraj, S. N. (2007). Protective role of *Hemidesmus indicus* R. Br. root extract against *Salmonella typhimurium*-induced cytotoxicity in Int 407 cell line. *Phytotherapy Research, 21*(12), 1209–1216. doi:10.1002/ptr.2241 PMID:17685386

Das, S., & Devaraj, S. N. (2008). Effect of *Hemidesmus indicus* R. Br. root extract against *Salmonella enterica* serovar *Typhimurium*-induced apoptosis in murine macrophage cell line (P388D1). *The Indian Journal of Medical Research, 128*(5), 647–657. PMID:19179687

Das, S., Prakash, R., & Devaraj, S. N. (2003b). Antidiarrhoeal effects of methanolic root extract of *Hemidesmus indicus* (Indian sarsaparilla)-an in vitro and in vivo study. *Indian Journal of Experimental Biology, 41*, 363–366. PMID:15255649

Das, S., & Singh Bisht, S. (2012). The bioactive and therapeutic potential of *Hemidesmus indicus* R. Br.(Indian Sarsaparilla) root. *Phytotherapy Research, 27*(6), 791–801. doi:10.1002/ptr.4788 PMID:22887725

Deepak, D., Srivastav, S., & Khare, A. (1995a). Indicusin-A pregnane diester triglycoside from *Hemidesmus indicus* R. Br. *Natural Product Letters, 6*(2), 81–86. doi:10.1080/10575639508044094

Deepak, D., Srivastav, S., & Khare, A. (1995b). *Structural studies of oligosaccharide segment of hemisine and desmisine.* Paper presented at the Trends in Carbohydrate Chemistry Developed From a Carbohydrate Conference, Lucknow, India.

Deepak, D., Srivastava, S., & Khare, A. (1997). Pregnane glycosides from *Hemidesmus indicus. Phytochemistry, 44*(1), 145–151. doi:10.1016/S0031-9422(96)00393-7 PMID:8983217

Desai, S., Babaria, P., Nakarani, M., Shah, K., & Paranjape, A. (2017). Antiosteoporotic effect of *Hemidesmus indicus* Linn. on ovariectomised rats. *Journal of Ethnopharmacology, 199*, 1–8. doi:10.1016/j.jep.2017.01.031 PMID:28119100

Dutta, A. T., Ghosh, S., & Chopra, R. N. (1938). Chemische Untersuchung der Wurzeln von *Hemidesmus Indicus*. I. Teil. *Archiv der Pharmazie, 276*(6), 333–340. doi:10.1002/ardp.19382760602

Dutta, M. K., Sen, T. K., & Sikdar, S. (1982). Some preliminary observations on anti-inflammatory properties of *Hemidesmus indicus* in rat. *Indian Journal of Pharmacology, 14*, 78.

Evans, D. A., Rajasekharan, S., & Subramoniam, A. (2004). Enhancement in the absorption of water and electrolytes from rat intestine by *Hemidesmus indicus* R. Br. root (water extract). *Phytotherapy Research, 18*(7), 511–515. doi:10.1002/ptr.1402 PMID:15305307

Fabricant, D. S., & Farnsworth, N. R. (2001). The value of plants used in traditional medicine for drug discovery. *Environmental Health Perspectives, 109*(1), 69–75. PMID:11250806

Farook, S. M., Atlee, W. C., Kannan, S., Kumar, S., & Davey, M. S. (2011). Assessment of analgesic, anti-pyretic and anti-inflammatory activity of hydro-alcoholic fraction of *Hemidesmus indicus* root in experimental animals. *Scholars Research Library Der Pharmacia Lettre, 3*(1), 442–447.

Ferruzzi, L., Turrini, E., Burattini, S., Falcieri, E., Poli, F., Mandrone, M., ... Hrelia, P. (2013). *Hemidesmus indicus* induces apoptosis as well as differentiation in a human promyelocytic leukemic cell line. *Journal of Ethnopharmacology, 147*(1), 84–91. doi:10.1016/j.jep.2013.02.009 PMID:23500881

Fimognari, C., Lenzi, M., Ferruzzi, L., Turrini, E., Scartezzini, P., Poli, F., ... Cantelli-Forti, G. (2011). Mitochondrial pathway mediates the antileukemic effects of *Hemidesmus indicus*, a promising botanical drug. *PLoS One*, *6*(6), 215–244. doi:10.1371/journal.pone.0021544 PMID:21738701

Fitwe, P., Rathod, V., Sarnaik, D. K., & Wakade, R. B. (2013). Sugandi the Healing-magical-spiritual dream herb. *Asian Journal of Biomedical and Pharmaceutical Sciences*, *3*(22), 6–8.

Gadge, N. B., & Jalalpure, S. S. (2011). Natriuretic and saluretic effects of *Hemidesmus indicus* R. Br. root extracts in rats. *Indian Journal of Pharmacology*, *43*(6), 714–717. PMID:22144781

Galhena, P. B., Samarakoon, S. R., Thabrew, M. I., Weerasinghe, G. A. K., Thammitiyagodage, M. G., Ratnasooriya, W. D., & Tennekoon, K. H. (2012). Anti-inflammatory activity is a possible mechanism by which the polyherbal formulation comprised of *Nigella sativa* (seeds), *Hemidesmus indicus* (root), and *Smilax glabra* (rhizome) mediates its antihepatocarcinogenic effects. *Evidence-Based Complementary and Alternative Medicine*, *2012*, 1–11. doi:10.1155/2012/108626 PMID:23243426

George, S., Tushar, K. V., Unnikrishnan, K. P., Hasim, K. M., & Balachandran, I. (2008). *Hemidesmus indicus* (L.) R. Br. A Review. *Journal of Plant Sciences*, *3*(2), 146–156. doi:10.3923/jps.2008.146.156

Gopi, S., & Setty, O. H. (2010). Beneficial effect of the administration of *Hemidesmus indicus* against bromobenzene induced oxidative stress in rat liver mitochondria. *Journal of Ethnopharmacology*, *127*(1), 200–203. doi:10.1016/j.jep.2009.09.043 PMID:19799985

Guerrini, A., Mancini, I., Maietti, S., Rossi, D., Poli, F., Sacchetti, G., ... Borgatti, M. (2014). Expression of pro-inflammatory interleukin-8 is reduced by ayurvedic decoctions. *Phytotherapy Research*, *28*(8), 1173–1181. doi:10.1002/ptr.5109 PMID:24395450

Gupta, M. M., Verma, R. K., & Misra, L. N. (1992). Terpenoids from *Hemidesmus indicus. Phytochemistry*, *31*(11), 4036–4037. doi:10.1016/S0031-9422(00)97582-4

Gupta, N. S. (2006). *The ayurvedic system of indian medicine*. New Delhi, India: Bharatiya Kala Prakashan.

Gupta, P. N. (1981). Antileprotic action of an extract from "Anantamul" (*Hemidesmus Indicus* R. Br.). *Leprosy in India*, *53*(3), 354–359. Retrieved from https://easyayurveda.com/2013/12/20/sariva-hemidesmus-indicus-benefits-usage-dose-side-effects/ PMID:7278140

Iddamaldeniya, S. S., Thabrew, M. I., Wickramasinghe, S. M. D. N., Ratnatunge, N., & Thammitiyagodage, M. G. (2006). A long-term investigation of the anti-hepatocarcinogenic potential of an indigenous medicine comprised of Nigella sativa, *Hemidesmus indicus* and Smilax glabra. *Journal of Carcinogenesis*, *5*(11), 1–7. PMID:16684351

Iddamaldeniya, S. S., Wickramasinghe, N., Thabrew, I., Ratnatunge, N., & Thammitiyagodage, M. G. (2003). Protection against diethylnitrosoamine-induced hepatocarcinogenesis by an indigenous medicine comprised of *Nigella sativa, Hemidesmus indicus* and *Smilax glabra*: A preliminary study. *Journal of Carcinogenesis*, *2*(6), 1–6. PMID:14613573

Jain, A., & Basal, E. (2003). Inhibition of Propionibacterium acnes-induced mediators of inflammation by Indian herbs. *Phytomedicine*, *10*(1), 34–38. doi:10.1078/094471103321648638 PMID:12622461

Jayaram, S., & Dharmesh, S. M. (2011). Assessment of antioxidant potentials of free and bound phenolics of *Hemidesmus indicus* (L) R. Br against oxidative damage. *Pharmacognosy Research, 3*(4), 225–231. doi:10.4103/0974-8490.89741 PMID:22224044

Kainthla, R. P., Kashyap, R. S., Deopujari, J. Y., Purohit, H. J., Taori, G. M., & Daginawala, H. F. (2006). Effect of *Hemidesmus indicus* (Anantmool) extract on IgG production and adenosine deaminase activity of human lymphocytes in vitro. *Indian Journal of Pharmacology, 38*(3), 190–193. doi:10.4103/0253-7613.25806

Kaur, A., Singh, S., Shirwaikar, A., & Setty, M. (2011). Effect of ethanolic extract of *Hemidesmus indicus* roots on Cisplatin induced Nephrotoxicity in rats. *Journal of Pharmacy Research, 4*(8), 2523–2525.

Kawlni, L., Bora, M., Upadhyay, S. N., Mukherjee, K., & Hazra, J. (2017). Pharmacological and therapeutic profile of Anantamula (*Hemidesmus indicus* (L.) R. Br.: A comprehensive review. *International Journal of Ayurveda and Pharma Research, 5*(11), 49–57.

Kennedy, D. O., & Wightman, E. L. (2011). Herbal extracts and phytochemicals: Plant secondary metabolites and the enhancement of human brain function. *Advances in Nutrition, 2*(1), 32–50. doi:10.3945/an.110.000117 PMID:22211188

Khandelwal, V. K. M., Balaraman, R., Ondrejcakova, M., Pancza, D., & Ravingerova, T. (2010). Effect of *Hemidesmus indicus* on ischemia-reperfusion injury in the isolated rat heart. *Pharmaceutical Biology, 48*(6), 611–614. doi:10.3109/13880200903218943 PMID:20645732

Khandelwal, V. K. M., Balaraman, R., Pancza, D., & Ravingerova, T. (2011). *Hemidesmus indicus* and *Hibiscus rosa-sinensis* affect ischemia reperfusion injury in isolated rat hearts. *Evidence-Based Complementary and Alternative Medicine, 2011*, 1–8. doi:10.1155/2011/802937 PMID:20953394

Khanna, V. G., & Kannabiran, K. (2008). Antimicrobial activity of saponin fraction from the roots of *Hemidesmus indicus. Research Journal of Medicinal Plant, 2*(1), 39–42. doi:10.3923/rjmp.2008.39.42

Khare, C. P. (2004). *Encyclopedia of indian medicinal plants: rational western therapy, ayurvedic and other traditional usage, botany*. New York, NY: Springer.

Kostova, I. (2005). Synthetic and natural coumarins as cytotoxic agents. *Current Medicinal Chemistry. Anti-Cancer Agents, 5*(1), 29–46. doi:10.2174/1568011053352550 PMID:15720259

Kotnis, M. S., Patel, P., Menon, S. N., & Sane, R. T. (2004). Renoprotective effect of *Hemidesmus indicus*, a herbal drug used in gentamicin-induced renal toxicity. *Nephrology (Carlton, Vic.), 9*(3), 142–152. doi:10.1111/j.1440-1797.2004.00247.x PMID:15189175

Kumar, G. S., Jayaveera, K. N., Kumar, A. C. K., Bharathi, T. B. M., Umachigi, P. S., & Vrushabendra, S. (2008). Evaluation of antioxidant and antiacne properties of terpenoidal fraction of *Hemidesmus indicus* (Indian sarsaparilla). *The Internet Journal of Aesthetic and Antiaging Medicine, 1*(1).

Kumar, G. S., Jayaveera, K. N., Kumar, C. K., Sanjay, U. P., Swamy, B. M., & Kumar, D. V. (2007). Antimicrobial effects of Indian medicinal plants against acne-inducing bacteria. *Tropical Journal of Pharmaceutical Research, 6*(2), 717–723. doi:10.4314/tjpr.v6i2.14651

Lakshman, K., Jayaprakash, B., & Joshi, H. (2005). Comparative anti-inflammatory activity studies of four species of Sariva. *Indian Journal of Pharmaceutical Sciences*, *67*(4), 481–482.

Lakshman, K., Shivaprasad, H. N., Jaiprakash, B., & Mohan, S. (2006). Anti-inflammatory and anti-pyretic activities of *Hemidesmus indicus* root extract. *African Journal of Traditional, Complementary, and Alternative Medicines*, *3*(1), 90–94.

Madhu, A., Gupta, G., Arali, B. K., Chellappan, D., & Dua, K. (2017). Anti-psychotic activity of aqueous root extract of *Hemidesmus indicus*: A time bound study in rats. *Recent Patents on Drug Delivery & Formulation*, *11*(1), 36–41. PMID:27993107

Madhu, A., Keerthi, P. H. V., Singh, J., & Shivalinge, G. K. P. (2009). To evaluate the antiepileptic activity of aqueous root extract of *Hemidesmus indicus* in rats. *Archives of Pharmaceutical Sciences and Research*, *1*(1), 43–47.

Mahalingam, G., & Kannabiran, K. (2008). Hypoglycemic activity of *Hemidesmus indicus* R. Br. on streptozotocin-induced diabetic rats. *International Journal of Diabetes in Developing Countries*, *28*(1), 6–10. doi:10.4103/0973-3930.41979 PMID:19902032

Mahalingam, G., & Kannabiran, K. (2009a). *Hemidesmus indicus* root extract ameliorates diabetes-mediated metabolic changes in rats. *International Journal of Green Pharmacy*, *3*(4), 314–318. doi:10.4103/0973-8258.59739

Mahalingam, G., & Kannabiran, K. (2009b). Antidiabetic activity of 2-hydroxy 4-methoxy benzoic acid isolated from the roots of *Hemidesmus indicus* on streptozotocin-induced diabetic rats. *International Journal of Diabetes and Metabolism*, *17*(2), 53–57. doi:10.1159/000497673

Malathi, M., & Maharani, B. (2011). Evaluation of anticonvulsant activity of ethanolic extract of roots of *Hemidesmus indicus* using adult albino rats. *Journal of Pharmacy Research*, *4*(10), 3345–3347.

Mallikarjun, N., Mesta, S. C., Sudharshan, S. J., & Vinayaka, K. S. (2010). Antimosquito (Insecticidal) activity of extracts of *Hemidesmus indicus* and *Swertia chirata* against *Aedes aegypti* mosquito larvae-A Comparative Study. *Drug Invent Today*, *2*, 106–108.

Mandal, S., Das, P. C., Joshi, P. C., Das, A., & Chatterjee, A. (1991). Hemidesminine-A New Coumarino-Lignoid from *Hemidesmus indicus* R. Br. *ChemInform*, *22*(37), 712–713.

Mary, N. K., Achuthan, C. R., Babu, B. H., & Padikkala, J. (2003). In vitro antioxidant and antithrombotic activity of *Hemidesmus indicus* (L) R. Br. *Journal of Ethnopharmacology*, *87*(2-3), 187–191. doi:10.1016/S0378-8741(03)00119-3 PMID:12860306

Mazumder, P. M., Das, S., Das, S., & Das, M. K. (2010). Cytotoxic activity of methanolic extracts of *Berberis aristata* DC and *Hemidesmus indicus* R.Br. in MCF7 Cell Line. *Journal of Current Pharmaceutical Research*, *1*(1), 12–15.

Mehta, A., Sethiya, N. K., Mehta, C., & Shah, G. B. (2012). Anti-arthritis activity of roots of *Hemidesmus indicus* R. Br. (Anantmul) in rats. *Asian Pacific Journal of Tropical Medicine*, *5*(2), 130–135. doi:10.1016/S1995-7645(12)60011-X PMID:22221757

Moideen, M. M., Varghese, R., Kumar, E. K., & Dhanapal, C. K. (2011). Wound healing activity of ethanolic extract of *Hemidesmus indicus* (Linn) R.Br leaves in rats. *Research Journal of Pharmaceutical, Biological and Chemical Sciences*, *2*(3), 643–651.

Murshed, S., Begum, R., & Nahar, N. (2005). Hypoglycemic and hypolipidemic effect of *Hemidesmus indicus* root on diabetic model rats. *Diabetes Research (Edinburgh, Lothian)*, *39*, 15–23.

Murti, P. B. R., & Seshadri, T. R. (1941). A study of the chemical components of the roots of *Decalepis Hamiltonii* (Makali Veru). In *Proceedings of the Indian Academy of Sciences-Section A*. India: Springer.

Nagarajan, S., Rao, J. M. L., & Gurudutt, K. N. (2001). Chemical composition of the volatiles of *Hemidesmus indicus* R. Br. *Flavour and Fragrance Journal*, *16*(3), 212–214. doi:10.1002/ffj.985

Oberai, K., Khare, M. P., & Khare, A. (1985). A pregnane ester diglycoside from *Hemidesmus indicus*. *Phytochemistry*, *24*(10), 2395–2397. doi:10.1016/S0031-9422(00)83049-6

Padhy, S. N., Mahato, S. B., & Dutta, N. L. (1973). Triterpenoids from the roots of *Hemidesmus indicus*. *Phytochemistry*, *12*(1), 217–218. doi:10.1016/S0031-9422(00)84656-7

Panchal, G. A., Panchal, S. J., & Patel, J. A. (2009). *Hemidesmus indicus*: A review. *Pharmacologyonline*, *2*, 758–771.

Pandey, G. S. (2013). *Bhavprakash Nighantu*. Varanasi, India: Chaukhamba Bharti Academy.

Patel, D. M., Kanzariya, N. R., Patel, N. J., Mehta, H. R., & Modh, H. A. (2010). Evaluation of the efficacy of methanolic extract of *Hemidesmus indicus* in urolithiasis in wistar rats. *Inventi Rapid: Ethnopharmacology*. Retrieved from http://inventi.in/journal/art_reader/3/9176/rapid/ethnopharmacology/pi

Prabakan, M., Anandan, R., & Devaki, T. (2000). Protective effect of *Hemidesmus indicus* against rifampicin and isoniazid-induced hepatotoxicity in rats. *Fitoterapia*, *71*(1), 55–59. doi:10.1016/S0367-326X(99)00120-3 PMID:11449471

Prakash, K., Sethi, A., Deepak, D., Khare, A., & Khare, M. P. (1991). Two pregnane glycosides from *Hemidesmus indicus*. *Phytochemistry*, *30*(1), 297–299. doi:10.1016/0031-9422(91)84141-E

Prasad, S., Kashyap, R. S., Deopujari, J. Y., Purohit, H. J., Taori, G. M., & Daginawala, H. F. (2007). Effect of *Fagonia arabica* (Dhamasa) on in vitro thrombolysis. *BMC Complementary and Alternative Medicine*, *7*(1), 36–41. doi:10.1186/1472-6882-7-36 PMID:17986325

Previati, M., Corbacella, E., Astolfi, L., Catozzi, M., Khan, M. T. H., Lampronti, I., ... Martini, A. (2007). Ethanolic extract from *Hemidesmus indicus* (Linn) displays otoprotectant activities on organotypic cultures without interfering on gentamicin uptake. *Journal of Chemical Neuroanatomy*, *34*(3-4), 128–133. doi:10.1016/j.jchemneu.2007.05.009 PMID:17640851

Rao, G. M., Venkateswararao, C., Rawat, A. K. S., Pushpangadan, P., & Shirwaikar, A. (2005). Antioxidant and antihepatotoxic activities of *Hemidesmus indicus* R. Br. *Acta Pharmaceutica Turcica*, *47*(2), 107–113.

Ravishankara, M. N., Shrivastava, N., Padh, H., & Rajani, M. (2002). Evaluation of antioxidant properties of root bark of *Hemidesmus indicus* R. Br. (Anantmul). *Phytomedicine*, *9*(2), 153–160. doi:10.1078/0944-7113-00104 PMID:11995949

Roy, S. K., Ali, M., Sharma, M. P., & Ramachandram, R. (2001). New pentacyclic triterpenes from the roots of *Hemidesmus indicus. Die Pharmazie*, *56*(3), 244–246. PMID:11265594

Roy, S. K., Ali, M., Sharma, M. P., & Ramachandram, R. (2002). Phytochemical investigation of *Hemidesmus indicus* R. Br. roots. *Indian Journal of Chemistry, Section B. Organic Chemistry Including Medicinal Chemistry*, *41B*(11), 2390–2394.

Samarakoon, S. R., Thabrew, I., Galhena, P. B., De Silva, D., & Tennekoon, K. H. (2010). A comparison of the cytotoxic potential of standardized aqueous and ethanolic extracts of a polyherbal mixture comprised of *Nigella sativa* (seeds), *Hemidesmus indicus* (roots) and *Smilax glabra* (rhizome). *Pharmacognosy Research*, *2*(6), 335–342. doi:10.4103/0974-8490.75451 PMID:21713135

Samarakoon, S. R., Thabrew, I., Galhena, P. B., & Tennekoon, K. H. (2012). Modulation of apoptosis in human hepatocellular carcinoma (HepG2 cells) by a standardized herbal decoction of *Nigella sativa* seeds, *Hemidesmus indicus* roots and *Smilax glabra* rhizomes with anti-hepatocarcinogenic effects. *BMC Complementary and Alternative Medicine*, *12*(1), 25. doi:10.1186/1472-6882-12-25 PMID:22458551

Samy, R. P., Thwin, M. M., Gopalakrishnakone, P., & Ignacimuthu, S. (2008). Ethnobotanical survey of folk plants for the treatment of snakebites in Southern part of Tamilnadu, India. *Journal of Ethnopharmacology*, *115*(2), 302–312. doi:10.1016/j.jep.2007.10.006 PMID:18055146

Sandeep, D., & Nair, C. K. K. (2010). Amelioration of cisplatin-induced nephrotoxicity by extracts of *Hemidesmus indicus* and *Acorus calamus. Pharmaceutical Biology*, *48*(3), 290–295. doi:10.3109/13880200903116048 PMID:20645815

Saravanan, N., & Nalini, N. (2007a). Antioxidant effect of *Hemidesmus indicus* on ethanol-induced hepatotoxicity in rats. *Journal of Medicinal Food*, *10*(4), 675–682. doi:10.1089/jmf.2006.110 PMID:18158840

Saravanan, N., & Nalini, N. (2007b). Effect of 2-hydroxy 4-methoxy benzoic acid on an experimental model of hyperlipidaemia, induced by chronic ethanol treatment. *The Journal of Pharmacy and Pharmacology*, *59*(11), 1537–1542. doi:10.1211/jpp.59.11.0011 PMID:17976265

Saravanan, N., Rajasankar, S., & Nalini, N. (2007). Antioxidant effect of 2-hydroxy-4-methoxy benzoic acid on ethanol-induced hepatotoxicity in rats. *The Journal of Pharmacy and Pharmacology*, *59*(3), 445–453. doi:10.1211/jpp.59.3.0015 PMID:17331349

Sastri, K. (2013). *Charak Samhita*. Varanasi, India: Chaukhambha Bharti Academy.

Sethi, A., Bhatia, A., Srivastava, S., Bhatia, G., Khan, M. M., Khanna, A. K., & Saxena, J. K. (2010). Pregnane glycoside from *Hemidesmus indicus* as a potential anti-oxidant and anti-dyslipidemic agent. *Natural Product Research*, *24*(15), 1371–1378. doi:10.1080/14786410802265084 PMID:20169502

Sethi, A., Srivastav, S. S., & Srivastav, S. (2006). Pregnane glycoside from *Hemidesmus indicus* R. Br. *Indian Journal of Heterocyclic Chemistry*, *16*(2), 191–192.

Shaikh, P. Z. (2011). Study of anti-inflammatory activity of ethanolic extract of *Hemidesmus indicus* roots in acute, subchronic & chronic inflammation in experimental animals. *International Journal of Pharmacy & Life Sciences*, *2*(10), 1154–1173.

Sharma, P., & Sharma, G. (1998). *Dhanvantari Nighantu*. Varanasi, India: Chaukhamba Oriental.

Sharma, P. C., Yelne, M. B., Dennis, T. J., Joshi, A., & Billore, K. V. (2000). *Database on medicinal plants used in Ayurveda*. New Delhi, India: Central Council for Research in Ayurveda and Siddha.

Sharma, P. V. (1998). *Dravyaguna vijnana*. Varanasi, India: Chaukhabha Bharti Academy.

Sharma, P. V. (2015). *Dravyaguna vijnana*. Varanasi, India: Chaukhabha Bharti Academy.

Shete, R. V., & Bodhankar, S. L. (2009). Neuropharmacology of ethanolic extract *of Hemidesmus indicus. Electronic Journal of Pharmacology and Therapy, 2*, 63–70.

Shete, R. V., & Bodhankar, S. L. (2010). *Hemidesmus indicus*: Evaluation of its nootropic effect in mice. *International Journal of Pharma and Bio Sciences, 1*(3), 1–10.

Shetty, T. K., Satav, J. G., & Nair, C. K. K. (2005). Radiation protection of DNA and membrane in vitro by extract of *Hemidesmus indicus. Phytotherapy Research, 19*(5), 387–390. doi:10.1002/ptr.1470 PMID:16106384

Siddique, N. A., Bari, M. A., Naderuzzaman, A. T. M., Khatun, N., Rahman, M. H., Sultana, R. S., ... Rahman, M. M. (2004). Collection of indigenous knowledge and identification of endangered medicinal plants by questionnaire survey in lin rind tract of Bangladesh. *The Journal of Biological Sciences, 4*(1), 72–80. doi:10.3923/jbs.2004.72.80

Sigler, P., Saksena, R., Deepak, D., & Khare, A. (2000). C21 steroidal glycosides from *Hemidesmus indicus. Phytochemistry, 54*(8), 983–987. doi:10.1016/S0031-9422(99)00611-1 PMID:11014302

Sivaraman, D., Kumar, P. S., Muralidharan, P., Rahman, H., Nagar, I., & Thoraipakkam, C. (2012). Effect of *Hemidesmus indicus* on cerebral infract Ischemia-reperfusion injured rats by four vessel occlusion method. *Pharmacologia, 3*(4), 91–102. doi:10.5567/pharmacologia.2012.91.102

Sowmia, C., & Kokilavani, R. (2007). Antidiabetic and antihypercholesterolemic effect of *Hemidesmus indicus* Linn. R. root in Alloxan induced diabetic rats. *Ancient Science of Life, 26*(4), 4. PMID:22557243

Subramanian, S. S., & Nair, A. G. R. (1968). Flavonoids of some asclepiadaceous plants. *Phytochemistry, 7*(9), 1703–1704. doi:10.1016/S0031-9422(00)88630-6

Sujatha, S., & Anusha, J. R. (2010). In vitro antibacterial activity on human pathogenic bacteria and larvicidal effect of root from *Hemidesmus indicus* (Linn.) on *Culex qinquifasciatus. International Journal of Phytomedicine, 2*(4), 418–424.

Sultana, S., Alam, A., Khan, N., & Sharma, S. (2003a). Inhibition of cutaneous oxidative stress and two-stage skin carcinogenesis by *Hemidesmus indicus* (L.) in Swiss albino mice. *Indian Journal of Experimental Biology, 41*(12), 1416–1423. PMID:15320495

Sultana, S., Khan, N., Sharma, S., & Alam, A. (2003b). Modulation of biochemical parameters by *Hemidesmus indicus* in cumene hydroperoxide-induced murine skin: Possible role in protection against free radicals-induced cutaneous oxidative stress and tumor promotion. *Journal of Ethnopharmacology, 85*(1), 33–41. doi:10.1016/S0378-8741(02)00360-4 PMID:12576200

Swathi, S., Amareshwari, P., & Venkatesh, K., R., R., A. (2019). Phytochemical and pharmacological benefits of *Hemidesmus indicus*: An updated review. *Journal of Pharmacognosy and Phytochemistry, 8*(1), 256–262.

Tabassum, Das. S., & Mazumder, A. (2015). Ethnobotanical review of *Hemidesmus indicus* R.Br. *International Journal of Current Research, 7*(12), 24251-24256.

Thabrew, M. I., Mitry, R. R., Morsy, M. A., & Hughes, R. D. (2005). Cytotoxic effects of a decoction of *Nigella sativa, Hemidesmus indicus* and *Smilax glabra* on human hepatoma HepG2 cells. *Life Sciences, 77*(12), 1319–1330. doi:10.1016/j.lfs.2005.01.022 PMID:15916774

Tirumani, P., Venu, S., Sridhar, G., Praveen Kumar, M., Rajashekhar, A. V., & Naga Raju, T. (2018). Delaying of cataract through intervention of *Hemidesmus indicus* in STZ induced diabetic rats. *Natural Product Research, 32*(11), 1295–1298. doi:10.1080/14786419.2017.1333991 PMID:28580798

Tripathi, I. (2003). *Raj Nighantu*. Varanasi, India: Bharati Acadamy.

Turrini, E., Calcabrini, C., Tacchini, M., Efferth, T., Sacchetti, G., Guerrini, A., ... Fimognari, C. (2018). In vitro study of the cytotoxic, cytostatic, and antigenotoxic profile of *Hemidesmus indicus* (L.) R. Br. (Apocynaceae) crude drug extract on t lymphoblastic cells. *Toxins, 10*(2), 70. doi:10.3390/toxins10020070 PMID:29415441

Turrini, E., Ferruzzi, L., Guerrini, A., Gotti, R., Tacchini, M., Teti, G., ... Fimognari, C. (2015). In vitro anti-angiogenic effects of *Hemidesmus indicus* in hypoxic and normoxic conditions. *Journal of Ethnopharmacology, 162*, 261–269. doi:10.1016/j.jep.2014.12.010 PMID:25560668

Verma, P. R., Joharapurkar, A. A., Chatpalliwar, V. A., & Asnani, A. J. (2005). Antinociceptive activity of alcoholic extract of *Hemidesmus indicus* R. Br. in mice. *Journal of Ethnopharmacology, 102*(2), 298–301. doi:10.1016/j.jep.2005.05.039 PMID:16081231

Vishali, K., Kavitha, K. N. V., Rajesh, V., & Perumal, P. (2011). Anti-ulcer activity of *Hemidesmus indicus* root extract on Indomethacin induced gastric ulcer in albino wistar rats. *Journal of Pharmacy Research, 4*(2), 391–392.

Warrier, P. K., Nambiar, V. P. K., & Ganapathy, P. M. (2000). Some important medicinal plants of the western ghats, India: a profile. New Delhi, India: Medicinal and Aromatic Plants Program in Asia (MAPPA), IDRC, SARO.

Weissner, W. (2014). Anantamul (*Hemidesmus indicus*) A Review of Biomedical Studies and U.S. Products. *Ayurveda Journal of Health, 12*(4), 40–52.

Xavier, T. F., Kannan, M., Lija, L., Auxillia, A., Rose, A. K. F., & Kumar, S. S. (2014). Ethnobotanical study of Kani tribes in Thoduhills of Kerala, South India. *Journal of Ethnopharmacology, 152*(1), 78–90. doi:10.1016/j.jep.2013.12.016 PMID:24393788

Zarei, M., & Javarappa, K. K. (2012). Anticarcinogenic and cytotoxic potential of *Hemidesmus indicus* root extract against Ehrlich Ascites tumor. *Der Pharmacia Lettre, 4*(3), 906–991.

Zhao, Z., Matsunami, K., Otsuka, H., Negi, N., Kumar, A., & Negi, D. S. (2013). A condensed phenylpropanoid glucoside and pregnane saponins from the roots of *Hemidesmus indicus*. *Journal of Natural Medicines, 67*(1), 137–142. doi:10.100711418-012-0659-6 PMID:22456894

Chapter 9
The Chemical Constituents, Anti-Inflammatory, Anti-Oxidant, and Ethnomedicinal Properties of *Aloe barbadensis*

Dickson Adom
Kwame Nkrumah University of Science and Technology, Ghana

Paul Appiah Sekyere
Kwame Nkrumah University of Science and Technology, Ghana

Mohan Kumar Krishnappa
Tamil Nadu Agricultural University, India

ABSTRACT

Aloe vera is a perennial, drought-resisting, succulent plant belonging to the Asphodelaceae family that has a long history of having many tremendous medicinal and anti-inflammatory as well as anti-bacterial properties. This chapter highlights the great potentials of the aloe vera species. Desk study and document analysis guided the systematic and extensive review of both published and unpublished resources on the Aloe vera Species. The chapter contends that horticulturists and plant scientists in the field of agriculture and medicine must pay attention to the Aloe vera as a medicinal plant that possesses the most powerful antibacterial and anti-inflammatory properties, which protects the body against constipation through enhancing the body metabolism, skin, and worm infestations.

INTRODUCTION

Aloe vera is a spiky cactus-like xerophytes clump forming perennial, drought-resisting, succulent plant with thick fibrous root (Ahlawat & Khatkar, 2011) belonging to the *Asphodelaceae* family. Specifically, the Aloe species is a member of the tree Lily family together with onion, garlic and turnip (Pandey

DOI: 10.4018/978-1-7998-1320-0.ch009

& Singh, 2016). In terms of physiology and phenology, Aloe vera belongs to a large class of plants known as "xeroids" characterized for its ability to close its stomata completely to avoid loss of water. This adaptation allows the Aloe vera plants to survive long and extreme drought periods (Akinyele & Odiyi, 2007). The Aloe vera plant has been mentioned throughout history and has been ranked as one of the leading and all-purpose plants in the world (Bai, Deva, Madan, & Sharma, 2013). Ahlawat and Khatkar (2011) contend that the popularity of Aloe vera is initiated by naturopaths, yog gurus, alternative medicine promoters and holistic healers. The name, Aloe, is derived from the Arabic "alloeh" or Hebrew "halal" meaning bitter shiny substance. *Aloe vera* is a species of Aloe that is particularly popular for its medicinal properties and as such is often referred to as the miracle plant (Shende & Telrandhe, 2017). In Ayurveda, Aloe is referred to as 'Kumari' meaning 'young girl' because it is believed among the Indians as giving users of products from them, youthful vigor, energy and vitality. Greek scientists regarded *Aloe vera* as *the universal panacea* and the Egyptians call Aloe "the plant of immortality" (Itrat & Zarnigar, 2013). It has a vast traditional role in indigenous system of medicine like homoeopathy (Baby & Justin, 2010). Aloe vera (*Aloe barbadensis* Miller) is a perennial plant of the *Xanthorrhoeaceae* family (USDA, 2012). It has been designated its own family, known as *Aloeaceae* (Eshun & He, 2004). This plant grows well in dry conditions which pretty much get sunlight (Kumar & Yadav, 2014). Aloe vera species cannot thrive in frost or freezing temperatures but only in warm tropical areas (Pandey & Singh, 2016). Aloe vera species can be found in Mexico, Pacific Rim countries, Central America, South America, The Caribbean, Australia, India and Africa.

According to the world checklist of selected plant species, the genus Aloe contains over 550 species (Royal Botanic Garden Kew, 2013). *Aloe barbadensis*, *Aloe arborescens,* and *Aloe chinensis* are the most popular. *Aloe barbadensis* Miller is considered the most biologically active species (Bozzi, Perrin, Austin, & Arcevera, 2007). However, only two species, *Aloe barbadensis* Miller and *Aloe aborescens* Miller are grown commercially (Sanjit & Tushar, 2018).

The leaves of the Aloe plant grow from the base in the rosette pattern. Mature plants can grow as tall as 2 and a half inches to 4 feet with the average being around 28 to 36 inches in length. Each plant usually has at least 12 leaves that, when fully matured, weighs up to three pounds (Pandey & Singh, 2016). Each leaf is composed of three layers: An inner clear gel that contains 99% water and rest is made of glucomannans, amino acids, lipids, sterols and vitamins. This Aloe vera gel is transparent and colorless, slippery mucilage that is produced by the thin-walled tubular cells in the thin-walled tubular cells in the inner central part (parenchyma) of the Aloe vera leaf ((Chatterjee, Chakraborty & Nandy, 2013). The middle layer contains the latex which is the bitter yellow sap and contains anthraquinones and glycosides. The outer thick layer of about 15 to 20 cells called rind has protective function and synthesizes carbohydrates and proteins (Nandal & Bhardwaj, 2012). The Aloe plant is said to be disease free though occasionally, black spots may occur in the upper surface because of fungal infection or soft rottening (Pardeshi & Tapar, 2016). In terms of phytochemistry, the dry matter basis of Aloe gel consists of lipid-4%, phenolic compounds-1%, proteins-7%, minerals-16%, polysaccharides-55%, and sugar-17% (Ahlawat & Khatkar, 2011). Acemannan is the major carbohydrate fraction in the gel and it is made up of water soluble long chain mannose polymer which demonstrates antineoplastic and antiviral effects (Bai et al., 2013).

The best time to cultivate Aloe vera in the year is from March to June and it grows well in coarse sandy loam soils with pH ranging from 7.0 to 8.5 (Maiti & Chandra, 2002). Aloe vera grows best in sunshine conditions with little amount of water for its establishment, growth and reproduction. The plant is highly adapted to areas with mean annual rainfall in the range 700 mm to 3000 mm. The plant thrives

Figure 1. Aloe barbadensis Miller

in areas with mean annual temperatures within the range 19°C -27°C, but can also tolerate temperatures within 10°C to 35°C. Aloe vera flourishes in rich soils, but can tolerate poor, saline, soil (PROTA, 2017).

The unique group of succulent plants of Aloe vera provokes wide ranging interest among both scientists and plant collectors (Smith & Van, 2009). It is believed that an immense indigenous knowledge has been found in local communities where Aloes are potentially grown (Asmelash, 2018). Using Aloes leaves exudates for medicinal purpose was dated back thousands of years in the history (Grace, Simmonds, Smith, & Van 2009). Aloe vera can be used alone or in combination with other substances of natural or synthetic origin to produce efficient pharmaceutical products that can serve as a strong buffer against diseases. Trade in wild and cultivated Aloes for health products is central to the livelihoods of many people in Arid and Semi-Arid lands where the cultivation of conventional crops is unfeasible (Cousins & Witkowski, 2012). Aloe vera is known to contain several pharmacologically active ingredients, including a carboxypeptidase that inactivates bradykinin in vitro, salicylates and a substance (s) that inhibits thromboxane formation in vivo. Scientific studies exist that support an antibacterial and antifungal effect for substance (s) in *A. vera* (Mantle, Gok, & Lennar, 2001). Aloe vera has been used in the production of cosmetic products because the polysaccharides that acts as moisturizers for the hydration of the skin. In home-made cosmetics where Aloe vera is used as an active ingredient, it is recommended that 20% to 40% of the Aloe vera gel is used. When it enters the dermis of the skin, Aloe stimulates the fibroblasts cells to reproduce at a relatively faster rate, producing collagen and elastin fibres that makes the skin

Figure 2: Aloe aborescens Miller
Source: Motaleb (2011)

more elastic and helps the skin in fighting against wrinkles (Bai et al., 2013). Aloe vera juice that has been well prepared can stimulate the immune system, heightening its robustness to withstand diseases. However, Aloe vera products are not to be used internally by pregnant and lactating women as well as those suffering from abdominal pain, intestinal problems and appendicitis (Alhawat & Khatkar, 2011). Economically, the raw material for the Aloe species generates over 125 million dollars while finished products from Aloe species also give the industry close to 110 million dollars (Pandey & Singh, 2016). The objective of this chapter is to briefly highlight the ethnomedical values of Aloe vera in the light of highlighting its chemical constituents, anti-inflammatory, anti-bacterial and medical properties as well as the preparation or formulation of products with Aloe gel as an active ingredient.

METHODS

The researchers employed document analysis (Bowen, 2009) and desk survey (Travis, 2016) for the systematic review of the ethnomedicinal values of Aloe vera species in traditional medicine in peer-reviewed journals, reports, and newspapers (Creswell, 2009). A systematic search using the keywords 'anti-inflammatory properties of Aloe vera', 'ethnobotany of Aloe vera', 'chemical constituents of Aloe vera', 'anti-bacterial, anti-viral, anti-fungal and medicinal uses of Aloe vera' from online databases such as Springer, JSTOR, PubMed, EBSCO, ProQuest and Google Scholar. The empirical studies that were included in the review were classical scholarships on Aloe vera though priority was given to recent studies on Aloe vera published from the last two decades 2000-2019. In all, a total of 85 articles were examined. The articles were carefully reviewed and analyzed to understand (Denzin & Lincoln, 1994) the ethnobotanical values of Aloe vera in traditional medicine. The key qualities in the interpretive document analysis that guided the review were authenticity, credibility, and representativeness (Hefferman, 2013). The documents were perused severally to understand thoroughly the contents of the papers in relation to the main theme of the paper (Creswell, 2009). The main ideas in the reviewed materials were summed up and discussed based on the research questions for the study. The new understanding was subjected to verification with the aim of validating the claims, assumptions, and theories made by scholars (Peshkin, 1993). Finally, a concise, persuasive discussion on the ethnobotanical values of Aloe vera was written.

Chemical Constituents of Aloe Vera

There are over 100 active biologic constituents that have been reported to be found in Aloe vera (Chatterjee et al., 2013). A wide variety of secondary compounds have been isolated from the Aloe vera latex (Reynolds, 2004). The isolated compounds are largely phenolic in nature, and many are anthraquinone *C* glycosides, anthrones, and free anthraquinones (Park et al., 1998). The levels of anthraquinone *C*-glycosides in Aloe vera latex are variable and may constitute up to 30% of the dry weight of the latex (Groom & Reynolds, 1987). Aloe vera latex contains four major C-glycosyl constituents: aloin A, aloin B, aloesin, and aloeresin A. Aloin A, a *C*-glycosylanthrone, also referred to as barbaloin, is the major component of aloe latex. Aloin A and its epimer, aloin B, also referred to as isobarbaloin, have a 9-anthrone skeleton and a β-D-glucopyranosyl substituent. Aloesin, also known as aloeresin B, is a 5-methyl chromone with an 8-β-D-glucopyranosyl substituent, and aloeresin A is a 5-methyl chromone with an 8-β-D-glucopyranosyl-2- O-trans-p-coumarol substituent. Several other C-glycosyl-chromones and anthrones have been isolated from Aloe vera, including aloe-emodin, the anthraquinone of barbaloin and isobarbaloin (Boudreau, Beland, Nichols & Pogribna, 2013). The main constituents of Aloe vera gel are mucopolysaccharides (glucomannans, polymannoses, about 10% of total solids), enzymes, anthranoids, lignin, saponins, vitamins, amino acids (almost 50% of the total amount consisting of 8 of the 10 essential amino acids) and minerals (quantities not given). Total solids are in the range of 1.3 to 2%, the rest being water (Lawless & Allen 2000). Choi and Chung (2003) report of tannins, sterols, salicyclic acid, organic acids, enzymes, saponins, vitamins (B_1, B_2, B_{12}, A, C), and niacin. Yeng, Yao and Fang (2010) isolated a new triglucosylated naphthalene derivative- aloveroside A along with two anthraquinone dimers and two 6-phenyl-2-pyrone derivatives from ethanol extracts. *Aloe excelsa*, also known as the Zimbabwe aloe have two compounds emodinandaloin which were isolated from the leaves of *A. excels* and tested for antibacterial activities against four gram negative and five gram positive bacterial strains of which emodin showed inhibitory activity against all test organisms with MIC ranging from 62.5 µg/ml in *B. subtilis* and *E. coli* to 250 µg/ml in *S. epidermidis* and *S. sonnei* (Coopoosamy & Magwa, 2006). The chemical constituents of the Aloe species are illustrated in Figure 3 below.

Medicinal Uses of Aloe Vera

Pharmaceutical industries globally are benefitting from the Aloe species in the treatment of ailments and diseases. The Aloe gel is used for the preparation of topical ointments, gel preparations, tablets and capsules (Ahlawat & Khatkar, 2011). Medicinal properties of Aloe vera range from external burn

Figure 3. Types of anthraquinones in Aloe vera gel
Source: Emias, Bisrat, Alvaro & Ben-Erik (2000)

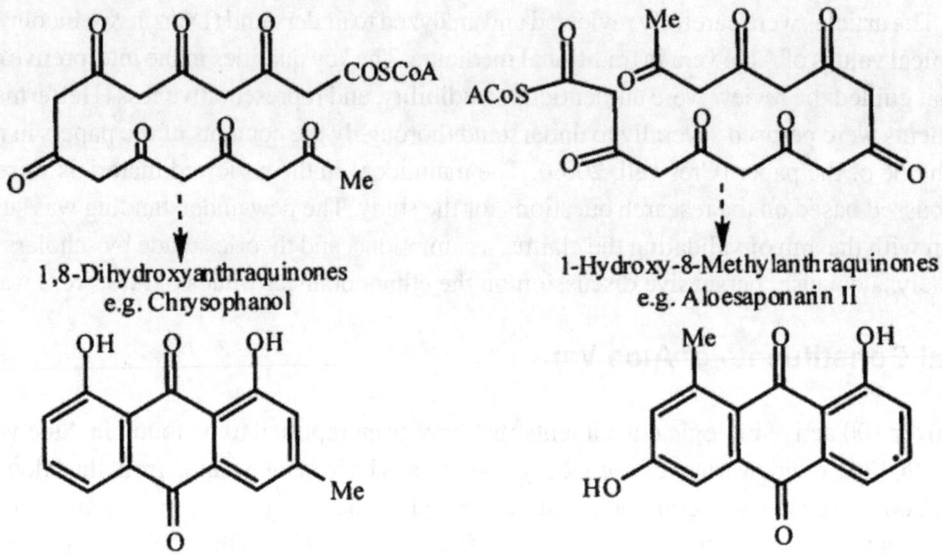

treatments to helping relieve constipation by consumption (Karkala & Bhushan, 2014). Aloe vera is the most efficacious natural plant used both externally and internally and there are numerous benefits that are derived from this wonderful plant. The health benefits of Aloe vera have been propagated throughout the world. There are two parts of Aloe vera that are commonly used. Anthraquinones present in latex are a potent laxative which increases intestinal water content, stimulates mucus secretion and increases intestinal peristalsis (Ishii, Tanizawa, & Takino, 1994). The bitter exudate is used as a natural drug for its cathartic effect and is widely employed as a bittering agent in alcoholic beverages and as a laxative (Karkala & Bhushan, 2014). The dried latex, with a high barbaloin content, is a strong laxative, but its use should be limited to no more than one week, and it should not be used during menstruation, pregnancy and nursing (Franz et al., 2005). Karkala and Bhushan (2014) outlined the five unique benefits of Aloe vera to the body as follows:

1. Penetration – aloe has the ability to reach deepest body tissues some 7 layers deep
2. Antiseptic – aloe has at least 6 antiseptic agents which kill bacteria, viruses and fungi
3. Stimulates cell growth – aloe stimulates the birth of new healthy tissue
4. Settles nerves – aloe has a clearing effect on the body's nervous system
5. Cleanses – aloe detoxifies and normalizes the body's metabolism

The Aloe gel stimulates cell growth and enhances and restores damaged skin in the field of dermatology (Shende & Telrandhe, 2017). This is due to the tannins, sterols, organic acids, enzymes, vitamins and minerals in Aloe vera plants, they can be used in healing wounds, burns and for the prevention of inflammation in seborrheic dermatitis (Rulz, Pleguezuelos, Murioz, & Gallardo, 2004).

Anti-Inflammatory Action and Immunity Activity of Aloe Vera

Because of salicylic acid, which is both analgetic and anti-inflammatory, the production of prostaglandins from arachidonic acid is inhibited, so Aloe has been used to help with arthritis and joint related problems (Efterpi & Panagiota, 2010). Again, Karkala and Bhushan (2014) reports that Aloe vera juice contains 12 essential nutrients that inhibits inflammation and improves joint and muscle mobility with no side effects. The gel of A. vera contains *bradykinase* and other antiprosta gland in compounds which relieve inflammation (Hamman, 2008). The Aloe vera sterol is anti-inflammatory in nature, helps in reducing the inflammation pain and act as natural analgesic (Gebremedhin, 2018). Aloe vera gel sterols were able to reduce inflammation by up to 37%, *Lupeol*, the most active anti- inflammatory sterol, reduced inflammation in a dose dependent manner (Jensen, Lenninger, Beaman, Taylor, & Benson, 2015).Results from a study on anti-inflammatory and analgesic effects of aqueous extract of Aloe Vera (*Aloe barbadensis*) in rats shows that the traditional use of *Aloe* barbadensis in the treatment of various types of pain and inflammatory conditions has a definite basis (Egesie, Chima, & Galam, 2011). Aloe vera inhibits the cyclooxygenase pathway from arachidonic acid (Hutter et al., 1996). Alprogen inhibit calcium influx into mast cells, thereby inhibiting the antigen-antibody- mediated release of histamine and leukotriene from mast cells (Ro et al., 2000).

Antiviral and Anti-Tumor Activity of Aloe Vera

Aloe vera facilitates the stimulation of immune system that in turn protects the body against viral and tumor related disorders. An induction of glutathione *S-transferase* and an inhibition of the tumor-promoting effects of phorbolmyristic acetate has been reported which suggest a possible benefit of using aloe gel in cancer chemoprevention (Kim, Kacew, & Lee, 1999). It has been reported by Radha, Gowrinath, Borra and Krishna (2011) that lectins inhibit aminopyrine uptake by parietal cells, thus the ability of the extract to inhibit gastric acid output may be as a result of direct action on the acid producing cells. An independent study showed that decolorized whole leaf Aloe vera with low levels of aloin did not cause adverse effects in rats, this reinforces the idea that aloin may be responsible for the adverse effects (tumor) seen in the colon of rats in the NTP study of non-decolorized whole leaf Aloe vera (National Center for Complementary and Integrative Health, 2016). The antiviral and anti-tumor actions may be direct or indirect. The indirect effect is due to stimulation of the immune system and the direct effect is due to the anthraquinones. The anthraquinonealoin renders viruses such as *herpes simplex*, *varicella zoster* and influenza inactive (Malik, 2013). Kim and Lee (1997) has reported that fractions of a polysaccharide has shown to inhibit the binding of benzopyrene to primary rat hepatocytes, thereby preventing the formation of potentially cancer initiating benzopyrene DNA adducts.

Anti-Fungal Properties of Aloe Vera

Bovine dermatophytosis causes economic loss due to hide damage and restrictions in showing and marketing infected cattle (Hany, El-sayed, Magda, Ahmed, & Magdy, 2004). Trichphytonverrucosum, Trichphytonmentagrophytes and Trichphytonmegnini are the most commonly fungal species that involved in ringworm (Quinn, Carter, Markey& Carter 1994). Aloe gel is used topically for infections that are caused by these fungal agents. This therapeutically benefit of Aloe vera gel leaves might be due to its properties and chemical components (Gebremedhin, 2018). Also the potency of A. vera in curing ringworm might be due to increasing the ability of internal immunity that strength the natural resistance by boosting T- lymphocyte cells that aid the immune system (Hany et al., 2004). Aloe Vera is used widely in dermatology, as it acts as an astringent, moisturizer, humidifier and cleanser. It softens the skin, diminishes wrinkles and cures acne, herpes, red spots, psoriasis, eczema, mycosis, fever blisters, skin irritation and provides protection to the skin against pollution. For its anti-fungal properties, Aloe vera is used as a fish tank water conditioner (Rajeswari et al., 2018).

Anti-Bacterial Properties of Aloe Vera

Because of the increasing resistance to antibiotics of many bacteria, plant extracts are of new interest as antiseptics and antimicrobial agents in medicine (Yisehak, Eyosyas, Cruz & Getachew, 2014). The bacteria that attack our body are rapidly combatted with the use of Aloe gel or juice. The combined antibacterial effects of the lignin, cinnamic acid, and chrysophanic acid are immediate. The cytotoxic characteristics owed to their anthracenic structure respond quickly on superficial cells, liberating these cells from unwanted guests (Alessandro & Stefano, 2005). The Aloe extract was potent against various species of bacteria including *S. aureus*, *S. agalctia*, *E. choli*, *Streptococcus pyogens*, *M. bacteriumtuberculosis*, *Pseudomonasaerogenosa*, *Klebsiella pneumonia*, etc. Whole leaf components of A. vera are proposed to have direct antibacterial properties include anthraquinones and saponins (Rubina, Priyanka, & Jeyakumar 2009). Mbajiuka, Obeagu and Okwandu (2014) demonstrated that ethanol and aqueous extracts of Aloe veragel had appreciable antimicrobial activity against *E scherichia coli*, *Staphylococcus aureus* and *Candida albicans* with methanol extract having the lowest activity on *Escherichia coli* but did not inhibit *Staphylococcus aureus* and *Candida albicans*. Many researchers mentioned that *Aloe vera* inhibits the growth of some microorganisms like *Str. pyogenes*, *Shigellaflexneri*, *Klebsiella*sp., especially against Gram-positive bacteria causing food poisoning or diseases in humans and animals (Ferro et al., 2003). Bhati, Jaidka and Somani (2015) showed that herbal dentifrices containing Aloe vera can be safely recommended as an alternative to ñuoridated dentifrices in terms of antimicrobial efficacy. They compared the number of bacteria in patients using a ñuoride-containing dentifrice, an Aloe vera containing dentifrice, and in patients, who did not use any dentifrice, a continuous reduction in microbial count was observed in the ñuoridated dentifrice group and Aloe vera dentifrice group. A similar trial was also conducted by Prabhakar, Karuna, Yavagal and Deepak (2015). They evaluated the efficacy of hand instruments in excavating dental caries and the role of Aloe vera as potential cavity disinfecting agent. The study included a control group and an Aloe vera group. Dentinal samples were collected three times from each carious tooth and subjected to microbiological analyses for total viable count. A statistically significant reduction in the bacterial count was observed in both groups post-excavation and post-cavity disinfection. In the Aloe vera group, a significant reduction in the number of bacterial, compared to control group, was observed. The results of this showed that Aloe vera extract can be used as a potential, natural disinfecting agent.

Aloe Vera as a Skin Care and Anti-Aging

Aloe Vera is used widely in dermatology, as it acts as an astringent, moisturizer, humidifier and cleanser. It softens the skin, diminishes wrinkles and cures acne, herpes, red spots, psoriasis, eczema, mycosis, fever blisters, skin irritation and provides protection to the skin against pollution. Also, it is ideal for sunburns, fragile skin, and for removal and repair of dead skin and cells (Karkala & Bhushan, 2014).

The nutrients and medicating substances contained in Aloe are widely used in the cosmetics industry, in skin creams, lotions, and special treatment preparations aimed at resolving a large number of skin problems. Topical application of the gel, on a daily basis, cleanses and regenerates the skin, supplying the correct amount of vitamins and stimulating blood circulation to guarantee improved oxygenation and rapid expulsion of toxins. Skin becomes smoother, moisturized, and more elastic because Aloe's antioxidants defend it against free radicals and their degenerative action on the epithelial cells. Its anti-aging effects are an added bonus. Mixture of a teaspoon of honey with one tablespoon of pure Aloe juice, the yolk of an egg, a teaspoon of castor oil, a teaspoon of lemon juice, and a teaspoon of black currant juice, a pinch of oatmeal to this mixture to give it a thicker consistency is an excellent bio-stimulant for the skin, cleansing it and feeding its deeper layers (Alessandro & Stefano, 2005). Paulsen, Korsholm and Brandrup (2005) conducted a randomized, double-blind, placebo-controlled study on 41 patients with stable plaque psoriasis. The study comprised a 2-week wash-out period followed by a 4-week treatment period with two applications daily, and follow-up visits after one and two months. Results of this study indicated that the score sum of erythema, infiltration and desquamation decreased in the *Aloe vera* extract-treated sites, compared with the placebo-treated areas with no serious side-effects. The efficacy of treatment with *Aloe vera* gel was lower than treatment with placebo gel.

Panahi et al. (2012) conducted randomized, double-blind trials on 32 infants in order to identify effective strategies to decrease the symptoms of nappy dermatitis in infants. Patients were treated with *Aloe vera* cream three times a day for 10 days. Severity of the nappy dermatitis was decreased in infants treated with *Aloe vera*. The results of these experiments show that topical application of *Aloe vera* gel can be used as a safe and well-functioning nappy dermatitis therapy. These beneficial effects could be attributed to the well-known anti-inñammatory properties of *Aloe vera* (Reuter et al., 2008). Rajar, Majeed, Parveen, Sheikh and Sushel (2008) conducted a trial to compare the efficacy of *Aloe vera* gel in the topical management of vulval lichen planus of which 34 female patients were randomized into two groups to receive *Aloe vera* gel or placebo for local application for eight (8) weeks. The results of this study showed that *Aloe vera* gel might be a safe and effective treatment for reducing symptoms, or even inducing remission in vulval lichen planus. Anthraquinone-rich *Aloe vera* extracts may function as absorbers of ultraviolet radiation in the sun, because anthraquinones absorb ultraviolet radiation (Committee of Experts on Cosmetic Products, 2008). *Aloe vera* supports the healing of first to second degree burns although the exact role is not well known. It suggests that *lectin* may be responsible for the therapeutic effect on skin exposure to UV and X-radiation (Maenthaisong, Chaiyakunapruk, & Niruntraporn, 2007).*Aloe vera* first gained popularity in the USA in the 1930s with reports of successful use of freshly cut leaves in treating X-ray burns. Both classes of *Aloe vera* leaf products, gel and latex are reported to possess a wide range of pharmaceutical activities. Topical usage of 50% ethanol extract of *Aloe vera*, three times per day for five consecutive days per week has been used for treatment of genital herpes and psoriasis vulgaris (Ulbricht et al., 2007).

Ethnobotanical Aspects and Bioactive Constituents of Aloe Vera

The Aloe vera leaf gel contains about 98% water. The total solid content of Aloe vera gel is 0.66% and soluble solids are 0.56% with some seasonal fluctuation. On dry matter basis aloe gel consists of polysaccharides 53%, sugars 17%, minerals 16%, proteins 7%, lipids 5% and phenolic compounds 2% (Sharrif & Verma, 2011). Aloe vera contains 200 potentially active constituents: vitamins, enzymes, minerals, sugars, lignin, saponins, salicylic acids and amino acids, which are responsible for the multifunctional activity of Aloe (Gautam & Awasthi, 2007). The bioactive compounds in Aloe vera are used as haemostatic, astringent, antiseptic, antibacterial, anti-inflammatory, antidiabetic, antiulcer, antioxidant and anticancer agents (Chatterjee et al., 2013). Aloe vera gel is used in fighting against all forms of skin diseases including pimples. In such instance, a heated leaf is applied on the face pimples, about two to three teaspoons jel from the leaves is mixed with two to three drops of lemon juice and the mixture is then applied on pimples for one week. In the case of abdominal pains, one large leaf of Aloe is cut into small pieces, put in a new bowl together with some lemon juice and ajwain seeds. It is then dried in the sun and crushed into powder for treating gas troubles and abdominal pains (Sultana, Mir, Mushtaq & Zafar, 2006). The raw gel is applied to inflamed skin and used on radiation burn to enhance the skin. Currently the plant is widely used in skin care, cosmetics and as nutraceuticals (Chatterjee et al., 2013).

In India (Ayurveda), Aloe vera is used as tonic for female reproductive system. It is believed to rejuvenate the energy in female and enhances their feminity (Kumar & Yadav, 2014). It is used in traditional Indian medicine for treating constipation, colic, skin diseases suc as eczema or psoriosis, intestinal worms, as a laxative, antihelmintic, haemorrhoid and as uterine stimulant (menstrual regulator). In Tami Nadu state, prepared curry of Aloe vera is taken along with Indian bread or rice (Heber, 2007) to improve the immune system. The skin and inner lining of aloe vera leaves are used as remedies for constipation in China (Babiei, Manafi & Tarafi, 2013). Also, Aloe vera is useful and is found in food, pharmaceutical products and cosmetics (Foster, Hunter, & Samman, 2011). Aloe vera is found in some drugs for treating type II diabetes mellitus (Coronado, Thompson, Tejeda & Godina, 2004).

Vitamins: Aloe vera contains Vitamins A (beta-carotene), C and E, which are antioxidants. It also contains Vitamins B12, B1, B2, B6, C, β-carotene, choline, folic acid, ß-carotene, choline, folic acid, α-tocopherol

Figure 4. The percentages of the bioactive constituents in Aloe vera

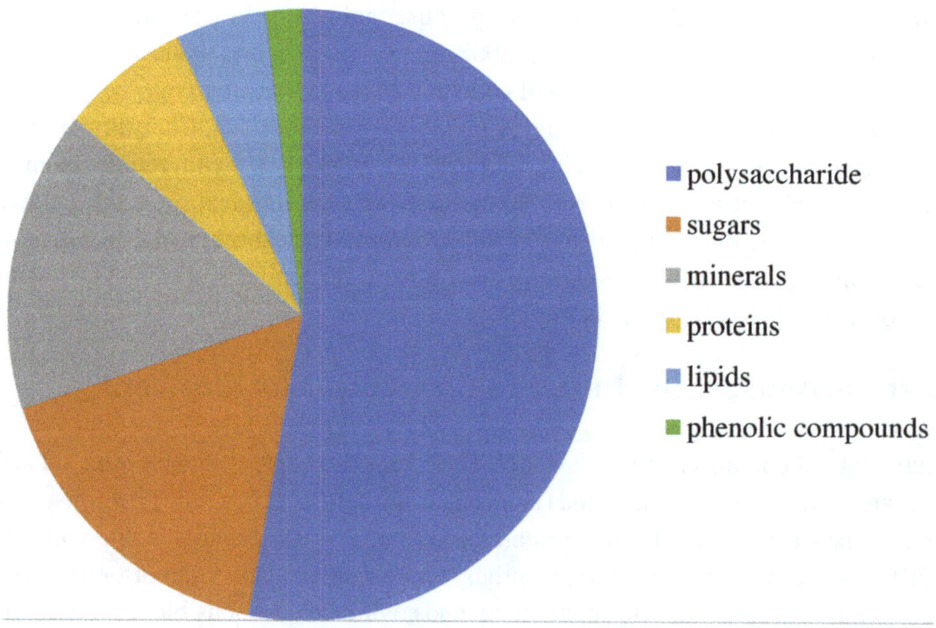

Enzymes: Aloe vera contains various enzymes such as bradykinase, catalase, aliiase, alkaline phosphatase, amylase, carboxypeptidase, cellulase, lipase, and peroxidasecyclooxidase, cyclooxigenase, lipase, oxidase, phosphoenolpyruvate carboxylase and superoxide dismutase. Bradykinase reduces excessive inflammation of the skin and the other enzymes help in the breakdown of fats and sugars (Bawankar, Singh, & Subramanian, 2014).

Inorganic Compounds: The inorganic compounds in the Aloe vera include calcium, chlorine, chromium, copper, iron, magnesium, manganese, potassium, phosphorous, sodium and zinc.

Organic Compounds: The organic compounds in Aloe vera are Arachidonic acid, Linolenic acid, triglycerides, triterpenoid, gibberellins, lignin, potassium sorbate, salicylic acid and uric acid (Senan, 2014).

Non- essential and essential amino acids: Aloe vera gel contains seven out of eight of the essential amino acids (isoleucine, leucine, lysine, methionine, phenylalanine, theonine and valine). These amino acids are required by the human body but cannot be synthesized. The nine non-essential amino acids (Hydroxyproline, threonine,proline, tyrosine, Histidine, Glycine, Glutamic acid, Arginine, Alaline) can also be found in aloe vera gel (Bhattacharya, Malik, & Singh, 2011).

Proteins: The proteins in Aloe vera include lectins and lectin-like substances (Misri, Fatema, Brishti, & Hoque, 2014).

Saccharides: The most essential sugars in Aloe vera are the long chain polysaccharides consisting of glucose and mannose, L-rhamnose and aldopentose. These are known as mucopolysaccharides. A glycoprotein with anti-allergic properties, called alprogen and anti-inflammatory compound, and C-glucosyl chromone has also been isolated from Aloe vera (Balasubramanian & Narayanan, 2013).

Carbohydrate: The carbohydrates in Aloe vera are pure mannan, acetylated mannan, acetylated glucomannan, glucogalactomannan, galatan, galactogalacturan, arabinogalactan, galactoglucoarabinomannan, pectic substance, xylan, and cellulose.

Minerals: The minerals in Aloe vera include calcium, sodium, manganese, magnesium, zinc, copper and chromium as well as other.

Anthraquinone: The anthraquinones in Aloe vera are Aloe-emodin, aloectic acid, anthranol, aloin A and B (or collectively known as barbaloin), isobarbaloin, emodin and ester of cinnamic acid.

Natural Products Containing Aloe vera

Aloe vera gel has been used for the production of various health and food products. Some of these products with Aloe vera gel as the main bioactive principle from some studies have been reviewed under this section.

CONCLUSION

Aloe vera is a traditional perennial plant with diverse benefits to mankind. This chapter has highlighted the unique relevance of the plant in pharmaceutical use especially by unveiling the plant's chemical constituents, anti-bacterial, anti-inflammatory and medicinal properties. Research institutions in the field of horticulture, agriculture, ethnobotany and other fields of plant science must investigate other benefits of the plant while highlighting how it could be branded in the production of modern pharmaceutical and/or agricultural products. In addition, international regulatory bodies such as the International Aloe Science Council must liaise with the Food and Drug Regulatory Boards of countries to set high standards for Aloe vera products approval and seal of quality.

Table 1. Natural products from Aloe vera

Product Description	Formulation/Preparation	Clinical Effects	References
Gel Beads	1. Collect, wash and rinse Aloe vera leaves. 2. Collect the Aloe gel by cutting the Aloe leaves in the middle section. 3. Pour 10ml of distilled water and add 300mg of sodium alginate solution. 4. Add 2gm of the Aloe gel at 1000 rpm using a magnetic stirrer for 15 minutes. 5. Extrude the mixture through a syringe (no-18) into 5% calcium chloride solution with gentle agitation at 37° C. 6. Allow the formed beads to stand for 5 minutes in the solution. 7. Separate the formed beads via filtration using a Wattman filter paper of 0.45 mm size.	1. It prevents photo suppression. 2. Effective treatment for first and second degree burns. 3. Prevents the formation of liver spots which is often found on aging skins. 4. Prevents the growth of microorganisms inside the oral cavity.	Bai et al. (2013)
Aloe vera Tooth Gel	1. Sodium saccharin, sodium fluoride, preservatives, colour, sodium salt and alum is added to the sufficient quantity of water. 2. Mix it properly and warm the water up to 45° C to 50° C 3. Add VTP gum as well as Sorbitol and poly-ethylene glycol. 4. Add silica powder (abrasive silica and hydrated silica) and mix it in 15 minutes. 5. Add the Aloe vera gel and flavor to it and mix it properly within 10-15 minutes. 6. Add sodium laurylsulphate to make a slurry. 7. Air bubbles are removed by the use of a vacuum. ***Preparation of the Aloe gel*** 1. Collect and wash the leaves of Aloe vera in a solution of water and a mild chlorine solution. 2. Cut the Aloe vera leaves transversely into pieces. 3. Remove the thick epidermis with a vegetable peeler while removing the inner gel-like pulp with a spoon, minced and homogenized in a mixer. 4. Different gellers such as sodium carboxy methylcellulose, carbomer 934, and carrageenam gum in different concentration of viscosity enhancers are used to optimize the concentration of the gelling agent to achieve consistency in the gel formulations.	1. It reduces plaque in the teeth and gives it a whitening effect. 2. It reduces the bad odor in the teeth. 3. It reduces inflammation in dental implants.	Pardeshi & Tapar (2016)
Aloe Juice	1. Use a sharp knife to remove the mucilage and fillet. 2. The Aloe vera gel fillet is washed with deionized water and transferred to the pulper. 3. The pulper is fitted to a refrigerated system to keep the temperature of the extracted juice lower to prevent decomposition. 4. The bioactivity of sensitive molecules in the Aloe vera juice is transferred into a protective tank. 5. The Aloe juice is mixed with other agents to produce a pharmaceutical, cosmetic or food product.	It has several therapeutic benefits such as wound healing, treatment of burns, protection against skin damage from X-rays, decreases blood glucose level, treatment of genital herpes and psoriases, lung cancer, and intestinal problems	Ahlawat & Khatkar (2011)
Aloe vera Skin Cream	1. The fresh leaves of Aloe vera is collected and washed in running water for 15 minutes. 2. Rinse it in sterile water and mild chlorine solution. 3. It is dissected longitudinally and the Aloe gel is removed using a sterile knife. 4. 5% Aloe vera gel is dissolved in 100% distilled water. 5. 23% olive oil is heated to 85° C while adding 5% of olivem 700 until it was melted. 6. After cooling, n-decane 0.3% was added with shaking and the aqueous phase is slowly added.	1. It is used in dermatology in enhancing the skin's smoothness. 2. It is used to heal wounds and burns.	Rulz et al. (2004)

REFERENCES

Ahlawat, K. S., & Khatkar, B. S. (2011). Processing Food Applications and Safety of Aloe vera Products: A Review. *Journal of Food Science and Technology*, *48*(5), 525–533. doi:10.100713197-011-0229-z PMID:23572784

Akinyele, B. O., & Odiyi, A. C. (2007). Comparative study of vegetative morphology and the existing taxonomic status of Aloe vera L. *Journal of Plant Sciences*, *2*(5), 558–563. doi:10.3923/jps.2007.558.563

Alessandro, B., & Stefano, S. (2005). *The great Aloe book, history, botany, composition, and pharmacological aspects of this legendary plant*. Trento: Ziccari Pty Ltd.

Asmelash, T. G. (2018). The Types of Aloe Species and Their Multi-Functions in Hammer District of Southern Ethiopia. *Annals of Reviews and Research*, *1*(2), 555–559.

Babaei, A., Manafi, M., & Tarafi, H. (2013). A study on Effects of Aloe vera leaf Extract on growth of *Aspergillus flavus*. *Annual Research & Review in Biology*, *3*, 1091–1097.

Baby, J., & Justin, S. R. (2010). Pharmacognostic and phytochemical properties of *Aloe vera* -an overview. *International Journal of Pharmaceutical Sciences Review and Research*, *4*, 106.

Bai, A., Ara, T., Deva, A. S., Madan, J., & Sharma, S. (2013). Preparation and Evaluation of Novel Aloe vera Gel Beads. *Journal of Global Biosciences*, *2*(6), 206–216.

Balasubramanian, J., & Narayanan, N. (2013). Natures Gift. *Species*, *2*(1), 11–13.

Bawankar, R., Singh, P., & Subramanian, B. (2014). Bioactive compounds and medicinal properties of Aloe vera L.: An update. *Journal of Plant Sciences*, *2*(3), 102–107.

Bhati, N., Jaidka, S., & Somani, R. (2015). Evaluation of antimicrobial efficacy of Aloe vera and Meswak containing dentifrices with ñouridated dentifrice: An in vivo study. *Journal of International Society of Preventive & Community Dentistry*, *5*(5), 394–399. doi:10.4103/2231-0762.165924 PMID:26539392

Bhattacharya, M., Malik, S., & Singh, A. (2011). Aloe vera barbedensis: A review on its Ethnopharmacological value. *Journal of Pharmacy Research*, *4*(1), 4507–4510.

Boudreau, M. D., Beland, F. A., Nichols, J. A., & Pogribna, M. (2013). Toxicology and carcinogenesis studies of a non-colorized whole leaf extract of Aloe barbadensis Miller (Aloe vera) in rats and mice (drinking water study). *National Toxicology Program Technical Report Series*, *577*(577), 1–266. PMID:24042237

Bowen, G. A. (2009). Document Analysis as a Qualitative Research Method. *Qualitative Research Journal*, *9*(2), 45–65. doi:10.3316/QRJ0902027

Bozzi, A., Perrin, C., Austin, S., & Arce, V. F. (2007). Quality and authenticity of commercial Aloe vera gel power. *Food Chemistry*, *103*(1), 22–30. doi:10.1016/j.foodchem.2006.05.061

Chatterjee, P., C.hakraborty, B., & Nandy, S. (2013). Aloe vera plant: Review with significant pharmacological activities. *Mintage Journal of Pharmaceutical and Medical Sciences*, *2*(3), 20–24.

Choi, S., & Chung, M. H. (2003). A review on the relationship between Aloe vera components and their biologic effects. *Seminar in integrative. Medicine*, *1*(1), 53–62.

Committee of Experts on Cosmetic Products. (2008). *Aloe extracts with anthraquinones. Active ingredients used in cosmetics: safety survey*. Strasbourg, France: Council of Europe Publishing.

Coopoosamy, R. M., & Magwa, M. L. (2006). Antibacterial activity of Aloe emodin and aloin A isolated from *Aloe excelsa*. *African Journal of Biotechnology*, *5*(11), 1092–1094.

Coronado, G. D., Thompson, B., Tejeda, S., & Godina, R. (2004). Attitudes and Beliefs among Mexican Americans about type II diabetes. *Journal of Health Care for the Poor and Underserved*, *15*(4), 576–588. doi:10.1353/hpu.2004.0057 PMID:15531816

Cousins, S. R., & Witkowski, E. T. F. (2012). African aloe ecology, a review. *Journal of Arid Environments*, *85*(1), 1–17. doi:10.1016/j.jaridenv.2012.03.022

Creswell, J. W. (2009). *Research Design* (3rd ed.). United States: Sage.

Denzin, N. K., & Lincoln, Y. S. (1994). Introduction: Entering the Field of Qualitative Research. In: N. K. Denzin, & Y. S. Lincoln (Eds.), Handbook of Qualitative Research. Thousand Oaks, CA: Sage.

Efterpi, V. C., & Panagita, C. F. P. (2010). Aloe vera, a plant for many uses. *Journal of Food Agriculture and Environment*, *8*(2), 245–249.

Egesie, U. G., Chima, K. E., & Galam, N. Z. (2011). Anti-inflammatory and Analgesic Effects of Aqueous Extract of Aloe Vera (*Aloe barbadensis*) in Rats. *African Journal of Biomedical Research*, *14*(1), 209–212.

Emias, D., Bisrat, D., Alvaro, V., & Ben-Erik, V. W. (2000). Chemistry of Aloe species. *Current Organic Chemistry*, *4*(10), 1055–1078. doi:10.2174/1385272003375932

Eshun, K., & He, Q. (2004). Aloe vera: A valuable ingredient for the food, pharmaceutical and cosmetic industries: A Review. *Critical Reviews in Food Science and Nutrition*, *44*(2), 91–96. doi:10.1080/10408690490424694 PMID:15116756

Ferro, V., Bradbury, B., Cameron, P., Shakir, E., Rahman, S., & Stimson, W. (2003). In vitro susceptibility of *Shigella flexneri* and *Streptococcus pyogenes* to inner gel of Aloe barbadensis Miller. *Antimicrobial Agents and Chemotherapy*, *3*(1), 1137–1139. doi:10.1128/AAC.47.3.1137-1139.2003 PMID:12604556

Foster, M., Hunter, L., & Samman, S. (2011). Evaluation of the nutritional and metabolic effects of Aloe vera. In I. F. F. Benzie, & S. Watchel-Galor (Eds.), Herbal Medicine: Biomolecular and Clinical Aspects (2nd ed.). Boca Raton, FL: CRC Press. doi:10.1201/b10787-4

Gautam, S., & Awasthi, P. (2007). Nutrient composition and physiochemical characteristics of *Aloe vera* (*Aloe barbadensis*) powder. *Journal of Food Science and Technology*, *44*(1), 224–225.

Gebremedhin, Y. (2018). Review on Medicinal Value of Aloe vera in Veterinary Practice. *Biomedical Journal of Science and Technical Research*, *6*(1), 1–6.

Grace, O. M., Simmonds, M. S. J., Smith, G. F., & Van, W. (2009). Documented utility and biocultural value of Aloe L. (Asphodelaceae) a review. *Economic Botany*, *63*(2), 167–178. doi:10.100712231-009-9082-7

Groom, Q. J., & Reynolds, T. (1987). Barbaloin in Aloe species. *Planta Medica*, *53*(4), 345–348. doi:10.1055-2006-962735 PMID:17269040

Hamman, J. (2008). Composition and application of Aloe vera leaf gel. *International Journal of Pharmacological Science*, *12*(8), 1599–1616.

Hany, Y. H., El-sayed, M., Magda, A. S., Ahmed, Z., & Magdy, A. A. (2004). Efficacy of Aloe vera gel leaves for treatment of skin infections among animals. In *Proceedings of the first anniversary conference of the Faculty of Veterinary Science.* Moshtohor: Semantic Scholar.

Heber, D. (2007). *Physicians' Desk Reference for Herbal Medicines* (4th ed.). Montvale: Thompson Health Care.

Hefferman, C. (2013). *Qualitative Research Approach.* [Online] Available from http://www.explorable.com

Hutter, J. A., Salmon, S., Stavinoha, W. B., Satsangi, N., Williams, R. F., & Streeper, R. T. (1996). Anti-inflammatory C-Glycosylchromone from Aloe Barbadensis. *Journal of Natural Products*, *59*(1), 541–543. doi:10.1021/np9601519 PMID:8778246

Ishii, Y., Tanizawa, H., & Takino, Y. (1994). Studies of Aloe vera: Mechanism of Cathartic Effect. *Biological & Pharmaceutical Bulletin*, *17*(1), 651–653. doi:10.1248/bpb.17.651 PMID:7920425

Itrat, M., & Zarnigar, K. (2013). *Aloe vera*: A review of its clinical effectiveness. *International Research Journal of Pharmacy*, *4*(1), 75–79.

Jensen, G. M., Lenninger, J., Beaman, R., Taylor, K., & Benson, N. (2015). Support of Joint Function, Range of Motion and Physical Activity Levels by Consumption of a Water-Soluble Egg Membrane Hydrolyzate. *Journal of Medicinal Food*, *18*(9), 1042–1048. doi:10.1089/jmf.2015.0041 PMID:26355952

Karkala, M., & Bhushan, B. (2014). Aloe vera: A wonder plant its history, cultivation and medicinal uses. *Journal of Pharmacognosy and Phytochemistry*, 2(5), 85–88.

Kim, H. S., Kacew, S., & Lee, B. M. (1999). In vitro chemopreventive effects of plant polysaccharides (*Aloe barbadensis Miller, Lentinusedodes, Ganoderma Lucidum and Cariolus Vesicolor*). *Carcinogenesis*, *20*(8), 1637–1871. doi:10.1093/carcin/20.8.1637 PMID:10426820

Kim, H. S., & Lee, B. M. (1997). Inhibition of benzopyrene-DNA adduct formation by *Aloe* barbadensis Miller. *Carcinogenesis*, *18*(1), 771–776. doi:10.1093/carcin/18.4.771 PMID:9111213

Kumar, S., & Yadav, J. P. (2014). Ethnobotanical and pharmacological properties of *Aloe vera*: A Review. *Journal of Medicinal Plants Research*, *8*(4), 1387–1398.

Lawless, J., & Allen, J. (2000). *Aloe vera-Natural wonder care*. Hammersmith: Harper Collins Publishers.

Maenthaisong, R., Chaiyakunapruk, N., Niruntraporn, S., & Kongkaew, C. (2007). The efficacy of *Aloe vera* for burn wound healing: A systematic review. *Burns*, *33*(1), 713–718. doi:10.1016/j.burns.2006.10.384 PMID:17499928

Maiti, S., & Chandra, R. (2002). *Cultivation of Aloe vera. National Research Centre for Medicinal and Aromatic plants*. Gujarat: Anand Press.

Malik, I. Z., & Zarnigar. (2013). Aloe vera, a review article on its clinical effectiveness. *International Research Journal of Pharmacy*, *4*(8), 75–79. doi:10.7897/2230-8407.04812

Mantle, D., Gok, M. A., & Lennard, T. W. (2001). Adverse and beneficial effects of plant extracts on skin and skin disorders. *Adverse Drug Reactions and Toxicological Reviews*, *20*(2), 89–103. PMID:11482001

Mbajiuka, C. S., Obeagu, E. I., & Okwandu, G. E. (2014). Antimicrobial effects of Aloe vera on some human pathogens. *International Journal of Current Microbiology and Applied Sciences*, *3*(3), 1022–1028.

Misri, J., Fatema, H., Brishti, M., & Hoque, M. (2014). Aloe vera gel as a novel edible coating for fresh fruits: A Review. *American Journal of Food Science and Technology*, *2*(3), 93–97. doi:10.12691/ajfst-2-3-3

Motaleb, M. A. (2011). *Selected Medicinal plants of Chittagong Hill Tracts*. Bangladesh: International Union for Conservation of Nature.

Nandal, U., & Bhardwaj, R. L. (2012). *Aloe vera*: A valuable wonder plant for food, medicine and cosmetic use - A review. *International Journal of Pharmaceutical Science Research*, *13*(1), 59–67.

National Center for Complementary and Integrative Health. (2016). *National toxicology program*. Available from https://nccih.nih.gov/health/aloevera

Panahi, Y., Sharif, M. R., Sharif, A., Beiraghdar, F., Zahiri Z, Amirchoopani, G., Eisar, T. M., & Sahebkar, A. (2012). A randomized comparative trial on the therapeutic efficacy of topical Aloe vera and *Calendula officinalis* on diaper dermatitis in children. *Scientific World Journal, 8*(10), 22-34.

Pandey, A., & Singh, S. (2016). Aloe vera: A Systematic Review of Its Industrial and Ethnomedical Efficacy. *International Journal of Pharmaceutical Research and Allied Sciences*, *5*(1), 21–33.

Pardeshi, M. D., & Tapar, K. K. (2016). Formulation and Evaluation of Aloe vera Gel with Active Salt and Alum: As a New Dentifrice. *International Journal of Research in Engineering and Social Sciences*, *6*(5), 297–310.

Paulsen, E., Korsholm, L., & Brandrup, F. (2005). A double-blind, placebo-controlled study of a commercial Aloe vera gel in the treatment of slight to moderate psoriasis vulgaris. *Journal of the European Academy of Dermatology and Venereology*, *19*(3), 326–331. doi:10.1111/j.1468-3083.2004.01186.x PMID:15857459

Peshkin, A. (1993). The Goodness of Qualitative Research. *Educational Researcher*, *22*(2), 23–29. doi:10.3102/0013189X022002023

Prabhakkar, A. R., Karuna, Y. M., Yavagal, C., & Deepak, B. M. (2015). Cavity disinfection in minimally invasive dentistry – comparative evaluation of Aloe vera and propolis: A randomized clinical trial. *Contemporary Clinical Dentistry*, *6*(1), 24–31. doi:10.4103/0976-237X.152933 PMID:25821369

PROTA. (2017). *Plant Resources of Tropical Africa*. Available from https://www.prota4u.org/

Quinn, P. J., Carter, M. E., Markey, B. K., & Carter, R. C. (1994). *Clinical veterinary microbiology* (1st ed.). United States: Mosby.

Radha, S., Borra, S. K., Krishna, S., & Gowrinath, R. M. (2011). Anti-ulcer effects of Aloe vera in non-steroidal and anti-inflammatory drug induced peptic ulcers in rats. *African Journal of Pharmacy and Pharmacology*, *5*(1), 1867–1871.

Rajar, U. D., Majeed, R., Parveen, N., Sheikh, I., & Sushel, C. (2008). Efficacy of Aloe vera gel in the treatment of vulval lichen planus. *Journal of the College of Physicians and Surgeons--Pakistan*, *18*(10), 612–614. PMID:18940117

Rajeswari, R., Umadevi, M., Sharmila, C. R., Pushpa, R., Selvavenkadesh, S., Sampath Kumar, K. P., & Debjit, B. (2018). Aloe vera: The Miracle Plant Its Medicinal and Traditional Uses in India. *Journal of Pharmacognosy and Phytochemistry*, *1*(4), 118–124.

Reuter, J., Jocher, A., Stump, J., Grossjohann, B., Franke, G., & Schempp, C. M. (2008). Investigation of the anti-inñammatory potential of Aloe vera gel (97.5%) in the ultraviolet erythema test. *Skin Pharmacology and Physiology*, *21*(2), 106–110. doi:10.1159/000114871 PMID:18253066

Reynolds, T. (2004). Aloe chemistry. In T. Reynolds (Ed.), Aloes: The genus Aloe (38th ed). Boca Raton, FL: CRC Press.

Ro, J. Y., Lee, B., Kim, J. Y., Chung, Y., Chung, M. H., Lee, S. K., ... Park, Y. I. (2000). Inhibitory mechanism of aloe single component (Alprogen) on mediator release in guinea pig lung mast cells activated with specific antigen-antibody reactions. *Journal of Pharmacological Experiment Therapeutics*, *292*(1), 21–114. PMID:10604937

Royal Botanical Garden Kew. (2013). *World Checklist of Aloe vera Species*. Available from www.londonexplorepass.com

Rubina, L., Priyanka, T., & Ebenezer, J. (2009). Isolation, purification and evaluation of antibacterial agents from Aloe vera. *Brazilian Journal of Microbiology*, *40*(4), 906–915. doi:10.1590/S1517-83822009000400023 PMID:24031440

Rulz, M. A., Pleguezuelos, M., Murioz, M., & Gallardo, V. (2004). Moisturizing Capacity of Aloe vera Gel in Skin Creams with Silicon-Based and Olive Oil-Based Latex Preparations. *Journal of Applied Cosmetology*, *22*(1), 25–33.

Sanjit, K. K., & Tushar, K. B. (2018). Phytochemical constituents of *Aloe vera* and their Multifunctional properties: A comprehensive review. *International Journal of Pharmaceutical Sciences*, *9*(4), 1416–1423.

Senan, V. P. (2014). Aloe vera: A miracle plant with biological action. *World Journal of Pharmaceutical Sciences*, *3*(11), 209–218.

Sharrif, M. M., & Verma, S. K. (2011). Aloe vera, their chemicals composition and applications: A review. *International Journal of Biology and Medical Research*, *2*(1), 466–471.

Shende, V., & Telrandhe, R. (2017). Formulation and Evaluation of Tooth Gel from Aloe vera Leaves Extract. *International Pharmaceutics and Drug Analysis*, *5*(10), 394–398.

Smith, G. F., & Van, W. B. (2009). The Conservation Status of Aloe in Southern Africa: An Updated Synopsis. *Bothalia*, *30*(2), 206–211. doi:10.4102/abc.v30i2.560

Sultana, S., Khan, M. A., Ahmad, M., & Zafar, M. (2006). Indigenous Knowledge of Folk Herbal Medicines by the Women of District Chakwal, Pakistan. *Ethnobotanical Leaflets*, *10*(1), 243–253.

Travis, D. (2016). *Desk Research: The What, Why, and How.* Available from https//:www.userfocus.co.uk

Ulbricht, C., Armstrong, J., Basch, E., Basch, S., Bent, S., Dacey, C., ... Weissner, W. (2007). An evidence-based systematic review of Aloe vera by the natural standard research collaboration. *Journal of Herbal Pharmacotherapy*, *7*(3-4), 279–323. doi:10.1080/15228940802153339 PMID:18928148

USDA. (2012). *Aloe barbadensis Miller. Beltville*. National Germplasm Resources Laboratory.

Vienna, C., Graz, R. B., Hohenhem, R. C., Milano, D. T., Trieste, A. T., & Wilen, T. (2007). Study of the assessment of plants/herbs, plant/herb extracts and their naturally or synthetically produced components as "additives" for use in animal production. *EFSA Supporting Publications*, *4*(1), 140–154.

Yang, Q. Y., Yao, C. S., & Fang, W. S. (2010). A new triglucosylated naphthalene glycoside from Aloe vera. *L. Fitoterapia*, *81*(1), 59–62. doi:10.1016/j.fitote.2009.07.006 PMID:19638305

Yisehak, T., Eyosyas, K., Cruz, C., & Getachew, G. (2014). Potential Antibacterial Activity of Crude Extracts from *Aloe vera, Zingiber officinale* and Vinca major Medicinal Plants. *International Journal of Microbiological Research*, *5*(3), 202–207.

Chapter 10
Ethnopharmacological Properties of Family *Asteraceae*

Neelesh Babu
Gurukula Kangri University, India

Navneet
Gurukula Kangri University, India

ABSTRACT

From ancient time, plants have been utilized as a great source of medicinal products for several types of diseases and disorders. Traditional knowledge is an important source for the development of new drugs. Several studies revealed that traditional knowledge of medicinal plants is being practised among several tribes throughout the world. Many researchers have been evaluated the authenticity of this information. Family Asteraceae got an important place among this medicinal heritage. This is one of the widely distributed families and large numbers of plants have been utilized in various skin-related problems. This chapter highlights the ethnopharmacological properties of this family.

INTRODUCTION

Therapeutic plants are widely significant as these are the source of essential raw materials to various pharmaceutical industries. As per World Health Organization (WHO) more than 21000 plants species are utilized throughout the world for medical reasons. Out of these, around 2000 plants have been utilized in Indian Systems of Medicine (ISM). In India around 650 - 700 species are utilized in about 4000 herbal and aromatic industries. There is no doubt that plants utilized in medication have been an incredible incentive for the medication advancement in the present ordinary restorative framework. Depending on the ethnobotanical data, the potential natural formulations of plants have been explored as per their customary use. In this scenario family Asteraceae got an important place among these medicinal heritages. This is known to be largest family of the flowering plants comprising 1600-1700 genera and more than 24000 species (Funk et. al., 2005). In India this family is represented by 167 genera and 900 species. Almost every life structure is represented by family Asteraceae including annual, biennial

DOI: 10.4018/978-1-7998-1320-0.ch010

or perennial herbs, under shrubs, bushes, some of the trees, a few scramblers and aquatics. Some are succulent, while others are prickly and some have smooth sap. Numerous species are adapted to endure the cool, dry winter season by underground stockpiling organs and creating annual stems in spring. The name Asteraceae is derived from the term Aster implies composite and alludes to the trademark inflorescence contains heads of flowers made out of some little flowers called florets and encompassed by bracts (Moreira-Munoz, 2007). Members of the family are cosmopolitan however, most common in the temperate areas even in tropical mountains.

TAXONOMY

Cassini in (1819) created the underlying order of the family gathering genera into different tribes. Several taxonomists have refined this innate idea in the manner that this is still a principal class over the genus level utilized for the classification of sunflowers (Panero & Funk, 2008). Carlquist (1976) and Wagenitz (1976) gathered these tribes into different subfamilies in Asteraceae systematics that started with characterizing the ideas of subfamilies Cichorioideae and Asteroideae. The morphological attributes used to outline these two groups are for the most part in light of discontinuities in the morphology of corolla, anther and style. As indicated by Bremer (1994), the Asteroideae are portrayed (with certain exemptions) by having true ray florets, disk corollas with short projections, caveate pollen, stigmatic surfaces of style branches isolated into two marginal lines sometimes intersecting at apices. Classification of the Asteraceae has been changed rapidly due to the incorporation of the results from the molecular phylogenetic studies. More precisely through the identification of the monophyletic groups which were traditionally included in Cichorioideae. Studies carried out by Jansen & Palmer (1987) and Bremer (1994) three different subfamilies were identified which includes Asteroideae, Barnadesioideae, and Cichorioideae with 17 tribes. Thorne & Reveal (2007) perceived the equivalent subfamily bunches utilizing the prior name Carduoideae for the Cichorioideae. They extended number of tribes up to 25, perceiving the three new tribes which were distinguished by molecular investigation. Tribal groupings of Panero and Funk (2002) were accepted by Kubitzki, Kadereit & Jeffrey (2007). At last they perceived 24 tribes and gathered them into five subfamilies. They perceived a monophyletic Barnadesioideae, a monophyletic Asteroideae and splits the grade of clades between these two groups into three polyphyletic subfamilies namely Mutisioideae, Carduoideae and Cichorioideae. Due to increase in the molecular studies, there is an increase in certainty and resolution in the phylogenetic analysis which leads to the identification of more major family lineages. In this scenario, genera *Corymbium*, *Gymnarrhena*, and *Hecastocleis* were discovered. Subfamily Asteroideae ruled the classification of the Asteraceae that contains over 70% of the species of the family. Molecular studies revealed three major lineage in the Asteroideae found by Kim & Jansen (1995) and Panero & Funk (2008) which have been recently perceived at the super tribe level as Asterodae, Helianthodae, and Senecionodae (Robinson, 2004; 2005).

BOTANY

Flowers

The characteristic features of this family are the presence of capitulum or head, an inferior, unilocular ovary with one ovule and exceptionally fused anthers surrounding the style. The capitulum is a specialized part of the flower that contains one to hundreds of individual flowers typically termed as florets. Capitulum has centripetal flowering sequence that is from outside to the center (Harris, 1995). The florets are present on the expanded shoot which could be flat, concave, convex, or rarely columnar. Leaf like structures called phyllaries surrounds the disc and florates collectively forming an involucre. The arrangement of phyllaries is in one row with equal and unequal length. Exceptionally most of the sunflowers have the series of phyllaries in which they could be in subequal with all phyllaries of equivalent length or imbricate. Another most common condition of phyllaries is involucres in which several series of phyllaries overlap each other. One of the best examples of involucres is *Cynara cardunculus*. Another structure called receptacle may be naked or sometime have bract like structure called paleae, scales or hairs.There are six kinds of corollas present in the Asteraceae among which two are actinomorphic and the rest four are zygomorphic (Bremer, 1994). Actinomorphic corollas are made out of five proportional lobes typically named disc corollas as they possess the greater part of the circle zone. They constitute five lobes and at the point when seen from over, the reflexed projections of the actinomorphic corolla look like a five-point star. Disc corollas may have fewer lobes with four being the most widely recognized; in spite of these corollas with three projections are additionally observed. Tubular corollas are narrow actinomorphic corollas, generally lacking stamens. Mostly zygomorphic corollas are restricted to the very first row of florets in the capitulum, albeit several rows of zygomorphic corollas are seen in few species. Bilabiate corollas are commonly present just in a few genera belongs to the early divergences of the family. The lobes are arranged in bilabiate corolla in 3+2 manner having the 3- lobe lamina facing outside and the 2-lobe lamina to the focal point of the capitulum. Sometime 2 lobes are separated and coiled. The pseudobilabiate corolla has a 4+1 arrangement.

Roots

Family Asteraceae have tap root system with several branched roots.

Stem

The stem in the family is erect, branched and cylindrical which might be woody or herbaceous and green in color. Another characteristic feature of the stem is pubescent appearance i.e covered with soft and short hairs.

Leaves

Leaves of this family are simple entire and spathulate. They might be cauline or ramal, alternate or exstipulate. They are also sessile or petiolate, acute or serrate having unicostate reticulate venation.

ECOLOGY AND DISTRIBUTION

Asteraceae or Compositae are cosmopolitan, however most regular in the temperate areas and tropical mountains. Asteraceae family is sometimes called as daisy family, sunflower family or thistle family. Considering more than 24000 - 30000 species and 1600 - 1700 genera worldwide it is known to be largest family of the flowering plants (Funk et. al, 2005). Most of the members belonging to this family are widely distributed in the temperate regions throughout the world. In India 955 taxa of the Asteraceae covering about 72.67% of the total Asteraceae in the country are known to be found in the temperate regions of Himalaya and the north-eastern region. The central focus of assorted variety of the Indian Asteraceae is the Himalayan biogeographic zone. This is because of its variable climatic conditions and elevations, which thus have brought about assorted living spaces, from the freezing deserts in the Trans-Himalayan biogeographic zone to the evergreen tropical and temperate lands of north-eastern side of India. 202 taxa of Asteraceae family are endemic to the Indian region. Entire Asteraceae which belongs to exotic taxa have been majorly divided into 18 categories. 30 taxa of this family are considered as rare, endangered and threatened.

Ethnobotany

Several plants of Asteraceae family have been widely utilized as a remedy for wound healing in several nations including with Asian, African and Europian countries. A large number of plants of this family have been utilized in various skin related issues. Some of the examples of these plants include *Achillea, Ageratum, Arctium, Artemisia, Caesulia, Centaurea, Chromolaena, Cichorium, Dendranthema, Eclipta, Gundelia, Helichrysum, Saussurea, Scorzonera, Sonchus, Tagetes, Tanacetum, Tridax* and *Vernonia* (Suntar, 2014). An account on herbs under this family which are utilized traditionally for skin related problems are given below (Table -1).

The medicinally important plant species of this family are described below:

Ageratum Conyzoides

Ageratum conyzoides has been widely utilized in Africa, Asia, and South America for several kinds of problems. The plant has been utilized as purgative, febrifuge, in ophthalmia, for the treatment of ulcers and as a wound healer. African countries widely utilized the plant for skin diseases, wound healing, headaches and dyspnea (Adjanohoun, 1998; Durodola, 1997). Traditionally it is utilized because of its anti-asthmatic, antispasmodic and haemostatic properties for the treatment of pneumonia (Okunade, 2002), onchocerciasis in Cameroon (Abbiw, 1990). It is widely used for the treatment of leprosy and purulent ophthalmia (Kasturi., et al 1973). Decoction is used in the fever and problems of stomach such as diarrhoea, dysentery, intestinal colic (Chopra, Nayar & Chopra, 2002), as anti-itching agent for the sleeping sickness as a mouthwash, and vermifuge (Kapur, 1993). Leaves are used for the treatment of cuts, sores, as anti-inflammatory agent, haemostatic, as an insecticide in the headache for the treatment of boils (Ahluwalia, 1968; Banerjee & Banerjee, 1986; Ramachandran & Nair, 1981; Siddiqui, Alam & Husain, 1989). On the other hand it is used as medicine for the treatment of typhoid, as an antidote for the snake poison and as an anti-tetanus agent (Dhasmana, 1985; Neogi, Prasad & Rao, 1989).

Table 1. Herbs under the family Asteraceae traditionally utilized for skin related problems

S.N	Plant	Common name	Part used	Reference
1	*Acanthospermum hispidum*DC	Kandlemullu	Leaves	Chakraborty, Gaikwad & Singh, 2012.
2	*Ageratum conyzoides*L.	Gindhoni	Leaves	Sharma, Gairola Sharma & Gaur, 2014.
3	*Achillea millefolium* L.	Yarrow	Aerial parts	Hemmati, Arzi & Amin, 2002.
4	*Achillea kellalensis* Bioss.	Golberrenjas	Flowers	Pirbalouti, Koohpayeh & Karimi, 2010.
5	*Arnica montana* L.	Leopards bane	Flowers, Rhizomes	Sabale, Bhimani, Prajapati & Sabale, 2012.
6	*Artemisia pallens* Walls	Davana	Leaves	Suresh, Singh, Vasavi, Ihsanulluah & Mary, 2011.
7	*Artemisia absinthium* L.	Wormwood	Whole plant	Lakshman, Yeasmin & Gabriel, 2014.
8	*Aspilia africana* C.D. Adams	Wild sunflower	Leaves	Komakech, Matsabisa & Kang, 2019.
9	*Achillea biebersteinii* Afan.	Sarıcicek	Aerial parts	Akkol, Koca, Pesin & Yilmazer, 2009.
10	*Blumea balsamifera* L.	Sambong	Leaves	Carvalho, Diniz & Suarez, 2018.
11	*Bidens pilosa*L.	Kumra	Leaves	Dharmananda, 2013.
12	*Chromolaena odorata* (L.) R.M. King & H. Rob.	Siam weed	Leaves	Logeeswari & Sripathi, 2012.
13	*Calendula officinalis* L.	Marigold flower	Flowers	Paul, Rahman, Mahmud, Uzzal & Das, 2017.
14	*Carthamus tinctorius* L.	Safflower	Leaves	Lakshman, Yeasmin & Gabriel,2014
15	*Caesulia axillaris*Roxb	Pink node flower	Flowers	Sharma, Gairola Sharma & Gaur, 2014.
16	*Chrysanthemum coronarium* L.	Crown daisy	Flowers	Dokuparthi & Manikanta, 2015,
17	*Elephantopus scaber* Linn	Elephant foot	Aerial parts	Singh, et al., 2005.
18	*Echinops echinatus* Roxb	Indian globe thistle	Roots	Maurya, Kushwaha & Seth, 2015.
19	*Eupatorium laevigatum* Lam.	Camara	Leaves	Sobrinho, de Morais, de Sauza & dos Santos, 2017.
20	*Echinacea pallida* Nutt	Pale purple coneflower	Roots	Zhai, et al., 2009.
21	*Flaveria trinervia* (Speng) C.Mohr	Speedy weed	Leaves	Umadevi, Mohanta, Kalaichelvan & Manavalan,2006; Logeeswari & Sripathi, 2012.
22	*Guizotia abyssinica* (L.f) Cass	Niger	Leaves	Lakshman, Yeasmin & Gabriel, 2014.
23	*Gynura procumbens* (Lour.) merr.	Sambung nyawa	Leaves	Zahra, et al., 2011.
24	*Tagetes patula* L.	Mexican marigold	Leaf, Root	Patel, 2012.
25	*Vicoa indica* DC.	Mugati	Roots	Lakshman, Yeasmin & Gabriel, 2014
26	*Xanthium strumarium* Linn.	Adhasis	Root, Fruit	Patel, 2012.

Ethnic Uses

In Bangladesh, its decoction is used for the treatment of diarrhea and dysentery. Its paste and poultice is used for the treatment of cuts, wounds and skin diseases. Mixture of its leaves with other herb is being utilized for the treatment of syphilis in Cameroon. Its leaves mixture with few herbs is utilized by Indians and Nepalies as an antiseptic for cuts and wounds. In Tanzania its leaves juice is used as eye drop (Kaur & Dogra, 2014).

Artemisia Absinthium

It is the well-known source of drug used in Unani medicine. Locally it is known as Afsanteen which has been widely utilized to cure chronic fever, swellings, and inflammation of liver. In European countries, tincture of this herb is consumed as a tonic, digestive supplement, febrifuge and anthelmintic. Traditionally it has been used in epilepsy, gastric problems, in the enlargement of spleen, urinary disorders and also for wound healing (Siddiqui, John & Paul, 1995; Srivastava, Badola, Shah & Gupta, 1984). Its flowers are used in diseases associated with stomach and as vermifuge (Durodola, 1977). Its leaves are also utilized in fever (Srivastava, Pathak & Hussain, 1980).

Ethnic Uses

Tribals of Kupwara district of Jammu and Kashmir, India uses this herb for several skin ailments. Its crushed leaves are boiled followed by frying in mustard oil with turmeric and salt is being used externally (Mir, 2014). Its dried leaves and flowers mixture is used with lukewarm milk for intestinal infection by Kashmiri peoples (Malik, Khuroo, Dar & Khan, 2011).

Bidens pilosa

B. pilosa is used as an herbal medicine and as an ingredient in teas. Its dried as well as fresh aerial part has been utilized as the ingredient in sauces and teas (Chiang, Chang, Chen, Ng & Lin, 2003; Rybalchenko, et al., 2010). Every part of this plant has been used as an ingredient in folk medicines as a frequently prepared dry powder, decoction, maceration or tincture (Redl, Breu, Davis & Bauer, 1994). Dry powder of this plant is used externally while as a decoction it is used as an internal remedy (Rybalchenko, et al., 2010). It has been reported that this plant has the capacity to treat more than 40 disorders including inflammation, immunological disorders, digestive disorders, infectious diseases, cancers, metabolic syndrome, wounds etc. (Pereira, Ibrahim, Lucchetti, da Silva & de Moraes, 1999; Tan, Dimo & Dongo, 2000; Dimo, et al., 2001; Wiart, 2000). Fresh plant has been utilized for the treatment of snake bites and wounds healing (Dharmananda, 2013).

Ethnic Uses

Traditionally, communities of Southern Brazil use this herb for various ailments. Its 20 g of aerial part boiled with 1 liter of water and consumed orally at the dose of 1 cup 3 times a day for urinary tract, kidney and bladder infections. Topically, this infusion is used 3 times a day for the treatment of skin eruptions, varicose veins and phlebitis. Half tablespoon of whole plant powder boiled in 250 ml of water utilized 1 cup twice a day orally for the treatment of sore throat, cough, stomach pain, diabetes and topically utilized for cleansing wound. Its leaves juice (3-4 leaves blended in 250 ml of water) utilized orally 1 cup a day for pain and inflammation (Borges, et al., 2013). In India, its leaves decoction is utilized for stomachache. Decoction of whole plant is utilized in colics, diarrhea, constipation, dysentery, conjunctivitis, otitis, pharynngitis and as an antirheumatic by African peoples (Bartolome, Villasenor & Yang, 2013).

Calendula Officinalis

Traditionally this herb is widely employed for the treatment of several problems including inflammations of internal organs, gastrointestinal ulcers, dysmenorrheal, as a diuretic and diaphoretic. It has also been utilized for inflammations of the oral, pharyngeal mucosa, wounds and burns (Yoshikawa, Murakami, Kishi, Kageura & Matsuda, 2001). It works as a cleansing and detoxifying herb; its infusion is very helpful in the treatment of chronic infections (Blumenthal, Goldberg & Brinckmann, 2000). Its dried flower heads are very good antipyretic, anti-tumor agents and also shows cicatrizing effects (Ukiya, et al., 2006). In current scenario its flowers infusion is being utilized as antifungal, antiseptic, freckles, sprain and conjunctivitis (Rehecho, Uriarte-Pueyo, Calvo, Vivas & Calvo, 2011). Tea made from this herb is utilized for eyewashes, gargles, rashes and sort of inflammatory skin (Safdar, et al., 2010). In homoeopathic medication its tincture is utilized for the mental tension and insomnia (Boericke, 1988). Ayurvedic and Unani system of medicine describes antipyretic, anti-inflammatory, antiepileptic and antimicrobial effect of its leaves and flowers (Kasiram, Sakharkar & Patil, 2000). It has also been used for the treatment of poor eyesight, menstrual irregularities, varicose veins, hemorrhoids and duodenal ulcers traditionally (Cetkovic, Djilas, Canadanovic-Brunet & Tumbas, 2004). It has been reported that its flowers were used for the treatment of liver disorders, snake bites and cardiological problems. In 18th century it has been used for headache, jaundice and red eyes. Moreover, it was used in the civil war for the treatment of wounds, measles, smallpox and jaundice.

Ethnic Uses

Its flowers are used for the treatment of wound, herpes, ulcers, scars and skin damage in India (Muley, Khadabadi & Banarase, 2009). People in Kishtwar, Jammu and Kashmir, India, uses its leaves and flowers paste for the skin diseases and burns (Kumar, Paul & Anand, 2009). In Bulgaria, it is used for prophylaxis, nerve problems, disorders of stomach, ulcers and wounds (Koleva, Dragoeva, Nanova, Koynova and Dashev, 2015). In European countries, its leaves are used as resolvent and diaphoretic whereas, its flowers are used as stimulant and antispasmodic. Decoction of its flowers is used for the treatment of measles and smallpox in England (Muley, Khadabadi & Banarase, 2009).

Eclipta Alba

It is said to be one of the promising herbs in Ayurveda for the treatment of hepatic disorders. According to Ayurveda this plant has been utilized as an antiaging agent as it has revitalizing property (Karthikumar, Vigneshwari & Jegatheesan, 2007; Uddin, et al., 2010; Vianna-da-Silva, 2003; Khanna & Kannabiran, 2009). It is one of the important herbs that have hair promoting compounds. Leaves extract of this herb has been utilized for the treatment of dandruff in Uttar Pradesh state of India and also has the property to make hair silky and shiny (Kumar, Agarwal, Singh & Deepak, 2012). Paste of whole plant with black pepper consumed orally by some of the tribes of Odisha state of India for the treatment of fever and jaundice (Sahu, Nayak & Dhal, 2013). Leaves juice at the dose of 5-10 ml is widely used for the treatment of hepatic disorders in Tripura state of India (Das, Chaudhary, Mandal & Talukdar, 2012). Its leaves and flowers are used to overcome asthma, cough, jaundice and urinary problem (Gautam & Batra, 2012). On the other hand leaf juice with honey is consumed to treat jaundice and only leaves extract with water is helpful in diarrhea (Abdel-Kader, et al., 1998). It has also been used for the treatment of wounds and cuts (Agnihotri & Gupta, 2013).

Ethnic Uses

In Uttar Pradesh West, India, its decoction three times a day is been utilized as acidity regulator. Its leaves extract is consumed orally for the treatment of alopecia. Decoction of whole plant with honey for 7 days is useful in bronchitis and pneumonia. Its leaves paste is utilized during burns. Its roots powder is consumed orally 3 times a day for the treatment of constipation. Fresh plant extract is very helpful in the treatment of jaundice if taken orally 3 times a day until cure. Its leaves paste is applied externally for weeks to treat eczema. Whole plants extract is consumed orally 2 times a day for 15 days for the treatment of urinary tract infection. Locals of Mount Abu, India, are also use its leaves paste for the treatment of urinary tract infection, jaundice and asthma. Santal tribe in Bangladesh boils its leaves with the leaves of *Scoparia dulcis* and *Cynodon dactylon* is consumed orally in morning and evening to overcome diabetes. Chakma tribe of Tripura utilizes its leaves juice for hepatic disorders. Local communities in Nepal utilizes its leaves extract for the treatment of cuts and wounds (Jahan, Al-Nahain, Majumder & Rahmatullah, 2014).

Grangea Maderaspatana

It is an important herb used in various diseases; according to previous studies this plant has immuno-potentiation activity. It has been widely used to treat few types of tumours in some region of China (Khare, 2008). It is used as antipyretic agent for the treatment of fever caused by deranged air and bile, remittent fever along with gastric irritability and depression. In India its juice is applied to overcome the burning sensation of the palms of hand as well as soles of the feet. Its juice with milk and sugar is utilized in acidity. For the treatment of remittent fever and heat eruptions its decoction is widely used. Its extract also has been used in jaundice and it also works as anthelmintic. In Indo China this plant is utilized as febrifuge.

Ethnic Uses

In South Orissa, India, its paste with cow milk is utilized during menstrual disorders. Its leaves paste is utilized with honey to treat disorders of stomach (Panda and Misra, 2011). In Nawarangpur district Orissa, India; its roots are chewed after meal up to 15 days for the treatment of dyspepsia (Dhal, Panda & Muduli, 2015).

Helianthus Annuus

It is a medicinal herb traditionally used for treatment of different types of illness. It is used as folk medicine in several parts of Asian and European countries. Its seeds are utilized for the treatment of laryngeal, pulmonary infections, bronchial, coughs and colds and in whooping cough. Infusion prepared from its seeds with rectified spirit of wine is helpful to overcome with ague and fever (Suo & Yang, 2014). Its tincture has also been used for the treatment of malarial fever. Generally, leaves are spread on a bed and covered with a material such as cloth which was further dampened with warm milk. Then patient is wrapped up in it. After some time sweat is produced and the process is rehashed until the fever has stopped. Mixture of its flowers and leaves are utilized for the treatment of bronchiectasis (Imran, Hussain, Gurmani, Zahid & Khan, 2008). It is a very potent medicine for soothing chest pains as well as consumed for food in Mexico. Its stems are useful for the treatment of cuts and wounds by applying its juice over the wounds and bandaged results in a speedy recovery. In Bangladesh its oil is utilized against bacterial infections in premature babies by rubbing on their skin several times a day. Its oil has also been utilized for the prevention and the treatment of acne (Ukiya, 2007).

Ethnic Uses

Traditionally in Russia, its flowers are utilized in rheumatic joints pain. Its flowers are soaked in vodka and soap chips in a sunny place for about one week and then rubbed on joints. In Southwest America, dried plant is utilized as tea and also added to bathing water to overcome with swelling of joints and arthritic pain. Decoction of its seeds is utilized in Russia for the treatment of jaundice, malaria and diarrhea. The fusion of its roasted seeds is useful in whooping cough. Its decoction of leaves is used in high fever by the people of v Southwest America (Rogers, 2015).

Parthenium Hysterophorus

Traditionally this plant is utilized for the treatment of fevers, migraine, headaches, rheumatoid arthritis, stomachaches, toothaches, insect bites, infertility, and problems with menstruation (Sharma & Gupta, 2012).This plant is widely employed for the treatment of ulcerated sores, wounds, fever, anaemia and heart related problems. Its roots extract is utilized for the treatment of dysentery and also been used as antifungal agent. It can be utilized externally as well as internally for a wide variety of ailments. It is an important medicine for the hepatic amoebiasis (Fazal, et al., 2011). It has huge number of bioactive compounds due to which it has multiple pharmacologic properties, which includes anticancer, anti-inflammatory, cardiotonic, antispasmodic etc. (Sharma & Gupta, 2012).

Ethnic Uses

In Ethiopia, dry roots powder of *P. hysterophorus* is taken with roots powder of *Prunus Africana* and consumed orally for 3 days to treat gonorrhea. Its leaves paste with fresh/dry roots of *Datura stramonium* applied externally for the treatment of wound (Mesfin, Demissew & Teklehaymanot, 2009). In Afar region of Ethiopia, its flowers are consumed orally for the treatment of somnambulism (Teklehaymanot, 2017).

Sonchus Oleraceus

This herb is credited with a huge number of therapeutic applications. It is being utilized by traditional medicinal systems in India as well as in several countries of the world and also as a folk medicine for the gastrointestinal tract disorder. Additionally, it is being used as food in few parts of Asian and African countries (Hussain, et al., 2010). Its juice is utilized in hemorrhage during labour (Bell, 1890). To expel placental worms its decoction with kopa, clover and salt is prescribed. With *Disphyma australe* it is being used traditionally for boils (Best, 1904). It also has been utilized as galactagogue, febrifuge, sedative, vermifuge, and in the liver disorder. Its ointment is very helpful for healing wounds and ulcers (Khare, 2008). It is consumed orally as a tonic to treat phthisis (Ahmad, Khan & Butt, 2015). Its decoction is administered to treat gastric disorders, hepatitis, infections, inflammation, headaches, general pain, rheumatism etc. (Vendruscolo & Mentz, 2006; Agra, Baracho, Nurit, Basilio & Coelho, 2007). Likewise, it is also being utilized for the treatment of vitiligo (Resende, de Aquino, do Nascimento, Aguiar & Fiorelli, 2015). Traditionally it has been utilized as a cathartic, sedative as well as in the treatment of cancer. It is also being used as a vermicide (Liu & Zhao, 2010).

Ethnic Uses

In Bengal, India, infusion of its leaves is utilized in high fever. Extract of its stems is utilized as sedative agent. It is being utilized as laxative in Italy. Its juice is utilized for the cleaning and the treatment of ulcers. It is used for the treatment of headache, pain, infection and rheumatism in Brazil. Its leaves are consumed orally as hepatoprotective in India.

Sphaeranthus Indicus

According to Ayurvedic literature, whole plant is medicinally important and can be used in the case of tuberculous meningitis, digestive disorders, bronchitis, spleen infection, anaemia, uterus pain, vaginal infection, piles, epilepsy, asthma, leukoderma, dysentery, vomiting, urinary discharges and rectal pain (Kirtikar & Basu, 1981). Traditionally it has been used for the treatment of epilepsy and mental disorders. Its dried leaves are useful in skin infections and nervous disorders at the dose of 20 grains twice a day (Nadkarni, 2007; Prajapati, Purohit, Sharma & Kumar, 2003). Its aqueous extract can be used as anthelminitic, diuretic and fish poison (Paranjape, 2001). It has also been used as an aphrodisiac (Kapoor & Kapoor, 1980). Its flowers are being used for the treatment of conjunctivitis. Reports suggested that it is helpful for strengthening of weak eyes (Agarwal, 1997). Its roots oil is useful for the treatment of scrofula. Its paste has been utilized for the treatment of edema, arthritis, filariasis, gout and cervical adenopathy (Sahu, 1984). It has been widely employed as laxative, for the treatment of loss appetite, enriches the blood, for the treatment of inflammation, eye infections, gleet, biliousness, boils, scabies and ringworm in Unani system of medication. Traditionally it is used for the treatment of diarrhea (Girach, Aminuddin, Siddiqui & Khan, 1994). Whole plant is being utilized in menstrual problems (Saha, Savini & Kasinathan, 1961). Its aqueous extract has been utilized in swollen glands of the neck and for jaundice (Ikram, 1981).

Ethnic Uses

In Rewa district, Madhya Pradesh, India, its leaves extract with black pepper powder is used to treat earache (Shukla, Srivastava & Rawat, 2010). Ayurvedic system of medicines claims that shade dried leaves powder at the dose of 20g 2 times a day is very useful for chronic skin diseases and also works as a nerve tonic. Its flowers extract is utilized for the treatment of conjunctivitis. Its paste is applied externally for the treatment of edema, arthritis, filariasis and gout (Galani, Patel & Rana, 2010).

Tridax Procumbens

It is another medicinally important species of Asteraceae which is employed as indigenous medicine for several problems. It is one of the best examples of Indian traditional medicine for wound healing purpose. It is extensively utilized as anticoagulant, antifungal, insect repellent, in diarrhea and dysentery (Ali, Ravinder & Ramachandran, 2001). In traditional medicines its leaves extract is being used to treat infectious skin diseases. It is well-known alternative of 'Bhringraj' which is well known medicine in Ayurvedic literature for liver disorders (Pathak, Saraf & Dixit, 1991). It has promising antioxidant activity and can also be used as hair growth promoter (Habila, Bello, Dzikwi, Musa & Abubakar, 2010; Saraf, Pathak & Dixit, 1991).

Ethnic Uses

Thottianaikan community of Tmail Nadu, India applied its leaves juice for cut wounds (Ganeshan, Venketeshan, & Banumathy, 2006). In Guatemala, its leaves juice is used to treat inflammation, cold, anemia, hepatopathies, vaginitis, stomach pain etc. Its dried leaves infusion and dried stem is very helpful to reduce inflammation, gastrointestinal and respiratory infections. Several Indian communities consumes its dried leaves with the mixture of other herbs to control diabetes, to treat diarrhea, jaundice and for wound healing. Decoction of this herb with *Phyllanthus amarus* is used as antimalerial, antibacterial and as a wound healer in Ghana. In Nigeria, whole dried plant is considered in fever, cough, back and stomachache, epilepsy and diarrhea (Beck, Mthison, Todorov, Calderon-Juarez & Kopp, 2018).

Tagetes Erecta

Traditionally different parts of this plant including flowers are used in several infectious diseases. Its leaves are utilized as antiseptic agent, in kidney disorders, piles, boils and also used in muscular pain. According to Ayurveda, its flowers can be used in epilepsy, fever and in eye diseases. It is actively utilized for the purification of blood, its flowers have been used for the treatment of bleeding piles, rheumatism, cold and bronchitis (Kirtikar, 1987). Several species of this genus shows antimicrobial, anti-inflammatory, hepatoprotective, wound healing, insecticidal and analgesic (Gopi, Elumalai & Jayasri, 2012; Rhama & Madhavan, 2011; Kiranami & Ibrahim, 2012).

Ethnic Uses

Juice of the flowers is infrequently used as the remedy for the purification of blood and its leaves are used in pain, inflammation and to stop bleeding in India. Its infusion is also used by Indian people to overcome cold, rheumatic pain and bronchitis. Brazilian people use its leaves infusion as vermifuge. Decoction of its leaves and flowers is used as carminative and diuretic by Mexican people. In Aztecs, whole plant is used to treat carbuncles and eye infections. Tribal communities of Kashipur, Uttarakhand, India, use its fresh leaves extract for eczematous condition and deafness (Joshi & Pant, 2012).

Vicoa Indica

Traditionally it is extensively utilized by northern eastern tribal population of India. It is being used as a contraceptive agent and female anti-fertility drug (Dhall & Dogra, 1988). The infusion of whole plant utilized in abortion (Tayade & Patil, 2005). Its roots are very useful remedy for cough and jaundice (Srinivasan, Natarajan, Mohanasundari, Venkatakrishnan & Nagamurugan, 2007).

Ethnic Uses

Thottianaikan community of Tiruchirappali district, Tmail Nadu, India, uses its leaves decoction to treat upset stomach and dysentery (Ganeshan, Venkateshan & Banumathy, 2006). Bagia tribe of Madhya Pradesh, India, uses whole plant extract for the treatment of fever and skin diseases. They use paste of its roots for healing and as an antidote for snake bite. They use boiled leaves decoction orally to cure dysentery and indigestion. Leaf juice is employed for 15 days to treat asthma (Tiwari, 2018).

CONCLUSION

Family Asteraceae is considered as the largest family of the angiosperm which is widely distributed all over the world and several species belongs to this family believed to possess various biological activities because of their chemical constituents (Funk, et al., 2005). Several studies showed that the plants of the family are being utilized traditionally for the treatment of several kinds of aliments in rural areas by many tribes all over the world. Majority of plants are used for the treatment of problems related to skin and most of them acts as a wound healer.

REFERENCES

Abbiw, D. K. (1990). Useful plants of Ghana, West African uses of wild and cultivated plants. England, Ltd., (28-34). Practical Action Publishing.

Abdel-Kader, M. S., Bahler, B. D., Malone, S., Werkhoven, M. C., van Troon, F., D, ... Kingston, D. G. I. (1998). DNA-damaging steroidal alkaloids from *Eclipta alba* from the Suriname rainforest. *Journal of Natural Products*, *61*(10), 1202–1208. doi:10.1021/np970561c PMID:9784152

Adjanohoun, E. J. (1988). Médecine Traditionnelle et Pharmacopée: Contribution aux études ethnobotaniques et floristiques en République Populaire du Congo. Paris (France Agence de Cooperation Culturelle et Technique. Agarwal, V. S. (1997). Drug plants of India. (1-656). New Delhi, India: Kalyani Publishers.

Agnihotri, N., & Gupta, A. K. (2013). Folkloremedicines for cuts and wounds in Kalyanpur block of Kanpur District, Uttar Pradesh, India. *Ph Tech Med*, *2*, 381–386.

Agra, M. F., Baracho, G. S., Nurit, K., Basílio, I. J. L. D., & Coelho, V. P. M. (2007). Medicinal and poisonous diversity of the flora of "Cariri Paraibano". *Brazilian Journal of Ethnopharmacology*, *111*(2), 383–395. doi:10.1016/j.jep.2006.12.007 PMID:17236731

Ahluwalia, K. S. (1968). Medicinal plants of Kerala-V. *Nagarjun*, *11*(7), 363–369.

Ahmad, S. S., Khan, S. M., & Butt, A. (2015). Exploring the medicinal plants wealth: An assessment of traditional medico-botanical knowledge of local communities in Changa Manga forest, Pakistan. *Pharmaceutical Chemistry Journal*, *2*(3), 24–32.

Akkol, E. K., Koca, U., Pesin, I., & Yilmazer, D. (2011). Evaluation of the wound healing potential of *Achillea biebersteinii* Afan.(Asteraceae) by in vivo excision and incision models. *Evidence-Based Complementary and Alternative Medicine*, *2011*, 1–7. doi:10.1093/ecam/nep039 PMID:19546149

Ali, M., Ravinder, E., & Ramachandran, R. (2001). Anticoagulant activity of *Tridax procumbens*. *Fitoterapia*, *72*, 313–315. doi:10.1016/S0367-326X(00)00296-3 PMID:11295316

Banerjee, A. K., & Banerjee, I. (1986). A survey of medicinal plants in Shevaroy hills. *Journal of Economic and Taxonomic Botany*, *8*, 271–290.

Bartolome, A. P., Villaseñor, I. M., & Yang, W. C. (2013). *Bidens pilosa* L.(Asteraceae): Botanical properties, traditional uses, phytochemistry, and pharmacology. *Evidence-Based Complementary and Alternative Medicine*, *2013*, 1–5. doi:10.1155/2013/340215 PMID:23935661

Beck, S., Mathison, H., Todorov, T., Calderon-Juarez, E. A., & Kopp, O. R. (2018). A review of medicinal uses and pharmacological activities of *Tridax procumbens* (L.). *Journal of Plant Studies*, *7*(1), 19–35. doi:10.5539/jps.v7n1p19

Bell, T. W. (1890). Medical notes on New Zealand. *New Zealand Medical Journal, 3*(65-83), 129-145.

Best, E. (1904). Maori medical lore. Notes on sickness and disease among the Maori people of New Zealand, and their treatment of the sick; together with some account of various beliefs, superstitions and rites pertaining to sickness, and the treatment thereof, as collected from the Tuhoe tribe. Part I. *The Journal of the Polynesian Society (NZ)*, *13*(52), 213–237.

Blumenthal, M., Goldberg, A., & Brinckmann, J. (2000). Herbal Medicine. Expanded Commission E monographs. Integrative Medicine Communications. *American Botanical Council*, *3*, 78–83.

Boericke, W. (1988). Pocket Manual of Homoeopathic Materia Medica. New Delhi, India. Homeopathic publication.

Borges, C. C., Matos, T. F., Moreira, J., Rossato, A. E., Zanette, V. C., & Amaral, P. A. (2013). *Bidens pilosa* L.(Asteraceae): Traditional use in a community of southern Brazil. *Revista Brasileira de Plantas Medicinais*, *15*(1), 34–40. doi:10.1590/S1516-05722013000100004

Bremer, K. R. (1994). *Asteraceae: cladistics & classification*. Portland, OR: Timber Press.

Carlquist, S. (1976). Tribal interrelationships and phylogeny of the Asteraceae. Aliso: *A Journal of Systematic and Evolutionary Botany, 8*(4), 465-492.

Carvalho, A. R. Jr, Diniz, R. M., Suarez, M. A., Figueiredo, C. S. S. S., Zagmignan, A., Grisotto, M. A. G., ... da Silva, L. C. N. (2018). Use of some Asteraceae plants for the treatment of wounds: From ethnopharmacological studies to scientific evidences. *Frontiers in Pharmacology*, *9*, 1–11. doi:10.3389/fphar.2018.00784 PMID:30186158

Cetkovic, G. S., Djilas, S. M., Canadanović-Brunet, J. M., & Tumbas, V. T. (2004). Antioxidant properties of marigold extracts. *Food Research International*, *37*(7), 643–650. doi:10.1016/j.foodres.2004.01.010

Chakraborty, A. K., Gaikwad, A. V., & Singh, K. B. (2012). Phytopharmacological review on *Acanthospermum hispidum*. *Journal of Applied Pharmaceutical Science*, *2*(1), 144–148.

Chiang, L. C., Chang, J. S., Chen, C. C., Ng, L. T., & Lin, C. C. (2003). Anti-Herpes simplex virus activity of *Bidens pilosa* and *Houttuynia cordata. The American Journal of Chinese Medicine*, *31*(3), 355–362. doi:10.1142/S0192415X03001090 PMID:12943167

Chopra, R. N., Nayar, S. L., & Chopra, I. C. (2002). *Glossary of Indian Medicinal Plants*. New Delhi, India: Niscair publisher.

Das, S., Choudhury, M. D., Mandal, S. C., & Talukdar, A. D. (2012). Traditional knowledge of ethnomedicinal hepatoprotective plants used by certain ethnic communities of Tripura State. *Indian Journal of Fundamental and Applied Life Sciences*, *2*, 84–97.

Dhal, N. K., Panda, S. S., & Muduli, S. D. (2015). Traditional uses of medicinal plants by native people in Nawarangpur district, Odisha, India. *Asian Journal of Plant Science and Research*, *5*(2), 27–33.

Dhall, K., & Dogra, M. (1988). Phase I and II clinical trials with *Vicoa indica* (Banjauri): Herbal medicine, as an antifertility agent. *Contraception*, *37*(1), 75–84. doi:10.1016/0010-7824(88)90150-3 PMID:3365984

Dharmananda, S. (2013). A Popular Remedy Escapes Notice of Western Practitioners. Retrieved from http://www.itmonline.org/arts/bidens.htm

Dhasmana, H. (1985). Medicinal plants of Pauri town (Garhwal) and adjacent forest regions (part 1: Family compositae). *Journal of Scientific Research in Plants and Medicines*, *6*, 45–47.

Dimo, T., Azay, J., Tan, P. V., Pellecuer, J., Cros, G., Bopelet, M., & Serrano, J. J. (2001). Effects of the aqueous and methylene chloride extracts of *Bidens pilosa* leaf on fructose-hypertensive rats. *Journal of Ethnopharmacology*, *76*(3), 215–221. doi:10.1016/S0378-8741(01)00229-X PMID:11448541

Dokuparthi, S. K., & Manikanta, P. (2015). Phytochemical and pharmacological studies on *Chrysanthemum coronarium* L.: A review. *Journal of Drug Discovery and Therapeutics*, *3*(27), 11–16.

Durodola, J. I. (1977). Antibacterial property of crude extracts from a herbal wound healing remedy–*Ageratum conyzoides*, L. *Planta Medica*, *32*(08), 388–390. doi:10.1055-0028-1097620 PMID:594219

Fazal, H., Ahmad, N., Ullah, I., Inayat, H., Khan, L., & Abbasi, B. H. (2011). Antibacterial potential in *Parthenium hysterophorus, Stevia rebaudiana* and *Ginkgo biloba. Pakistan Journal of Botany*, *43*(2), 1307–1313.

Funk, V. A., Bayer, R. J., Keeley, S., Chan, R., Watson, L., Gemeinholzer, B., & Jansen, R. K. (2005). Everywhere but Antarctica: Using a super tree to understand the diversity and distribution of the Compositae. *Biologiske Skrifter*, *55*, 343–374.

Galani, V. J., Patel, B. G., & Rana, D. G. (2010). *Sphaeranthus indicus* Linn.: A phytopharmacological review. *International Journal of Ayurveda Research*, *1*(4), 247–253. doi:10.4103/0974-7788.76790 PMID:21455454

Ganesan, S., Venkateshan, G., & Banumathy, N. (2006). Medicinal plants used by ethnic group Thottianaickans of Semmalai hills (reserved forest), Tiruchirappalli district, Tamil Nadu. *Indian Journal of Traditional Knowledge*, *5*(2), 245–252.

Gautam, A., & Batra, A. (2012). Ethnomedicinal plant product used by the local traditional practitioners in Mount Abu. *World Journal of Pharmaceutical Sciences*, *1*, 10–18.

Girach, R. D., Aminuddin, Siddiqui, P. A., & Khan, S. A. (1994). Traditional plant remedies among the Kondh of District Dhenkanal (Orissa). *International Journal of Pharmacognosy*, *32*(3), 274–283. doi:10.3109/13880209409083005

Gopi, G., Elumalai, A., & Jayasri, P. (2012). A concise review on *Tagetes erecta*. *International Journal of Phytopharmacy Research*, *3*(1), 16–19.

Habila, J. D., Bello, I. A., Dzikwi, A. A., Musa, H., & Abubakar, N. (2010). Total phenolics and antioxidant activity of *Tridax procumbens* Linn. *African Journal of Pharmacy and Pharmacology*, *4*(3), 123–126.

Harris, E. M. (1995). Inflorescence and floral ontogeny in Asteraceae: A synthesis of historical and current concepts. *Botanical Review*, *61*(3), 93–131. doi:10.1007/BF02887192

Hemmati, A. A., Arzi, A., & Amin, M. (2002). Effect of *Achillea millefolium* extract in wound healing of rabbit. *Journal of Natural Remedies*, *2*(2), 164–167.

Hussain, J., Muhammad, Z., Ullah, R., Khan, F. U., Khan, I. U., Khan, N., & Jan, S. (2010). Evaluation of the chemical composition of *Sonchus eruca* and *Sonchus asper*. *The Journal of American Science*, *6*(9), 231–235.

Ikram, M. (1981). A review on the medicinal plants. *Hamdard*, *24*(1/2), 102–129.

Imran, M., Hussain, A., Gurmani, Z. A., Zahid, M. S., & Khan, S. (2008). Cultivation of sunflower and its utilization in livestock. *Science Journal of Technology Dev*, *27*, 1–2.

Jahan, R., Al-Nahain, A., Majumder, S., & Rahmatullah, M. (2014). Ethnopharmacological significance of *Eclipta alba* (L.) Hassk. (Asteraceae). *International Scholarly Research Notices*, *2014*, 1–22. doi:10.1155/2014/385969 PMID:27355071

Joshi, B., & Pant, S. C. (2012). Ethnobotanical study of some common plants used among the tribal communities of Kashipur, Uttarakhand. *Indian Journal of Natural Products and Resources*, *3*(2), 262–266.

Kapoor, S. L., & Kapoor, L. D. (1980). Medicinal plant wealth of the Karimnagar district of Andhra Pradesh. *Bulletin of Medical Ethnobotany and Research*, *1*, 120–144.

Kapur, S. K. (1993). Ethno-medico plants of Kangra valley (Himachal Pradesh). *Journal of Economic and Taxonomic Botany*, *17*, 395–408.

Karthikumar, S., Vigneswari, K., & Jegatheesan, K. (2007). Screening of antibacterial and antioxidant activities of leaves of *Eclipta prostrata* (L). *Scientific Research and Essays*, (4): 101–104.

Kasiram, K., Sakharkar, P. R., & Patil, A. T. (2000). Antifungal activity of *Calendula officinalis*. *Indian Journal of Pharmaceutical Sciences*, *62*(6), 464–466.

Kasturi, T. R., Thomas, M., & Abraham, E. M. (1973). Essential oil of *Ageratum conyzoides*: Isolation and structure of two new constituents. *Indian Journal of Chemistry*, 91–95.

Kaur, R., & Dogra, N. K. (2014). A review on traditional uses, chemical constituents and pharmacology of *Ageratum conyzoides* L.(Asteraceae). *International Journal of Pharmaceutical and Biological Archives*, *5*(5), 33–45.

Khanna, V. G., & Kannabiran, K. (2009). Anticancer-cytotoxic activity of saponins isolated from the leaves of *Gymnema sylvestre* and *Eclipta prostrata* on Hela cells. *International Journal of Green Pharmacy*, *3*(3), 227–229. doi:10.4103/0973-8258.56280

Khare, C. P. (2008). *Indian medicinal plants: an illustrated dictionary*. Springer Science & Business Media.

Kim, K. J., & Jansen, R. K. (1995). ndhF sequence evolution and the major clades in the sunflower family. *Proceedings of the National Academy of Sciences of the United States of America*, *92*(22), 10379–10383. doi:10.1073/pnas.92.22.10379 PMID:7479788

Kiranmai, M., & Ibrahim, M. (2012). Antibacterial potential of different extracts of *Tagetes erecta* Linn. *International Journal of Pharmaceutics*, *2*(1), 90–96.

Kirtikar, K. R. (1987). *Indian Medicinal Plants. Lalit Mohan Basu, Allahabad. 2, (146)*. New Delhi, India: Jayyd Press.

Kirtikar, K. R., & Basu, B. D. (1981). ICS. Indian medicinal plants. Allahabad, India. *Lalit Mohan Basu*, *2*(2), 1346–1348.

Koleva, V., Dragoeva, A., Nanova, Z., Koynova, T., & Dashev, G. (2015). An ethnobotanical study on current status of some medicinal plants used in Bulgaria. *International Journal of Current Microbiology and Applied Sciences*, *4*(4), 297–305.

Komakech, R., Matsabisa, M. G., & Kang, Y. (2019). The wound healing potential of *Aspilia africana* (Pers.) CD Adams (Asteraceae). *Evidence-Based Complementary and Alternative Medicine*, *2019*, 1–12. doi:10.1155/2019/7957860 PMID:30800171

Kubitzki, K., Kadereit, J. W., & Jeffrey, C. (2007). The families and genera of vascular plants: *Flowering plants. Eudicots*, *8*, 1–645.

Kumar, A., Agarwal, S., Singh, A., & Deepak, D. (2012). Medicobotanical study of some weeds growing in Moradabad District of Western Uttar Pradesh in India. *Indian Journal of Social Research*, *3*(1), 107–111.

Kumar, M., Paul, Y., & Anand, V. K. (2009). An ethnobotanical study of medicinal plants used by the locals in Kishtwar, Jammu and Kashmir, India. *Ethnobotanical Leaflets*, *13*, 1240–1256.

Lakshman, H. C., Yeasmin, T., & Gabriel, K. P. (2014). Herbs of Asteraceae and their ethano medicinal uses in dermatological problems. *Journal of Bio-Science*, *22*, 127–129. doi:10.3329/jbs.v22i0.30016

Liu, Z., & Zhao, X. (2010). Effects of different extraction methods on extraction rate of polysaccharide from *Sonchus oleraceus* L. *Medicinal Plant*, *1*(5), 31–35.

Logeeswari, K., & Sripathi, S. K. (2012). Wound healing medicinal plants: A review. *International Journal of Chemical Environmental and Pharmaceutical Research*, *3*(3), 199–218.

Malik, A. H., Khuroo, A. A., Dar, G. H., & Khan, Z. S. (2011). Ethnomedicinal uses of some plants in the Kashmir Himalaya. *Indian Journal of Traditional Knowledge*, *10*(2), 362–366.

Maurya, S. K., Kushwaha, A. K., & Seth, A. (2015). Ethnomedicinal review of Usnakantaka (*Echinops echinatus* Roxb.). *Pharmacognosy Reviews*, *9*(18), 149–154. doi:10.4103/0973-7847.162138 PMID:26392713

Mesfin, F., Demissew, S., & Teklehaymanot, T. (2009). An ethnobotanical study of medicinal plants in Wonago Woreda, SNNPR, Ethiopia. *Journal of Ethnobiology and Ethnomedicine*, *5*(1), 28–46. doi:10.1186/1746-4269-5-28 PMID:19821994

Mir, M. Y. (2014). Indigenous knowledge of using medicinal plants in treating skin diseases by tribal's of Kupwara, J&K, India. *International Journal of Herbal Medicine*, *1*(6), 62–68.

Moreira-Muñoz, A. (2007). The Austral floristic realm revisited. *Journal of Biogeography*, *34*(10), 1649–1660. doi:10.1111/j.1365-2699.2007.01757.x

Muley, B. P., Khadabadi, S. S., & Banarase, N. B. (2009). Phytochemical constituents and pharmacological activities of *Calendula officinalis* Linn (Asteraceae): A review. *Tropical Journal of Pharmaceutical Research*, *8*(5), 455–465. doi:10.4314/tjpr.v8i5.48090

Nadkarni, K. H. (2007). Indian Materica Medica. (1-739). Bombay Popular Publisher, Bombay.

Neogi, B., Prasad, M. N. V., & Rao, R. R. (1989). Ethnobotany of some weeds of Khasi and Garo hills, Meghalaya, Northeastern India. *Economic Botany*, *43*(4), 471–479. doi:10.1007/BF02935921

Okunade, A. L. (2002). *Ageratum conyzoides* L. (Asteraceae). *Fitoterapia*, *73*(1), 1–16. doi:10.1016/S0367-326X(01)00364-1 PMID:11864757

Panda, A., & Misra, M. K. (2011). Ethnomedicinal survey of some wetland plants of South Orissa and their conservation. *Indian Journal of Traditional Knowledge*, *10*(2), 296–303.

Panero, J. L., & Funk, V. A. (2002). Toward a phylogenetic subfamilial classification for the Compositae (Asteraceae). *Proceedings of the Biological Society of Washington*, *115*, 909–922.

Panero, J. L., & Funk, V. A. (2008). The value of sampling anomalous taxa in phylogenetic studies: Major clades of the Asteraceae revealed. *Molecular Phylogenetics and Evolution*, *47*(2), 757–782. doi:10.1016/j.ympev.2008.02.011 PMID:18375151

Paranjape, P. (2001). *Indian medicinal plants. Forgotten healer: a guide to Ayurvedic herbal medicine*. Delhi, India: Chaukhamba Sanskrit Pratisthan.

Patel, D. K. (2012). Study on medicinal plants with special reference to family Asteraceae, Fabaceae and Solanaceae in GGV-Campus, Bilaspur (CG) in central India. *Current Biotica*, *3*(4), 34–38.

Pathak, A. K., Saraf, S., & Dixit, V. K. (1991). Hepatoprotective activity potential of *Tridax procumbens*. *BMC Complementary and Alternative Medicine*, *10*, 10–15.

Paul, S., Rahman, A. N. M. A., Al Mahmud, M. A., Uzzal, A. R., & Das, Z. C. (2017). Wound healing by marigold (*Calendula officinalis*) and turmeric (*Curcuma longa*) paste: A comparative approach. *Journal of Advanced Veterinary and Animal Research*, *4*(4), 333–342. doi:10.5455/javar.2017.d229

Pereira, R. L., Ibrahim, T., Lucchetti, L., da Silva, A. J. R., & de Moraes, V. L. G. (1999). Immunosuppressive and anti-inflammatory effects of methanolic extract and the polyacetylene isolated from *Bidens pilosa* L. *Immunopharmacology*, *43*(1), 31–37. doi:10.1016/S0162-3109(99)00039-9 PMID:10437654

Pirbalouti, A. G., Koohpayeh, A., & Karimi, I. (2010). The wound healing activity of flower extracts of *Punica granatum* and *Achillea kellalensis* in Wistar rats. *Acta Poloniae Pharmaceutica*, *67*(1), 107–110. PMID:20210088

Prajapati, N. D., Purohit, S. S., Sharma, A. K., & Kumar, T. (2003). *A Handbook of Medicinal Plants* (pp. 1–1155). Jodhpur, India: Agrobios.

Pratibha, J., & Sahu, T. R. (1993). An ethnobotanical study of Noradehi Sanctuary Park of Madhya Pradesh, India: Native plant remedies for scorpion sting and snake bite. *Journal of Economic and Taxonomic Botany*, *17*(2), 315–328.

Ramachandran, V. S., & Nair, N. C. (1981). Ethnobotanical observations on irulars of Tamil Nadu (India). *Journal of Economic and Taxonomic Botany*, *2*, 183–190.

Redl, K., Breu, W., Davis, B., & Bauer, R. (1994). Anti-inflammatory active polyacetylenes from *Bidens campylotheca. Planta Medica*, *60*(1), 58–62. doi:10.1055-2006-959409 PMID:8134419

Rehecho, S., Uriarte-Pueyo, I., Calvo, J., Vivas, L. A., & Calvo, M. I. (2011). Ethnopharmacological survey of medicinal plants in Nor-Yauyos, a part of the Landscape Reserve Nor-Yauyos-Cochas, Peru. *Journal of Ethnopharmacology*, *133*(1), 75–85. doi:10.1016/j.jep.2010.09.006 PMID:20837126

Resende, J. H. C., De Aquino, G. S. T., Do Nascimento, F. R. F., Aguiar, M. M., & Fiorelli, R. K. A. (2015). Oral use of an infusion of leaves of *Solanum paniculatum* L., *Jacaranda brasiliensis* and *Sonchus oleraceus* for Treatment of Vitiligo. Journal of Cosmetics. *Dermatological Sciences and Applications*, *5*(4), 317–331. doi:10.4236/jcdsa.2015.54039

Rhama, S., & Madhavan, S. (2011). Antibacterial activity of the flavonoid, patulitrin isolated from the flowers of *Tagetes erecta* L. *International Journal of Pharm Tech Research*, *3*(3), 1407–1409.

Robinson, H. (2004). New supertribes Helianthodae and Senecionodae, for the subfamily Asteroideae (Asteraceae). *Phytologia*, *86*(3), 116–120. doi:10.5962/bhl.part.28428

Robinson, H. (2005). Validation of the supertribe Asterodae. *Phytologia*, *87*, 72–88.

Rogers, M. (2015). Herbalpedia: Herbs and everything about them, Sunflower. Retrieved from https://www.herbalpedia.com/blog

Rybalchenko, N. P., Prykhodko, V. A., Nagorna, S. S., Volynets, N. N., Ostapchuk, A. N., Klochko, V. V., ... Avdeeva, L. V. (2010). In vitro antifungal activity of phenylheptatriyne from *Bidens cernua* L. against yeasts. *Fitoterapia*, *81*(5), 336–338. doi:10.1016/j.fitote.2009.10.007 PMID:19837144

Sabale, P., Bhimani, B., Prajapati, C., & Sabale, V. (2012). An overview of medicinal plants as wound healers. *Journal of Applied Pharmaceutical Science*, *2*(11), 143–150.

Safdar, W., Majeed, H., Naveed, I., Kayani, W. K., Ahmed, H., Hussain, S., & Kamal, A. (2010). Pharmacognostical study of the medicinal plant *Calendula officinalis* L. (family Compositae). *International Journal of Cell Science & Molecular Biology*, *1*, 108–116.

Saha, J. C., Savini, E. C., & Kasinathan, S. (1961). Ecbolic properties of Indian medicinal plants. Part 1. *The Indian Journal of Medical Research*, *49*, 130–151. PMID:14496025

Sahu, C. R., Nayak, R. K., & Dhal, N. K. (2013). Traditional herbal remedies for various diseases used by tribals of Boudh district, Odisha, India for sustainable development. *International Journal of Herbal Medicine*, *1*(1), 12–20.

Sahu, T. R. (1984). Less known uses of weeds as medicinal plants. *Ancient Science of Life*, *3*(4), 245–249. PMID:22557414

Saraf, S., Pathak, A. K., & Dixit, V. K. (1991). Hair growth promoting activity of *Tridax procumbens*. *Fitoterapia*, *62*, 495–498.

Sharma, J., Gairola, S., Sharma, Y. P., & Gaur, R. D. (2014). Ethnomedicinal plants used to treat skin diseases by Tharu community of district Udham Singh Nagar, Uttarakhand, India. *Journal of Ethnopharmacology*, *158*, 140–206. doi:10.1016/j.jep.2014.10.004 PMID:25448505

Sharma, S., & Gupta, N. (2012). Antimicrobial potential of a weed plant *Parthenium hysterphorus*: An in vivo study. *International Journal of Pharmaceutical Research Development*, *4*, 112–118.

Shukla, A. N., Srivastava, S., & Rawat, A. K. S. (2010). An ethnobotanical study of medicinal plants of Rewa district, Madhya Pradesh. *Indian Journal of Traditional Knowledge*, *9*(1), 191–202.

Siddhiqui, M. A. A., John, A. Q., & Paul, T. M. (1995). Status of some important medicinal and aromatic plants of Kashmir Himalaya. *Advances in Plant Sciences (Muzaffarnagar)*, *8*, 134–139.

Siddiqui, M. B., Alam, M. M., & Husain, W. (1989). Traditional treatment of skin diseases in Uttar Pradesh, India. *Economic Botany*, *43*(4), 480–486. doi:10.1007/BF02935922

Singh, S. D. J., Krishna, V., Mankani, K. L., Manjunatha, B. K., Vidya, S. M., & Manohara, Y. N. (2005). Wound healing activity of the leaf extracts and deoxyelephantopin isolated from *Elephantopus scaber* Linn. *Indian Journal of Pharmacology*, *37*(4), 238–242. doi:10.4103/0253-7613.16570

Sobrinho, A. C. N., de Morais, S. M., de Souza, E. B., & dos Santos Fontenelle, R. O. (2017). The genus *Eupatorium* L.(Asteraceae): A review of their antimicrobial activity. *Journal of Medicinal Plants Research*, *11*(3), 43–57. doi:10.5897/JMPR2016.6313

Srinivasan, K., Natarajan, D., Mohanasundari, C., Venkatakrishnan, C., & Nagamurugan, N. (2007). Antibacterial, preliminary phytochemical and pharmacognostical screening on the leaves of *Vicoa indica* (L.) DC. *Iranian Journal of Pharmacology and Therapeutics*, *6*(1), 109–110.

Srivastava, T. N., Badola, D. P., Shah, D. C., & Gupta, O. P. (1984). Ethno-medico-botanical exploration of Figurez Valley Kashmir. *Bulletin of Medical Ethnobotany and Research*, *5*, 15–54.

Srivastava, T. N., Pathak, N. N., & Hussain, S. M. S. (1980). Medicinal plants of Kailash (J & K). *Bulletin of Ethno Medicine and Botanical Research*, *2*(2), 145–166.

Suntar, I. (2014). The medicinal value of Asteraceae family plants in terms of wound healing activity. FABAD. *Journal of Pharmaceutical Sciences*, *39*(1), 21–31. PMID:24450391

Suo, M., & Yang, J. (2014). Ceramides isolated from *Helianthus annuus* L. *Helvetica Chimica Acta*, *97*(3), 355–360. doi:10.1002/hlca.201300194

Suresh, J., Singh, A., Vasavi, A., Ihsanullah, M., & Mary, S. (2011). Phytochemical and pharmacological properties of *Artemisia pallens*. *International Journal of Pharmaceutical Sciences and Research*, *2*(12), 3081.

Tan, P. V., Dimo, T., & Dongo, E. (2000). Effects of methanol, cyclohexane and methylene chloride extracts of *Bidens pilosa* on various gastric ulcer models in rats. *Journal of Ethnopharmacology*, *73*(3), 415–421. doi:10.1016/S0378-8741(00)00290-7 PMID:11090994

Tayade, S. K., & Patil, D. A. (2005). Ethnomedicinal traditions of tribals of Nandurbar District (Maharashtra). *Journal of Phytological Research*, *18*(2), 251–254.

Teklehaymanot, T. (2017). An ethnobotanical survey of medicinal and edible plants of Yalo Woreda in Afar regional state, Ethiopia. *Journal of Ethnobiology and Ethnomedicine*, *13*(1), 40–65. doi:10.118613002-017-0166-7 PMID:28679438

Thorne, R. F., & Reveal, J. L. (2007). An updated classification of the class Magnoliopsida ("Angiospermae"). *Botanical Review*, *73*(2), 67–181. doi:10.1663/0006-8101(2007)73[67:AUCOTC]2.0.CO;2

Tiwari, V. J. (2018). Validity of Ethnopharmacological uses of *Vicoa indica* (L.) DC., Family–Asteraceae. *Research Journal of Pharmacognosy and Phytochemistry*, *10*(3), 207–210. doi:10.5958/0975-4385.2018.00033.X

Uddin, M. N., Rahman, M. A., Ahmed, N. U., Rana, M. S., Akter, R., & Chowdhury, A. M. A. (2010). Antioxidant, cytotoxic, and antimicrobial properties of *Eclipta alba* ethanol extract. *International Journal of Biological and Medical Research*, *1*(4), 341–346.

Ukiya, M., Akihisa, T., Yasukawa, K., Koike, K., Takahashi, A., Suzuki, T., & Kimura, Y. (2007). Triterpene glycosides from the flower petals of sunflower (*Helianthus annuus*) and their anti-inflammatory activity. *Journal of Natural Products*, *70*(5), 813–816. doi:10.1021/np078002l PMID:17480100

Ukiya, M., Akihisa, T., Yasukawa, K., Tokuda, H., Suzuki, T., & Kimura, Y. (2006). Anti-inflammatory, anti-tumor-promoting, and cytotoxic activities of constituents of marigold (*Calendula officinalis*) flowers. *Journal of Natural Products*, *69*(12), 1692–1696. doi:10.1021/np068016b PMID:17190444

Umadevi, S., Mohanta, G. P., Kalaichelvan, V. K., & Manavalan, R. (2006). Studies on wound healing effect of *Flaveria trinervia* leaf in mice. *Indian Journal of Pharmaceutical Sciences*, *68*(1), 106–108. doi:10.4103/0250-474X.22979

Vendruscolo, G. S., & Mentz, A. Z. (2006). Ethnobotanical survey of the medicinal plants used by the community of Ponta Grossa neighborhood, Porto Alegre, Rio Grande do Sul, Brazil. *Iheringia Botânica. Series*, *61*, 83–103.

Vianna-da-Silva., N. M. (2003). Moraes ram, Da silva, A. J. M., Costa, P. R. R., Melo, P. A. Antivenom effect of a new synthetic coumestan analog of wedelolactone. *The Journal of Venomous Animals and Toxins*, *9*(2), 381.

Wagenitz, G. (1976). Systematics and phylogeny of the Compositae (Asteraceae). *Plant Systematics and Evolution*, *125*(1), 29–46. doi:10.1007/BF00986129

Wiart, C. (2000). *Medicinal Plants of Southeast Asia*. Kuala Lumpur: Pelanduk Publisher.

Yoshikawa, M., Murakami, T., Kishi, A., Kageura, T., & Matsuda, H. (2001). Medicinal flowers. III. Marigold. (1): Hypoglycemic, gastric emptying inhibitory, and gastroprotective principles and new oleanane-type triterpene oligoglycosides, calendasaponins A, B, C, and D, from Egyptian *Calendula officinalis*. *Chemical & Pharmaceutical Bulletin*, *49*(7), 863–870. doi:10.1248/cpb.49.863 PMID:11456093

Zahra, A. A., Kadir, F. A., & Mahmood, A. A., Al hadi, A. A., Suzy, S. M., Sabri, S. Z., Latif, I. I., & Ketuly, K. A. (2011). Acute toxicity study and wound healing potential of *Gynura procumbens* leaf extract in rats. *Journal of Medicinal Plants Research*, *5*(12), 2551–2558.

Zhai, Z., Haney, D. M., Wu, L., Solco, A. K., Murphy, P. A., Wurtele, E. S., & Cunnick, J. E. (2009). Alcohol extract of *Echinacea pallida* reverses stress-delayed wound healing in mice. *Phytomedicine*, *16*(6-7), 669–678. doi:10.1016/j.phymed.2009.02.010 PMID:19303756

Section 3
Traditional Practices of Ethnobotanical Plants

Chapter 11
Ethnobotanical Uses of Medicinal Plants for the Treatment of Herniated Disc in Turkish Traditional Medicine

Mustafa Eray Bozyel
https://orcid.org/0000-0001-5556-4880
Canakkale Onsekiz Mart University, Turkey

Elif Merdamert Bozyel
https://orcid.org/0000-0001-5470-2372
Canakkale Onsekiz Mart University, Turkey

Kerem Canli
Dokuz Eylül University, Turkey

ABSTRACT

Humans have known the therapeutic properties and plant origin poisons of many plant species since ancient times. Ancient people who had no real and scientific knowledge on diseases and treatments tried to find solutions to these problems by the opportunities they had and the theories they could produce. Plants are the most important source of traditional folk medicine. People around the world have benefited from plants grown in their environment for this purpose. In these drugs, whole plant or various plant organs or secondary substances were used. Over the years, several therapies have emerged to treat herniated disc, ranging from conservative treatment to minimally invasive and percutaneous techniques to open surgical methods. This chapter examines the recent studies to form a compilation on ethnobotanical uses of medicinal plants for the treatment of herniated disc and their local names, parts, and usage forms in Turkish traditional medicine.

DOI: 10.4018/978-1-7998-1320-0.ch011

INTRODUCTION

Medicinal plant culture is an indispensable accumulation of knowledge depending on observations, and trial and errors for centuries. It started with garlic, castor oil and myrrh in the Egyptian Ebers papyrus in 1500 BC. It is a traditional culture that has reached today with the help of Indian, Chinese and Arabic medicine (Emre, 2012).

In general, the branch of science that investigates the folk culture of plants is called ethnobotany. Local studies on this subject are also studies that contribute to the recognition of world culture. In this respect, ethnobotanical studies aiming to provide a permanent accumulation of the folk culture that humanity has developed in the historical process and transfer to the next generations are carried on by caring about the world.

It is very clear that our country, which has a very rich environment in terms of plants, has also a very important position in terms of ethnobotany. Anatolia, where there is a very rich cultural accumulation, carries the traces of various civilizations. Therefore, Turkey is a very rich source for ethnobotanical research. It is very important to evaluate this source and to contribute to the recognition of our country in this respect with scientific studies. Ethnobotanical studies also allow our culture to be detected in time. Ethnobotanical research on the scientific validity has shown much increase over the past 40 years and especially the last 20 years in Turkey.

Wild plants used as herbal folk remedies and nutrients have a special importance for public health within ethnobotanical studies. With the increasing researches on these subjects, the situation in our country is tried to be recognized scientifically.

As it is known, the most important condition of being healthy is healthy nutrition. Plants are one of our most valuable nutrients because they contain vitamins, minerals and a variety of health-related substances, and they help to maintain body resistance against diseases.

Plants that are both cultivated and grown in nature make important contributions to our health. Among the wild plants, there are many plants where the people gather from nature and eat raw or cook in various ways. The use of wild plants as a source of food is especially common in rural areas of our country.

Plants used as folk medicine have traditionally played a role in the elimination of various health problems of the people for centuries. It is a fact that the people living in rural areas have widely used traditional methods in the treatment of various diseases. The role of herbal folk medicines in the treatment of simple and frequent health problems requiring rapid intervention (wound, cut, etc.) is important.

One of the interesting features of our folk culture is that the use of plants in traditional treatment is reflected in the local name of the plants (Stomach grass, Wound grass, etc.). Recognition and introduction of plants according to their use in treatment is an indication of the importance given to traditional treatment with plants in our country. The information traditionally transmitted from generation to generation is a very important source for scientific research in the field of medicine and new drugs can be developed by using these sources (Tuzlacı, 2016).

Due to their easy adaptation to our body, medicinal plants have remained on the agenda for centuries. Medicinal plants can have more than one beneficial effect at the same time, unlike drugs produced to treat a single disease. For example, a plant may be good for acne, headache, indigestion, constipation and fatigue. The main effect of this plant is to increase bile release. Fat excretion is regulated by bile and the skin heals. In addition, bile increase improves constipation. When constipation disappears and the intestines begin to work regularly, there is no accumulation of toxins in the body. This causes loss of complaints like headaches and fatigue (Emre, 2012).

The Uses of Natural Products in Modern Time

The use of natural substances is as old as human history. Thus, after the formation of the active substance concept, naturally occurring substances are presented in at least as much as synthetic compounds or in a greater proportion of medical preparations.

In recent years, the toxic effects seen in the direct use of active substances have led to the use of natural products that are effective but not harmful in a shorter period of time. This situation became a treatment method. The interest of the Western world, especially in the treatment methods applied in India and China, has started to increase.

Until the 19th century, the only methods of treatment were the plants and the galenic preparations prepared from them. Plant chemistry became important in the beginning of this century and pure substances were isolated from the plants. Morphine in 1816, Kinin in 1820, Digitoxin in 1868 were obtained as pure. Thus, a new period of treatment started to begin. Galenic preparations prepared by using herbal drugs were replaced with active substance use. With the reflection of industrialization on treatment, the use of galenic preparations has inevitably decreased since the beginning of the 20th century.

The presence of alkaloids in the 19th century, and later the emergence of antibiotics and sulfonamides increased the interest in synthetic materials in the pharmaceutical industry until the 1960s. However, the expensive costs and side effects of such drugs have brought about the use of natural remedies as in the previous centuries.

Today, plants are the sources of both active substances and starting material of semi-synthesis products. Direct plant use in treatment decreased in countries where technology was advanced and increased in countries where the economy was based on agriculture. In addition, societies' life philosophy and social habits also play an active role in this situation. In the USA, where consumption economics dominates, the trend towards synthetic drugs is higher, whereas in European countries, which are more rigorous in preserving traditional values, the direct use of plants for treatment has not lost its importance. In addition, the concerns of the countries for assessing their own resources are clearly evidents for the pharmacopoeias of Eastern European countries. Many herbal drugs are registered in the pharmacopoeias of Hungary, a country with suitable climatic conditions for the production of medicinal plants (Demirhan Erdemir, 2007).

In recent years, the importance given to the plants used for medicinal purposes has increased in our country and the studies on the medicinal uses of plants have increased (Başer, Honda, & Miki, 1986; Bozyel et al., 2019; Bozyel, Merdamert Bozyel, Canli, & Altuner, 2019; Bozyel & Merdamert, 2018; Canli, Yetgin, Akata, & Altuner, 2016; Canli, Bozyel, & Altuner, 2017; Canli, Simsek, Yetgin, & Altuner, 2017; Canli, Yetgin, Akata, & Altuner, 2017a; Canli, Yetgin, Akata, & Altuner, 2017b; Canli, Yetgin, Benek, Bozyel, & Altuner, 2019; Çömlekçioğlu & Karaman, 2008; Malyer, Aydın, Tümen, & Er, 2004; Polat, Satıl, & Çakılcıoğlu, 2011; Selvi, Satıl, Polat, & Çakılcıoğlu, 2012; Tulukçu & Sağdıç, 2011; Yetgin, Canli, & Altuner, 2018).

The main reasons for this are:

1- Developing countries, which do not have a sufficient level of chemical industry, want to obtain an easy and inexpensive treatment opportunity by taking advantage of the plants in their nature. In this area, countries such as Egypt, India and Pakistan are making great efforts and receiving positive results.
2- Dangerous side effects seen in some of the new synthetic substances have been introduced into the treatment area. Since the herbal drugs have been used for a long time, their side effects are well

known. On the other hand, the newly introduced synthetic substances do not have sufficient time to control, so some dangerous side effects are understood only after they are used. This situation leads to irreparable damages.

3- Some drug primary substances are obtained from herbal drugs are cheaper and easier to reach than synthetic ones. Steroid compounds, Cinchona alkaloids, Opium alkaloids, Ergot alkaloids, Atropa alkaloids, Rauvolfia alkaloids, Strychnos alkaloids, Digitalis glycosides are examples of such applications.

4- Another advantage of herbal drugs is that they have several effects. Synthetic compounds usually have a single effect. Some, like antibiotics, require some other medication to prevent side effects. There is no such situation in herbal drugs.

Medicinal Plants for The Treatment of Herniated Disc

The spine, which is one of the most basic and integrating part of the human skeleton, also acts as the protector of the spinal cord where the nerves that provide communication and management of the human body are gathered and reach the brain. In medical terminology, the spine consists of four main sections (Cervical, Thoracic, Lumbar and Sacral) and 29 vertebrae. The first three of these sections also refer to the curved areas of the spine (Budak & Bozkurt, 2013).

Back pain is one of the most common disorders worldwide with significant impact on public health. It is estimated that about 80% of adults in western world will suffer back pain at some points of their lives. One of the most frequent causes of back pain is herniated disc, which triggers pain pathways via mechanical compression together with immunologic and inflammatory changes (Marcia et al., 2018).

Herniated disc is the protrusion of the liquid part in the middle of the intervertebral disc from the tear formed in the surrounding structure (Hammer, 1999).

The intervertebral disc consists of 3 zones: an outer zone made up of fibrocartilage attaching the other 2 zones to each other; the vertebral body consisting of the central nucleus pulposus (i.e. a fibro-gelatinous mass composed of 80%-90% water, collagen, and a mucopolysaccharide matrix); and the peripheral annulus fibrosus (formed by the concentric alternating lamellae of obliquely oriented collagenous fibers). The annulus fibers run obliquely between vertebrae and are arranged primarily in concentric layers. The annulus is the primary disc structure that resists rotational forces through the orientation of the lamellae. Resistance to forward bending occurs due to the relatively greater thickness of the posterior lamellae (Urban & McMullin, 1985; Marchand & Ahmed, 1990; Tsuji et al., 1993; Saal, 1996; Weinstein, Herring, & Cole, 1998). The main function of the intervertebral discs is to shock absorption. Primarily, the annulus acts as a shock absorber, not the nucleus, which is predominantly liquid (and incompressible). When an axial load occurs, the increased force on the incompressible nucleus pushes on the annulus and stretches its fibers. If the fibers break, then a herniated nucleus pulposus occurs (Barr & Harrast, 2007).

The herniated disc, which protrudes outward, exerts pressure on the nerves that pass through the spinal canal or posterior-lateral (Hertling & Kessler, 2006). The nerves in the spine are very sensitive even to a very weak pressure, so this condition causes hernia (Liebenson, 2007; Magee, 2013).

The common factor in risk factors of herniated disc is primarily advanced age. While the rate of water in a fetus disc is 90%, it decreases to 80% in children and 50-60% in adults. This means that the discs lose their elasticity with age (Waddell, 1998).

A vertebra trauma, an inactive work and lifestyle, working more by sitting, obesity, lifting heavy things, fighting sports, doing unconscious sports, wrong sitting and posture habit, disliking his/her own profession, living in a restless environment and stress, smoking and alcohol are some of the other risk factors in developing herniated disc (Miller et al., 1988; Vlok & Hendrix, 1991; Jensen et al., 1994)

Over the years, several therapies have emerged to treat herniated disc, ranging from conservative treatment to minimally invasive and percutaneous techniques to open surgical methods. Although conservative treatment is the mainstay of management for most patients, there is a growing concern regarding the use of opioid therapy in this and other populations. Beyond that, persistent pain (more than 6 months) often leads to consideration of minimally invasive techniques and/or open surgery. Although open surgery can be effective (49–95%), it also carries the risks of general anaesthesia, high cost and significant postoperative complications, including new or worsening of neurological deficit of 1–3%, and re-operation rates of 4–10% (Marcia et al., 2018).

According to the World Health Organization (WHO) data, 80% of the world's population is treated with herbal drugs (Sarı et al., 2010). In this case, ethnobotanical knowledge transferred from the ancient times to today comes to the fore and the use of medicinal plants gains importance (Bozyel & Merdamert, 2018).

This study examines the recent studies to form a compilation on ethnobotanical uses of medicinal plants for the treatment of herniated disc and their local names, parts and usage forms in Turkish traditional medicine. As a result of the study, only the most used 31 plant taxa from 17 families are listed (Table 1).

Since there are many such plants, the examples provided below were selected amongst the most widely used species.

Liquidambar orientalis Mill. (Altingiaceae) (Günlükağacı)

Description: Tree up to 20 m. Leaves palmatifid, 5-10 x 6-13 cm, divided into 5 lobes, each lobe undivided or sparsely lobulate, margin finely sinuate-dentate or -serrate, glabrous or rarely with a sparse tuft of short hairs beneath at the base of the main veins. Fruiting head 2,5-3 cm diameter, pendulous. Flowering; March to April. Flood plains, marshy places, valley sides near streams, sea level to 800 m. Eastern Mediterranean element, of Boreal-Tertiary origin (Davis, 1965-1985).

Origanum saccatum P.H.Davis (Lamiaceae) (Tahtacı kekiği)

Description: Subshrub to 80 cm, ± hirsute at base, otherwise glabrous. Branches to 1-2 pairs per stem, to 7 cm; branches of 2nd order sometimes present. Leaves subsessile, cordate to ovate, 6-40 x 4-24 mm, acute or obtuse, glaucous. Spicules 10-35 x 6-11 mm. Bracts ovate, 6-11 x 2-6 mm, acuminate or acute. Calyx 1-lipped to circa 2/5, circa 5 mm; upper lip subentire; lower lip consisting of 2 very small lobes. Corolla pink, circa 12 mm. Flowering; July to August. Calcareous rocks and slopes, sometimes in *Pinus* woods, circa 1000 m. Endemic. Eastern Mediterranean element (Davis, 1965-1985).

Laurus nobilis L. (Lauraceae) (Defne)

Description: Aromatic evergreen shrub or tree, 2-15 m; dioecious. Leaves 3-10 x 2-4 cm, narrowly oblong-lanceolate to broadly ovate, acute or acuminate, somewhat wavy-margined, coriaceous. Male flowers with 8-12 stamens, filaments with stipitate glands near base. Female flowers often with 4 staminodes. Fruit 10-12 mm, globular to ellipsoid, black. Flowering; March to May. Coastal macchie, dense bushes mixed with *Myrtus*, *Phillyrea* and *Erica arborea*, scattered as underwood in *Pinus brutia* forest, rocky slopes, damp gorges, ancient settlements; naturalized and cultivated for ornament, sea level to 1200 m. South Europe, Northwest Africa, Cyrenaica, West Syria, Crimea; often subspontaneous in Western Mediterranean. Mediterranean element (Davis, 1965-1985).

Cedrus libani A.Rich. (Pinaceae) (Katranağacı)

Description: Columnar tree, taller than broad. Branches ± horizontal, the youngest sparsely pubescent to glabrous. Leaves glaucous, frequently with silvery lines, 10-40 mm, somewhat flattened, acute. Ripe cones ovoid, 6-9 cm. Dominant tree, or in mixed forest, 1000-2000 m. Lebanon, Cyprus, Northwest Africa. A disjunct Mediterranean element (Davis, 1965-1985).

Triticum aestivum L. (Poaceae) (Ekmeklik buğday)

Description: Annual. Stems solitary or tufted, 35-140 cm, thin-walled and hollow, rarely partly filled with pith, smooth, glabrous throughout or nodes glabrescent. Leaf sheaths glabrous or puberulent. Ligule circa 1 mm. Leaf blades linear-acuminate, 6-15 mm broad, scabridulous, green or ± glaucous, glabrous or glabrescent, margins smooth. Spikes lax to dense, 5-10 ern, ±square in cross-section. Rachis tough, glabrous or ciliate, upper internodes 4-8 mm. Spikelets 3-6-flowered. Glumes 6-10 mm, coriaceous, glabrous, pubescent or villous, yellow to blackish-purple; keel crested at apex or in upper half only, scabridulous, terminating in 2-3 mm acute tooth or 4-10 mm scabrid awn. Fertile lemma 12-15 mm, unawned or with 6-12 cm scabrid awn. Caryopsis ellipsoid, pale coloured or reddish. Endosperm mealy to flinty. 2n=42. Flowering; April to May. Volcanic rock, hillslopes, chalky steppe, dry plains, by roadsides, growing with Aegilops speltoides, *A. umbellulata*, *A. triuncialis*, *Triticum turgidum*, *Secale montanum* and *S. cereale*, or cultivated in solid stands, 10-2500 m. *T. aestivum*, the bread wheat, is the most widespread wheat in cultivation today (Davis, 1965-1985).

Agrimonia eupatoria L. (Rosaceae) (Fıtıkotu)

Description: Erect perennial 20-120 cm; basal leaves often in a rosette. Leaves pinnate, 8-25 cm; leaflets 4-13-paired, serrate-crenate, glandular below, unequal, larger pairs 2-10 cm, alternating with smaller pairs, 1 cm or less. Inflorescence, a lax, many-flowered, spike-like raceme, 10-35 cm; pedicels 1-3 mm. Flowers erect. Sepals ovate, 3-nerved, 1,5-2 mm, glandular. Petals yellow, 3-6 mm. Mature hypanthium obconical to turbinate, 7-10 x 5-8 mm, basal part hairy and deeply grooved. Inner bristles erect, the outer ascending, patent or slightly deflexed. Flowering; May to September. Damp meadows, by streams, ea level to 2300 m. Europe, Atlantic Islands, North Africa, West, Central & East Asia (Davis, 1965-1985).

Ruta chalepensis L. (Rutaceae) (Kokar sedefotu)

Description: Stem rigid, somewhat woody to suffruticose, 18-40 cm, glabrous, densely foliose. Leaves bi- or subtripinnatisect, ultimate segments narrowly oblonglanceolate or obovate. Inflorescence lax, pedicels equalling or exceeding the capsules; branches and pedicels glabrous, rarely a few minute glands above. Bracts glabrous, very broad, the lowest cordate-ovate, much wider than the subtended branch. Sepals glabrous, deltoid-ovate. Petals oblong, fringed with long cilia. Capsule glabrous, each segment acuminate-pointed. Flowering May to June. Rocky limestone slopes and fields, sea level to 300 m. Mediterranean area, Arabia, Somalia; cultivated elsewhere (Davis, 1965-1985).

Urtica dioica L. (Urticaceae) (Isırgan)

Description: Coarse perennial herb with an extensively spreading, matted root system, forming clumps, 30-150 cm tall. Leaves broadly ovate, 4-11 x 3-10 cm, sharply and coarsely dentate, acuminate. Dioecious, rarely monoecious. Male and female inflorescences similar in form, to circa 8 cm, much-branched. Female flowers with prominent purplish penicillate stigma, inner pair of perianth segments finally accrescent to circa 1,5 mm, ovate, pilose over entire surface. Flowering; June to September. Forests, shaded ravines and rocks, margins of streams, 500-2700 m. Europe, North Africa, West Asia, Siberia. Euro-Siberian element (Davis, 1965-1985).

FUTURE RESEARCH DIRECTIONS

The use of medicinal plants in traditional medicine is mentioned above. Today, the medicinal uses of these plants take their place in modern medicine with traditional complementary/alternative medicine applications. According to the results, the uses of medicinal plants which are great interests under the headings of phytotherapy and aromatherapy in general, are promising for future studies (Karadeniz et al., 2017, Ceylan et al., 2009). There isn't present detailed information about the herniated disc and the treatment of the herniated disc in Turkey. In particular, the adaptation of medicinal plants used in the treatment of herniated disc to modern medicine has great importance. Priority should be given to research on this subject.

CONCLUSION

This review includes 31 medicinal plant taxa representing 17 families and two of them are endemic. The highest number of medicinal plants were observed in the Lamiaceae family. Herbal medicine used by the local people are prepared as infusion, decoction, crushed, etc. The most commonly used plant parts are leaves, fruits and capitula. In addition, other aerial and underground plant parts such as roots, seeds and stems can also be used in the treatment of herniated disc.

However, it is known that some plants traditionally used for therapeutic purposes among local populations actually contain toxic compounds which cause various disorders, and in some cases even death. Therefore, studies that demonstrate the medicinal characteristics and uses of different plants are just for informational purposes. It is also known that the various compounds present in plants may have different effects on different individuals. Although herbal medicines have fewer side effects than synthetic medicinal medicines, the application of such plants for medicinal purposes must be carried out under the supervision of specialists.

Furthermore, this review includes some of the herbal medicines that have therapeutic activity for various types of herniated disc. This review can help others to explore herbs to further extent and its use in various other disease and toxicity studies along with clinical trials.

REFERENCES

Akan, H., & Bakır Sade, Y. (2015). Kâhta (adıyaman) merkezi ve narince köyü'nün etnobotanik açıdan araştırılması. *BEÜ Fen Bilimleri Dergisi*, *4*(2), 219–224.

Akaydin, G., Şimşek, I., Arituluk, Z. C., & Yeşilada, E. (2013). An ethnobotanical survey in selected towns of the Mediterranean subregion (Turkey). *Turkish Journal of Biology*, *37*(2), 230–247.

Altundağ, E. (2009). *Iğdır ilinin (doğu anadolu bölgesi) doğal bitkilerinin halk tarafından kullanımı* (Unpublished doctoral dissertation). İstanbul Üniversitesi, İstanbul.

Altundag, E., & Ozturk, M. (2011). Ethnomedicinal studies on the plant resources of east Anatolia, Turkey. *Procedia: Social and Behavioral Sciences*, *19*, 756–777. doi:10.1016/j.sbspro.2011.05.195

Arı, S. (2014). *Afyonkarahisar ve civarında halk tarafından kullanılan bazı bitkilerin etnobotanik özellikleri* (Unpublished doctoral dissertation) Afyon Kocatepe Üniversitesi, Afyon.

Bağcı, Y. (2000). Aladağlar (yahyalı, kayseri) ve çevresinin etnobotanik özellikleri. *Ot Sistematik Botanik Dergisi*, *7*(1), 89–94.

Barr, K. P., & Harrast, M. A. (2007). Low back pain. In: Braddom RL, editor. Physical medicine and rehabilitation. Philadelphia, CN: Saunder & Elsevier, 883-929.

Başer, K. H. C., Honda, G., & Miki, W. (1986). Herb drugs and herbalists in turkey. Institute for the Study of Languages and Cultures of Asia and Africa, Studia Culturae Islamicae, 27, Tokyo.

Baytop, T. (1984). Türkiye'de bitkiler ile tedavi. İstanbul, Turkey: İ. Ü. Yayınları. 3255 Ecz. Fak. No: 40.

Baytop, T. (1999). *Türkiye'de bitkiler ile tedavi, 2. baskı*. İstanbul, Turkey: Nobel Tıp Kitapevleri.

Bozyel, M. E., & Merdamert, E. (2018). Antiurolithiatic activity of medicinal plants in turkey. In R. Efe (Ed.), *Science, ecology, and engineering research in the globalizing world* (pp. 152–167). Sofia: St. Kliment Ohridski University Press.

Bozyel, M. E., Merdamert Bozyel, E., Canli, K., & Altuner, E. M. (2019). Anticancer Uses of Medicinal Plants in Turkish Traditional Medicine. *KSU J. Agric Nat*, *22*(Suppl 2), 465–484.

Bozyel, M. E., Şenturan, M., Benek, A., Merdamert Bozyel, E., Canli, K., & Altuner, E. M. (2019). In vitro antimicrobial activity screening of heliotropium europaeum against wide range of microorganisms and multi drug resistant (mdr) bacteria. *European Journal of Biomedical and Pharmaceutical Sciences*, *6*(3), 113–117.

Budak, E. Ç., & Bozkurt, M. R. (2013). Vertebra lomber disklerde meydana gelen bozulmaların manyetik rezonans görüntüleme (mrg) ile analizi. *AJIT-e: Online Academic Journal of Information Technology*, *4*(12), 125–144.

Bulut, G., & Tuzlaci, E. (2013). An ethnobotanical study of medicinal plants in turgutlu (Manisa-turkey). *Journal of Ethnopharmacology*, *149*(3), 633–647. doi:10.1016/j.jep.2013.07.016 PMID:23933313

Canli, K., Bozyel, M. E., & Altuner, E. M. (2017c). In vitro antimicrobial activity screening of maclura pomifera fruits against wide range of microorganisms. *International Journal of Pharmaceutical Science Invention*, *6*(8), 19–22.

Canli, K., Simsek, O., Yetgin, A., & Altuner, E. M. (2017d). Determination of the chemical composition and antimicrobial activity of frankenia hirsuta. *Bangladesh Journal of Pharmacology*, *12*(4), 463–469. doi:10.3329/bjp.v12i4.33652

Canli, K., Yetgin, A., Akata, I., & Altuner, E. M. (2016). In vitro antimicrobial screening of aquilaria agallocha roots. *African Journal of Traditional, Complementary, and Alternative Medicines*, *13*(5), 178–181. PMID:28487909

Canli, K., Yetgin, A., Akata, I., & Altuner, E. M. (2017a). Antimicrobial activity and chemical composition screening of epilobium montanum root. *Indian Journal of Pharmaceutical Education and Research*, *51*(3s), 239–243. doi:10.5530/ijper.51.3s.21

Canli, K., Yetgin, A., Akata, I., & Altuner, E. M. (2017b). Antimicrobial activity and chemical composition screening of anacyclus pyrethrum root. *Indian Journal of Pharmaceutical Education and Research*, *51*(3s), 244–248. doi:10.5530/ijper.51.3s.22

Canli, K., Yetgin, A., Benek, A., Bozyel, M. E., & Altuner, E. M. (2019). In vitro antimicrobial activity screening of ethanol extract of lavandula stoechas and investigation of its biochemical composition. *Advances in Pharmacological Sciences, 2019. Article ID, 3201458*, 1–6. doi:10.1155/2019/3201458

Ceylan, S., Azal, Ö., Taşlipinar, A., Türker, T., Açikel, C. H., & Gulec, M. (2009). Complementary and alternative medicine use among Turkish diabetes patients. *Complementary Therapies in Medicine*, *17*(2), 78–83. doi:10.1016/j.ctim.2008.07.003 PMID:19185265

Çömlekçioğlu, N., & Karaman, Ş. (2008). The medicinal plants found in the local herbal markets in the city of kahramanmaraş in turkey. *KSU Journal of Science and Engineering*, *11*(1), 23–32.

Dağlı, M. (2015). *Şanlıurfa merkez ve bağlı köylerde etnobotanik bir araştırma* (Unpublished master's thesis). Harran Üniversitesi, Şanlıurfa.

Davis, P. H. (1965-1985). *Flora of turkey and the east Aegean islands*. Edinburgh at the University Press, Vol. 1-9, Edinburgh.

Demirci, S., & Özhatay, N. (2012). An ethnobotanical study in kahramanmaraş (turkey); wild plants used for medicinal purpose in andirin, kahramanmaraş. *Turk J Pharm Sci*, *9*(1), 75–92.

Demirhan Erdemir, A. (2007). *Şifalı bitkiler: doğal ilaçlarla geleneksel tedaviler*. İstanbul, Turkey: Alfa Yayınları.

Emre, A. (2012). *Şifalı bitkiler ve vitaminler*. İstanbul, Turkey: Alfa Yayınları.

Gençay, A. (2007). *Cizre (şırnak)'nin etnobotanik özellikleri* (Unpublished master's thesis). Yüzüncü Yıl Üniversitesi, Van.

Gençler, Ö. A. M., & Koyuncu, M. (2005). Traditional medicinal plants used in pınarbaşı area (kayseri-türkiye). *Turkish J. Pharm. Sci.*, *2*(2), 63–82.

Güner, A., Aslan, S., Ekim, T., Vural, M., & Babaç, M. T. (Eds.). (2012). *Türkiye bitkileri listesi damarlı bitkiler*. İstanbul, Turkey: Nezahat Gökyiğit Botanik Bahçesi ve Flora Araştırmaları Derneği Yayını.

Hammer, W. (1999). *Functional soft tissue examination and treatment by manual methods* (2nd ed.). Gaithersburg, MD: Aspen.

Hertling, D., & Kessler, R. M. (2006). *Management of common musculoskeletal disorders: physical therapy principles and methods*. Lippincott Williams & Wilkins.

İzgi, M. N. (2017). Intent-to-treat medicinal plants found ın the flora of mardin province of turkey and neighborhood. In *International Symposium of Medicinal, Aromatic and Dye Plants (REYHAN 2017) Proceedings Book* (pp. 1-8). Malatya.

Jensen, M. C., Brant-Zawadzki, M. N., Obuchowski, N., Modic, M. T., Malkasian, D., & Ross, J. S. (1994). Magnetic resonance imaging of the lumbar spine in people without back pain. *The New England Journal of Medicine*, *331*(2), 69–73. doi:10.1056/NEJM199407143310201 PMID:8208267

Karadeniz, C., Pınarlı, F. G., Oğuz, A., Gürsel, T., & Canter, B. (2007). Complementary/alternative medicine use in a pediatric oncology unit in Turkey. *Pediatric Blood & Cancer*, *48*(5), 540–543. doi:10.1002/pbc.21012 PMID:16900485

Karakurt, E. (2014). *Kelkit (gümüşhane) ilçesinin etnobotanik özellikleri* (Unpublished master's thesis), Erzincan Üniversitesi, Turkey.

Korkmaz, M., & Karakurt, E. (2014). Kelkit (gümüşhane) aktarlarında satılan tıbbi bitkiler. *Journal of Natural and Applied Science*, *18*(3), 60–80.

Liebenson, C. (2007). *Rehabilitation of the spine: a practitioner's manual*. Lippincott Williams & Wilkins.

Magee, D. J. (2013). *Orthopedic physical assessment*. Elsevier Health Sciences.

Malyer, H., Aydın, Ö. A., Tümen, G., & Er, S. (2004). Tekirdağ ve çevresindeki aktarlarda satılan bazı bitkiler ve tıbbi kullanım özellikleri. *Dumlupınar Üniversitesi Fen Bilimleri Enstitüsü Dergisi*, *7*, 103–112.

Marchand, F., & Ahmed, A. M. (1990, May). Investigation of the laminate structure of lumbar disc anulus fibrosus. *Spine*, *15*(5), 402–410. doi:10.1097/00007632-199005000-00011 PMID:2363068

Marcia, S., Bellini, M., Hirsch, J. A., Chandra, R. V., Piras, E., Marras, M., ... Saba, L. (2018). Efficacy of an ethyl alcohol gel in symptomatic disc herniation. *European Journal of Radiology*, *109*, 101–107. doi:10.1016/j.ejrad.2018.10.029 PMID:30527290

Miller, J. A., Schmatz, C., & Schultz, A. B. (1988). Lumbar disc degeneration: Correlation with age, sex, and spine level in 600 autopsy specimens. *Spine*, *13*(2), 173–178. doi:10.1097/00007632-198802000-00008 PMID:3406837

Ozturk, M., & Altay, V. (2017). Endemic medicinal and aromatic plants of turkiye -conservation and sustainable use. In *1st International Congress on Medicinal and Aromatic Plants: "Natural and Healthy Life" Proceedings Book* (pp. 11-39). Konya, Turkey.

Polat, R., Satıl, F., & Çakılcıoğlu, U. (2011). Medicinal plants and their use properties of sold in herbal market in bingöl (turkey) district. *Biological Diversity and Conservation*, *4*(3), 25–35.

Saal, J. A. (1996). Natural history and nonoperative treatment of lumbar disc herniation. *Spine, 21*(24Suppl), 2–9. doi:10.1097/00007632-199612151-00002 PMID:9112320

Sağıroğlu, M., Topuz, T., Ceylan, K., & Turna, M. (2013). An ethnobotanical survey from yahyalı (kayseri) and tarsus (mersin). *Sakarya University J Sci and Lit, 2*, 13–37.

Sakinoğlu Oruç, F. Ç., & Oruç, S. H. (2017). Ethnobotonical aspects of some species in düzce and its vicinity. In *1st International Congress on Medicinal and Aromatic Plants: "Natural and Healthy Life" Proceedings Book* (pp. 391-399). Konya, Turkey.

Sargın, S. A. (2013). *Alaşehir ve çevresinde (manisa) tarımsal biyoçeşitlilik ve etnobotanik araştırmaları* (Unpublished doctoral dissertation). Balıkesir Üniversitesi, Balıkesir.

Sargin, S. A. (2015). Ethnobotanical survey of medicinal plants in Bozyazı District of Mersin, Turkey. *Journal of Ethnopharmacology, 173*, 105–126. doi:10.1016/j.jep.2015.07.009 PMID:26190351

Sargın, S. A., Selvi, S., & López, V. (2015). Ethnomedicinal plants of Sarigöl District (Manisa), Turkey. *Journal of Ethnopharmacology, 171*, 64–84. doi:10.1016/j.jep.2015.05.031 PMID:26026370

Sarı, A. O., Oğuz, B., Bilgiç, A., Tort, N., Güvensen, A., & Şenol, S. G. (2010). Ege ve güney marmara bölgelerinde halk ilacı olarak kullanılan bitkiler. *Anadolu Ege Tarimsal Arastirma Enstitüsü Dergisi, 20*(2), 1–21.

Selvi, S., Satıl, F., Polat, R., & Çakılcıoğlu, U. (2012). Kazdağlarından (balıkesir-edremit) toplanan ve bölgedeki aktarlarda satılan tıbbi bitkiler üzerine bir araştırma. In Kazdağları III. Ulusal Sempozyumu Bildiri Kitabı (pp. 505-513). Edremit, Balıkesir.

Tsuji, H., Hirano, N., Ohshima, H., Ishihara, H., Terahata, N., & Motoe, T. (1993). Structural variation of the anterior and posterior anulus fibrosus in the development of human lumbar intervertebral disc. A risk factor for intervertebral disc rupture. *Spine, 18*(2), 204–210. doi:10.1097/00007632-199302000-00006 PMID:8441935

Tulukçu, E., & Sağdıç, O. (2011). Konya'da aktarlarda satılan tıbbi bitkiler ve kullanılan kısımları. *Erciyes Üniversitesi Fen Bilimleri Enstitüsü Dergisi, 27*(4), 304–308.

Tuzlacı, E. (2005). *Bodrum'da bitkiler ve yaşam*. İstanbul, Turkey: Güzel Sanatlar Matbaası.

Tuzlacı, E. (2006). *Şifa niyetine, türkiye'nin bitkisel halk ilaçları*. İstanbul, Turkey: Alfa Yayıncılık.

Tuzlacı, E. (2016). *Türkiye bitkileri geleneksel ilaç rehberi*. İstanbul, Turkey: İstanbul Tıp Kitabevleri.

Urban, J. P. G., & McMullin, J. F. (1985). Swelling pressure of the inervertebral disc: Influence of proteoglycan and collagen contents. *Biorheology, 22*(2), 145–157. doi:10.3233/BIR-1985-22205 PMID:3986322

Vlok, G. J., & Hendrix, M. R. G. (1991). The lumbar disc: Evaluating the causes of pain. *Orthopedics, 14*(4), 419–425. PMID:1828114

Waddell, G. (1998). *The back pain revolution*. London, UK: Churchill Livingstone.

Weinstein, S. M., Herring, S. A., & Cole, A. J. (1998). Rehabilitation of the patient with spinal pain. In Rehabilitation medicine principles and practice. 3rd ed. Philadelphia, PA: Lippincot Williams & Wilkins, 1423-1453.

Yeşilada, E., Honda, G., Sezik, E., Tabata, M., Goto, K., & Ikeshiro, Y. (1993). Traditional medicine in Turkey iv. Folk medicine in the Mediterranean subdivision. *Journal of Ethnopharmacology*, *39*(1), 31–38. doi:10.1016/0378-8741(93)90048-A PMID:8331960

Yetgin, A., Canli, K., & Altuner, E. M. (2018). Comparison of antimicrobial activity of allium sativum cloves from China and Taşköprü, Turkey. *Advances in pharmacological sciences, 2018. Article ID, 9302840*, 1–5. doi:10.1155/2018/9302840

KEY TERMS AND DEFINITIONS

Cervical vertebrae: In human anatomy, cervical vertebrae are the vertebrae of the neck, immediately below the skull.

Decoction: It is a method of extraction by boiling plant material to dissolve the chemicals of the material, which may include stems, roots, bark and rhizomes.

Ethnomedicine: It is a study or comparison of the traditional medicine based on bioactive compounds in plants and animals and practiced by various ethnic groups, especially those with less access to western medicines.

Infusion: It is the process of extracting chemical compounds or flavors from plant material in a solvent such as water, oil or alcohol, by allowing the material to remain suspended in the solvent over time.

Lumbar vertebrae: In human anatomy, lumbar vertebrae are the five vertebrae between the rib cage and the pelvis.

Medicinal plants: They are known to be therapeutic due to the various molecules and compounds they contain and are used by public to treat many diseases.

Thoracic vertebrae: In human anatomy, thoracic vertebrae compose the middle segment of the vertebral column, between the cervical vertebrae and the lumbar vertebrae.

Traditional medicine: It refers to the knowledge, skills and practices based on the theories, beliefs and experiences indigenous to different cultures, used in the maintenance of health and in the prevention, diagnosis or treatment of physical and mental illness.

Turkey: Republic of Turkey is a Eurasian country that connects Asia and Europe with the Bosphorus bridges.

Table 1. Medicinal plants for the treatment of herniated disc in Turkish traditional medicine

Plant species	Local name*	Parts	Usage form	Disease type	References
Altingiaceae					
Liquidambar orientalis Mill.	Günlük ağacı	Roots	Mixed with olive oil	Herniated thoracic/ lumbar disc	Akan & Bakır Sade, 2015
Apocynaceae					
Nerium oleander L.	Zakkum	-	Oil	Herniated thoracic/ lumbar disc	Sargin et al., 2015
Asteraceae					
Inula viscosa (L.) Aiton	Sümenit	Leaves	-	Herniated thoracic/ lumbar disc	Tuzlacı, 2005, Tuzlacı, 2006
Onopordum heteracanthum C.A.Mey.	Kavlanık	Capitula	Infusion	Herniated thoracic/ lumbar disc	Tuzlacı, 2016
***Scorzonera tomentosa* L.	Alabent	Leaves Roots	Crushed	Herniated thoracic/ lumbar disc	Bağcı, 2000, Gençler & Koyuncu, 2005
Cannabaceae					
Celtis tournefortii Lam.	Dardağan	Fruits	Crushed	Herniated cervical disc	İzgi, 2017
Lamiaceae					
***Origanum saccatum* P.H.Davis	Tahtacı kekiği	Aerial parts	Infusion	Herniated thoracic/ lumbar disc	Sargin, 2015, Ozturk & Altay, 2017
Thymbra spicata L.	Zahter	Aerial parts	Infusion	Herniated thoracic/ lumbar disc	Sargin, 2015
Thymus fallax Fisch. & C.A.Mey.	Catri	Aerial parts	Decoction Boiled	Herniated lumbar disc	Altundağ, 2009, Tuzlacı, 2016
Thymus kotschyanus Boiss. & Hohen. ssp. *kotschyanus*	Kekik	Aerial parts	Decoction Boiled	Herniated thoracic/ lumbar disc Herniated lumbar disc	Altundağ, 2009, Tuzlacı, 2016
Thymus migricus Klokov & Des.-Shost.	Peynir kekiği	Aerial parts	Decoction	Herniated thoracic/ lumbar disc	Altundağ, 2009
Thymus transcaucasicus Ronniger	Kır kekiği	Aerial parts	Decoction Boiled	Herniated thoracic/ lumbar disc Herniated lumbar disc	Altundağ, 2009, Tuzlacı, 2016
Lauraceae					
Laurus nobilis L.	Defne	Seeds Fruits	Oil	Herniated thoracic/ lumbar disc	Akaydin et al., 2013, Tuzlacı, 2016
Malvaceae					
Hibiscus trionum L.	Kerkede	Fruits	Poultice	Herniated thoracic/ lumbar disc	Dağlı, 2015
Moraceae					
Ficus carica L.	İncir	Fruits Leaves Pedicels Branches	Raw eaten - - -	Herniated thoracic/ lumbar disc	Sargın *et al.*, 2015
Ficus carica ssp. *rupestris* (Hausskn.) Browicz	İt inciri	Leaves Fruits	Poultice	Herniated thoracic/ lumbar disc	Sargın, 2013
Pinaceae					
Cedrus libani A.Rich.	Katran ağacı	Stems	Terebinth	Herniated thoracic/ lumbar disc	Tuzlacı, 2016
Pinus nigra J. F. Arnold ssp. *pallasiana* (Lamb.) Holmboe var. *pallasiana*	Karaçam	Resin+Tar	Wrapping	Herniated thoracic/ lumbar disc	Arı, 2014
Pinus sylvestris L.	Sarıçam	Resin	Wrapping	Herniated thoracic/ lumbar disc	Karakurt, 2014
Piperaceae					
Piper nigrum L.	Karabiber	Fruits	Oil	Herniated cervical disc	Korkmaz & Karakurt, 2014
Platanaceae					
Platanus occidentalis L.	Londra çınarı	Leaves	Infusion	Herniated thoracic/ lumbar disc Herniated cervical disc	Sakinoğlu Oruç & Oruç, 2017
Poaceae					
Elymus repens (L.) Gould	Sabankıran	Whole plant	Infusion	Herniated thoracic/ lumbar disc	Karakurt, 2014
Triticum aestivum L.	Ekmeklik buğday	Stems Aerial parts	Decoction	Herniated thoracic/ lumbar disc Herniated lumbar disc	Gençay, 2007, Sağıroğlu et al., 2013, Tuzlacı, 2016
Rosaceae					
Agrimonia eupatoria L.	Fıtıkotu	Whole plant	Infusion Crushed	Herniated thoracic/ lumbar disc	Baytop, 1984, Altundag & Ozturk, 2011

continued on following page

Table 1. Continued

Plant species	Local name*	Parts	Usage form	Disease type	References
Sanguisorba minor ssp. *balearica* (Bourg. ex Nyman) Muñoz Garm. & C.Navarro	Kelekayağı	Whole plant	Decoction	Herniated thoracic/ lumbar disc	Yeşilada et al., 1993
Rutaceae					
Ruta chalepensis L.	Kokar sedefotu	Flo. branches Leaves+ flo. branches	Mixed with olive oil Chopped, boiled, cooked with olive oil and wrapping	Herniated thoracic/ lumbar disc	Tuzlacı, 2005 Tuzlacı, 2016
Santalaceae					
Viscum album sp. *abietis* (Wiesb.) Abromerit	Göknar güveliği	Leaves	-	Herniated thoracic/ lumbar disc	Çömlekçioğlu & Karaman, 2008
Typhaceae					
Typha latifolia L.	Cil	Capitula	-	Herniated thoracic/ lumbar disc	Baytop, 1999
Typha laxmannii Lepechin	Papur	Capitula	-	Herniated thoracic/ lumbar disc	Altundağ, 2009
Urticaceae					
Urtica dioica L.	Isırgan	Leaves Aerial parts Seeds	Infusion Infusion Crushed	Herniated thoracic/ lumbar disc	Bulut & Tuzlacı, 2013, Tuzlacı, 2016, Demirci & Özhatay, 2012
Urtica pilulifera L.	Dalağan	Fruits	Infusion with *Salvia* leaves	Herniated thoracic/ lumbar disc	Sargın, 2013

* Güner et al., 2012; ** Endemic taxon; "-": No information

Chapter 12
Turkish Herbal Medicine in the Treatment of Back Pain

Mustafa Eray Bozyel
https://orcid.org/0000-0001-5556-4880
Canakkale Onsekiz Mart University, Turkey

Elif Merdamert Bozyel
https://orcid.org/0000-0001-5470-2372
Canakkale Onsekiz Mart University, Turkey

Kerem Canli
Dokuz Eylül University, Turkey

ABSTRACT

The number of plants used for treatments has been steadily increasing since ancient times. There were around 250 herbal drugs used during the Mesopotamian civilization. During the time of the Greeks, 60 medicinal plants were known. During the Arab-Persian civilization, this number increased to about 4,000. At the beginning of the 19th century, the number of known medicinal plants reached 13,000. These herbal drugs are used to treat many diseases, from diabetes to cancer. Back pain and related symptoms rank as the second most common affliction of humankind. One of the proposed methods to control and treat back pain symptoms is traditional folk medicine. This chapter examines the recent ethnobotanical studies to form a compilation on medicinal plants used as Turkish herbal medicine in the treatment of back pain and their local names, parts, and usage forms.

INTRODUCTION

The therapeutic properties of many plant species have been known by humans since ancient times. On the other hand, poisons of plant origin have been known and used by people since the first ages of the history. Native African used *Strophanthus hispidus* and *Strophanthus kombe* seeds against their enemies as arrow poisons. In addition, the ancient Greeks used *Conium maculatum* was used as an official poi-

DOI: 10.4018/978-1-7998-1320-0.ch012

son substance. Researchers know that the wild and domestic animals benefit from the healing effects of plants and do not eat the harmful plants.

The ancient people, who have no real and scientific knowledge on diseases and treatments, tried to find solutions to health problems through the opportunities they had and the theories they could produce. As the archaeological findings come to light, medical tools of the Roman period, votive materials in the temples, drug recipes, wall paintings, and others reveal a more specific picture of this period's medicine.

Plants are the most important source of traditional folk medicine. People living in various parts of the world have benefited from the plants grown in their environment for this purpose. In these drugs, individuals used the whole plant, various plant organs (roots, leaves, flowers, fruits, seeds, etc.), or plant-derived substances (milk, terebinth, resin, balsam, etc.). Individuals use plants directly or in various forms (pills, powder, infusion, decoction, ointment, cataplasm). In Turkey, the most commonly used traditional medicine preparation forms are infusions and decoctions.

In folk medicine, plants are often used alone or with each other. Individuals create special prescriptions in the preparation of these drugs, and they adjuste the amounts accordingly. The effect of a plant can be increased when used with other plants (Sağında, 2014).

In a report published by the World Health Organization (WHO) published in 2000, authors state that 50% of people in Europe, Australia, and North America reported using one of the alternative supportive treatment methods in addition to the drug for the treatment of diseases. Herbal supplements were the most frequently used among these methods (Ersöz, 2012).

In today's world, medicinal plants have become an area of interest in contemporary scientific medicine in addition to being the main treatment tool of alternative medicine. The interest for the use of these plants on the way to preventing diseases or living healthier is becoming more and more widespread. Due to this increase in interest, the production, processing, and sales of medicinal plants are emerging as a growing sector (Kıran, 2006).

It is of great importance to recognize and use the plants that are still used today and that are the basis of many of synthetic drugs. In the last quarter of the 20th century, researhcers found that synthetic drugs have many side effects and potential for damage in addition to their benefits. Due to this finding, there has been a return to herbal medicine in the treatment of many diseases. In the Green Stream, called "Return to Nature", herbal medicines became the first choice in the treatment of simple daily diseases (Kızılarslan, 2008).

Traditional Uses of Medicinal Plants in Turkey

Turkey is located at the intersection of three different phytogeographical regions. These phytogeographical regions are European-Siberian (in North Anatolia), the Mediterranean (in Western and Southern Anatolia), and Irano-Turanian (in Central and South-Eastern Anatolia) (Erik & Tarikahya, 2004).

These phytogeographical regions of Turkey are the main cause of the evolution of different species of plants. More than 12,000 plant taxa grow naturally in Turkey. Approximately 3,649 (3/1 ratio) of them are endemic taxa (Güner et al., 2012). However, the endemic richness of Turkey is not limited to this number: the average number of plants is increasing by the discovery of a new endemic taxa presence every 10 days (Güner et al., 2012; Torlak et al., 2010).

Turkey, thanks to its important geographical location, is one of the leading countries in the aromatic and medicinal plant trade. The superiority provided by its geographical location has brought wealth to Turkey in terms of aromatic and medicinal plants compared to other countries, and this has enabled the creation of many industrial inputs (Bayram et al., 2010). Instead of ignoring the benefits that can be obtained from plants used for centuries researchers have thought that it would be more appropriate to investigate local formulas or recipes with the studies performed (Faydaoğlu & Sürücüoğlu, 2011).

Due to its cultural and floristic richness, Turkey is an important country, where reserachers have conducted many traditional folk medicine studies. Historically, the oldest written source about plants grown in the land and used as folk medicine is Materia Medica, written by Dioscorides in A.D. 78 (Sargın et al., 2015). In Turkey, the scientific studies on traditional folk medicine began in 1923, and until today, in many parts of the country reserachers have conducted an increasing number of studies on the subject (Paksoy et al., 2016). In recent years, the importance given to the plants used for medicinal purposes has increased in Turkey, and the studies on the medicinal uses of plants have increased (Başer et al.,1986; Bozyel et al., 2019; Bozyel & Merdamert, 2018; Canli et al., 2016; Canli et al., 2017a; Canli et al., 2017b; Canli et al., 2017c; Canli et al., 2017d; Canli et al., 2019; Çömlekçioğlu & Karaman, 2008; Malyer et al., 2004; Polat et al., 2011; Selvi et al., 2012; Tulukçu & Sağdıç, 2011; Yetgin et al., 2018).

Examples of medicinal plants and their use in Turkey are as follows; *Achillea millefolium*: diuretic, appetizing, wound healing; *Brassica nigra*: appetizing, analgesic; *Capparis spinosa*: diuretic, relieving constipation, strengthening; *Ecballium elaterium*: laxative, relieving sinusitis; *Equisetum arvense*: antiurolithiatic, hemostatic, wound healing; *Hypericum perforatum*: antispasmodic, antiseptic, wound healing; *Origanum onites*: sedative, antiseptic, worm reducer; *Pinus brutia*: antiseptic, wound healing; *Urtica urens*: blood purifier, diuretic, antirheumatic (Baytop, 1999).

The Importance of Herbal Medicine in Back Pain Treatment

Back pain and related symptoms rank as the second most common affliction of humankind. Moreover, pain management for back ailments encompasses a number of diverse treatments, with medication routinely being prescribed to alleviate immediate pain and suffering or enduring discomfort. The variety of drugs used (e.g., antidepressants, anti-inflammatory medications, muscle relaxants, corticosteroids, and narcotic analgesics) suggests that there is no consensus among physicians as to which drug, or combination of drugs, is the most effective. Much controversy also exists in the lack of selection criteria for prescribing narcotic analgesics. Without clear-cut standards for back pain, clinical judgment on the part of medical care providers may be susceptible to sociocultural and non-clinical influences (Kposowa & Tsunokai, 2002).

The most common type of back pain is lower back pain (LBP). LBP is a major public health problem that is one of the most serious pain problems (Andersson, 1997; Picavet et al., 1999; Van Tulder et al., 1995; Volinn, 1997). The lifetime prevalence of LBP ranges from 30% to 85%, with a point prevalence of 20% to 40% in developed countries (Bergenudd & Nilsson, 1988; Biering-Sørensen et al., 1989; Lee et al., 2001; Walsh et al., 1992). About two thirds of the adult population suffers from LBP at any given time (Deyo & Phillips, 1996). LBP is most common to those between the ages of 35 and 64 years (Andersson, 1997; Deyo & Tsui-Wu, 1987). Additionally, LBP causes loss of working days; outpatient attendance and number of days of incapacity also increase for individuals with LBP (Hansson & Hansson, 2000). Little is known about the prevalence and determinants of LBP among populations from developing countries (Gilgil et al., 2005) as most reports come from industrialized countries (Volinn, 1997; Sjølie, 1998).

Researchers usually define LBP as a pain, muscle tension, or stiffness localized below the costal margin and above the inferior gluteal folds. Moreover, reserachers typically classify LBP as being "specific" or "nonspecific." Specific LBP is defined as symptoms caused by a specific pathophysiologic mechanism, such as hernia nuclei pulposi, infection, inflammation, osteoporosis, rheumatoid arthritis, fracture, or tumor. Nonspecific LBP is defined as symptoms without clear specific cause, that is, LBP of unknown origin. Approximately 90% of all LBP patients have nonspecific LBP (Van Tulder et al., 2002).

In Turkey, community-based trials on the prevalence of LBP are scarce. One of these trials, the Turkish National Burden of Disease study, found the 30-day period prevalence of LBP to be 34.2%. Although the study by Gilgil et al. does not represent the whole country, as it was conducted in one urban area in Turkey, the lifetime, 12-month, reserachers found the point prevalence of LBP to be 46.6%, 35.9%, and 20.1%, respectively (Gilgil et al., 2005).

One of the proposed treatment methods to control and treat back pain symptoms is use of traditional folk medicine. Nowadays, in the age of new technologies and the considerable advances in science, patients benefit from the inclusive development in the field of traditional folk medicine and herbal/natural products. In many countries, several herbal medicines traditionally have been used to control and treat for various types of pain along with the chemical drugs. In addition, some of these herbal medicines have been clinically verified to aid with relief from symptoms of back pain (Malek et al., 2017).

This study examines recent ethnobotanical studies to form a compilation on medicinal plants used as Turkish herbal medicine in the treatment of back pain. Authors include local names, parts, and usage forms of Turkish herbal medicines.

As a result of the study, the authors found that 66 plant taxa (Table 1) and one multi-herbal formula (Table 2) are used as Turkish herbal medicine in the treatment of back pain.

Since there are many so many plants, the authors selected the examples provided below amongst the most widely-used species.

Petroselinum Crispum (Mill.) A.W.Hill (Apiaceae) (Maydanoz)

Description: This plant is a glabrous biennial. Its stems are erect, terete, and striate and measure 50-80 cm. This plant contains glabrous leaves that ar triangular-ovate in outline and that measure three to 10 x two to seven cm. Segments are cuneate-obovate to lanceolate-linear and usually deeply lobed with irregularly crenate lobes. Inflorescence paniculate-corymbose, umbels long-peduncled, rays 7-20, subequal, 15-4 cm; bracts 1-3, subulate, 3-7 mm, entire or 3-fid; bracteoles 5-8, subulate, small. Mericarps 2,5-3 x 0,5 mm, slightly arcuate at maturity. The plant flowers from June to August. Additionally, the plant is cultirvated in fields, vineyards, from sea level to 2000 m. Moreover, the plant is cultivated all over the world as a vegetable. The origin of this species is still unknown (Davis, 1965-1985).

Drimia Maritima (L.) Stearn (Asparagaceae) (Kum örümcekotu)

Description: The bulb of this plant is five to 15 cm diameter. Its leaves are flat and measure 17-50 x two to six cm. Its scape is 100-150 cm. Additionally, its inflorescence is a dense raceme with numerous flowers. Pedicels are 15-20 mm and are patent during anthesis, but become erect in fruit. The plant's perianth segments are white with greenish fascia, measuring seven to eight mm. In additiona, its anthers are medifixed, and measure 1.5 to two mm. 2n = 50. The plant flowers from September to November. This plant is grown in coastal habitats from sea level to 300 m, Often, the plant flowers before the autumn rains. The plant grows in the Mediterranean area of South Iran. Mediterranean element (Davis, 1965-1985).

Table 1. Herbal medicine for the treatment of back pain in Turkey

Plant species	Local name*	Parts	Usage form	Back pain type	References
Amaranthaceae					
Spinacia oleracea L.	Ispanak	Leaves	-	Upper	Sargın, 2013
Amaryllidaceae					
Allium sativum L.	Sarmısak	Bulbs	- Raw eaten Crushed	Lower	Kültür, 2007, Karcı et al., 2017, Tuzlacı, 2016
Apiaceae					
Petroselinum crispum (Mill.) A.W.Hill	Maydanoz	Aerial parts Leaves	Raw eaten (mixed with olive oil) Boiled with corn flour and sugar	Upper Lower	Bulut et al., 2014, Tuzlacı, 2016
Asparagaceae					
Asparagus acutifolius L.	Tilkişen	Roots	Wrapping (fresh)	Lower	Tuzlacı, 2016
Drimia maritima (L.) Stearn	Kum örümcekotu	Leaves Bulbs	Wrapping half cooked leaves or olive oiled leaves, wrapping cutted bulbs Divided into 2 pieces	Lower	Uysal, 2008, Tuzlacı, 2016
Scilla bifolia L.	Orman sümbülü	Bulbs	Crushed and keeped in olive oil	Lower	Tuzlacı, 2016
Asteraceae					
Achillea arabica Kotschy	Hanzabel	Aerial parts	Infusion	Upper	Sezik et al., 2001
Achillea pannonica Scheele	Kurpotu	Aerial parts	Decoction	Upper	Kültür, 2007; Tuzlacı, 2016
Arctium platylepis (Boiss. & Bal.) Sosn. ex Grossh.	Baldikeni	Leaves Basal leaves	Wrapping Wrapping (fresh) Poultice	Upper	Altundağ, 2009, Tuzlacı, 2016, Altundag & Ozturk, 2011
Artemisia annua	Kabe süpürgesi	Flowers	Decoction mixed with flour	Upper	Tuzlacı, 2006
Scorzonera cinerea Boiss.	Bozkonak	Latex	Crushed Raw eaten	Upper	Altundag & Ozturk, 2011, Polat et al., 2013
Scorzonera latifolia (Fisch & Mey.) DC.	Dağsakızı	Roots, Latex	Crushed	Upper	Altundag & Ozturk, 2011
Scorzonera rigida DC.	Serttekesakalı	Latex	Crushed	Upper	Altundag & Ozturk, 2011
Scorzonera dzhawakhetica Grossh.	Bozkulak sakızı	Latex	Crushed	Upper	Altundag & Ozturk, 2011
Inula helenium L. ssp. *pseudohelenium* Grierson	Andızotu	Whole plants, Roots	Infusion	Upper	Altundag & Ozturk, 2011
Inula viscosa (L.) Aiton	Sümenit	Leafy shoots	Chopped, boiled, cooked with olive oil and wrapping (fresh)	Lower	Tuzlacı, 2016
Matricaria aurea Sch.Bip.	Çiçeğe zer	Aerial parts	Raw eaten Infusion	Upper	Akgul et al., 2018
Brassicaceae					
Brassica oleracea	Lahana	Leaves	Wrapping	Lower	Ertuğ, 2002
Brassica nigra (L.) W.D.J.Koch	Karahardal	Leaves Stems	-	Lower	Güneş, 2010
Raphanus sativus L.	Turp	Tubers	Inside of tuber scooped out and filled with honey; after one night, water that spilled over from the pit is drunk.	Lower	Sarı et al., 2010
Cactaceae					
Opuntia ficus-barbarica A.Berger	Firenkinciri	Stems	Spines are heated after burning.	Lower	Tuzlacı, 2016
Cannabaceae					
Celtis tournefortii Lam.	Dardağan	Fruits	Crushed	Upper	İzgi, 2017
Capparaceae					
Capparis spinosa L.	Kebere	Roots	Infusion	Lower	Furkan, 2016
Cistaceae					
Cistus laurifolius L.	Karağan	Flo. branches Flowers Leaves	Cataplasm Decoction Boiled and Wrapping (fresh)	Lower	Ertuğ et al., 2003, Günbatan et al., 2016 Oral, 2007, Tuzlacı, 2016
Cucurbitaceae					
Cucurbita moschata Duchesne	Bal kabağı	Fruit fibers	Wrapping	Upper	Karakurt, 2014
Ecballium elaterium (L.) A.Rich.	Eşekhıyarı	Fruits	Keeped in kerosene Juice mixed with Olive oil and absolute alcohol	Upper Lower	Tuzlacı, 2016
Cupressaceae					
Juniperus excelsa Bieb.	Boz ardıç	Seeds	Maseration	Upper	Paksoy et al., 2016

continued on following page

Table 1. Continued

Plant species	Local name*	Parts	Usage form	Back pain type	References
Equisetaceae					
Equisetum arvense L.	Atkuyruğu	Aerial parts	Infusion	Lower	Akan & Bakır Sade, 2015
Gentianaceae					
Centaurium erythraea Rafn.	Kırmızı kantaron	Branches	Mixed with olive oil	Lower	Polat, 2000
Hypericaceae					
Hypericum perforatum L.	Kantaron	Flo. branches	Infusion	Upper	Kültür, 2007
Juglandaceae					
Juglans regia L.	Ceviz	Leaves	Decoction	Upper Lower	Aktan, 2011
Lamiaceae					
Mentha longifolia ssp. *typhoides* (Briq.) Harley	Dere nanesi	Leaves	- - Boiled and mixed with flour	Lower	Uğurlu & Seçmen, 2008, Karcı et al., 2017, Tuzlacı, 2016
Origanum onites L.	Bilyalı kekik	Aerial parts	Infusion Essential oil	Upper	Sadıkoğlu, 2003, Tuzlacı, 2016
Origanum vulgare L.	Karakınık	Aerial parts	-	Lower	Doğanoğlu et al., 2006
Teucrium polium L.	Acıyavşan	Aerial parts	-	Lower	Doğanoğlu *et al.*, 2006
Thymus fallax Fisch. & C.A.Mey.	Catri	Aerial parts Whole plant	Decoction Infusion Decoction	Lower Upper	Altundağ, 2009 Altundag & Ozturk, 2011
Thymus kotschyanus Boiss. & Hohen.	Kekik	Aerial parts	Infusion	Upper	Polat et al., 2017
Thymus kotschyanus Boiss. & Hohen. ssp. *kotschyanus*	Kekik	Aerial parts Whole plant	Decoction Infusion Decoction	Lower Upper	Altundağ, 2009 Altundag & Ozturk, 2011
Thymus migricus Klokov & Des.-Shost.	Peynir kekiği	Aerial parts Whole plant	Decoction Infusion Decoction	Lower Upper	Altundağ, 2009 Altundag & Ozturk, 2011
Thymus transcaucasicus Ronniger	Kır kekiği	Whole plant	Infusion Decoction	Upper	Altundag & Ozturk, 2011
Vitex agnus-castus L.	Hayıt	Leaves Seeds	Cataplasm	Lower	Ertuğ et al., 2003
Lauraceae					
Laurus nobilis L.	Defne	Seeds Fruits	Oil Crushed, boiled and obtained oil (fresh)	Lower	Uysal, 2008, Tuzlacı, 2016
Malvaceae					
Abelmoschus esculentus (L.) Moench	Bamya	Fruits	Poultice	Upper	Akan & Bakır Sade, 2015
Malva neglecta Wallr.	Çobançöreği	Whole plant Aerial parts	Raw eaten, Poultice, Decoction Poultice	Lower	Özgen et al., 2012, Tuzlacı, 2016
Malva pusilla Sm.	Kabalık	Whole plant	Raw eaten	Lower	Arıturuk, 2010
Malva sylvestris L.	Ebegümeci	Aerial parts Leaves	Infusion Crushed (fresh)	Lower	Aktan, 2011, Tuzlacı, 2016
Moraceae					
Ficus carica L.	İncir	Latex	-	Lower	Sargin *et al.*, 2015
Nitrariaceae					
Peganum harmala L.	Üzerlik	Fruits	-	Upper	Altundağ, 2009
Oleaceae					
Olea europaea L. ssp. *europaea*	Zeytin	Oil	Wrapping oil mixed with lemon juice and Aspirin	Upper	Metin, 2009
Pinaceae					
Cedrus libani A.Rich	Katranağacı	Tar	Wrapping	Upper	Eşen, 2008, Saday, 2009, Güneş, 2010
Pinus brutia Ten.	Kızılçam	Branc. bark Resin Shoots Stems	- - - Terebinth mixed with honey	Lower	Metin, 2009, Tuzlacı & Erol, 1999, Tuzlacı, 2016
Platanaceae					
Platanus orientalis L.	Çınar	Leaves	Infusion	Lower	Aktan, 2011
Ranunculaceae					
Helleborus orientalis Lam.	Çöpleme	Rhizomes	-	Lower	Eryaşar, 1998
Rosaceae					
Laurocerasus officinalis M.Roem.	Karayemiş	Leaves	Crushed and wrapping	Upper	Tuzlacı, 2016

continued on following page

Table 1. Continued

Plant species	Local name*	Parts	Usage form	Back pain type	References
Rubiaceae					
Rubia tinctorum L.	Kökboyası	Flo. branches Leaves+flo. branches	Crushed with grape fruits	Lower	Oral, 2007, Tuzlacı, 2016
Rutaceae					
Ruta chalepensis L.	Kokar sedefotu	Aerial parts	Decoction	Upper	Güzel et al., 2015
Santalaceae					
Viscum album sp. *abietis* (Wiesb.) Abromerit	Göknar güveliği	-	-	Lower	Güneş, 2010
Solanacaeae					
Hyoscyamus niger L.	Banotu	Seeds	-	Lower	Güneş, 2010
Physalis alkekengi L.	Güveyfeneri	Seeds Fruits	- Poultice	Lower	Sadıkoğlu & Alpınar, 2001, Tuzlacı, 2016
Thymelaeaceae					
Daphne oleoides Schreb.	Gövçek	Aerial parts Flo. branches	Cooked with butter and applied on the back side - -	Lower	Yesilada et al., 1995, Yesilada et al., 1997, Güneş, 2010
Daphne oleoides ssp. *oleoides* Schreb.	Gövçek	Whole plant	-	Lower	Yesilada et al., 1995,
Tiliaceae					
Tilia rubra DC.	Kafkas ıhlamuru	Flowers Leaves	Infusion	Lower	Korkmaz & Karakurt, 2014
Urticaceae					
Urtica sp.	Isırgan	Aerial parts	Chopped and wrapping	Lower	Tuzlacı, 2016
Urtica dioica L.	Isırgan	Whole plant Seeds+ leaves Leaves	Dec. with ethanol Decoction Infusion	Lower	Erol, 1995; Tuzlacı, 2016, Uysal, 2008, Bulut & Tuzlacı, 2013
Urtica urens L.	Cılağan	Whole plant	Dec. with ethanol	Lower	Tuzlacı & Erol, 1999; Tuzlacı, 2016
Urtica pilulifera L.	Dalağan	Fruits	Infusion with Salvia leaves	Lower	Sargın, 2013

* Güner et al., 2012; "-": No information

Achillea Arabica Kotschy (Asteraceae) (Hanzabel)

Description: *Achillea arabica* (syn: *Achillea biebersteinii*) has a height that ranges from 10-100 cm. Its stems are terete, longitudinally striped, and ± densely spreading-pubescent. Leaves loosely or ± densely pilose, 2-3-pinnatisect, cauline similar to basal, median cauline linear to linear-lanceolate, 2.5-12 x 0.5-2.5 cm, rachis entire, 0.5-1 mm, primary segments numerous, ultimate segments setaceous to linear, acute to acuminate, 1-4 x 0.2-1 mm. It has 30-200 or more capitula, corymbs that are two to 10 cm broad, peduncles that measure 0.5 to four mm. Its involucre is oblong to broadly ovoid and measures three to four x two to three mm. Additionally, its phyllaries are ovate-triangular to oblong, ± obtuse, pale, pubescent, and inner with scarious tip. Ligules 4-5, golden yellow, 1-2 mm; disc flowers 10-30. The plant flowers from May to September. It is grown in coniferous forest, steppe, dry meadows, rocky slopes, and fallow fields, 350-3450 m. This plant can be found in South Bulgaria, Southwest, and Central Asia. Irano-Turanian element (Davis, 1965-1985).

Matricaria Aurea Sch.Bip. (Asteraceae) (Çiçeğe zer)

Description: This plant has glabrous stems that measure five to 20 cm. It has two to three leaves that are pinnatisect, linear-oblanceolate in outline, and measure up to six cm. The ultimate segments are linear, circa 0.2 mm broad. capitula solitary or two to three at branch ends. Its involucre is four to five mm broad, its phyllaries are ovate, ± obtuse and measure 2.5-3.5 mm, and its margins are brown. Ray flowers absent. Disc flowers measure one mm. Additionally, its achenes is brown, measures 0.75 mm, contains three whitish ribs on posterior surface, and is ecoronate. The plant flowers from March to May.

Table 2. Multi-herbal formula for the treatment of back pain in Turkey

Plant species	Local name*	Parts	Usage form	Back pain type	References
Ficus carica+ *Olea europaea* ssp. *europaea*+ *Abelmoschus esculentus*+ *Mentha x piperita*	İncir+ Zeytin+ Bamya+ Nane	Fruits+ Oil+ Fruits+ Leaves	Wrapping	Lower	Çilden, 2011

* Güner et al., 2012

It is a weed of roadside and cultivated ground, found from sea level to 800 m. The plant is native to North Africa and Southwest Asia eastwards to West Himalayas (Davis, 1965-1985).

Brassica Nigra (L.) W.D.J.Koch (Brassicaceae) (Karahardal)

Description: This is an annual herb that grows up to 1.5 m or more. It is usually branched from about the middle. Its stem is covered below with stiff bristly hairs and is glabrous in region of inflorescence. Its basal leaves are petiolate, lyrate-pinnatisect with a large terminal lobe, and hispid on both surfaces. Its upper leaves are oblong-linear, entire, and glabrous. Additionally, its petals are bright yellow and meausre eight x three to four mm. Its siliquae meausres 10-17 x 1.5 to two mm and includes a very narrow beak measuring 1.5 to three mm. Seeds are not mucilaginous. This plant flowers from March to May. It is found in waste places and fields in the lowlands. The plant grows throughout Europe, North Africa, Southwest Asia, and most temperate parts of the world (Davis, 1965-1985).

Capparis Spinosa L. (Capparaceae) (Kebere)

Description: Its leaves are round or very broadly ovate. It is usually without a mucro, but sometimes with a very small, much recurved mucro. The plant is occasionally emarginate and often glabrous. Its flowers are weakly zygomorphic with whitish petals. The plant flowers in July, 100-150 m (Davis, 1965-1985).

Ecballium Elaterium (L.) A.Rich. (Cucurbitaceae) (Eşekhıyarı)

Description: This is a perennial herb. Its stems are prostrate, stout, hispid, pale green, and measure up to 25 cm. Its leaves are rather thick, stiff, deep green, and scabrid above, pale green and densely hispid-tomentose beneath, and ovate-triangular, cordate, sinuate and undulate at the margins. The plant is rather concave above, shortly five-lobed with the basal lobes ± erect, measuring 4-14.5 x 3.5-17 cm. Its peduncles are erect and hispid, measuring from two to 10 cm. Its male flowers measure 3.5-3.7 cm and are racemes. Additionally, its petals measure from eight to16 x four to seven mm. Its female flowers are often co-axillary with male flowers. Its peduncles measure from one to eight cm, and its petals measure from six to 12 x three to six mm. Its fruit are pendulous, measuring from three to five x 1.5-2.5 cm. Its pulp is exceedingly bitter. Its seeds are about four mm. The plant flowers from April to October. It grows in waste places, roadsides, and riverbanks, from sea level to 600m. Azores; It is native to the Mediterranean area, South Russia, Crimea, and Caucasia. Mediterranean element (Davis, 1965-1985).

Equisetum Arvense L. (Equisetaceae) (Atkuyruğu)

Description: This plant has stems of two types: the fertile, non-green stems are produced before the green sterile stems. Its sterile stems measure 20-80 cm. They are green, branched, and slightly rough with six to 19 deep grooves. Its sheaths are green, have teeth as many as grooves, are one-ribbed, are acute, and contain blackish tips. Its central hollow is somewhat less than half the stem's diameter. Its branches are spreading andthree to four grooved. The lowest internode is longer than the stem sheath. Its fertile stems measure up to 25 em. They are simple, brownish, and contain four to six distant leaf-sheaths. Its cones are one to four cm. This plant grows in waste places, stream sides, etc., lip to 1700 m. It is found in Europe, North Temperate Asia, and North America (Davis, 1965-1985).

Hypericum Perforatum L. (Hypericaceae) (Kantaron)

Description: This plant has stems of 10110 cm. The stems are erect, sometimes rooting, two-lined, and contains branches ± ascending. Its leaves measure from five to 35 mm. The leaves are narrowly ovate or lanceolate to elliptic-oblong or linear. They are occasionally oblanceolate, sessile or subsessile, plane, and always with large pellucid dots. Its sepals lanceolate to oblong or elliptic, acute to acuminate or shortly aristate, entire, and without or with a few superficial black dots. Next, its petals measure from eight to 15 mm, with a few marginal black dots and sometimes superficial black streaks. Its capsule meausres from five to nine mm, is ovoid to pyramidal, and contains dorsal vittae and lateral vesicles. The plant flowers from May to August. It grows in dry habitats in mesophytic regions, near water elsewhere, from sea Ievel to 2500 m. The plant is native to Europe, North Africa, Caucasia, Siberia, Central Asia, Iran, North Iraq, Cyprus, and West Syria, and it has been introduced elsewhere (Davis, 1965-1985).

Origanum Onites L. (Lamiaceae) (Bilyalı kekik)

Description: This plant is subshrub to 6.5 cm and hirsute. It branches to 10 pairs per stem up to 13 cm. Its leaves are petiolate to subsessile, cordate, ovate or elliptic. The leaves measure from three to 22 x two to 19 mm. They are ± acute or acuminate, remotely serrulate or entire, and contain veins ± raised on lower surface. Is spicules are in corymbiform inflorescence, circa three to 17 x four mm. Its bracts are obovate or elliptic, measuring two to five x 1.5 to four mm. The bracts are obtuse to acuminate and entire or denticulate. Its calyx measures from two to three mm. Its corolla is white, measuring from three to seven mm. The plant flowers from April to August. Its is grown on stony hills and rocky slopes, usually on limestone, sometimes in partial shade, from sea level to 1400 m. The plant is native to Sicily, Greece, and Aegean. East Mediterranean element (Davis, 1965-1985).

Vitex Agnus-Castus L. (Lamiaceae) (Hayıt)

Description: *Vitex agnus-castus L.* (syn: *Vitex pseudo-negundo*) is a shrub or rarely a small tree, measuring 1-3 m. The plant is much-branched and shortly tomentose-canescent. Its leaves are digitately five-partite, leaflets usually entire, 3.5-15 x 0.5-2.8 cm, occasionally. broader or distinctly dentate, acute, narrowed to both ends, sessile or at least terminal petiolulate, dull green above, white-tomentose beneath; Its petioles are long and are those of lower leaves to c. four cm. Its inflorescence is rather dense. Its cymes are compact and frequently subglobose, sessile or subsessile. Its calyx measures three mm and is densely tomentose outside and obscurely nerved. Its teeth containt a broadly triangular inner surface with prominent primary nerves. One or two intermediate nerves are present below each sinus with a mainly looping reticulum of secondary venation in upper third of tube or less, chiefly below each tooth. Its corolla is pale lilac to blue, circa eight mm. Its tube is distinctly exserted. Its drupe is globose and black or reddish, measuring three to four mm. The plant flowers from June to September. Its is grown mostly on sandy ground, parched alluvial soils, and rocky areas near sea. It is also grown on limestone slopes from sea level to 750 m. The plant is native to the Mediterranean area, extending to West Caucasia. Mediterranean element (Davis, 1965-1985).

Pinus Brutia Ten. (Pinaceae) (Kızılçam)

Description: This plant is a tree with a straight trunk and branches, up to 25 m. Its twigs are glabrous and usually reddish in the first year. They become greyish brown, rarely all greyish. Its buds are non-resinous, and the scales are recurved and fimbriate. Its leaves meausre up to 180 x 0.5-1.5 mm and are flaccid and light green. Its cones are elongate, brown, and usually more than two borne together. Its scales are usually with a depression containing the umbo. Its peduncles are very short to almost absent. The cones are erect or ascending and never deflexed. This plant is frequently a dominant forest tree, growing from sea level to 1200 m. The plant is native to South Italy, South Aegean, Crimea, Cyprus, North Iraq, West Syria, and West Caucasus. In the broad sense, an East Mediterranean element (Davis, 1965-1985).

Urtica Urens L. (Urticaceae) (Cılağan)

Description: This plant is an annual herb and is simple or sometimes quite bushy. This plant measures 10-60 cm. Its leaves are ovate or elliptic, sharply and coarsely dentate, and obtuse to shortly acuminate. The plant is monoecious. Its spikes are dense, to circa one cm, axillary and congested above but lower remote withnumerous male and female flowers. Female flowers have an inner pair of perianth segments accrescent in fruit, are pilose around margins, ± glabrous centrally, cordate-ovate, and measure two mm. The plant flowers from April to June. It grows in waste ground, near sea level to circa 1000 m. The plant is native to Europe, North Africa, and Southwest Asia. It is widely adventive as a weed in many other regions (Davis, 1965-1985).

FUTURE RESEARCH DIRECTIONS

The authors have mentioned the use of medicinal plants in traditional medicine above. Today, the medicinal uses of these plants take their place in modern medicine with traditional complementary/alternative medical applications. According to the results of this research, the uses of medicinal plants, which are of great interest under the headings of phytotherapy and aromatherapy in general, are promising for future studies (Ceylan et al., 2009; Karadeniz et al., 2017). Gilgil et al. (2005) and Oksuz (2006) presented detailed information about back pain and the treatment of the lower back pain in Turkey. In particular, the adaptation of medicinal plants used in the treatment of back pain to modern medicine is of great importance. Priority should be given to research on this subject.

CONCLUSION

This review includes 66 medicinal plant taxa representing 33 families and also 1 multi-herbal formula. The authors observed the highest number of medicinal plants in the Asteraceae family. Herbal medicine used by the local people are prepared as decoction, infusion, wrapping, etc. The most commonly used plant parts are leaves, fruits, and seeds. In addition, individuals can use other aerial plant parts, such as flowering branches, latex, stems, and roots, in the treatment of back pain.

Back pain is second health problem in both developed and developing countries. In many countries, the several herbal medicines traditionally have been used to control and treat various types of pain along with the chemical drugs. In addition, researchers have clinically verified some of these herbal medicine types for the relief of symptoms of back pain. Furthermore, this review includes some of the herbal medicines that have therapeutic activity for various types of back pain. This review can help others to explore herbs to a further extent. It can also help other so explore the use of herbs in various other diseases and toxicity studies along with clinical trials.

REFERENCES

Akan, H., & Bakır Sade, Y. (2015). Kâhta (adıyaman) merkezi ve narince köyü'nün etnobotanik açıdan araştırılması. *BEÜ Fen Bilimleri Dergisi*, *4*(2), 219–224.

Akgul, A., Akgul, A., Senol, S. G., Yildirim, H., Secmen, O., & Dogan, Y. (2018). An ethnobotanical study in Midyat (Turkey), a city on the silk road where cultures meet. *Journal of Ethnobiology and Ethnomedicine*, *14*(12), 1–18. PMID:29415748

Aktan, T. (2011). *Yenişehir (bursa) köylerinin etnobotanik özellikleri* (Unpublished master's thesis). Celal Bayar Üniversitesi, Turkey.

Altundağ, E. (2009). *Iğdır ilinin (doğu anadolu bölgesi) doğal bitkilerinin halk tarafından kullanımı* (Unpublished doctoral dissertation), İstanbul Üniversitesi, Turkey.

Altundag, E., & Ozturk, M. (2011). Ethnomedicinal studies on the plant resources of East Anatolia, Turkey. *Procedia: Social and Behavioral Sciences*, *19*, 756–777. doi:10.1016/j.sbspro.2011.05.195

Andersson, G. B. J. (1997). The epidemiology of spinal disorders. In J. W. Frymoyer (Ed.), *The Adult Spine: Principles and Practice* (2nd ed., pp. 93–141). New York, NY: Raven Press.

Arıturuk, Z. C. (2010). *Tefenni (burdur) ilçesinin florası ve halk ilaçları* (Unpublished master's thesis), Hacettepe Üniversitesi, Turkey.

Başer, K. H. C., Honda, G., & Miki, W. (1986). Herb drugs and herbalists in turkey. Institute for the Study of Languages and Cultures of Asia and Africa, Studia Culturae Islamicae, 27, Tokyo.

Bayram, E., Kırıcı, S., Tansı, S., Yılmaz, G., Arabacı, O., Kızıl, S., & Telci, İ. (2010). Tıbbi ve aromatik bitkiler üretiminin arttırılması olanakları. In Türkiye ziraat mühendisliği vii. teknik kongresi bildiriler kitabı 1 (pp. 437-456). Ankara, Turkey.

Baytop, T. (1999). *Türkiye'de bitkiler ile tedavi: geçmişte ve bugün*. İstanbul, Turkey: Nobel Tıp Kitabevleri.

Bergenudd, H., & Nilsson, B. O. (1988). Back pain in middle age: Occupational workload and psychologic factors: An epidemiologic survey. *Spine*, *13*(1), 58–60. doi:10.1097/00007632-198801000-00014 PMID:2967994

Biering-Sørensen, F., Thomsen, C. E., & Hilden, J. (1989). Risk indicators for low back trouble. *Scandinavian Journal of Rehabilitation Medicine*, *21*(3), 151–157. PMID:2529630

Bozyel, M. E., & Merdamert, E. (2018). Antiurolithiatic activity of medicinal plants in turkey. In R. Efe (Ed.), *Science, ecology and engineering research in the globalizing world* (pp. 152–167). Sofia, Bulgaria: St. Kliment Ohridski University Press.

Bozyel, M. E., Şenturan, M., Benek, A., Merdamert Bozyel, E., Canli, K., & Altuner, E. M. (2019). In vitro antimicrobial activity screening of heliotropium europaeum against wide range of microorganisms and multi drug resistant (mdr) bacteria. *European Journal of Biomedical and Pharmaceutical Sciences*, *6*(3), 113–117.

Bulut, G., & Tuzlaci, E. (2013). An ethnobotanical study of medicinal plants in Turgutlu (Manisa-Turkey). *Journal of Ethnopharmacology*, *149*(3), 633–647. doi:10.1016/j.jep.2013.07.016 PMID:23933313

Bulut, G., Tuzlacı, E., Doğan, A., & Şenkardeş, İ. (2014). An ethnopharmacological review on the Turkish apiaceae species. *Journal of Faculty Pharmacy of Istanbul University*, *44*(2), 163–179.

Canli, K., Bozyel, M. E., & Altuner, E. M. (2017c). In vitro antimicrobial activity screening of maclura pomifera fruits against wide range of microorganisms. *International Journal of Pharmaceutical Science Invention*, *6*(8), 19–22.

Canli, K., Simsek, O., Yetgin, A., & Altuner, E. M. (2017d). Determination of the chemical composition and antimicrobial activity of frankenia hirsuta. *Bangladesh Journal of Pharmacology*, *12*(4), 463–469. doi:10.3329/bjp.v12i4.33652

Canli, K., Yetgin, A., Akata, I., & Altuner, E. M. (2016). In vitro antimicrobial screening of aquilaria agallocha roots. *African Journal of Traditional, Complementary, and Alternative Medicines*, *13*(5), 178–181. PMID:28487909

Canli, K., Yetgin, A., Akata, I., & Altuner, E. M. (2017a). Antimicrobial activity and chemical composition screening of epilobium montanum root. *Indian Journal of Pharmaceutical Education and Research*, *51*(3s), 239–243. doi:10.5530/ijper.51.3s.21

Canli, K., Yetgin, A., Akata, I., & Altuner, E. M. (2017b). Antimicrobial activity and chemical composition screening of anacyclus pyrethrum root. *Indian Journal of Pharmaceutical Education and Research*, *51*(3s), 244–248. doi:10.5530/ijper.51.3s.22

Canli, K., Yetgin, A., Benek, A., Bozyel, M. E, & Altuner, E. M. (2019). In vitro antimicrobial activity screening of ethanol extract of lavandula stoechas and investigation of its biochemical composition. *Advances in Pharmacological Sciences, 2019*, Article ID 3201458, 6 pages. . doi:10.1155/2019/3201458

Çilden, E. (2011). *Paşayaylası (aydın) florası ve etnobotanik özellikleri* (Unpublished master's thesis), Hacettepe Üniversitesi, Turkey.

Çömlekçioğlu, N., & Karaman, Ş. (2008). The medicinal plants found in the local herbal markets in the city of Kahramanmaraş in Turkey. *KSU Journal of Science and Engineering, 11*(1), 23–32.

Davis, P. H. (1965-1985). *Flora of Turkey and the East Aegean Islands* (Vols. 1-9). Edinburgh, UK: Edinburgh at the University Press.

Deyo, R. A., & Phillips, W. R. (1996). Low back pain: A primary care challenge. *Spine, 21*(24), 2826–2832. doi:10.1097/00007632-199612150-00003 PMID:9112706

Deyo, R. A., & Tsui-Wu, Y. J. (1987). Descriptive epidemiology of low-back pain and its related medical care in the United States. *Spine, 12*(3), 264–268. doi:10.1097/00007632-198704000-00013 PMID:2954221

Doğanoğlu, Ö., Gezer, A., & Yücedağ, C. (2006). Göller bölgesi-yenişarbademli yöresi'nin önemli bazı tıbbi ve aromatik bitki taksonları üzerine araştırmalar. *Süleyman Demirel Üniversitesi Fen Bilimleri Enstitüsü Dergisi, 10*(1), 66–73.

Erik, S., & Tarikahya, B. (2004). Türkiye florası üzerine. *Kebikeç, 17*, 139–163.

Erol, M. K. (1995). *Eğirdir (ısparta) yöresinin geleneksel halk ilacı olarak kullanılan bitkileri* (Unpublished master's thesis), Marmara Üniversitesi, Turkey.

Ersöz, T. (2012). Bitkisel ilaçlar ve gıda takviyeleri ile ilgili genel yaklaşım ve sorunlar. *MİSED, 27-28*, 9–19.

Ertuğ, F., Tümen, G., & Çelik, A. (2003). Buldan (denizli) etnobotanik alan araştırma raporu 2002 yılı çalışması. *Türkiye Bilimler Akademisi, TÜBA-TÜKSEK Türkiye Kültür Envanteri Pilot Bölge Çalışmaları*, 76-87.

Eryaşar, P. (1998). *Gönen (balıkesir) yöresinin geleneksel halk ilacı olarak kullanılan bitkileri* (Unpublished master's thesis). Marmara Üniversitesi, Turkey.

Eşen, B. (2008). *Aydınlar köyü ve çevresinin (erdemli/mersin) etnobotanik özellikleri* (Unpublished master's thesis), Selçuk Üniversitesi, Turkey.

Faydaoğlu, E., & Sürücüoğlu, M. S. (2011). Geçmişten günümüze tıbbi ve aromatik bitkilerin kullanılması ve ekonomik önemi. *Kastamonu Üniversitesi Orman Fakültesi Dergisi, 11*(1), 52–67.

Furkan, M. K. (2016). *Adıyaman ilinde yetişen bazı bitkilerin etnobotanik özellikleri* (Unpublished master's thesis), Adıyaman Üniversitesi, Adıyaman.

Gilgil, E., Kaçar, C., Bütün, B., Tuncer, T., Urhan, S., Yildirim, Ç., ... Dündar, Ü. (2005). Prevalence of low back pain in a developing urban setting. *Spine, 30*(9), 1093–1098. doi:10.1097/01.brs.0000161007.46849.4c PMID:15864165

Günbatan, T., Gürbüz, İ., & Özkan, A. M. G. (2016). The current status of ethnopharmacobotanical knowledge in çamlıdere (Ankara, Turkey). *Turkish Journal of Botany, 40*(3), 241–249. doi:10.3906/bot-1501-37

Güner, A., Aslan, S., Ekim, T., Vural, M., & Babaç, M. T. (Eds.). (2012). *Türkiye bitkileri listesi damarlı bitkiler*. İstanbul, Turkey: Nezahat Gökyiğit Botanik Bahçesi ve Flora Araştırmaları Derneği Yayını.

Güneş, S. (2010). *Karaisalı (adana) ve köylerinde halkın kullandığı doğal bitkilerin etnobotanik yönden araştırılması* (Unpublished master's thesis), Niğde Üniversitesi, Niğde.

Güzel, Y., Güzelşemme, M., & Miski, M. (2015). Ethnobotany of medicinal plants used in antakya: A multicultural district in hatay province of turkey. *Journal of Ethnopharmacology*, *174*, 118–152. doi:10.1016/j.jep.2015.07.042 PMID:26239155

Hansson, T. H., & Hansson, E. K. (2000). The effects of common medical interventions on pain, back function, and work resumption in patients with chronic low back pain: A prospective 2-year cohort study in six countries. *Spine*, *25*(23), 3055–3064. doi:10.1097/00007632-200012010-00013 PMID:11145817

İzgi, M. N. (2017). Intent-to-treat medicinal plants found ın the flora of mardin province of turkey and neighborhood. In *International Symposium of Medicinal, Aromatic and Dye Plants (REYHAN 2017)* (pp. 1-8). Malatya. Academic Press.

Karakurt, E. (2014). *Kelkit (gümüşhane) ilçesinin etnobotanik özellikleri* (Unpublished master's thesis), Erzincan Üniversitesi, Erzincan.

Karcı, E., Gürbüz, İ., Akaydın, G., & Günbatan, T. (2017). Folk medicines of bafra (Samsun-Turkey). *Turkish Journal of Biochemistry*, *42*(4), 381–399. doi:10.1515/tjb-2017-0172

Kıran, Ö. (2006). *Kozan yöresindeki tıbbi bitkiler ve bunların halk arasında kullanılışı* (Unpublished master's thesis), Çukurova Üniversitesi, Adana.

Kızılarslan, Ç. (2008). *İzmit körfezi'nin güney kesiminde etnobotanik bir araştırma* (Unpublished master's thesis), İstanbul Üniversitesi, Turkey.

Korkmaz, M., & Karakurt, E. (2014). Kelkit (gümüşhane) aktarlarında satılan tıbbi bitkiler. *Journal of Natural and Applied Science*, *18*(3), 60–80.

Kposowa, A. J., & Tsunokai, G. T. (2002). Searching for relief: Racial differences in treatment of patients with back pain. *Race and Society*, *5*(2), 193–223. doi:10.1016/j.racsoc.2004.01.004

Kültür, Ş. (2007). Medicinal plants used in Kırklareli Province (Turkey). *Journal of Ethnopharmacology*, *111*(2), 341–364. doi:10.1016/j.jep.2006.11.035 PMID:17257791

Lee, P., Helewa, A., Goldsmith, C. H., Smythe, H. A., & Stitt, L. W. (2001). Low back pain: Prevalence and risk factors in an industrial setting. *The Journal of Rheumatology*, *28*(2), 346–351. PMID:11246674

Malek, V. G., Abbasian, A., Karimi, A., Khanlarkhani, N., & Parvari, S. (2017). 5 medicinal plants used in iranian traditional medicine for low back pain: A quick review. *European Journal of Experimental Biology*, *7*(4, 24), 1-4.

Malyer, H., Aydın, Ö. A., Tümen, G., & Er, S. (2004). Tekirdağ ve çevresindeki aktarlarda satılan bazı bitkiler ve tıbbi kullanım özellikleri. *Dumlupınar Üniversitesi Fen Bilimleri Enstitüsü Dergisi*, *7*, 103–112.

Metin, A. (2009). *Mut ve çevresinde yetişen bitkilerin (mersin) etnobotanik özellikleri* (Unpublished master's thesis), Selçuk Üniversitesi, Turkey.

Oksuz, E. (2006). Prevalence, risk factors, and preference-based health states of low back pain in a Turkish population. *Spine*, *31*(25), E968–E972. doi:10.1097/01.brs.0000247787.25382.3c PMID:17139213

Oral, D. Ç. (2007). *Konya ilinde kullanılan halk ilaçları üzerinde etnobotanik araştırmalar* (Unpublished master's thesis), Gazi Üniversitesi, Turkey.

Özgen, U., Kaya, Y., & Houghton, P. (2012). Folk medicines in the villages of Ilıca District (Erzurum, Turkey). *Turkish Journal of Biology*, *36*(1), 93–106.

Paksoy, M. Y., Selvi, S., & Savran, A. (2016). Ethnopharmacological survey of medicinal plants in Ulukışla (Niğde-Turkey). *Journal of Herbal Medicine*, *6*(1), 42–48. doi:10.1016/j.hermed.2015.04.003

Picavet, H. S. J., Schouten, J. S. A. G., & Smit, H. A. (1999). Prevalence and consequences of low back problems in the netherlands, working vs non-working population, the MORGEN-study. *Public Health*, *113*(2), 73–77. doi:10.1016/S0033-3506(99)00122-5 PMID:10355306

Polat, R. (2010). *Havran ve burhaniye (balıkesir) çevresinde tarımsal biyoçeşitlilik ve etnobotanik araştırmaları* (Unpublished doctoral dissertation), Balıkesir Üniversitesi, Balıkesir.

Polat, R., Cakilcioglu, U., & Satıl, F. (2013). Traditional uses of medicinal plants in Solhan (Bingöl-Turkey). *Journal of Ethnopharmacology*, *148*(3), 951–963. doi:10.1016/j.jep.2013.05.050 PMID:23770029

Polat, R., Güner, B., Babacan, E. Y., & Çakılcıoğlu, U. (2017). Survey of wild food plants for human consumption in Bingöl (Turkey). *Indian Journal of Traditional Knowledge*, *16*(3), 378–384.

Polat, R., Satıl, F., & Çakılcıoğlu, U. (2011). Medicinal plants and their use properties of sold in herbal market in Bingöl (Turkey) district. *Biological Diversity and Conservation*, *4*(3), 25–35.

Saday, H. (2009). *Güzeloluk köyü ve çevresinin (erdemli/mersin) etnobotanik özellikleri* (Unpublished doctoral dissertation). Selçuk Üniversitesi, Turkey.

Sadıkoğlu, E. (2003). *Koçarlı (aydın) yöresinin geleneksel halk ilacı olarak kullanılan bitkileri* (Unpublished master's thesis). Marmara Üniversitesi, Turkey.

Sadıkoğlu, N., & Alpınar, K. (2001). Etnobotanik açıdan bartın. In XIII. bitkisel ilaç hammaddeleri toplantısı bildiri kitabı (pp. 87-100). İstanbul, Turkey.

Sağında, A. (2014). *Silifke bölgesindeki bitkisel halk ilaçlarının araştırılması* (Unpublished doctoral dissertation). Selçuk Üniversitesi, Turkey.

Sargın, S. A. (2013). *Alaşehir ve çevresinde (manisa) tarımsal biyoçeşitlilik ve etnobotanik araştırmaları* (Unpublished doctoral dissertation). Balıkesir Üniversitesi, Turkey.

Sargin, S. A. (2015). Ethnobotanical survey of medicinal plants in bozyazı district of Mersin, Turkey. *Journal of Ethnopharmacology*, *173*, 105–126. doi:10.1016/j.jep.2015.07.009 PMID:26190351

Sargın, S. A., Selvi, S., & López, V. (2015). Ethnomedicinal plants of Sarigöl District (Manisa), Turkey. *Journal of Ethnopharmacology*, *171*, 64–84. doi:10.1016/j.jep.2015.05.031 PMID:26026370

Sarı, A. O., Oğuz, B., Bilgiç, A., Tort, N., Güvensen, A., & Şenol, S. G. (2010). Ege ve güney marmara bölgelerinde halk ilacı olarak kullanılan bitkiler. *Anadolu Ege Tarimsal Arastirma Enstitüsü Dergisi*, *20*(2), 1–21.

Selvi, S., Satıl, F., Polat, R., & Çakılcıoğlu, U. (2012). Kazdağlarından (balıkesir-edremit) toplanan ve bölgedeki aktarlarda satılan tıbbi bitkiler üzerine bir araştırma. In Kazdağları iii. ulusal sempozyumu bildiri kitabı (pp. 505-513). Edremit, Balıkesir.

Sezik, E., Yeşilada, E., Honda, G., Takaishi, Y., Takeda, Y., & Tanaka, T. (2001). Traditional medicine in Turkey x. folk medicine in Central Anatolia. *Journal of Ethnopharmacology*, *75*(2-3), 95–115. doi:10.1016/S0378-8741(00)00399-8 PMID:11297840

Sjølie, A. N. (1998). The epidemiology of low back pain in the rest of the world. *Spine*, *23*(8), 960–961. PMID:9580967

Torlak, H., Vural, M., & Aytaç, Z. (2010). *Türkiye'nin endemik bitkileri*. Ankara, Turkey: Kültür ve Turizm Bakanlığı.

Tulukçu, E., & Sağdıç, O. (2011). Konya'da aktarlarda satılan tıbbi bitkiler ve kullanılan kısımları. *Erciyes Üniversitesi Fen Bilimleri Enstitüsü Dergisi*, *27*(4), 304–308.

Tuzlacı, E. (2016). *Türkiye bitkileri geleneksel ilaç rehberi*. İstanbul, Turkey: İstanbul Tıp Kitabevleri.

Tuzlacı, E., & Erol, M. K. (1999). Turkish folk medicinal plants. part II: Eğirdir (ısparta). *Fitoterapia*, *70*(6), 593–610. doi:10.1016/S0367-326X(99)00074-X

Uğurlu, E., & Seçmen, Ö. (2008). Medicinal plants popularly used in the villages of Yunt Mountain (Manisa-Turkey). *Fitoterapia*, *79*(2), 126–131. doi:10.1016/j.fitote.2007.07.016 PMID:17878061

Uysal, G. (2008). *Köyceğiz (muğla) ilçesinin etnobotaniği* (Unpublished master's thesis). Muğla Üniversitesi, Turkey.

Van Tulder, M. W., Koes, B. W., & Bouter, L. M. (1995). A cost-of-illness study of back pain in The Netherlands. *Pain*, *62*(2), 233–240. doi:10.1016/0304-3959(94)00272-G PMID:8545149

Volinn, E. (1997). The epidemiology of low back pain in the rest of the world: A review of surveys in low-and middle-income countries. *Spine*, *22*(15), 1747–1754. doi:10.1097/00007632-199708010-00013 PMID:9259786

Walsh, K., Cruddas, M., & Coggon, D. (1992). Low back pain in eight areas of Britain. *Journal of Epidemiology and Community Health*, *46*(3), 227–230. doi:10.1136/jech.46.3.227 PMID:1386619

Yesilada, E., Honda, G., Sezik, E., Tabata, M., Fujita, T., Tanaka, T., ... Takaishi, Y. (1995). Traditional medicine in Turkey. v. folk medicine in the inner taurus mountains. *Journal of Ethnopharmacology*, *46*(3), 133–152. doi:10.1016/0378-8741(95)01241-5 PMID:7564412

Yesilada, E., Üstün, O., Sezik, E., Takaishi, Y., Ono, Y., & Honda, G. (1997). Inhibitory effects of turkish folk remedies on inflammatory cytokines: Interleukin-1α, interleukin-1β and tumor necrosis factor α. *Journal of Ethnopharmacology*, *58*(1), 59–73. doi:10.1016/S0378-8741(97)00076-7 PMID:9324006

Yetgin, A., Canli, K., & Altuner, E. M. (2018). Comparison of antimicrobial activity of allium sativum cloves from China and Taşköprü, Turkey. *Advances in Pharmacological Sciences, 2018*, Article ID 9302840, 5 pages. . doi:10.1155/2018/9302840

KEY TERMS AND DEFINITIONS

Decoction: Decoction is a method of extraction by boiling plant material to dissolve the chemicals of the material, which may include stems, roots, bark, and rhizomes.

Herbal Medicine: Finished, labelled medicinal products that contain active ingredients that include aerial or underground parts of plants, other plant material, or combinations thereof, whether in the crude state or as plant preparations.

Infusion: Infusion is the process of extracting chemical compounds or flavors from plant material in a solvent such as water, oil, or alcohol, by allowing the material to remain suspended in the solvent over time.

Lower Back: The lower back starts at the bottom of the rib cage and is called the lumbar region.

Medicinal Plants: Medicinal plants are known to be therapeutic due to the various molecules and compounds they contain, and they are used by the public to treat many diseases.

Turkey: The Republic of Turkey is a Eurasian country that connects Asia and Europe with the Bosphorus bridges.

Upper Back: The upper back is the area between the base of the neck and the bottom of the rib cage.

WHO: World Health Organization (WHO) is an organization of the United Nations that conducts international studies on public health.

Chapter 13
A Review of Traditional Medicinal Plants Used for Treatment of Leucorrhoea in India

Vartika Jain
https://orcid.org/0000-0002-8490-2495
Government Meera Girls' College, India

ABSTRACT

Leucorrhoea is a common disorder that is prevalent in females. Ethnic communities still treat leucorrhoea with plants that grow in their surroundings due to lack of sufficient medical facilities and/or costly modern treatment. Therefore, the author tried to look for plants used in traditional ethno-medicine and Ayurvedic systems of medicine for treatment of leucorrhoea in India. This review reveals the use of 345 plants by various folk communities and 31 unique plants from Ayurveda, which gives a figure of 376 plant species prevalent in Indian traditional medicine for treatment of leucorrhoea. Further analysis revealed maximum use of roots followed by the leaves and the stem bark of these plants. The chapter indicates 29 highly credible ethnomedicinal plant species with a rating of 5. This review shall prove helpful in screening of potential ethnomedicinal plant species for carrying out further scientific validation studies to assess their efficacy in leucorrhoea, which may contribute to the development of novel phyto-therapeutic molecules for effective treatment.

INTRODUCTION

Traditional medicine is considered to be materials and practices employed by healers for treatment of an ailment that originated from a specific region, descended over generations, and further spread to a larger region (Jain & Jain, 2017). As per this definition, ethnomedicine or folk medicine is mostly communicated through oral mediums among generations and is a part of traditional medicine. There is neither any written literature nor any institute to impart a degree course in ethnomedicine. Ayurveda,

DOI: 10.4018/978-1-7998-1320-0.ch013

on the other hand, is a well-known Indian traditional system of medicine that possesses prescribed literature to study, diagnoses and treats, the patient, and confers a degree is before its commercial practice. Interestingly, the roots of Ayurveda are also thought to be in ethnomedicine (Jain & Jain, 2015). These traditional medicinal systems employ various plants/animals/minerals for treatment of different diseases/disorders of humans and animals that are safe, effective, and growing mostly in nearby places of healers and/or patients.

The relationship between humans and plants is age-old. Humans have been using plants for fulfilling basic needs of daily life as well as for performing rituals as a part of cultural relationship. Plants are being utilized for medicinal purposes mostly in the form of crude drug formulations, such as tinctures, poultices, powders etc. (Balick & Cox, 1997). From such crude formulations, recent advances have led to the isolation of active drug molecules, and researchers have developed techniques for standardization of natural drugs to ensure the purity of drugs through analytical marker compounds (Balunas & Kinghorn, 2005). Ethnobotanical leads have played a significant role in drug discovery; hence, documentation of ethnobotanical plant species that are useful as medicine is very important.

Leucorrhoea, locally known as "shwet pradar" or "safed pani", is the term for abnormal, excessive vaginal discharge of white/yellow/green color that is associated with itching and sometimes inflammation (Abid et al., 2016; Trollope-Kumar, 2001). It is one of the most common problems among females, and many times it is ignored rather than treated. Moreover, in severe conditions, leucorrhoea becomes a reason for infertility. Women who have anemia, constipation, abortion, or are taking excessive medicines are more prone to suffer from leucorrhoea (Sarkar, 1983). Additionally, poor nutrition, hygiene, overwork, stress, and sexual anxiety are also considered to be among some of the causes of leucorrhoea (Trollope-Kumar, 2001). Leucorrhoea is told to be a culturally-shaped disorder that could be better understood through an anthropological lens than biomedical lens (Trollope-Kumar, 2001). Yet scientific etiology of leucorrhoea has its roots in physiological, pathological, and/or sometimes psychological reasons. Physiological leucorrhoea is a result of hormonal stimulation; pathological reasons involve infection by any pathogen, such as bacteria, fungi, virus etc.; and factors such as age and psychosocial indicators, such as financial losses, family conflicts, socio-economic status etc. are among psychological causes of the disease (Abid et al., 2016; Trollope-Kumar, 2001). However, in the long term, leucorrhoea reduces the quality of life and feeling of wellbeing in women.

India is a country that is rich in ethnic and floristic diversity, and here, history of traditional medicine could be traced back to even before the Vedic period. Due to the high cost of modern medical treatment or even the lack of such facilities available in remote areas, ethnic communities used to treat their various ailments with the natural resources available in their region (Jain & Jain, 2017). Besides the high cost, side effects associated with modern drugs is another important issue, and therefore, the need for a safe, effective, and economical treatments for various ailments always exists. In view of this, plants are wonderful natural gifts that may provide safe and effective treatment at a very reasonable cost. Therefore, the author made an attempt to look for plants that are employed by various ethnic communities of India as well as plants which are incorporated in the traditional Ayurvedic system of medicine for treatment of leucorrhoea.

For this purpose, the author conducted an exhaustive literature survey on various databases of research papers and review articles, like Pubmed, Scopus, Elsevier, Google Scholar and two major databases on ethnobotany of India [i.e., *Dictionary of Indian Folk Medicine and Ethnobotany* (Jain, 1991) and *Compendium of Indian Folk Medicine and Ethnobotany* (Jain & Jain, 2016)]. The author thoroughly scrutinized these sources along with relevant Ayurvedic literature (Bhutya, 2011; Joshi et al., 2008).

The author adopted correct botanical identities of plants from a well-known international database (The Plant List). For the convenience of readers, the author has not adopted recent name changes of some traditionally well-known families, such as Mimosaceae, Fabaceae, Caesalpiniaceae, Tiliaceae etc. The author has presented an alphabetical list of these plant species by their botanical names in italics, authority along with family, plant part used, and corresponding reference have in Table 1 and 2, respectively.

MAIN FOCUS OF THE CHAPTER

The present study reveals that 345 plant species of 267 genera are being utilized by ethnic communities for the treatment of leucorrhoea in India (Table 1). Out of these 345 ethnomedicinal plants, seven plants belong to Pteridophyta, two belong to Gymnosperm, and the remaining 336 are Angiosperms, of which 302 are Dicot and 34 are Monocot plant species. Further analysis shows that the genus that has the most species is *Euphorbia* with eight species, followed by *Ficus* with six species, *Acacia* with five species, *Sida* and *Terminalia* each with four species, and rest of the genera have three, two or one single species. Among the 109 families, some dominant families of these 345 plant species are Leguminosae, which has 42 plants (including Mimosaceae, Caesalpiniaceae, and Fabaceae) followed by Euphorbiaceae, which has 23 plants (including Phyllanthaceae), Malvaceae, which has 22 plants (including Sterculiaceae and Tiliaceae), Asteraceae, which has 15 plants, and Rubiaceae, which has 11 plants. Analysis of plant part used from Table 1 reveals that roots were the most frequently utilized plant part (36%) followed by leaf (26%), stem bark (19.7%) whole plant (12.46%), seed (12.17%), fruit (7.53%), flower (5.21%), root bark (3.76%), stem (3.18%), rhizome (2.9%), tuber (2.32%), gum (1.74%), and latex (1.16%). The rest parts were utilized for less than one percent.

The author also screened these 345 plants for their credibility rating by a previously established methodology (Jain, 2004), which is well known to assess the credibility of folk medicinal claims based on criteria of reporting of a claim from different locations and different ethnic groups. The author observed that there are 29 plant species with the highest credibility score of 5 (folk claim reported from more than two different states and ethnic groups). These species have been indicated with a dagger (†) symbol in Table 1. Out of these 29 plant species, there are six potential plant species, namely *Abrus precatorius, Byttneria herbacea, Dalbergia sissoo*, *Prosopis cineraria*, *Firmiana simplex* and *Xanthium strumarium,* which have yet not been included in the traditional Ayurvedic system of medicine for treatment of leucorrhoea and thus could be further analyzed for their phytochemistry and pharmacology to search for novel phyto-therapeutic molecules for treatment.

Analysis on the basis of time scale of ethnomedicinal data on 345 plant species, reveals that there were 42 plant species that were reported until 1991 for treatment of leucorrhoea. After 1991, due to spread of ethnobotany as a discipline in the entire country, rapid documentation of ethnomedicinal plant knowledge started, and sources recorded an increase of 303 plant species for treatment of leucorrhoea. Not only the addition of knowledge about new plants, but also loss of previously reported knowledge about any plant species is a part of dynamism in ethnobotany (Jain, 2005). Additionally, in the present study, it was interesting to observe that there are 17 plant species that have been reported for leucorrhoea treatment before 1991, but none have been again reported from any geographical region of country after 1991, which indicates the importance of documentation of ethnobotanical knowledge at the earliest before the plant species or the informant is lost in the trough of time.

Ayurveda, one of India's major traditional medicinal systems, describes medicinal uses of about 1500-2000 plant species, out of which about 500 plants at most are utilized on a regular basis (Bhutya, 2011; Jain & Jain, 2015). Scrutiny of these 500 plants revealed that there are 80 major plants mentioned in Ayurveda that are being used either singly or in combination for the treatment of leucorrhoea. Notably, 41 plants and their parts are common to both ethnomedicine and Ayurvedic literature (Table 1). However, there are 31 plant species reported from Ayurvedic literature have yet not been reported in ethnomedicine for treatment of leucorrhoea (Table 2). Combining these 31 plant species with 345 plants reported in ethnomedicine gives a total of 376 plant species being used for treatment of leucorrhoea in Indian traditional medicine. This could be considered a significant number of plant wealth, which could be further screened for their efficacy in treatment of leucorrhoea.

Another important observation is the use of different plant parts of eight ethnomedicinal plants for treatment of leucorrhoea in Ayurveda. These are as follows: stem bark of *Acacia nilotica,* leaf of *Aegle marmelos,* seed of *Areca catechu,* root of *Argyeria nervosa,* gum of *Bombax ceiba,* whole plant of *Corchorus depressus,* Bark of *Picrorhiza kurrooa,* and leaf of *Nelumbo nucifera*. If it is considered that Ayurveda has its origin in ethnomedicine, then it is appreciable that the use of 41 plant parts are common in both medicinal systems. However, the question arises of how the use of 31 plant species and plant parts of eight plant species for treatment of leucorrhoea have not been reported in ethnomedicine yet when the database of ethnobotanical plant species in India covers about 6000 plant species (Jain & Jain, 2015).

FUTURE RESEARCH PERSPECTIVES

The importance of diet and exercise is well-mentioned in various types of health literature to improve conditions of any disease. Additionally, plants are not only used as a dietary component, but they also act as beneficial drugs. Humans have been using the therapeutic potential of plants since long ago. Due to the ever-increasing drug resistant strains of pathogens and newer diseases, there is always a need to discover drugs compatible to present-day ailments, and plants have always been supportive in this. Any drug discovery program demands a high amount of money, skilled labor, and instrumentation, so it is better to start a drug discovery project on a plant with some previously reported efficacy for that disease rather than undertaking a random screening of plants. Previously reported efficacy may either be in the form of folk claims or from traditional systems of medicine. In the past, many of the wonderful drugs have been developed from ethnobotanical leads. Examples include Ajmaline, Aspirin, Codeine, Colchicine, Ephedrine, Eugenol, Pilocarpine, Quinine, and Toxiferine (Cox, 1994).

Reverse pharmacology that means the application of clinical knowledge of a healer/traditional medical practitioner for developing drugs in a laboratory. Reverse pharmacology is currently a frequently utilized approach (Patwardhan et al., 2015). India is a land of geographical, cultural, ethnic, and floristic diversity, and thus, ethnobotanical knowledge may lead towards the opening of new vistas for development of drugs. In this regard, author has reported a large number of plants to be used for treatment of leucorrhoea in traditional medicine. A thorough phytochemical and pharmacological profiling of these plant species would help in scientific validation and characterization of bioactive molecules, which may be useful not only in treatment of leucorrhoea but also in management of other diseases.

Joshi et al. (2008) have provided a scientific basis behind the efficacy of eight Ayurvedic medicinal plants in treatment of leucorrhoea. These plant species are *Albizia lebbeck, Asparagus racemosus, Azadirachta indica, Cyperus rotundus, Ficus religiosa, Glycyrrhiza glabra, Nelumbo nucifera,* and *Woodfordia*

Table 1. List of plant species used by ethnic communities of India for treatment of leucorrhoea

S. No.	Botanical Name	Family	Plant part[#]	References
1.	*Abelmoschus esculentus* (L.) Moench	Malvaceae	rt	Jain & Jain 2016
2.	*Abelmoschus moschatus* Medik.	Malvaceae	rt	Jain & Jain 2016
3.	*Abroma augusta* (L.) L.f.	Sterculiaceae	rt, st, lf	Jain & Jain 2016
4.	*Abrus precatorius*† L.	Fabaceae	rt, sd, lf px	Jain & Jain 2016 Jain 1991
5.	*Abutilon hirtum* (Lam.) Sweet	Malvaceae	lf	Jain & Jain 2016
6.	*Abutilon indicum* (L.) Sweet	Malvaceae	rt,lf,sb,sd	Jain & Jain 2016
7.	*Acacia catechu*† (L.f.) Willd.	Mimosaceae	bk,hw bk	Jain & Jain 2016 Jain 1991
8.	*Acacia farnesiana* (L.) Willd.	Mimosaceae	bk	Balamurugan et al. 2018
9.	*Acacia leucophloea* (Roxb.) Willd.	Mimosaceae	bk	Vishnuvardhan et al 2018
10.	*Acacia nilotica* (L.) Willd.ex Delile	Mimosaceae	fr,lf	Jain & Jain 2016
11.	*Acacia nilotica* subsp. *indica* (Benth.) Brenan	Mimosaceae	sb,bk,gm	Jain & Jain 2016
12.	*Achyranthes aspera* L.	Amaranthaceae	rt	Jain & Jain 2016
13.	*Achyranthes aspera* L. var. *porphyristachya* (Wall. ex Moq.) Hook.f.	Amaranthaceae	rt	Jain & Jain 2016
14.	*Aconitum heterophyllum* Wall. ex Royle	Ranunculaceae	rt	Jain & Jain 2016
15.	*Actiniopteris radiata* (Sw.) Link	Pteridaceae	lf,wp	Jain & Jain 2016
16.	*Adenanthera pavonina* L.	Mimosaceae	sd	Jain & Jain 2016
17.	*Aegle marmelos* (L.) Correa	Rutaceae	rt	Jain & Jain 2016
18.	*Aerva lanata* (L.) Juss.	Amaranthaceae	wp rt	Jain & Jain 2016 Hemadri & Rao 1983
19.	*Aerva sanguinolenta* (L.) Blume.	Amaranthaceae	rt,lf	Jain & Jain 2016
20.	*Aesculus indica* (Wall. ex Cambess) Hook.	Hippocastanaceae	sd rt	Jain & Jain 2016 Jain 1991
21.	*Ailanthus excelsa* Roxb.	Simaroubaceae	bk	Jain & Jain 2016
22.	*Ajuga parviflora* Benth.	Lamiaceae	lf	Jain & Jain 2016
23.	*Albizia lebbeck* (L.) Benth.	Mimosaceae	bk	Jain & Jain 2016
24.	*Aloe vera*† (L.) Burm.f.	Xanthorrhoeaceae	lf	Jain & Jain 2016
25.	*Alstonia scholaris* (L.) R. Br.	Apocynaceae	st,rb,la	Jain & Jain 2016
26.	*Alternanthera sessilis* (L.) R.Br. ex DC.	Amaranthaceae	lf	Jain & Jain 2016
27.	*Amaranthus hybridus* L.	Amaranthaceae	lf,rt	Jain 1991
28.	*Amaranthus spinosus* L.*	Amaranthaceae	wp rt	Jain & Jain 2016 Saheb et al. 2017
29.	*Amaranthus viridis* L. syn. *Amaranthus gracilis* Desf.	Amaranthaceae	rt	Hemadri & Rao 1983
30.	*Ampelocissus latifolia* (Roxb.) Planch	Vitaceae	tu	Jain & Jain 2016
31.	*Andrographis nallamalayana* J.L.Ellis	Acanthaceae	rt	Jain & Jain 2016
32.	*Andrographis paniculata* (Burm.f.) Wall. ex Nees	Acanthaceae	lf,wp rt	Jain & Jain 2016 Saheb et al. 2017
33.	*Anogeissus latifolia* Wall. ex Guillem. & Perr.	Combretaceae	gm	Jain 1991
34.	*Aphanamixis polystachya* (Wall.) R. Parker	Meliaceae	fr,bk,lf,rt	Jain & Jain 2016
35.	*Aquilaria malaccensis* Lam. syn. *Aquilaria agallocha* Roxb.	Thymelaeaceae	resin	Jain & Jain 2016
36.	*Areca catechu* L. *	Arecaceae	rt,lf lf	Jain & Jain 2016 Jain 1991
37.	*Arenga wightii* Griff.	Arecaceae	st	Jain & Jain 2016
38.	*Argemone mexicana* L.	Papaveraceae	lf,rt	Jain & Jain 2016
39.	*Argyreia nervosa* (Burm.f.) Bojer	Convolvulaceae	wp	Jain & Jain 2016
40.	*Argyreia strigosa* Roberty	Convolvulaceae	wp	Jain & Jain 2016
41.	*Aristolochia bracteolata* Lam.	Aristolochiaceae	st	Jain & Jain 2016
42.	*Aristolochia indica* L.	Aristolochiaceae	lf	Jain & Jain 2016
43.	*Arnebia benthamii* (Wall. ex G. Don) I. M. Johnst.	Boraginaceae	rt	Jain & Jain 2016
44.	*Artocarpus heterophyllus* Lamk.	Moraceae	rt	Jain & Jain 2016
45.	*Asparagus adscendens* Roxb.*	Asparagaceae	rt	Jain & Jain 2016
46.	*Asparagus filicinus* Buch.-Ham. ex D. Don	Asparagaceae	rt	Jain & Jain 2016
47.	*Asparagus racemosus*† Willd.*	Asparagaceae	rt,lf,wp rt	Jain & Jain 2016 Jain 1991
48.	*Astragalus rhizanthus* subsp. *candolleanus* (Benth.) Podlech.	Fabaceae	rt	Rana et al. 2013 Jain & Jain 2016
49.	*Averrhoa carambola* L.	Oxalidaceae	fr	Jain & Jain 2016
50.	*Azadirachta indica* A.Juss.*	Meliaceae	lf fr,bk	Sarkar 1983 Balamurugan et al. 2018
51.	*Barleria prionitis* L.	Acanthaceae	rt	Jain & Jain 2016

continued on following page

Table 1. Continued

S. No.	Botanical Name	Family	Plant part#	References
52.	*Barringtonia acutangula* (L.) Gaertn.	Lecythidaceae	fr	Jain & Jain 2016
53.	*Bauhinia purpurea* L.	Caesalpiniaceae	bk	Jain & Jain 2016
54.	*Bauhinia racemosa* Lamk.	Caesalpiniaceae	sb	Jain & Jain 2016
55.	*Bauhinia variegata* L.	Caesalpiniaceae	rb	Jain & Jain 2016
56.	*Benkara fasciculata* (Roxb.) Ridsdale	Rubiaceae	rt	Jain & Jain 2016
57.	*Bergenia ciliata* (Haw.) Sternb.*	Saxifragaceae	rt	Singh et al. 2017
58.	*Bergia suffruticosa* (Delile.) Fenzl.	Elatinaceae	wp	Jain & Jain 2016
			wp	Jain 1991
59.	*Blepharispermum subsessile* DC.	Asteraceae	lf	Jain & Jain 2016
60.	*Blumea axillaris* (Lam.) DC.	Asteraceae	wp	Jain & Jain 2016
61.	*Boehmeria macrophylla* Hornem.	Urticaceae	rt	Jain & Jain 2016
62.	*Boerhavia chinensis* (L.) Rottb.	Nyctaginaceae	rt	Jain & Jain 2016
63.	*Boerhavia diffusa*† L.	Nyctaginaceae	rt,lf	Jain & Jain 2016
			wp,rt	Balamurugan et al. 2018
64.	*Bombax ceiba*† L.*	Malvaceae	fl,bk,rt	Jain & Jain 2016
65.	*Breynia retusa* (Dennst.) Alston	Euphorbiaceae	rt,sb	Jain & Jain 2016
66.	*Buchanania cochinchinesis* (Lour.) M.R.Almeida	Anacardiaceae	bk	Jain & Jain 2016
67.	*Butea monosperma*† (Lamk.) Taub.*	Fabaceae	gm,sb,lf,fl	Jain & Jain 2016
68.	*Butea superba* Roxb.	Fabaceae	rt	Jain & Jain 2016
69.	*Byttneria herbacea*† Roxb.	Malvaceae	rt,lf	Jain & Jain 2016
70.	*Cajanus cajan* (L.) Millsp.	Fabaceae	lf	Jain & Jain 2016
71.	*Callicarpa arborea* Roxb.	Lamiaceae	rt,bk	Jain & Jain 2016
72.	*Calotropis gigantea* (L.) Dryand.	Asclepiadaceae	rt	Jain & Jain 2016, Sarkar 1983
73.	*Calotropis procera* (Aiton.) Dryand.	Asclepiadaceae	rt,gs	Jain & Jain 2016
74.	*Canscora alata* (Roth) Wall.	Gentianaceae	wp	Jain & Jain 2016
75.	*Canthium parviflorum* Lam.	Rubiaceae	lf	Saheb et al. 2017
76.	*Cardamine impatiens* L.	Brassicaceae	wp	Jain & Jain 2016
77.	*Cardiospermum halicacabum* L.	Sapindaceae	rt	Jain & Jain 2016
78.	*Careya arborea* Roxb.	Lecythidaceae	bk	Jain & Jain 2016
79.	*Carum carvi* L.	Apiaceae	sd	Jain & Jain 2016
80.	*Cassia fistula* L.	Caesalpiniaceae	sd	Sinha 1991
81.	*Cassytha filiformis* L.	Lauraceae	wp	Jain & Jain 2016
82.	*Catharanthus roseus* (L.) G. Don.	Apocynaceae	fl,lf	Jain & Jain 2016
83.	*Catunaregam spinosa* (Thunb.) Tirveng	Rubiaceae	fl	Jain & Jain 2016
84.	*Celastrus paniculatus* Willd.	Celastraceae	rt,bk,rb	Jain & Jain 2016
85.	*Celosia argentea* L. syn. *Celosia cristata* L.	Amaranthaceae	wp	Jain & Jain 2016
			rt	Hemadri & Rao 1983
86.	*Centella asiatica* (L.) Urb.	Apiaceae	wp,lf	Jain & Jain 2016
87.	*Cheilanthes albomarginata* C.B. Clarke	Pteridaceae	wp	Jain & Jain 2016
88.	*Chlorophytum tuberosum* (Roxb.) Baker	Asparagaceae	tu	Jain & Jain 2016
			tu	Jain 1991
89.	*Cicer arietinum* L.	Fabaceae	sd	Jain & Jain 2016
90.	*Cinnamomum tamala* (Buch.-Ham.) T. Nees & Eberm.	Lauraceae	bk,lf	Jain & Jain 2016
91.	*Cirsium verutum* (D. Don) Spreng.	Asteraceae	rt	Jain & Jain 2016
92.	*Cissampelos pareira* L.	Menispermaceae	fl	Jain & Jain 2016
93.	*Cissus repanda* (Wight. & Arn.) Vahl.	Vitaceae	rt	Jain & Jain 2016
94.	*Clitoria ternatea* var. *pilosula* Wall. ex Baker	Fabaceae	rt	Jain & Jain 2016
95.	*Coccinia grandis* (L.) Voigt.	Cucurbitaceae	rt	Jain & Jain 2016
96.	*Cocculus hirsutus*† (L.) W. Theob	Menispermaceae	lf,st,rt	Jain & Jain 2016
			lf	Jain 1991
97.	*Cocculus pendulus* (J.R.Forst. & G. Forst.) Diels.	Menispermaceae	lf	Jain & Jain 2016
98.	*Cocos nucifera* L.	Arecaceae	infl	Jain & Jain 2016
99.	*Cochlospermum religiosum* (L.) Alston*	Bixaceae	gm	Jain & Jain 2016
100.	*Coldenia procumbens* L.	Boraginaceae	rt	Hemadri & Rao 1983
101.	*Coleus forskohlii* (Willd.) Briq.	Lamiaceae	rt	Rana et al. 2013
102.	*Corchorus depressus* (L.) Stocks.	Tiliaceae	fr	Jain & Jain 2016
103.	*Cordia macleodii* Hook.f. & Thomson	Boraginaceae	bk	Jain & Jain 2016
104.	*Corydalis meifolia* Wall.	Fumariaceae	rt	Rana et al. 2013
105.	*Crateva religiosa* G.Forst.	Capparaceae	sb	Jain & Jain 2016
106.	*Crotalaria medicaginea* Lamk.	Fabaceae	lf	Jain & Jain 2016
107.	*Crotalaria verrucosa* L.	Fabaceae	lf	Jain & Jain 2016

continued on following page

Table 1. Continued

S. No.	Botanical Name	Family	Plant part#	References
108.	*Croton scabiosus* Bedd.	Euphorbiaceae	sb	Jain & Jain 2016
109.	*Cryptolepis dubia* (Burm.f) M.R. Almeida	Apocynaceae	sh	Jain & Jain 2016
110.	*Cuminum cyminum* L.*	Apiaceae	sd,sb	Jain & Jain 2016
111.	*Curculigo orchioides*† Gaertn.*	Hypoxidaceae	tu rt	Jain & Jain 2016 Jain 1991
112.	*Curcuma longa* L.*	Zingiberaceae	rh	Jain & Jain 2016
113.	*Cyanthillium albicans* (DC.) H. Rob. syn. *Vernonia albicans* DC.	Asteraceae	rt	Jain 1991
114.	*Cyanthillium cinereum* (L.) H.Rob.	Asteraceae	lf	Jain & Jain 2016
115.	*Cynodon dactylon* (L.) Pers.*	Poaceae	lf,wp	Jain & Jain 2016
116.	*Cyperus pangorei* Rottb.	Cyperaceae	rh	Jain & Jain 2016
117.	*Dactylorhiza hatagirea* (D. Don.) Soo	Orchidaceae	rh,rt	Jain & Jain 2016
118.	*Dalbergia sissoo*† DC.	Fabaceae	lf	Jain & Jain 2016
119.	*Dendrophthoe falcata* (L.f.) Ettingsh.	Loranthaceae	lf,rt,wp	Jain & Jain 2016
120.	*Desmodium oojeinense* (Roxb.) H. Ohashi syn. *Ougeinia oojeinensis* Hochr.	Fabaceae	bk	Jain & Jain 2016
121.	*Desmostachya bipinnata* (L.) Stapf	Poaceae	rt	Jain & Jain 2016
122.	*Dichrostachys cinerea* (L.) Wight & Arn.	Mimosaceae	lf lf	Jain & Jain 2016 Jain 1991
123.	*Dicliptera chinensis* (L.) Juss.	Acanthaceae	wp	Jain & Jain 2016
124.	*Dioscorea belophylla* (Prain) Voight ex Haines	Dioscoreaceae	tu	Jain & Jain 2016
125.	*Dioscorea bulbifera* L.	Dioscoreaceae	tu	Jain & Jain 2016
126.	*Dioscorea oppositifolia* L.	Dioscoreaceae	tu	Jain & Jain 2016
127.	*Diospyros melanoxylon* Roxb.	Ebenaceae	fl	Jain & Jain 2016
128.	*Dolichandrone falcata* (Wall. ex DC.) Seem	Bignoniaceae	sb	Jain & Jain 2016
129.	*Dolichos trilobus* L.	Fabaceae	rt	Jain & Jain 2016
130.	*Drimia indica* (Roxb.) Jessop	Asparagaceae	bb	Jain & Jain 2016
131.	*Echinops echinatus* Roxb.	Asteraceae	fl head	Jain & Jain 2016
132.	*Eclipta prostrata* (L.) L.	Asteraceae	lf	Jain & Jain 2016
133.	*Eleusine coracana* (L.) Gaertn.	Poaceae	sd	Hemadri & Rao 1983
134.	*Elytraria acaulis* (L.F.) Lindau	Acanthaceae	rt	Jain & Jain 2016
135.	*Enicostema axillare* (Poir. ex Lam.) A. Raynal	Gentianaceae	sh,lf,wp	Jain & Jain 2016
136.	*Ensete superbum* (Roxb.) Cheesman	Musaceae	sd	Jain & Jain 2016
137.	*Eriolaena candollei* Wall.	Sterculiaceae	bk	Jain & Jain 2016
138.	*Erythrina stricta* Roxb.	Fabaceae	bk,fl	Jain & Jain 2016
139.	*Eulophia herbacea* Lindl.	Orchidaceae	rh	Jain & Jain 2016
140.	*Eulophia spectabilis* (Dennst.) Suresh	Orchidaceae	tu	Jain & Jain 2016
141.	*Euphorbia heyneana* Spreng.	Euphorbiaceae	wp	Jain & Jain 2016
142.	*Euphorbia hirta*† L.	Euphorbiaceae	wp,lf,rt	Jain & Jain 2016
143.	*Euphorbia hypericifolia* L.	Euphorbiaceae	lf	Jain 1991
144.	*Euphorbia indica* Lam.	Euphorbiaceae	lf	Jain & Jain 2016
145.	*Euphorbia neriifolia* L.	Euphorbiaceae	bk	Jain & Jain 2016
146.	*Euphorbia prostrata* Aiton	Euphorbiaceae	wp	Jain & Jain 2016
147.	*Euphorbia thymifolia* L.	Euphorbiaceae	wp	Jain & Jain 2016
148.	*Evolvulus alsinoides* (L.) L.	Convolvulaceae	wp wp	Jain & Jain 2016 Jain 1991
149.	*Feronia elephantum* Corr.	Rutaceae	la	Balamurugan et al. 2018
150.	*Feronia limonia* (L.) Swingle	Rutaceae	lf	Jain & Jain 2016
151.	*Ficus auriculata* Lour.	Moraceae	la	Jain & Jain 2016
152.	*Ficus benghalensis* L.*	Moraceae	fr,rt,bk,lf aerial rt	Jain & Jain 2016 Jain 1991
153.	*Ficus racemosa*† L.	Moraceae	fr,bb	Jain & Jain 2016
154.	*Ficus religiosa*† L.*	Moraceae	bk,fr,sd,lf	Jain & Jain 2016
155.	*Ficus semicordata* (Buch.-Ham.) ex Sm.	Moraceae	fr	Jain & Jain 2016
156.	*Ficus virens* Aiton	Moraceae	bk	Jain & Jain 2016
157.	*Firmiana simplex*† (L.) W. Wight syn. *Sterculia urens* Roxb.	Sterculiaceae	sb,gm	Jain & Jain 2016
158.	*Flacourtia indica* (Burm.f.) Merr.	Flacourtiaceae	sb	Jain & Jain 2016
159.	*Flemingia strobilifera* (L.) W.T. Aiton	Fabaceae	rt	Jain 1991
160.	*Flueggea leucopyrus* Willd.	Phyllanthaceae	lf	Jain & Jain 2016
161.	*Geranium nepalense* Sweet	Geraniaceae	wp	Jain & Jain 2016
162.	*Givotia moluccana* (L.) Sreem.	Euphorbiaceae	sb	Jain & Jain 2016
163.	*Glaphyropteridopsis erubescens* (Wall. ex Hook) Ching	Thelypteridaceae	rh	Jain & Jain 2016
164.	*Glochidion multiloculare* (Rottler ex Willd.) Voigt.	Phyllanthaceae	bk,sd	Jain & Jain 2016

continued on following page

Table 1. Continued

S. No.	Botanical Name	Family	Plant part[#]	References
165.	*Glossocardia bosvallia* (L.f.) DC.	Asteraceae	lf	Jain & Jain 2016
166.	*Glycosmis mauritiana* (Lam.) Tanaka	Rutaceae	sb	Jain & Jain 2016
167.	*Glycyrrhiza glabra* L.*	Fabaceae	rt,lf	Jain & Jain 2016
168.	*Goniothalamus sesquipedalis* Hook.f. & Thomson.	Annonaceae	rt	Jain & Jain 2016
169.	*Gossypium herbaceum* L.	Malvaceae	rt	Jain & Jain 2016
170.	*Grewia flavescens* Juss.	Malvaceae	rt	Jain & Jain 2016
171.	*Grewia hirsuta* Vahl	Malvaceae	rt	Jain & Jain 2016
172.	*Gymnosporia senegalensis* (Lam.) Loes.	Celastraceae	rt,lf	Jain & Jain 2016
173.	*Haldina cordifolia* (Roxb.) Ridsdale	Rubiaceae	sb	Jain & Jain 2016
174.	*Helicteres isora* L.	Sterculiaceae	rt	Jain & Jain 2016
175.	*Hemidesmus indicus* (L.) R.Br.ex Schult.*	Asclepiadaceae	rt,st,lf	Jain & Jain 2016
176.	*Heracleum lanatum* Michx.	Apiaceae	rt	Jain & Jain 2016
177.	*Hibiscus micranthus* L.f.	Malvaceae	lf,fr	Jain & Jain 2016
178.	*Hibiscus rosa-sinensis*[†] L.*	Malvaceae	fl,lf,wp	Jain & Jain 2016
			fl	Sarkar 1983
179.	*Holoptelea integrifolia* Planch.	Ulmaceae	lf	Jain & Jain 2016
180.	*Holostemma ada-kodien* Schult.	Asclepiadaceae	tu	Jain & Jain 2016
181.	*Hygrophila auriculata* (Schumach.) Heine.*	Acanthaceae	sd,lf	Jain & Jain 2016
182.	*Hymenodictyon obovatum* Wall.	Rubiaceae	bk	Jain & Jain 2016
183.	*Hyptis suaveolens* (L.) Poit.	Lamiaceae	sd	Jain & Jain 2016
184.	*Ichnocarpus frutescens* (L.) W.T.	Apocynaceae	rt	Jain & Jain 2016
185.	*Imperata cylindrica* (L.) P. Beauv.	Poaceae	rt	Jain & Jain 2016
186.	*Indigofera cassioides* DC.	Fabaceae	bk,fl,lf	Jain & Jain 2016
187.	*Indigofera tinctoria* L.	Fabaceae	wp	Jain & Jain 2016
188.	*Ipomoea digitata* L.	Convolvulaceae	sd	Jain & Jain 2016
189.	*Ipomoea marginata* (Desr.) Verdc.	Convolvulaceae	rt	Jain & Jain 2016
190.	*Ixora coccinea* L.	Rubiaceae	rt,lf,fl	Jain & Jain 2016
191.	*Jatropha curcas* L.	Euphorbiaceae	tw	Jain & Jain 2016
192.	*Jatropha gossypiifolia* L.	Euphorbiaceae	rt	Jain & Jain 2016
193.	*Juglans regia* L.	Juglandaceae	lf	Jain & Jain 2016
194.	*Juniperus communis* L.	Cupressaceae	lf	Jain & Jain 2016
195.	*Juniperus excelsa* M. Bieb.	Cupressaceae	fr,wd	Jain & Jain 2016
196.	*Jurinea dolomiaea* Boiss.	Asteraceae	rt	Jain & Jain 2016
197.	*Justicia gendarussa* Burm.f.	Acanthaceae	rt,lf	Jain & Jain 2016
198.	*Lasia spinosa* (L.) Thwaites.	Araceae	rh,lf	Jain & Jain 2016
199.	*Leea macrophylla* Roxb. ex Hornem.	Vitaceae	lf	Jain 1991
200.	*Leucas aspera* (Willd.) Link	Lamiaceae	lf	Jain & Jain 2016
201.	*Leucas biflora* (Vahl.) Sm.	Lamiaceae	lf	Jain & Jain 2016
202.	*Leucas cephalotes* (Roth.) Spreng	Lamiaceae	lf	Jain & Jain 2016
203.	*Linum usitatissimum* L.	Linaceae	sd	Jain & Jain 2016
204.	*Litsea glutinosa* (Lour.) C.B.Robins	Laurace	sb	Jain & Jain 2016
205.	*Lonicera zeylanica* Gaertn.	Caprifoliaceae	rt	Jain & Jain 2016
206.	*Ludwigia adscendens* (L.) H. Hara	Onagraceae	rt	Jain & Jain 2016
207.	*Ludwigia perennis* L.	Onagraceae	rt	Jain & Jain 2016
208.	*Lycopersicon esculentum* Mill.	Solanaceae	fl	Jain & Jain 2016
209.	*Madhuca longifolia* var. *latifolia* (Roxb.) A. Chev.	Sapotaceae	bk	Jain & Jain 2016
210.	*Mallotus philippensis* (Lam.) Muell.-Arg.	Euphorbiaceae	lf	Jain & Jain 2016
211.	*Mangifera indica* L.*	Anacardiaceae	sb	Jain & Jain 2016
			sd	Balamurugan et al. 2018
212.	*Melastoma malabathricum* L.	Melastomataceae	lf	Jain & Jain 2016
213.	*Memecylon umbellatum* Burm.f.	Melastomataceae	rb	Saheb et al. 2017
214.	*Merremia tridentata* (L.) Hallier.f.	Convolvulaceae	px	Jain & Jain 2016
215.	*Mesua ferrea* L.	Calophyllaceae	Stamen fr,sd	Jain & Jain 2016 Sinha 1991
216.	*Mimosa hamata* Willd	Mimosaceae	fl	Jain & Jain 2016
217.	*Mimosa pudica*[†] L.	Mimosaceae	wp,rt,rb wp	Jain & Jain 2016 Jain 1991
218.	*Mimusops elengi* L.	Sapotaceae	sb	Jain & Jain 2016
219.	*Mirabilis jalapa* L.	Nyctaginaceae	lf,rt	Jain & Jain 2016
220.	*Mitragyna parvifolia* (Roxb.) Korth	Rubiaceae	Bk	Jain 1991
221.	*Momordica charantia* L.	Cucurbitaceae	sd	Jain & Jain 2016

continued on following page

Table 1. Continued

S. No.	Botanical Name	Family	Plant part#	References
222.	*Momordica dioica* Roxb. ex willd.	Cucurbitaceae	rt	Jain & Jain 2016
223.	*Morinda citrifolia* L.	Rubiaceae	rt	Jain & Jain 2016
224.	*Morinda pubescens* Sm.	Rubiaceae	rt,lf,fr	Jain & Jain 2016
225.	*Moringa concanensis* Nimmo	Moringaceae	sb	Jain & Jain 2016
226.	*Mucuna pruriens*† (L.) DC.	Fabaceae	rt,lf,sd	Jain & Jain 2016
227.	*Murraya koenigii* Sprin	Rutaceae	rt	Jain & Jain 2016
228.	*Musa paradisiaca* L.*	Musaceae	fr,lf,st	Jain & Jain 2016
229.	*Myrica nagi* Thunb.	Myricaceae	rb	Sinha 1991
230.	*Myristica fragrans* Houtt.	Myristicaceae	fr,sd	Jain & Jain 2016
231.	*Nelumbium speciosum* Willd.	Nymphaeaceae	sd	Balamurugan et al. 2018
232.	*Nelumbo nucifera* Gaertn.*	Nelumbonaceae	fr	Jain & Jain 2016
233.	*Nymphaea nouchali* Burm.f.*	Nymphaeaceae	sd	Jain & Jain 2016
234.	*Nymphaea pubescens* Willd.	Nymphaeaceae	fl,rh	Jain & Jain 2016
235.	*Ocimum basilicum* L.*	Lamiaceae	px	Jain & Jain 2016
236.	*Oroxylum indicum* (L.) Kurz.	Bignoniaceae	sb,fl,st	Jain & Jain 2016
237.	*Oryza sativa* L.	Poaceae	sd	Jain & Jain 2016
238.	*Paederia foetida* L.	Rubiaceae	lf	Jain & Jain 2016
239.	*Paeonia emodi Wall. ex Royle*	Paeonaceae	lf	Rana et al. 2013
240.	*Paracalyx scariosus* (Roxb.) Ali.	Papaveraceae	rb	Jain & Jain 2016
241.	*Parthenocissus semicordata* (Wall.) Planch	Vitaceae	ap	Jain & Jain 2016
242.	*Pedalium murex* L.*	Pedaliaceae	lf	Jain & Jain 2016
			fr	Jain 1991
243.	*Phoenix acaulis* Roxb.	Arecaceae	rt,lf	Jain & Jain 2016
244.	*Phyllanthus emblica*† L.*	Phyllanthaceae	fr,sd,sb	Jain & Jain 2016
245.	*Phyllanthus fraternus* G.L.Webster	Phyllanthaceae	rt,sd	Jain & Jain 2016
246.	*Phyllanthus niruri* L.	Phyllanthaceae	rt	Jain & Jain 2016
247.	*Physalis minima* L.	Solanaceae	lf	Jain & Jain 2016
248.	*Picrorhiza kurrooa* Royle	Plantaginaceae	rh	Sinha 1991
249.	*Piper betle* L.	Piperaceae	lf	Jain & Jain 2016
250.	*Piper nigrum* L.	Piperaceae	sd	Jain & Jain 2016
251.	*Pithecellobium dulce* (Roxb.) Benth	Mimosaceae	sb	Jain & Jain 2016
252.	*Platostoma hispidum* (L.) A.J. Paton	Lamiaceae	rt	Jain & Jain 2016
253.	*Plumbago indica* L.	Plumbaginaceae	rt	Jain & Jain 2016
254.	*Plumbago zeylanica* L.	Plumbaginaceae	rt,rb	Jain & Jain 2016
255.	*Pogostemon elsholtzioides* Benth.	Lamiaceae	lf	Jain & Jain 2016
256.	*Polyalthia longifolia* (Sonn.) Thwaites	Annonaceae	lf,bk	Jain & Jain 2016
257.	*Polygonatum cirrhifolium* (Wall.) Royle	Asparagaceae	rt	Jain & Jain 2016
258.	*Polygonatum verticillatum* (L.) All.	Asparagaceae	rt	Jain & Jain 2016
259.	*Polygonum polystachyum* Wall. ex Meissn.	Polygonaceae	rt	Rana et al. 2013
260.	*Polygonum viviparum* L.	Polygonaceae	rt	Jain 1991
261.	*Pongamia pinnata* (L.) Pierre	Fabaceae	sb	Jain & Jain 2016
262.	*Prosopis cineraria*† (L.) Druce.	Mimosaceae	lf,rb	Jain & Jain 2016
263.	*Psidium guajava* L.	Myrtaceae	rb	Jain & Jain 2016
264.	*Pterocarpus marsupium* Roxb.	Fabaceae	sb	Jain & Jain 2016
265.	*Pueraria tuberosa* (Willd.) DC.	Fabaceae	rh	Jain & Jain 2016
266.	*Punica granatum* L.	Punicaceae	lf	Jain & Jain 2016
267.	*Quercus infectoria* G.Olivier *	Fagaceae	gall	Jain & Jain 2016
268.	*Rheum australe* D. Don	Polygonaceae	rt	Jain & Jain 2016
269.	*Rheum moocroftianum* Royle	Polygonaceae	rt	Rana et al. 2013
270.	*Rhus coriaria* L.	Anacardiaceae	gall	Sinha 1991
271.	*Robinia pseudoacacia* L.	Fabaceae	sd	Jain & Jain 2016
272.	*Rubia cordifolia* L.	Rubiaceae	rt	Jain & Jain 2016
273.	*Rubus ellipticus* Smith	Rosaceae	rt	Jain & Jain 2016
274.	*Rubus niveus* Thunb.	Rosaceae	rt	Jain & Jain 2016
275.	*Rumex nepalensis* Spreng	Polygonaceae	lf	Jain & Jain 2016
			rt	Rana et al. 2013
276.	*Santalum album* L.*	Santalaceae	hw	Jain & Jain 2016
277.	*Saraca asoca*† (Roxb.) Willd.*	Caesalpiniaceae	bk	Jain & Jain 2016
278.	*Satyrium nepalense* D. Don	Orchidaceae	rt	Jain & Jain 2016
279.	*Saurauia roxburghii* Wall.	Actinidiaceae	rt	Jain & Jain 2016
280.	*Saussurea gossypiphora* D. Don	Asteraceae	rt,fl	Rana et al. 2013

continued on following page

Table 1. Continued

S. No.	Botanical Name	Family	Plant part[#]	References
281.	*Saussurea simpsoniana* (Fielding & Gardner) Lipsch.	Asteraceae	fl,rt	Jain & Jain 2016
282.	*Scolopia crenata* Clos	Salicaceae	lf	Jain & Jain 2016
283.	*Selaginella bryopteris* Baker	Selaginellaceae	wp	Jain & Jain 2016
			wp	Jain 1991
284.	*Selaginella sellowii* Hieron.	Selaginellaceae	wp	Jain & Jain 2016
285.	*Selaginella tamariscina* (P. Beauv.) Spring	Selaginellaceae	lf	Jain & Jain 2016
286.	*Semecarpus anacardium* L.f.	Anacardiaceae	bk	Jain & Jain 2016
287.	*Senna auriculata* (L.) Roxb.	Caesalpiniaceae	lf,fr,sb,rt	Jain & Jain 2016
288.	*Senna montana* (Roth.) V. Singh	Caesalpiniaceae	lf	Jain & Jain 2016
289.	*Shorea robusta* Gaertn.*	Dipterocarpaceae	sb	Jain & Jain 2016
290.	*Sida acuta*[†] Burm.f.	Malvaceae	sd,rt,lf,wp	Jain & Jain 2016
291.	*Sida cordata* (Burm.f.) Borss. Waalk.	Malvaceae	lf	Jain & Jain 2016
			wp	Jain 1991
292.	*Sida cordifolia*[†] L.*	Malvaceae	rt,rb,sd	Jain & Jain 2016
			sd	Jain 1991
			wp	Saheb et al. 2017
293.	*Sida rhombifolia* L.	Malvaceae	rb	Jain & Jain 2016
294.	*Sigesbeckia orientalis* L.	Asteraceae	rt	Jain & Jain 2016
295.	*Sinopodophyllum hexandrum* (Royle) T.S. Ying	Berberidaceae	rt	Jain & Jain 2016
296.	*Smilax aspera* L.	Smilacaceae	wp	Jain & Jain 2016
			rt	Rana et al. 2013
297.	*Smilax ovalifolia* Roxb. ex D. Don	Smilacaceae	rt	Jain & Jain 2016
298.	*Smilax zeylanica* L.	Smilacaceae	rt	Jain & Jain 2016
			rt	Jain 1991
299.	*Solanum americanum* Mill.	Solanaceae	lf	Jain & Jain 2016
300.	*Soymida febrifuga* (Roxb.) A. Juss.	Meliaceae	bk	Jain & Jain 2016
301.	*Spilanthes acmella* (L.) L.	Asteraceae	lf	Jain & Jain 2016
302.	*Spiranthes sinensis* (Pers.) Ames.	Orchidaceae	rt	Jain & Jain 2016
303.	*Stephania japonica* (Thunb.) Miers	Menispermaceae	lf	Jain & Jain 2016
304.	*Sterculia villosa* Roxb.	Sterculiaceae	sb	Saheb et al. 2017
305.	*Streblus asper* Lour.	Moraceae	la,sb	Jain & Jain 2016
306.	*Strychnos potatorum* L.f.	Loganiaceae	sd	Saheb et al. 2017
307.	*Swertia chirayta* H. Karst	Gentianaceae	lf	Jain & Jain 2016
308.	*Swertia paniculata Wall.*	Gentianaceae	lf	Rana et al. 2013
309.	*Swietenia mahogani* L.	Meliaceae	sd	Jain & Jain 2016
310.	*Symplocos racemosa* Roxb.*	Symplocaceae	sb	Sinha 1991
311.	*Syzygium cumini*[†] (L.) Skeels.	Myrtaceae	bk,sd,fr	Jain & Jain 2016
312.	*Syzygium salicifolium* (Wight) J. Graham	Myrtaceae	bk	Jain & Jain 2016
313.	*Tamarindus indica* L.*	Caesalpiniaceae	sd	Jain & Jain 2016
314.	*Tecomella undulata* (Sm.) Seem.*	Bignoniaceae	rt,rb,br	Jain & Jain 2016
315.	*Tectaria cicutaria* (L.) Copel.	Tectariaceae	rh	Jain & Jain 2016
316.	*Tectona grandis* L.f.	Verbenaceae	sb	Jain & Jain 2016
317.	*Tephrosia purpurea* (L.) Pers.	Fabaceae	sh	Jain & Jain 2016
318.	*Tephrosia villosa* (L.) Pers.	Fabaceae	rt	Jain & Jain 2016
319.	*Terminalia arjuna* (Roxb.ex DC.) Wight. & Arn.*	Combretaceae	bk	Jain & Jain 2016
320.	*Terminalia bellirica* (Gaertn.) Roxb.	Combretaceae	fr,bk gm	Jain & Jain 2016 Saheb et al. 2017
321.	*Terminalia cuneata* Roth	Combretaceae	rt,sb	Jain & Jain 2016
322.	*Terminalia elliptica* Willd.	Combretaceae	bk	Jain & Jain 2016
323.	*Thalictrum reniforme* Wall.	Ranunculaceae	wp	Jain & Jain 2016
324.	*Tinospora cordifolia*[†] (Willd.) Miers ex Hook. f. & Thomson*	Menispermaceae	lf,st,wp	Jain & Jain 2016
325.	*Tolypanthus lageniferus* Tiegh.	Loranthaceae	wp	Jain & Jain 2016
326.	*Trachyspermum ammi* (L.) Sprague syn.	Apiaceae	sd	Jain & Jain 2016
327.	*Trapa natans* L.*	Trapaceae	sd	Jain & Jain 2016
328.	*Trapa natans* var. *bispinosa* (Roxb.) Makino	Trapaceae	cotyledon	Jain & Jain 2016
329.	*Trianthema portulacastrum* L.	Aizoaceae	lf	Jain & Jain 2016
330.	*Tribulus terrestris*[†] L.*	Zygophyllaceae	fr,wp	Jain & Jain 2016
			fr	Jain 1991
331.	*Tridax procumbens* (L.) L.	Asteraceae	rt,lf	Jain & Jain 2016
332.	*Trigonella foenum-graecum* L.*	Fabaceae	fr	Jain & Jain 2016
333.	*Triticum aestivum* L.	Poaceae	sd	Jain & Jain 2016

continued on following page

Table 1. Continued

S. No.	Botanical Name	Family	Plant part#	References
334.	*Vallisneria natans* (Lour.) H. Hara	Hydrocharitaceae	px	Jain & Jain 2016
335.	*Vallisneria spiralis* L.	Hydrocharitaceae	wp	Jain 1991
336.	*Vigna mungo* (L.) Hepper	Fabaceae	sd	Jain & Jain 2016
337.	*Vigna radiata* (L.) R.Wilczek.	Fabaceae	sd	Jain & Jain 2016
338.	*Viola canescens* Wall.	Violaceae	wp	Rana et al. 2013
339.	*Vitex negundo* L.	Lamiaceae	rt	Jain & Jain 2016
340.	*Waltheria indica* L.	Sterculiaceae	rt	Jain & Jain 2016
341.	*Withania somnifera*† (L.) Dunal	Solanaceae	rt,st,fr	Jain & Jain 2016
342.	*Woodfordia fruticosa*† (L.) Kurz*	Lythraceae	fl	Jain & Jain 2016
343.	*Xanthium strumarium*† L.	Asteraceae	lf,wp,px	Jain & Jain 2016
344.	*Ziziphus mauritiana* Lam.	Rhamnaceae	fr	Jain & Jain 2016
345.	*Ziziphus nummularia* (Burm.f) Wight. & Arn.*	Rhamnaceae	rt	Jain & Jain 2016

*Plants also mentioned in Ayurveda

†Plants with highest credibility score-5

#Symbols for Plant parts: ap: aerial part; bb: bulb; bk: bark; br: branch; fl: flower; fr: fruit; gm: gum; gs: gynostegium; hw: hardwood; infl: inflorescence; la: latex; lf: leaf; px: part not specified; rb: root bark; rh: rhizome; rt: root; sb: stem bark; sd: seed; sh: shoot; st: stem; tw: twig; tu: tuber; wd: wood; wp: whole plant

species. All these plants species possess antimicrobial and anti-inflammatory properties, besides other health-beneficial activities. If researchers look at the high credibility rating of ethnomedicinal plant species, as discussed in previous section, they find that three plant species – namely *A. racemosus, F. religiosa* and *Woodfordia* spp. – are also part of the Ayurvedic medicinal system. Thus, the scientific basis of these plant species, as discussed by Joshi et al. (2008), seems to be quite relevant, and the author further emphasizes the importance of traditional knowledge in the discovery of novel drug molecules. Besides, properly framed clinical studies on 41 common plant species of ethnomedicine and Ayurveda for their efficacy in leucorrhoea as well as pharmacognostical characterization of these herbal drugs in order to avoid adulteration may establish them in global scenario.

Leucorrhoea is associated many times with inflammation of vaginal epithelium. Ethnic communities treat leucorrhoea with the help of various plants either in form of oral intake, through washing the affected area, or through local application. If these plant species possess anti-inflammatory potential, one could co-relate it with relieving of leucorrhoea symptoms. Researchers are required to understand not only its anti-inflammatory mechanism, but the complete mechanism of action in order to fully utilize the plant potential, as the beneficial effect of any plant might be a synergistic effect of many pharmacological activities. Therefore, research on these over 300 plant species should be initiated at the earliest time possible. Detailed scientific studies may prove beneficial not only in treatment of leucorrhoea and associated symptoms but also in some other diseases/disorders.

CONCLUSION

Traditional medicine is focus of research almost all over the world due to its safety, efficacy, and cost-effectiveness. The present chapter provides a list of 376 traditional medicinal plants out of which 345 are ethnomedicinal plant species and 31 are Ayurvedic medicinal plants being used for treatment of leucorrhoea in India. This chapter further analyzes the highest credibility score of ethnomedicinal plant species and presents a total of six potential plant species that have not yet been incorporated in Ayurveda and could be a source of novel therapeutic molecules for treatment of leucorrhoea. This large database

Table 2. List of major plants useful for treatment of leucorrhoea (singly/combination) in Ayurveda system of medicine

S. No.	Botanical name	Family	Plant part	References
1.	*Amomum subulatum* Roxb.	Zingiberaceae	sd	Bhutya 2011
2.	*Bambusa bambos* (L.) Voss.	Poaceae	gm	Bhutya 2011
3.	*Breonia chinensis* (Lam.) Capuron. syn. *Anthocephalus chinensis* Hassk.	Rubiaceae	bk	Bhutya 2011
4.	*Caesalpinia sappan* L.	Caesalpiniaceae	hw	Bhutya 2011
5.	*Cedrus deodara* (Roxb. ex D. Don) G. Don	Pinaceae	st pith	Bhutya 2011
6.	*Commiphora wightii* (Arn.) Bhandari	Burseraceae	gm	Bhutya 2011
7.	*Crocus sativus* L.	Iridaceae	Dried stigma	Bhutya 2011
8.	*Cucumis sativus* L.	Cucurbitaceae	sd	Bhutya 2011
9.	*Cyperus rotundus* L.	Cyperaceae	rt	Joshi et al. 2008
10.	*Dactylorhiza incarnata* (L.) Soó	Orchidaceae	tu	Bhutya 2011
11.	*Delonix regia* (Hook.) Raf.	Caesalpiniaceae	lf	Bhutya 2011
12.	*Desmodium gangeticum* (L.) DC.	Fabaceae	rb	Bhutya 2011
13.	*Diospyros malabarica* (Desr.) Kostel.	Ebenaceae	fr	Bhutya 2011
14.	*Elettaria cardamomum* Maton	Zingiberaceae	sd	Bhutya 2011
15.	*Eulophia dabia* Hochr.	Orchidaceae	tu	Bhutya 2011
16.	*Euryale ferox* Salisb.	Nymphaeaceae	sd	Bhutya 2011
17.	*Ficus arnottiana* Miq.	Moraceae	gm	Bhutya 2011
18.	*Ficus lacor* Buch.-Ham.	Moraceae	bk	Bhutya 2011
19.	*Gossypium herbaceum* L.	Malvaceae	rt	Bhutya 2011
20.	*Holarrhena pubescens* Wall. ex G.Don	Apocynaceae	bk	Bhutya 2011
21.	*Ocimum gratissimum* L.	Lamiaceae	lf	Bhutya 2011
22.	*Ocimum tenuiflorum* L.	Lamiaceae	lf	Bhutya 2011
23.	*Peristrophe bicalyculata* Nees.	Acanthaceae	wp	Bhutya 2011
24.	*Piper longum* L.	Piperaceae	fr	Bhutya 2011
25.	*Ricinus communis* L.	Euphorbiaceae	rb	Bhutya 2011
26.	*Rosa centifolia* L.	Rosaceae	fl	Bhutya 2011
27.	*Saccharum bengalense* Retz.	Poaceae	rt	Bhutya 2011
28.	*Solanum surattense* Burm.f.	Solanaceae	st	Bhutya 2011
29.	*Teramnus labialis* (L.f.) Spreng.	Fabaceae	wp	Bhutya 2011
30.	*Thespesia populnea* Sol. ex Correa	Malvaceae	bk	Bhutya 2011
31.	*Uraria picta* (Jacq.) DC.	Fabaceae	wp	Bhutya 2011

of over 350 plant species at a single place shall be useful for present day ethnobotanists to look for these plant species and their uses in the currently developing smart ecosystems and cities as well as for pharmacologists to obtain clues for producing new therapeutic drugs.

ACKNOWLEDGMENT

This research received no specific grant from any funding agency in the public, commercial, or not-for-profit sectors. The author extends her thanks to Dr. R. K. Bhutya, who gave details about some Ayurvedic medicinal plants.

REFERENCES

Abid, M. J., Kumar, K., Khan, R., Ali, S., Chandra, P., Rani, R. & Khan, N. A. (2016). Assessment of Leucorrhea diseases in female students. *Journal of Scientific and Innovative Research*, *5*(4), 116–118.

Balamurugan, S., Vijayakumar, S., Prabhu, S., & Yabesh, J. E. M. (2018). Traditional plants used for the treatment of gynaecological disorders in Vedaranyam Taluk, South India: An ethnomedicinal survey. *Journal of Traditional and Complementary Medicine*, *8*(2), 308–323. doi:10.1016/j.jtcme.2017.06.009 PMID:29736387

Balick, M. J., & Cox, P. A. (1997). *Plants, people, and culture: The Science of ethnobotany*. New York, NY: Scientific American Library.

Balunas, M. J., & Kinghorn, A. D. (2005). Drug discovery from medicinal plants. *Life Sciences*, *78*(5), 431–441. doi:10.1016/j.lfs.2005.09.012 PMID:16198377

Bhutya, R. K. (2011). *Ayurvedic medicinal plants of India* (Vol. 1-2). Jodhpur, Rajasthan: Scientific Publishers.

Cox, P. A. (1994). The ethnobotanical approach to drug discovery: Strengths and limitations. *Ciba Foundation Symposium*, *185*, 25–41. PMID:7736859

Hemadri, K., & Rao, S. S. (1983). Leucorrhoea and Menorrhagia: Tribal medicine. *Ancient Science of Life*, *3*(1), 40–41. PMID:22557376

Jain, S. K. (1991). *Dictionary of Indian folk medicine and ethnobotany*. New Delhi, India: Deep Publications.

Jain, S. K. (2004). Credibility of traditional knowledge: The criteria of multilocational and multiethnic use. *Indian Journal of Traditional Knowledge*, *3*(2), 137–153.

Jain, S. K. (2005). Dynamism in ethnobotany. *Ethnobotany*, *17*, 3–13.

Jain, S. K., & Jain, V. (2015). Evolution of ethnobotany in last six decades in India: An overview. *Ethnobotany*, *27*, 1–6.

Jain, S. K., & Jain, V. (2017). *Methods and approaches in ethnobotany (Concepts, practices and prospects)*. New Delhi, India: Deep Publications.

Jain, V., & Jain, S. K. (2016). *Compendium of Indian folk medicine and ethnobotany (1991-2015)*. New Delhi, India: Deep Publications.

Joshi, J., Vaidya, R., Shah, R., & Vaidya, A. B. (2008). The phytochemical and pharmacological basis of Ayurvedic plants used in treatment of Leucorrhoea and Menorrhagia. *The Journal of Research and Education in Indian Medicine*, *14*, 1–16.

Patwardhan, B., Mutalik, G., & Tillu, G. (2015). Drug discovery and Ayurveda. In Patwardhan & ... (Eds.), *Integrative approaches for health: Biomedical research, Ayurveda and Yoga* (pp. 229–258). Amsterdam, The Netherlands: Academic Press, Elsevier. doi:10.1016/B978-0-12-801282-6.00009-7

Rana, C. S., Ballabha, R., Sharma, A., Dangwal, L. R., & Tiwari, J. K. (2013). Herbal remedies for leucorrhoea: A study from the Garhwal Himalaya, India. *Global Journal of Research on Medicinal Plants and Indigenous Medicine*, *2*(10), 685–691.

Saheb, T. S., Zakia, S., Swamulu, M., & Mahesh, M. (2017). Medicinal plants used against gynaecological disorders by the tribal people of Nallamalais Andhra Pradesh. *Journal of Pharmacognosy and Phytochemistry*, *6*(2), 242–244.

Sarkar, P. R. (1983). *Yogic treatment and natural remedies. Tiljala*. Kolkata, India: AMPS Publications.

Singh, A., Nautiyal, M. C., Kunwar, R. M., & Bussmann, R. W. (2017). Ethnomedicinal plants used by local inhabitants of Jakholi block, Rudraprayag District, Western Himalaya, India. *Journal of Ethnobiology and Ethnomedicine*, *13*(1), 49. doi:10.118613002-017-0178-3 PMID:28836997

Sinha, R. K. (1991). Herbal remedies prescribed by the street herbal vendors (tribal medicineman) in the treatment of some common human ecological disorders (gynaecological disorders in women). *Ancient Science of Life*, *10*(3), 207–210. PMID:22556535

The Plant List. A working list of all plant species. Retrieved from www.theplantlist.org

Trollope-Kumar, K. (2001). Cultural and biomedical meanings of the complaint of leukorrhea in South Asian women. *Tropical Medicine & International Health*, *6*(4), 260–266. doi:10.1046/j.1365-3156.2001.00699.x PMID:11348516

Vishnuvardhan, Z., Jyothirmayi, G. L. V., & Jyothi, D. (2018). Medicinal plants of tribal traditional system from Guntur District, Andhra Pradesh, India. *International Journal of Life Sciences (Kathmandu)*, *6*(1), 194–204.

ADDITIONAL READING

Charak, S. (1983). Yonivyapad– Chikitsasthan 30th Chapter, Shree Kashi Sanskrit Granthamala 198, Choukhambha Sanskrit Sansthan, Part II, 2nd ed., 751-799.

Chaudhuri, M., Chatterjee, B. D., & Banerjee, M. (1998). A clinicobacteriological study on leucorrhoea. *Journal of the Indian Medical Association*, *96*(2), 46–50. PMID:9715025

Copeman, E. (1855). The pathology and treatment of leucorrhœa. *The British and Foreign Medico-Chirurgical Review*, *16*(31), 62–70. PMID:30163799

Dhiman, K. (2014). Leucorrhoea in Ayurvedic literature - A review. *Ayurpharm International Journal of Ayurveda and Allied Sciences*, *3*(3), 73–78.

Ghorbani, A., Naghibi, F., & Mosaddeh, M. (2006). Ethnobotany, ethnopharmacology and drug discovery. *Indian Journal of Pharmaceutical Sciences*, *2*(2), 109–118.

Joshi, J. V., Rege, V., Vaidya, R. A., Nabar, N. S., Lulla, M., & Sundar, S. (2005). Allopathic or Ayurvedic treatment of leucorrhoea. *Journal of National Integrated Medical Association*, *47*, 9–12.

Satyavati, C., & Reddy, D. B. (1957). Leucorrhoea: Review based on a study of 100 cases. *Journal of the Indian Medical Association*, *28*(5), 229–232. PMID:13416606

Shelley, B. C. L. (2009). Ethnobotany & the process of drug discovery: A laboratory exercise. *The American Biology Teacher*, *71*(9), 541–547. doi:10.2307/20565377

KEY TERMS AND DEFINITIONS

Anti-Inflammatory: A drug or substance that has the potential to reduce inflammation and pain in the body.

Ethnobotany: A branch of science that deals with all kinds of relationships existing between humans and plants.

Ethnomedicine: A branch of science that deals with plants employed for treatment of various human ailments.

Indigenous Medicine: Knowledge about various materials and practices employed for treatment of disease in a specific region and popular in a localized area.

Leucorrhoea: Excessive vaginal discharge of a white/yellow/green color.

Modern Medicine: Also known as Allopathic or western medicine, modern medicine is the most advanced, scientifically based, and statistically approved therapeutic system and diagnostic methodology.

Traditional Medicine: Knowledge about various materials and practices employed for treatment of diseases that have descended over generations.

Chapter 14
Wild Plants Used as a Forest Ecosystem Services by the Local Inhabitants of Champawat District of Uttarakhand to the Subsistence of Their Livelihood

Kapil Kharkwal
India Post, India

Dalip Kumar Mansotra
Gurukul Kangri University, India

Bhupendra Singh Bhandari
HNB Garhwal University, India

ABSTRACT

This chapter deals with scientific information about the wild socioeconomic plants used by the local inhabitants of the Champawat district of Uttarakhand for the enhancement of their livelihood. During this entire study, it was observed and documented that poor rural people of this area are educationally and financially backward, and they do not have any permanent source of income; due to this reason they are unable to fulfill their necessary requirements. Therefore, they are totally dependent on various forest ecosystem services, viz., selling prepared herbal medicines, wild edible fruits, and wild ornamental plants, etc., which shares approximately 45-50% of their total annual income and plays a vital role for the upliftment of their livelihood. During this research work, a total of 63 plant species belonging to 58 genera under 46 families have been documented, which are being used as livelihood resources by the locals of this area.

DOI: 10.4018/978-1-7998-1320-0.ch014

INTRODUCTION

India is one of the great resources of biodiversity in the world. In India, the Central Himalaya is a great reservoir of various flora and fauna. The Himalayan region possesses great diversity of vegetation, which is important from aesthetic, medicinal, and nutritional aspects. According to an estimate, between 35,000 and 70,000 plant species are used in medicine worldwide. It is believed that out of these, 1,600 species of medicinal plants are being traditionally used in India. Himalaya is recognized as one of the hotspots of biodiversity, harboring nearly 8,000 species of flowering plants, including 25.3% that are endemic. About 1,748 plants are reported as used for medicinal purposes in Indian Himalaya. A large number of wild plants growing in this zone are used for food, medicine, edible fruits, fodder, fuel, timber, and many other purposes by local communities. Thus, a very close relationship and dependence have been established between the forest and inhabitants of the hill. The Himalayan vegetation ranges from sub-tropical deciduous to evergreen forests across the foothills to alpine meadows above the timberline (Singh & Attri, 2014). **The present study area district, Champawat, comes under Kumaun Himalaya of Uttarakhand, which was earlier known as Kali Kumaun**. During the 11th to 12th century, it was a capital of "Chanddynasty" (a ruler of the Kumaun region). Since that ancient time, this area was lush and green with different kinds of vegetation. This district lies between 28.9529^0N and 29.5212^0N latitude and 79.8132^0 E and 80.3178^0E longitude, covering an area of about 1,766 square kilometers. It is ringed by Udham Singh Nagar to the south, Nainital to the west, Pithoragarh and Almora to the north, and Nepal to the east. The Kali River acts as an international border with Nepal. Some perennial rivers and rivulets, viz., Jabgura, Kali, Ladhiya, Lohawati, and Ramganga, constitute the drainage system of the district. This area falls in the Shivalik hills and lesser Himalayan zone, and is very rich in flowering and non-flowering plants because of varied vegetations, climate, topography, soils, rivers, and valleys, etc. Until now, there has been no previous work done by any researchers, particularly for the assessment of wild plants used as a livelihood subsistence, although some fragmentary reports are presented earlier by several workers (Kumari et al., 2011; Osmaston, 1978; Pande & Pande, 2002, 2003; Pant & Samant, 2010; Pusalkar & Srivastava, 2018) regarding different plant groups, but the information about wild plants and their uses for livelihood subsistence is completely lacking in this area. There are some interior areas in the Champawat district of Uttarakhand where people are financially and educationally backward, and they are not engaged in any kind of permanent jobs. Therefore, they are totally dependent on the forest ecosystem for the subsistence of their livelihood. So in this research paper an attempt has been made to document the wild plants that are being used as livelihood subsistence by the locals of this smallest border district of Uttarakhand.

MATERIALS AND METHODS

The research work was carried out in the Champawat district of Kumaun Himalaya, Uttarakhand during the years 2017–2018. In this research work, different places in the Champawat district and its nearby areas, namely, Pancheshwar, Munch, Tamli, Mudiyani, Pokhari, Kharhi, and Pati were regularly surveyed for the documentation of economic plant resources of this area. During this survey period, a number of local fruit vendors, natives (older household members and women), farmers, and experienced personnel in the forest department were interviewed. Group discussion and other participatory methods were also used to learn about the economic uses of these plants used by local peoples. Necessary permissions have been taken from the forest department for the collection and documentation of plant specimens. Plants collected during the survey period have been properly processed as per the standard method given by Jain and Rao (1976). Some plant specimens were identified in the field, and others were identified by matching them with authenticated specimens housed at the BSD herbarium, Dehradun. Asterisk marks (*) indicate that information was gathered by local peoples.

Local Inhabitants

The indigenous people of Uttarakhand primarily belong to the pahari community (including Garhwali and Kumauni people from Garhwal and Kumaon, respectively), with a predominantly Hindu population. According to the Census of India (2011), the population of Uttarakhand is 101 million, with an average population density of 189 people per square kilometer and a decadal growth rate of 18.8% (Directorate of Economics and Statistics, 2014). More than 90% of the population is living in rural areas, and their livelihoods have been dependent on forest-based resources for generations (Sandhu & Sandhu, 2015). The economy is primarily based on agro-pastoralism (Rao & Pant, 2001; Sarin, 2001), with approximately 16% of the total area under cultivation (Singh & Singh, 1992), and almost all households have cattle, either a few for fulfilling household needs of milk and dairy, or large herds of goat, sheep, or buffalo (Semwal et al., 2007). With an increase in population, there is a demand to expand the area under agriculture to meet livelihood needs. However, the fragile landscape, prone to landslides, earthquakes, and erosion, limits the scope for agricultural expansion for the people of the Himalayan region (Rao & Pant, 2001). Apart from the settled residents, these mountain ecosystems also plays a vital role in the lives of the pastoral transhumance communities like the gaddis, bakarwalas, and gujjars who travel with their cattle and goats to nutrient-rich foraging grounds every year (Rao & Pant, 2001; Rawat, 1998). According to Guha (1985), British administrators introduced some of these pastoral communities to avoid fuel load accumulation in the forest understory of economically valuable pine and cedar forests.

Floristic Diversity and Description of the Study Area

Forests constitute one of the most important natural resources. They play a vital role in the social, cultural, historical, economic, and industrial development of any country and in maintaining its ecological balance. Forests are considered not only the resource base for sustenance of the population but also as the storehouses of biodiversity. Forests are solely responsible to maintain and improve the moisture regime and provide clear air. They moderate floods and make the streams perennial. They also produce humus and thereby maintain soil fertility. The diversity in the climatic and physical setting produces a markedly diverse flora and fauna. With increasing catastrophes (mass wasting, cloud bursts, avalanches) forests have been severely fragmented, and at many places degraded, causing a threat of local extinction for many wild species of plants. The sustainability of the forest ecosystem is an essential component of the environmental conservation efforts, and any degradation of forests will have an adverse impact on various systems, such as water resources, agriculture bio diversity, environment, climate, and human health, besides its impact on the subsistence living of tribal and other communities living in and around forest areas. Therefore, the function with respect to the conservation of soil, water, forests, and biodiversity are vital for the welfare of present and future generations. Climatic extremes like cloudbursts, hailstorms, and earthquakes play a critical role as an environmental constraint. Global warming is affecting ice and glacier cover in the region. Furthermore, due to inaccessibility of the region, it is difficult to quantify and assess the damage caused by these natural events. The environmental constraints against the exploitation of natural resources also include other natural phenomena like mass wasting, high seismic activity, landslides, glacial lake outburst floods, erosion, and sedimentation. Soil erosion is one of the major environmental constraints, which results in frequent flooding in the plains downstream and damage to agriculture, life, and infrastructure. The livestock population in the region has also increased during the recent past and the problem of grazing in the high altitudes has acquired serious dimension, as a result, the vegetative cover is decreasing, which has resulted in the loss of topsoil due to excessive erosion. The relatively soft rocks and favorable climatic conditions are the factors behind quick weathering, followed Workshop on Himalayan Ecology by mass wasting and landslides. Intense monsoon rainfall also accelerates soil erosion in the region. The Himalayan rivers also carry a very heavy sediment load, especially during summer and the rainy season, which provides conditions for river shifting. A cloudburst is a natural phenomenon in the Himalayas and it has become frequent in the recent years. Geomorphic features like cirques, hanging valleys with a broad valley, erosional glacial features with an open valley on one side, and dense forests (especially if oak on another side) are present, with an average height of 5,000–6,500 feet. It always occurs in areas where the southern face has agriculture land. These artificial deposits are less compact and loosely laid on the surface of various types of pre-existing landforms and are prone to erosion lands transportation in bed load due to a mixture of finer sand, silt, and coarse gravel, and even boulders, and get a lot of direct sunlight, and the northern face lies behind the area and has thick forests. Cloud always move south to north in the Himalayas, and when they come up against cirque-like features, where the valley is closed, they become stagnant and get piled up, and, after some time, clouds precipitate suddenly, with a heavy discharge in rivers, usually causing massive damage to forests and other property.

Among the natural resources of Uttarakhand, forests are the most important, both economically and environmentally. The forest area is reported to be 3,466 thousand hectares and accounts for around 62.27% of the area of Uttarakhand. The alpine and tropical rainforests that cover most parts of the state make natural habitats for some of the best-known wildlife creatures. The Jim Corbett National Park is home to Royal Bengal tigers. Another rainforest in the region is Rajaji National Park, famous for its large number of pachyderms. Alpine forests in the region include the Valley of Flowers National Park (known for its amazing variety of flowers), Nanda Devi National Park, Govind National Park, and Gangotri National Park. Uttaranchal Himalaya is very rich in forest resources and diversity. The plant diversity is extremely rich, from the valley regions to the highly elevated alpine meadows, locally known as kharaks or bugyals. The main forest types of the state according to altitude are described as follows:

1. **Deodar forests (*Cedrus deodara*)**

   They are found in the Chakrata region, Tons Valley, and Uttarkashi, and the highly elevated regions of water parting, mainly Alaknanda, Nandakini, Pindar, and Mandakini (1,650~2,300 m). The Deodar is a tall coniferous tree used for house construction and paneling.
2. **Blue Pine forests (*Pinus wallichiana*).**

   The Blue Pine is also known as Kail. It is found in the Chakrata, Tons Valley, Uttarkashi, and Joshimath areas. These are found mostly mixed with Deodar forests (1,650~2,300 m).
3. **Chir forests (*Pinus roxburghii*)**

   The Chir pine is found in the whole of the Uttaranchal Hills. The forests exist mostly at the height of 1,000 to 1,650 m. This pine is used for making packing cases and paneling in interior decoration and is also used as firewood.

4. **Teak forests (*Tectona grandis*)**
   The forests have been artificially planted in the Terai belt of the Kumaon region. The plant is a broad-leaved tree and flourishes in the plains. It is the best timber, with great demand for construction work as well as for furniture making.
5. **Bamboo forests (*Dendro calamus*)**
   Bamboo forests are found abundant in the Lansdowne and Kalagarh areas of the Garhwal region (325~1,000 m). Bamboo is known as the poor man's timber. Its versatility of use makes it a force of livelihood for a large number of people. It is a broad-leaved species and helpful for reducing soil erosion.
6. **Oak forests (*Quercus* sp.)**
   Oak forests are found in the whole area of Uttaranchal between 1,325 m and 1,925 m. It is used for firewood and charcoal manufacturing. It is a broad-leaved tree.
7. **Fir (*Abies pindrow*) & Spruce (*Picea smithiana*) forests**
   The two species of trees are found mostly between 2,300 m and 2,950 m. The forests exist in Chakrata, the Tons Valley, Uttarkashi, Pauri, and Pithoragarh.
8. **Sal forests (*Shorea robusta*)**
   The Sal is found in the foothills of the Himalaya in the Terai and Bhabar areas (500 m~1,000 m). The forests are rich in species, such as Rohini, Amultas, Kanju, and Amla.

Table 1 shows forest diversity based on altitude (below 1,000 m to snow line) in the state. In the present study area, the forests are depleting at a large scale. The depletion of forests is causing severe impediments for the stability of the landscape. Most of the forest patches are clearing for agricultural fields due to mounting population pressure on the hilly slopes of the state. Therefore, the rate of deforestation is high. As it is reported that the highest rate of deforestation in any biome is in tropical upland forest, i.e., 1.1% per year (FAO, 1993), the similar situation is applied with the state. The number of people that depend on the forest for firewood is high, resulting in the degradation of the forest and the deterioration of the environment. A case study of the five villages is done on the basis of their elevation and distance from the road head. Per day firewood consumption (in kg) both in summers and winters is calculated as an average in Table 2, which reveals that the elevation and distance form the road play a crucial role in determining per day firewood consumption in the villages. The author collected the data related with per day fire consumption and forest types after an extensive field visit and household survey with a sampling method. The highly elevated villages consume more firewood than the villages in low-lying areas. Firewood consumption also depends on the availability of forestland. In the areas with dense vegetation, more firewood is collected than that in the other areas. The degradation of natural forests is a global problem (Guppy, 1984; Sayer & Whitmore, 1991). People have been destroying forests for millennia, ever since agriculture was started (William 1989). In the Himalaya, deforestation is also argued to be not a recent phenomenon. It has a long history, being well established in late 18th century at least (Mahat et al., 1986). The heavy pressure of the population on the agricultural land due to the high growth rate of the human population compelled the inhabitants of the region to solve the problem through merging forestland either for settlement or for agricultural practices. However, the extent of impairment of various processes attributed to vegetal degradation depends upon a range of other factors, including past histories, intensity of removal of natural vegetation, patterns of natural regeneration, and/or other human interferences (Alford, 1992; Gilmour et al., 1989; Ramakrishnan et al., 1992; Valdiya & Bartarya, 1989, 1991). The management of the forest resource in the mainland of Uttaranchal is facing setbacks due to two reasons: first, an inaccessibility of the temperate forestland and second, the depletion of the forest in surrounding settlements. The instability of the land due to natural phenomena (terrestrial and atmospheric), such as slope failure, landslides, mass movements, and rock fall, further accentuate the problem of the depletion of the forest. There may be two solutions for the sustainable management of forest resources: (1) The proper utilization of the waste woods, which are found in the inaccessible areas of the region through small-scale, village-based forest industries. That will help the economic development of the region as well as the sustainable livelihood of the populace there. (2) Afforestation, around the settlements and the areas, where soil erosion is more prone, should be done to stabilize hill slopes.

Uttarakhand is a storehouse of a rich variety of herbs and medicinal and aromatic plant species. The government intends to exploit this advantage. Special emphasis is on research and development An integrated action plan has been drawn up for this purpose in coordination with the Government of India and other concerned agencies in the state and elsewhere in the country. Research and development work in the area of medicinal plants and the commercial production of applications and formulations will be developed in conjunction with research institutions and reputed companies. A Medicinal and Aromatic Plants Export Zone has been set up, covering seven districts of Uttarakhand, and Specialized Herbal Parks are in the offing. **Medical and health care:** The salubrious climate, pollution-free environment, and the availability of a wide range of flora and fauna in the mountainous terrain make Uttarakhand an ideal location for developing centers for alternative medicine and health care facilities. Possibilities for establishing gyms, saunas, and related facilities, which would motivate health care enthusiasts, exist in plenty. **Forest product—herbs and spices:** A significant portion of Uttarakhand is under forest cover (almost 70%). Thus, there is excellent potential for the development of forest resources–based industries in the state. In addition, there is ample scope to develop industries based on forest and agro-wastes such as lantana, pineneedles, plant, and vegetative fibers such as Rambans, etc. **Horticulture and floriculture:** Uttarakhand has almost all the different Agro-geo climatic zones, making it particularly conducive to commercial horticulture and floriculture. Horticulture schemes are being promoted in a big way through adequate incentives and facilities. Given the constraints on the productivity of field crops, a shift from the cultivation of low-value field crops to high-value crops, such as fruits and vegetables, appears to be the most obvious option for the state. The Uttarakhand region has proved to be suitable for growing different types of temperate, sub-tropical, and tropical fruits. Also, the region has wide scope for growing different kinds of vegetables, flowers, ornamental plants, mushrooms, and medicinal plants in its different climatic zones. Temperate fruits, such as apples, pears, peaches, plums, apricots, cherries, and walnuts, are grown in the places of 1,000 m~3,000 m (Sati, 2004). The places from 300 m to 1,400 m are planted with citrus, mango, litchi, banana, guava, papaya, strawberry, and different vegetables and crops. Market forces and institutional setup created for gearing the horticultural development led to economic growth, but at the cost of equity. Prosperous farmers benefited more than small and marginal farmers (Swarup & Sikka, 1987). The land under fruit cultivation is 190,192 ha and under vegetable 80,332 ha, accounting for 29% and 15% of the total area sown, respectively. The total production of fruits and vegetables in Uttarakhand is estimated to be 345,339 metric tons and 492,785 metric tons, respectively, according to data available in 2003. Apples are the most important among the various fruits grown in the region, with cultivated land of 54,000 ha. Among vegetables, potatoes are the most important crop. In the state, fruits are grown in the different altitudinal zones, and among them apples, peaches, citrus, mangoes, plums, apricots, walnuts, litchis, and other fruits are important. The land used for production of various vegetables in the region was almost twice increased and their production was increased by 2.5 times during the past two decades. This clearly indicates that the growing of fruits and vegetables has spread over larger areas of land. However, a lack of proper marketing facilities, and an absence of post-harvest technologies and storage are serious constraints to a more rapid and systematic development of fruit and vegetable cultivation and marketing in the region. Cultivation of offseason vegetables in different altitudinal zones of Uttarakhand in summer offers extensive market opportunities (Mehta, 1990). Herbs are naturally grown in the meadows (kharaks). Fortunately, the state has extensive meadows along the Great Himalayan Ranges. These herbs are locally utilized for medicinal purposes, with positive results. Simultaneously, a tea garden practice is centuries old, but presently its extension is available in very limited areas. Following the creation of Uttarakhand State, the government has launched a scheme for establishing tea gardens in different geographically suitable areas. Floriculture is being developed in a big way in order to supply the domestic market and penetrate foreign markets. Floriculture can help generate employment opportunities as well as augment incomes of farmers. With the climate being ideal for growing flowers nearly year round, it is proposed to establish Floriculture Parks with common infrastructure facilities for sorting, pre-cooling, cold chain, processing, grading, and packing/marketing facilities.

Environmental Conditions and Soil Types

Climate

The Central Himalayan state of Uttarakhand experiences three major seasons: summer (April–June), monsoon (July–September), and winter (December–February), along with a brief spring season preceding summer and an autumn season just prior to the beginning of winter (FSI, 2011). Depending on the elevation, climatic patterns in Uttarakhand can resemble the indogangetic plains in the foothills (2,000 m). In the foothills, the lowest and highest mean summer temperatures are around 16°C and 36°C respectively; however, temperatures up to 45°C have also been recorded in peak summer months. Winters can be moderately cold with mean minimum and maximum temperatures of 5°C to 18°C, respectively (Dutta et al., 2015). However, temperatures as low as -2°C, accompanied with fog and frost, have also been recorded in the peak winter months (December–January) in the foothills. In the hill regions (data from Nainital observatory: altitude >1,900 m) the mean monthly maximum temperature ranges between 18°C (January) and 24°C (May) and the mean monthly minimum temperature ranges between 4°C (January) and 17°C (July). Winters can be harsh, with temperatures reaching sub-zero levels, accompanied by snowfall and hailstorms at higher altitudes. Rainfall in the region is primarily monsoonal and most of the precipitation occurs in the months of July and August, with an average annual precipitation of nearly 1,558 mm (Kumar & Ram, 2005).

Soil

Soil types in the Central Himalaya vary from rich alluvial soil in the flat tarai region at the foothills to the poor-quality, bare soil of the mountains to the desert-like soils of the trans-Himalayan region (Jina et al., 2011). A detailed soil survey by Kumar (1988) reveals the presence of four orders of soil in the region: (1) Inceptisols (moderately developed soil), (2) Alfisols (soil with high base content), (3) Entisols (soil developed in organic material) and (4) Mollisols (soils with dark A horizon). In a broader sense, most of the forest soils in the region belong to the brown earth soil category (or Eurochrepts), which has calcium as the dominant exchangeable base (Singh & Singh, 1992). Based on information collated from

30 studies on the soil of the Central Himalaya, Singh and Singh (1992) give a comprehensive overview of Himalayan soil. In terms of texture, the soil varies from coarse, loamy sand in the lower elevations to clay loam at higher elevation (~2,000 m). Soil pH is slightly acidic to neutral (6–7) at a lower elevation.

RESULT AND DISCUSSION

Western Himalaya is a reservoir of many natural resources, of which the vegetational aspect is predominant. Its unique setting within the Himalayan region possesses luxuriant and varied vegetation, most of which is important from nutritional, aesthetic, and medicinal viewpoints. Locals of the Western Himalayan region are highly dependent on natural resources for products such as fuel fodder, timber, and NTFPs. Among these resources, herbal as well as ethno-medicinal plants are one of the most important NTFPs, which have been playing an important role in improving the health of locals for a long time. Through field surveys, it was observed that a total of 63 plant species belonging to 58 genera under 46 families have commonly been used as a livelihood subsistence by the local inhabitants of the district of Champawat. These plant species include Angiosperms, Gymnosperms, Pteridophytes, Fungi, and Lichens. During this entire study, it was also observed and documented that the poor rural peoples of this area are totally dependent on the forest ecosystem, which shares approximately 45–50% of their total annual income and plays an important role in their livelihood. (plate.1 a to f).

Almost all poor families depend on traditional ethno-medicinal plants due to the lack of medical facilities in rural areas. Today about 65% of the Indian population depends on the traditional system of medicine. They diagnose and cure different diseases through their own traditional knowledge. Extensive surveys and literature reveals that considerable work has been done on medicinal plants and their ethno-medicinal uses in the prospective of the Himalayan region, including studies on ethnobotanical uses of plants in Garhwal Himalaya, ethnobotany of Kumaun Himalaya, the indigenous knowledge of medicinal plants and wild edibles among three tribal sub-communities of the Central Himalayas, and ethnomedicinal plant diversity in Kumaun Himalaya. Similarly, the indigenous knowledge system of the high altitude society has been reported of Kumaun Himalaya, threat status of rare species of Kedarnath Wild life sanctuary with economic uses and medicinal plants traditional health care knowledge of Vaidhyas, palsi and others, recorded ethnobotanical uses of plants by Bhotiya tribal in Niti valley · Endorsing the declining indigenous ethnobotanical knowledge system of Seabuckthorn in Central Himalaya, But in the present scenario, there is continuous exploitation of these herbal reserves or natural resources, and traditional knowledge is depleting day by day in the younger generation due to lack of proper interaction with Vaidhyas and other knowledgeable persons. Therefore, the present study has been undertaken to explore the more traditional knowledge on ethno-medicinal plants and their uses, with the help of Vaidhyas and other experienced persons, and the expected outcome will be shared with the younger generation, to conserve traditional knowledge as well as natural and herbal reservoir from a future perspective.

Descriptions of Some of the Most Important Plants

1. ***Diploknema butyracea*** (Roxb.) H.J.Lam
   Indian butter tree is locally known as **"Chyuri"** in Kumauni. In Hindi it is called "Madhuca," and in Sanskrit, "Madhupushpa." Vegetable ghee is obtained from the seed, which is very in demand and yields high prices.

2. ***Myrica esculenta*** Buch.-Ham. ex D. Don

It is commonly known as kaphal. The tree has great importance since it is from the ancient Ayurveda and Yunani system of medicines, where most of the part of the tree is utilized as a medicine. The bark of the tree is traditionally used as an antiseptic, used for washing putrid sores, against fish poisoning, and in an external plaster for rheumatism. It is also used for tanning and dying (yellow-colored dye) (Jeeva et al., 2011).

3. *Rhododendron arboreum* Smith.

It is commonly known as Burash. Flowers are eaten raw or made into juice to cure stomach diseases (Tiwari et al., 2010b; Uniyal et al., 2006).

4. **Diplazium esculentum** Retz.

It is commonly known as Lingura. The rhizomes are kept in the granaries to check them from insect and pests. Young fronds are used as green vegetables and also used as salad or cooked as vegetables (Upreti et al., 2009).

5. ***Dioscorea bulbifera*** Linn.

It is commonly known as Genthi. Tubers are cooked as vegetables. The tubers are used for piles, dysentery, and syphilis and are also applied to ulcers after being dried and powdered. The rhizomes are considered anorexiant, and an ethyl alcohol (50%) extract of its aerial parts is used as diuretic (Asolkar et al., 1992).

6. ***Bauhinia purpurea*** L.

It is commonly known as Khairwal. The flowers are used as a laxative; the buds (dried) are anthelmintic, useful in piles and blood dysentery; the bark is used for diarrhea; the root-bark is mixed with curd found efficacious in hemorrhoids and its paste can be mixed with dried ginger and applied internally in the treatment of a goiter; its root is carminative.

7. ***Selaginella bryopteris*** Baker

It is locally known as Ryun. Young fronds are sold in the market as an ornamental plant.

8. ***Origanum vulgare*** L.

It is locally known as Ban Tulsi. Its dried leaves are sold as an herbal powder to cure fever as well as other health problems.

9. ***Morchella esculenta*** L.

It is locally known as Chyo. It fruiting body is sought after by shopkeepers and yields very high prices.

10. ***Drepanostachyum falcatum*** (Nees) Keng f.

It is locally known as Ringal. The wood is used for basket making by locals.

11. ***Boenninghausenia albiflora*** (Hook.) Rchb. ex Meisn.

It is locally known as Pissu Booti. The dried powder of its leaves is sold with cows' ghee and yields high prices.

12. ***Cedrus deodara*** (Roxb. ex D.Don) G.Don

It is commonly known as Deodar. Wood Oil is sold as an eczema medicine.

13. *Bergenia ciliata* (Haw.) Sternb.

It is commonly known as Pasan Bhed. The dried powder of its root is used to treat piles.

14. *Adiantum edgeworthii* Hook.

It is commonly known as Ryun. Its rachis is sold as antiseptic medicine during Upnayansanskar in hindu dharma to protect ear boils.

15. *Justicia adhatoda* L.

It is locally known as Basing. The leaf powder is used for asthmatic patients.

Wild Edible Plants with Nutraceutical Activity

Wild edible plants have traditionally occupied an important position in the socio-cultural, spiritual, and health arena of rural and tribal lives of India. India has one of the oldest, richest, and most diverse cultural traditions associated with the use of medicinal plants in the form of traditional systems of medicine (Government of India, 2000). Plants are the invaluable, incredible, and traditional sources for the curability of various diseases in the form of medicines (Guerra et al., 2003). Plants' secondary metabolites have been implicated for most plants' therapeutic activities (Timothy et al., 2008). Accordingly, the plant and its products are safe, and as a result there is continuous use of plant product as a drug, as it is found to be an alternative way to cure patients; this approach was in practice from ancient times (Archana et al., 2011). Traditional treatments may provide valuable clues for the development of new oral hypoglycemic agents and simple dietary adjuncts. One important plant *Urtica dioica* has been used for centuries for food and medical purposes (Mehri et al., 2011). Some plants have been shown to have protective antioxidant effects and are therefore hepatoprotective. Such plants are *Parkia clappertoniana* and *Xylopiaa ethiopica* (Patrick-Iwuanyanwu et al., 2010). There are following wild edible plants that play a critical role in neutraceutics.

Conservation Strategies

Conservation is the process of managing the biosphere in order to obtain the greatest benefit for the present generation and maintaining the potential for the future. The conservation of plant resources is of global concern because we don't know what we are losing and what we will need in the future. Conservation methods vary, with many biological and environmental factors (Rajasekharan & Ganeshan, 2002). Small isolated populations, and endemic and rare species in particular, are subjected to genetic drift and inbreeding, and their genetic variation is consequently expected to be low compared to that of larger populations, which may lead to a decrease in species' ability to survive environmental changes and demographic fluctuations, both in the short and long term (Bilington, 1991; Gaston & Cunin, 1997a; Karron, 1997). Hence, the maintenance of genetic variation is essential for the long-term protection of a taxon (Hamrick & Godt, 1989; Simberloff, 1988).

Strategies for Plants

The two main strategies are ex situ (protection of species outside their natural habitats) and in situ (in their natural surroundings) conservation. There is a need for coordinated conservation efforts based on these strategies (Figure 1). More information is required on medicinal plant production, utilization, trade, monitoring the stock of medicinal plants, development of sustainable harvesting practices, preservation of traditional knowledge, and intellectual property rights. The World Conservation Union (formerly known as the International Union for Conservation of Nature and Natural Resources) categorized plants into "Red Data List Categories" (IUCN, 2001) based on the detailed knowledge of the population dynamics and genetics of the species, viz., extinct, extinct in the wild, threatened (critically endangered, endangered, and vulnerable), and low risk (conservation dependent, near threatened, and least concern), and indeterminate where the data are insufficient. Conservationists focus their attention exclusively on species extinction rather than genetic erosion within individual gene pools, and the latter may be of equal importance in terms of loss of biodiversity (Maxted, 2001). Hence, it is imperative that implementing viable strategies to conserve the populations and genetic resources of medicinally important species is a must to avoid further loss. Ongoing efforts in India include both in situ and ex situ conservation measures, viz., plant tissue culture, introduction of new crop genetic resources, research in habitat restoration, pollution abatement, seed storage, and tissue banking, etc. (Jackson & Sutherland, 2000).

In Situ Conservation

In situ or onsite conservation involves maintaining genetic resources in their natural habitats, i.e., within the ecosystem to which it is adapted, whether as wild or crop cultivar in a farmer's field as components of the traditional agricultural systems (Damania, 1996). The key operational steps for establishing in situ gene banks for the conservation of prioritized medicinal plants include: threat assessment; establishment of a network of medicinal plant forest reserves; involving local stakeholders; botanical, ecological, trade, and ethno-medical surveys; assessing intraspecific variability of prioritized species; designing species recovery programs; establishment of a medicinal plant seed center, etc. Conclusively, no in situ conservation project can succeed without the complete cooperation and involvement of local people (Srinivasamurthy & Ghate, 2002).

Ex Situ Conservation

Ex situ conservation involves the conservation of biodiversity outside the native or natural habitat where the genetic variation is maintained away from its original location. The ex situ genetic conservation fulfills the requirement of present or future economic, social, and environmental needs. Conservation also includes the propagation and assessment of molecular diversity (Olorode, 2004). The conservation of medicinal plants include a combination of methods, depending on factors such as geographic sites, biological characteristics of plants, available infrastructure, and network having an access to different geographical areas, human resources, and number of accessions in a given collection (Rajasekharan & Ganeshan, 2002).

Need for Conservation of Plant Species

To meet the requirements of expanding regional and international markets' healthcare products and the needs of growing populations, large quantities of medicinal plants are harvested from forests (Desilva, 1997). In India, a large number of medicinal plants are extracted from the wild to meet the increasing demand for raw material needed for domestic consumption and for export. As a result, the natural sources are rapidly depleting. Medicinal plants contribute to health, income, agroforestry system, cultural identity, and livelihood security. Hence there is a need for the conservation, cultivation, maintenance, and assessment of germplasm for future use. The conservation of biological diversity involves protecting, restoring, and enhancing the variety of life in an area so that the abundance and distribution of species and communities contributes to sustainable development. The ultimate goal of conservation biology is to maintain the evolutionary potential of species by maintaining natural levels of diversity, which is essential for species and populations to respond to long- and short-term environmental changes in order to overcome stochastic factors failing, which would result in extinction

Figure 1.

Ecosystem Services

The concept of ecosystem services is seen as a promising approach communicating the links between ecosystems and human well-being. Although the term "ecosystem services" was primary introduced by Ehrlich and Ehrlich (1981), the concept's origin in modern history dates back to the late 1960s and 1970s, highlighting the societal value on nature's functions. Ecosystem services are the direct and indirect contributions to human well-being by ecosystems; these benefits mainly refer to provisioning, regulating, cultural, and supporting services needed to maintain other services. First, provisioning services refer to all economic activities derived from physical goods provided by an ecosystem like food, raw materials, and drinking water. Second, regulating services refer to the indirect benefits provided by an ecosystem like climate regulation, prevention of soil erosion, and air purification. Third, cultural services refer to the non-material services resulting from experiences, providing personal and cultural enrichment. And finally, supporting services refer to the rest of functions that make the ecosystems possible to provide all the above-mentioned benefits. These diverse services are interrelated in many different ways, ranging from synergistic to tolerant, conflictive, and mutually exclusionary ones. The concept of ecosystem services is multidisciplinary by nature and exhaustive analyses require collaboration between different disciplines, inherently combining the natural and social sciences.

The field of ecosystem services began to grow, following studies by Daily and Costanza et al. Initiatives such as the Millennium Ecosystem Assessment project, the Economics of Ecosystems and Biodiversity study, and the Intergovernmental Science-Policy Platform on Biodiversity and Ecosystem Services are milestones that turned the concept into a political instrument to achieve the sustainable use of natural resources. The evaluation of ecosystem services is aimed at providing useful knowledge for policy-making, defining strategies, and managing ecosystems by stakeholders. Regions that are highly susceptible to the supply of ecosystem services and human well-being need to be informed of all aspects of trends in the flow of ecosystem services, including the impact of government intervention and social and ecological pressures. The theoretical framework of ecosystem services has the potential to support informed decision-making with evidence-based knowledge. At the time, it contributes to face up to challenges to ensure sustainability and it can be applied within forest ecosystems. Forests are widely acknowledged as being principal ecosystem service providers. Some national classifications consider up to 100 different forest services, including food, timber, and fuel production; water conservation and regulation; nutrient retention; carbon sequestration; biodiversity protection; climate regulation; ecotourism; and spiritual and traditional values. There are several drivers of change that threaten worldwide forest system sustainability. Among them, we should highlight the expansion of farming land and urbanization areas, as well as the adverse effects of climate change. Food security is a global concern due to the continuously growing population, particularly in disadvantaged areas. Economic development and the expansion of cultivation surfaces are the principal factors driving changes in forest land use in the last decade. The 2016 FAO State of the World's Forests Report estimated the total area of global forests to be 30.74% of the planet's surface. Between 2000 and 2010, the net deforestation registered seven million hectares annually in tropical countries and a six million hectares net increase in the agricultural land per year. Often land managers consider farming and forest uses as rivals. Further ways of sustainable and complementary management of land are hereby ignored. It must be acknowledged that adequate forest system management, such as forestry and multifunctional systems, not only directly contributes to human well-being in the form of provisioning services, but it also produces synergies with agricultural systems capable of increasing food production and other complementary service such as pollination services, biological control of pests, and prevention of soil erosion.

Climate change threatens forests and their service flow. Some of the consequences of global climate change are expected to be long-term droughts and imbalances in water supplies, a decrease in carbon sequestering capabilities, soil erosion, loss of soil fertility, desertification, reduction in food production capacity, and the threat posed by the prevalence of certain species and communities. An increase in arid land, principally due to the decrease of precipitation, but also due to higher temperatures, is one of the main threats to diversity and the survival of certain forests. All this can generate a wide range of large-scale and difficult impacts, which can be costly or irreversible for ecosystems and biodiversity.

This would reduce the ability to mitigate adverse effects and, consequently, reduce human well-being, particularly for the poorest populations of low-income countries and arid climate regions.

A bibliometric analysis has been implemented to study the diverse research fields related to forests. Malesios and Arabatzis assess journals on forestry; Aleixandre-Benavent et al. analyze research trends on worldwide deforestation; Song and Zhao make use of the bibliometric method to analyze the global research on forestry ecology in a 10 year-period. Regarding ecosystem services, the bibliometric analysis has also been implemented. For instance, Zhang et al. assess the global research on ecosystem services and Tancoigne et al. analyze the role of Agricultural Sciences in the study of ecosystem services. However, no bibliometric studies were found for forest ecosystem services. The bibliometric analysis pursues the identification, organization and analysis of the main components within a specific research area Garfield was a forerunner of this methodology in the mid-twentieth century. The application of this method has been extended to areas like medicine, engineering, biology, and administration. Bibliometry is based on statistical and mathematic analysis tools. Thanks to them, the evolution and research trends of a topic can be studied, as well as the relevance of its publications. Furthermore, this methodology allows the identification of the main driving forces of a study field, that is, authors, journals, institutions, and countries. As traditional approaches in the bibliometric analysis, we can find co-occurrence analysis, co-quotation, and bibliographic coupling applied to database metadata. These metadata usually refer to the year of publication; authors' names, institutions, and countries; topic categories; article citations; and keywords. The technical evolution in recent years has given rise to the development of viewing tools through text extraction and data mining, development of analysis frames to assess innovation, overlaying maps and variable associations tool developments based on routine types of automatized software, and methods to identify and view the evolution of specific scientific topics within a time segment. Durieux and Gevenois defined diverse types of indicators within the bibliometric analysis. Such indicators refer to productivity, quality, and structure of scientific production. Productivity indicators show the amount of works published on a field and can also be applied to authors, journals, and institutions. As far as qualitative indicators are concerned, some of them, like the number of citations, H-index, and impact factor-based journal rankings, are well known. Structure indicators measure established links between agents. It has a special relevance in those areas where international cooperation is required by the own research topic. The obtained results by the bibliometric analysis are very useful for diverse users. It may help junior researchers approach a topic, get to know the main research lines within a topic, identify the most productive agents, and learn about future researching trends. Senior researchers can, at the time, keep updated about a specific subject. The bibliometric analysis of rapidly growing areas related to the development of new technologies and innovations may also show investment options for analysts and business agents. Furthermore, data compilation on the evolution and new trends of subjects related to the management of natural resources and ecosystems, environmental protection, climate change, or sustainability can be very useful for research agencies, government institutions, and private research centers when trying to understand collaboration strategies and identify key collaborators to strengthen up potential weaknesses (Hassan et al., 2014). The consequences of climate change, the deterioration of the environment, and the demands of an increasing world population are the driving factors behind this increase of publications on forest ecosystem service research. All this has generated a new and growing corpus of scientific literature that needs to be analyzed, organized, and systematized. However, there is no knowledge of studies devoted to the dynamic analysis of this global research field. This study aims to fill this knowledge gap through the analysis of the worldwide dynamics of research on forest ecosystem services during last 20 years. The main objective of this work is to identify the evolution of

Table 1. List of wild plants used as forest ecosystem services

Sr. No	Plant Name	Family	Altitude (in m.)	Habit	Used as livelihood subsistence
1.	*Selaginella bryopteris* Baker	Selaginellaceae	805m.	Fern (Herb)	*Young fronds are sold in the market as an ornamental plant.
2.	*Rhododendron arboreum* Sm.	Ericaceae	1612m.	Small Tree	*Juice of its flowers is sold in the market yield high rate.
3.	*Myrica esculenta* Buch.-Ham. ex D. Don	Myricaceae	1610m.	Tree	*Ripened fruits (edible) are supplied by the poor rural people of village areas for marketing purposes.
4.	*Diploknema butyracea* (Roxb.) H.J.Lam	Sapotaceae	818m.	Tree	*Vegetable ghee is obtained from seeds, which are very in demand and yield high prices.
5.	*Asparagus racemosus* Willd.	Asparagaceae	1735m.	Shrub	*Young plant is sold as a vegetable.
6.	*Origanum vulgare* L.	Lamiaceae	1905m.	Herb	*Its dried leaves are sold as an herbal powder to cure fever as well as other health problems.
7.	*Diplazium esculentum*(Retz.) Sw.	Athyriaceae	2115m.	Fern	Young fronds are sold as a green vegetable and are beneficial for anaemic patients.
8.	*Bauhinia vahlii* Wight &Arn.	Fabaceae	580m.	Climber	Leaves are used for religious purposes, viz., as a worship and shradha in hindu dharma .
9.	*Drepanostachyum falcatum*(Nees) Keng f.	Poaceae	1625m.	Woody herb	*Wood is used in basket making by locals.
10.	*Aesculus indica* (Wall. ex Cambess.) Hook.	Sapindaceae	1920m.	Tree	*Mature fruits are used in soap manufacturing.
11.	*Usnea pectinata* Taylor	Usneaceae	1710m.	Lichen	*Whole plant is used in dye making.
12.	*Ramaria botrytis* (Pers.) Ricken	Gomphaceae	1520m.	Fungi	*Fruiting body is sold and used as a vegetable.
13.	*Morchella esculenta* L.	Morchellaceae	1815m.	Fungi	*Fruiting body is sought after by shopkeepers and yield a very high price.
14.	*Agave americana* L.	Asparagaceae	1825m.	Shrub	Leaf extract is used in fishing.
15.	*Ageratum conyzoides* (L.) L.	Asteraceae	580m.	Herb	Used as animal fodder.
16.	*Boenninghausenia albiflora* (Hook.) Rchb. ex Meisn.	Rutaceae	1922m.	Herb	Dried powder of its leaves is sold with cows ghee and yields high prices.
17.	*Aster thomsonii* C.B.Clarke	Asteraceae	1616m.	Herb	Leaves are used as fodder.
18.	*Alnus nepalensis* D.Don	Betulaceae	1516m.	Tree	Wood is used in furniture making.
19.	*Artemisia nilagirica* (C.B.Clarke) Pamp.	Asteraceae	1505m.	Herb	Dried leaves are sold to prepare dhoop for worship.
20.	*Azadirachta indica* A.Juss.	Meliaceae	510m.	Tree	Leaf extract is used against skin allergy.
21.	*Bauhinia purpurea* L.	Fabaceae	430m.	Tree	Leaves are used for religious purposes.
22.	*Bauhinia variegata* L.	Fabaceae	420m.	Tree	Leaves are used as fodder.
23.	*Cedrus deodara* (Roxb. ex D.Don) G.Don	Pinaceae	1910m.	Tree	Wood Oil is sold as an eczema medicine.
24.	*Pinus roxburghii* Sarg.	Pinaceae	1715m.	Tree	Wood is used for furniture making.
25.	*Berberis aristata* DC.	Berberidaceae	1782m.	Shrub	Root is sold by localized as a diabetes medicine.
26.	*Bergenia ciliata* (Haw.) Sternb.	Saxifragaceae	1716m.	Herb	Dried powder of its root is used to treat piles.
27.	*Adiantum edgeworthii* Hook.	Pteridaceae	1615m.	Fern (Herb)	Its rachis is sold as antiseptic medicine during Upnayansanskar in hindu dharma to protect ear boils.
28.	*Cannabis sativa* L.	Cannabaceae	2015m.	Herb	Dried seeds are sold as spices and yield very high prices.
29.	*Celtis australis* L.	Cannabaceae	1719m.	Tree	Leaves are used as fodder.
30.	*Chenopodium album* L.	Amaranthaceae	1515m.	Herb	Young leaves are sold by locals and cooked as vegetables.
31.	*Chenopodium ambrosioides* L.	Amaranthaceae	1516m.	Herb	Extract of its leaves is used as insect repellent.
32.	*Cinnamomum tamala* (Buch.-Ham.) T.Nees&Eberm.	Lauraceae	615m.	Tree	Dried leaves and bark are used as spices; yield very high cost.

continued on following page

Table 1. Continued

Sr. No	Plant Name	Family	Altitude (in m.)	Habit	Used as livelihood subsistence
33.	*Centella asiatica* (L.) Urb.	Apiaceae	1812m.	Herb	Extract of its leaves is used in the treatment of fever as well as other mental problems.
34.	*Cornus capitata* Wall.	Cornaceae	1856m.	Tree	Fruits are edible.
35.	*Cynoglossum glochidiatum* Wall. ex Benth.	Boraginaceae	1912m.	Herb	Leaves are used as ethno-veterinary medicine for animal fever.
36.	*Cuscuta reflexa* Roxb.	Convolvulaceae	1729m.	Climber	Extract of this plant is used in paralysis.
37.	*Datura alba* L.	Solanaceae	1503m.	Herb	Dried leaves are used to cure asthma.
38.	*Dioscorea deltoidea* Wall. ex Griseb.	Dioscoreaceae	1535m.	Climber	Boiled tubers are sold to enhance vigor of human body.
39.	*Hedychium spicatum* Sm.	Zinegiberacea	1521m.	Herb	Dried powder of its rhizome is mostly used in the treatment of various gastric problems.
40.	*Holboellia latifolia*Wall.	Berberidaceae	1515m.	Woody Climber	Fruits are used to cure liver problem.
41.	*Juglans regia* L.	Juglandaceae	1716m.	Tree	Fruits are sold, which are beneficial for health.
42.	*Justicia adhatoda* L.	Acanthaceae	980m.	Shrub	Leaf powder is used for asthmatic patients.
43.	*Leptodermis lanceolata* Wall.	Rubiaceae	1536m.	Shrub	Leaves are used as fodder.
44.	*Melothria heterophylla* (Lour.) Cogn.	Cucurbitaceae	1643m.	Climber	*Dried powder of its roots is used in various urino-genital diseases.
45.	*Morus serrata* Roxb.	Moraceae	1515m.	Tree	Fruits are sold in the market.
46.	*Plantago himalaica* Pilg.	Plantaginaceae	1715m.	Herb	Dried seeds of this plant are used for diarrhea and liver problems.
47.	*Prinsepia utilis* Royle	Rosaceae	1615m.	Shrub	Seed oil is used for bone fractures.
48.	*Pyracantha crenulata* (Roxb. ex D.Don) M.Roem.	Rosaceae	1615m.	Shrub	Fruits are used for heart problems.
49.	*Quercus leucotrichophora* A. Camus	Fagaceae	1471m.	Tree	*Leaves are used as fodder.
50.	*Quercus semecarpifolia* Sm.	Fagaceae	2215m.	Tree	*Leaves are used as fodder.
51.	*Ranunculus diffusus* DC.	Ranunculaceae	1710m.	Herb	1-2 drops of its leaf extract is used to cure migraine
52.	*Ricinus communis* L.	Euphorbiaceae	816m.	Shrub	Seed oil is sold in the market as a medicine for bone fractures.
53.	*Sapindus mukorossi* Gaertn.	Sapindaceae	615m.	Tree	Fruits are used in soap making.
54.	*Thymus linearis* Benth.	Lamiaceae	1610m.	Herb	Dried powder of its leaves is sold as a noise, fever medicine.
55.	*Tagetes minuta* L.	Asteraceae	1416m.	Herb	Leaves are used as insect repellent.
56.	*Tectona grandis* L.f.	Lamiaceae	415m.	Tree	Wood is sold for furniture making.
57.	*Terminalia arjuna* (Roxb. ex DC.) Wight &Arn.	Combretaceae	340m.	Tree	Extract of its bark is used for heart problems.
58.	*Terminalia chebula* Retz.	Combretaceae	650m.	Tree	Its fruits are in very high demand and are sold by local people as a digestive medicine.
59.	*Toona serrata* (Royle) M. Roem.	Meliaceae	1615m.	Tree	Leaves are used as fodder.
60.	*Urtica dioica* L.	Urticaceae	1515m.	Herb	*Leaves are cooked as vegetable, a good source of iron.
61.	*Thalictrum foliolosum* DC.	Ranunculaceae	1715m.	Herb	*Dried root powder is sold by local vaidhyas, hakims as a jaundice medicine.
62.	*Tinospora cordifolia* (Willd.) Miers	Menispermaceae	815m.	Climber	Dried powder of its leaves is sold in the market to improve immune system and fever.
63.	*Shorea robusta* Gaertn	Dipterocarpaceae	425m.	Tree	*Wood is sold for furniture making.

this research line and assess the relative significance of studies devoted to ecosystem services within the forest research area.

ACKNOWLEDGMENT

The authors are thankful to the Director, Botanical Survey of India, Kolkata for providing the necessary facilities and to the staff of the forest Department of Champawat district, Uttarakhand for their kind cooperation during this survey work.

REFERENCES

Archana, S. J., Paul, R., & Tiwari, A. (2011). Indian medicinal plants: A rich source of natural immuno-modulator. *International Journal of Pharmacology*, *7*(2), 198–205. doi:10.3923/ijp.2011.198.205

Arora, R. K., & Anjula, P. (1996). *Wild edible plants of India: Diversity conservation and use*. New Delhi, India: ICAR.

Asolkar, L.V., Kakkar, K. K., & Chakra, O. J. (1992). Second supplement to glossary of Indian medicinal plants with active principles part-1 (A-K) 1965-1981. NISC, CSIR, New Delhi, India, pp. 265-266.

Chadha, Y. R. (1995). The wealth of India, raw materials. *Publications and Information Directorate*, *10*(Sp-W), 591.

Guerra, R. N., Pereira, H. A., Silveria, L. M., & Olea, R. S. (2003). Immunomodulatory property of AlternantheratenellaColla aqueous extracts in mice. *Braz. J. Med. Biol. Res.*, 36, 1,215-1,219.

Harborne, J. B. (1962). Plant polyphenols. 5: Occurrence of azalein and related pigments in flowers of *Plumbago* and *Rhododendron* species. *Archives of Biochemistry and Biophysics*, *96*(1), 171–178. doi:10.1016/0003-9861(62)90467-8 PMID:13904580

Hassan, S. U., Haddawy, P., & Zhu, J. (2014). A bibliometric study of the world's research activity in sustainable development and its sub-areas using scientific literature. [CrossRef]. *Scientometrics*, *99*(2), 549–579. doi:10.100711192-013-1193-3

Jain, S. K., & Rao, R. R. (1976). *A handbook of field and herbarium methods*. New Delhi, India: Today and Tomorrow's Printers and Publishers.

Jeeva, S., Lyndem, F. G., Sawian, J. T., Laloo, R. C., & Mishra, B. P. (2011). MyricaesculentaBuch.- Ham. ex D. Don. - a potential ethnomedicinal species in a subtropical forest of Meghalaya, northeast India. *Asian Pacific Journal of Tropical Biomedicine*, *1*(2), S174–S177. doi:10.1016/S2221-1691(11)60150-0

Kumari, P., Joshi, G. C., & Tiwari, L. M. (2011). Diversity and status of ethnomedicinal plants of Almora district in Uttarakhand, India. *International Journal of Biodeversity and Conservation*, *3*(7), 298–326.

Osmaston, A. E. (1978). *Forest flora for Kumaun*. Dehra Dun, India: International Book Distributors.

Pande, H. C., & Pande, P. C. (2002). *An illustrated fern flora of the Kumaon Himalaya* (Vol. II). Dehra Dun, India: Bishen Singh Mahendra Pal Singh.

Pande, H. C., & Pande, P. C. (2003). *An illustrated fern flora of the Kumaon Himalaya* (Vol. I). Dehra Dun, India: Bishen Singh Mahendra Pal Singh.

Pant, S., & Samant, S. S. (2010). Ethnobotanical observation in the Mornaula Reserve Forest of Kumoun, West Himalaya, India. *Ethnobotanical Leaflets*, *14*, 193–217.

Pusalkar, P. K., & Srivastava, S. K. (2018). *Flora of Uttarakhand* (Vol. 1). Kolkata, India: Botanical Survey of India.

Singh, P., & Attri, B. L. (2014). Survey on traditional uses of medicinal plants of Bageshwar Valley (Kumaun Himalaya) of Uttarakhand, India. *International Journal of Conservation Science*, *5*(2), 223–234.

Tiwari, J. K., Dangwal, L. R., Rana, C. S., Tiwari, P., & Ballabha, R. (2010). Indigenous uses of plant species in Nanda Devi Biosphere Reserve, Uttarakhand, India. *Rep. Opin*, *2*, 67–70.

Uniyal, S. K., Singh, K. N., Jamwal, P., & Lal, B. (2006). Traditional use of medicinal plants among the tribal communities of ChhotaBhangal, Western Himalaya. *Journal of Ethnobiology and Ethnomedicine*, *2*(1), 14. doi:10.1186/1746-4269-2-14 PMID:16545146

Upreti, K., Jalal, J. S., Tewari, L. M., Joshi, G. C., Pangtey, Y. P. S., & Tewari, G. (2009). Ethnomedicinal uses of pteridophytes of Kumaun Himalaya, Uttarakhand, India. *The Journal of American Science*, *5*, 167–170.

Upreti, K., Jalal, J. S., Tiwari, L. M., Joshi, G. C., Pangtey, Y. P. S., & Tiwari, G. (2009). Ethnomedicinal uses of pteridophytes of Kumaun Himalaya, Uttarakhand, India. *The Journal of American Science*, *5*(4), 167–170.

Chapter 15
Plant Species Used by Locals as Ethno-Medicine in Gulmarg Wildlife Sanctuary, Kashmir Himalaya (India)

Zishan Ahmad Wani
Baba Ghulam Shah Badshah University, Rajouri, India

Shreekar Pant
Baba Ghulam Shah Badshah University, Rajouri, India

ABSTRACT

Villages in the Indian Himalayan Region have a rich repository of the indigenous knowledge and practices. Documentation of such knowledge is urgently required in view of the advent of modernization. Therefore, an attempt is made to document the indigenous uses and practices of the plants utilized by the local inhabitants of the Gulmarg Wildlife Sanctuary, Jammu and Kashmir., India. A total of 48 plant species belonging to 46 genera and 25 families are used traditionally to cure various diseases. Various plant parts of these species are used to cure cold, cough, fever, liver disorder, kidney stones, skin diseases, and eye and ear complaints. Such kinds of studies will help in developing a comprehensive database of the plants used traditionally, strengthening the health care system in the villages, and conserving traditional knowledge for the future.

INTRODUCTION

The discovery of plants for medicinal purposes has a long tradition extending back to old civilizations (Noman, 2003). Numerous wild and cultivated plants play a vital role in the culture, customs and traditional health care system (with no or few side effects) of tribal communities and this interrelationship has evolved over generations of experience and practice (Hassan. Ahmad and Mohi-ud-din, 2013). This is the reason that people in developing countries where many resources are not available, especially for

DOI: 10.4018/978-1-7998-1320-0.ch015

health care (Bangladesh, Nepal, Sri Lanka, Indonesia, and India), have strong convictions towards this system (Malik, Siddique, Sofi and Butola, 2011). A total of 7,500 plant species are estimated to be used by 4,635 ethnic communities for human and animal health care across the country (Baba et al., 2012). As they have no side effects. now they have become valuable in the development of healthcare and conservation programs in different parts of the world (Balick, 1996). Further, documentation of such knowledge provides clues for the identification of novel drug discovery (Cox, 2000). However, while great deal of information about the traditional uses and practices is still intact with tribal, the native healers are often reluctant to share their knowledge with outsiders (Lone, 2003; Rajoriya, Choudhary, Shah, Rawat and Jat, 2016) and solely transmit it orally from one generation to another. Due to this fact, such knowledge is disappearing from the society.

India occupies a premier position in the uses of herbal drugs and nearly 2,500 plant species are used in botanically different medicinal health care formulations. The estimated number of medicinal drug manufacturing units in India is over 7,800 which consume about 20,000 tons of herbs annually (Ramakrishnappa, 2002). Ayurveda, Unani,and other medicinal systems are the classical examples of such information (Fabricant & Farnsworth, 2001).

In Himalaya, approximately 8,000 species of flowering plants occur, out of which 1748 plants are mostly used for ethnomedicinal purpose for curing different disease ((Adhikari, Babu, Saklani, & Rawat,2010; Akash & Navneet, 2018; Akash, Navneet & Bhandari, 2018 a,b,c; Akash, Navneet & Bhandari, 2019; Akash, Navneet, Bhandari & Bijalwan, 2019; Wani, Kumar & Akash, 2015, 2016). At the same time, this region is severely affected by various anthropogenic pressures like grazing, trampling, scraping etc. which causes severe threat to these medicinal plants (Akash & Navneet, 2019).

The Kashmir Himalaya, often referred as *Terrestrial Paradise on Earth*, is located at the northwestern tip (Mittermeier, Gil, Hoffman, Pilgrim, Brooks, Mittermeier, Lamoreux and Da Fonseca, 2005), and supports a rich and spectacular biodiversity of great scientific curiosity and significance (Hussain, 2001). It alone contributes nearly 20% of the total Himalayan plant wealth within just 2.15% (15,948 km^2) of the total land area (Dar, Bhagat and Khan, 2002). Several ethnobotanical studies have been carried out throughout the Kashmir Himalaya like Ara and Naqshi, (1992); Bhat, Nigam and Majaz, (2012a, 2012b); Bhat, Mahajan, Sayyed and Bhat, (2014); Chak, Agarwal and Kak, (2009); Dad and Khan, (2010); Jan and Khare, (2013); Kapahi, Srivastava and Sarin, (1993); Kaul, (2010); Khan, (2016); Khan, Khuroo and Dar,(2004); Khuroo, Rashid, Reshi, Dar and Wafai, (2007); Lone, (2003); Lone and Bhardwaj, (2013a, 2013b); Lone and Pandit (2007); Lone, Bhardwaj, Shah and Tabassum (2014); Mala, Lone, Lone and Arya, (2012) Mir, (2014a, 2014b, 2014c, 2014d); Mir and John, (2014); Navchoo and Bhat, (1994); Shapoo, Kaloo, Ganie and Singh, (2013); Singh, (1995); Singh and Bedi, (2017); Wagay, (2014) etc. But a very few studies have been conducted on different aspects of traditional uses of the plants present in the Gulmarg Wildlife Sanctuary (GWLS). Therefore, the present study has been conducted to document the indigenous knowledge and practices of the species used to cure various diseases.

STUDY AREA

The Gulmarg Wildlife Sanctuary (GWLS) has a rich heritage of traditional medicinal plants and is home to ethnic groups like Gujjars and Bakerwals, which have a rich heritage of traditional knowledge especially related to medicinal plants. It falls 26 km to the southwest of District Baramulla, stretching between 74°.17’ to 74°.79’ N latitude and 34°.55’ to 34°.60’ E longitude, at an altitude of 2400-4300m amsl, and is divided into 20 compartments, numbered from 31 to 41 and 50 to 58 (Fig. 1). In 1981, it was proposed to be notified as a Biosphere Reserve, but in 1987, it was declared as a Wildlife Sanctuary (*Notified vide S.R.O.147, Dated: 14-03-1987*). Gulmarg Wildlife Sanctuary covers an area of 180 km^2 including the top catchment of Ferozpur Nallah and forests surrounding the Gulmarg Bowl. Further, the area is surrounded in the north by the Jhelum Valley Forest Division-Baramulla, and in the south by the Forest Division of Poonch and Pir-Panjal. The east is flanked by the village of Drang and Badrakoot forests of Special Forest Division – Tangmarg and on the west by the Special Forest Division Tangmarg and Baba Reshi. Local inhabitants of the GWLS depend on cattle rearing and collecting non-timber forest produce from the surrounding areas for their livelihood.

The GWLS is a paradise for various ethnic groups like the Gujjars and Bakerwal, which have a rich heritage of traditional knowledge, particularly regarding the use of plants for human and animal health care, and they transmitted this knowledge orally from the times immemorial. But due to lack of awareness, industrialization, and urbanization, such valuable knowledge is now limited to elderly persons only. Therefore, it is the need of the hour to document this knowledge well before it is gone.

MATERIALS AND METHODS

This study is based on the extensive and intensive survey conducted in the GWLS between 2018 and 2019. The information regarding the ethnomedicinal plants used for curing various diseases and ailments was collected from the local communities through personal interviews and discussions. Semi-structured, open ended questionnaires were used for the collection of data. *Hakeems* (local herbal healers) were also consulted regarding the traditional and indigenous practices. The fresh samples of the species were collected and brought to the department for identification, with the help of research papers and local floras (Bhat, 1984; Javeid, 1968; Navchoo & Kachroo, 1995; Singh & Kachroo, 1994). Herbarium was deposited in the Department of Botany, BGSB University, Rajouri. The nativity of the plant species was identified following the methods of Samant, Dhar and Palni, (1998) and Khuroo et al, (2007).

RESULTS AND DISCUSSION

The study recorded 48 plant species belonging to 46 genera and 25 families (Fig. 2) distributed within different life forms i.e., trees (5 species), shrubs (4 species), herbs (37 species) and climbers (2 species) {(Fig. 3)}, ranging from 1600 to 4000m amsl altitude. These species are used by the locals for curing various diseases like gastric disorders, skin diseases, toothache, respiratory diseases, etc., {(Fig. 5)}. The utilization pattern of the species indicated that fruits, seeds, roots, leaves, flowers, etc. are mainly used (Fig. 6). Further, an altitudinal wise analysis of the species has been presented (Fig. 4). The indigenous uses and practices of the species used to cure various diseases have been presented in Table-1. Out of the total collected species, only 11 species are native to the Himalayan Region, while 37 species are non-native {(Fig. 7)}. The non-native species have been introduced for different purposes, like food (4 species), horticulture (2 species), medicinal (4 species), landscaping (1 species), fodder (2 species), and some have been introduced unintentionally (24 species) (Fig. 8). Further, out of the 48 species, 22 species have been identified as threatened species, (2 critically endangered, 5 endangered and 15 vul-

Figure 1. Map of the study area

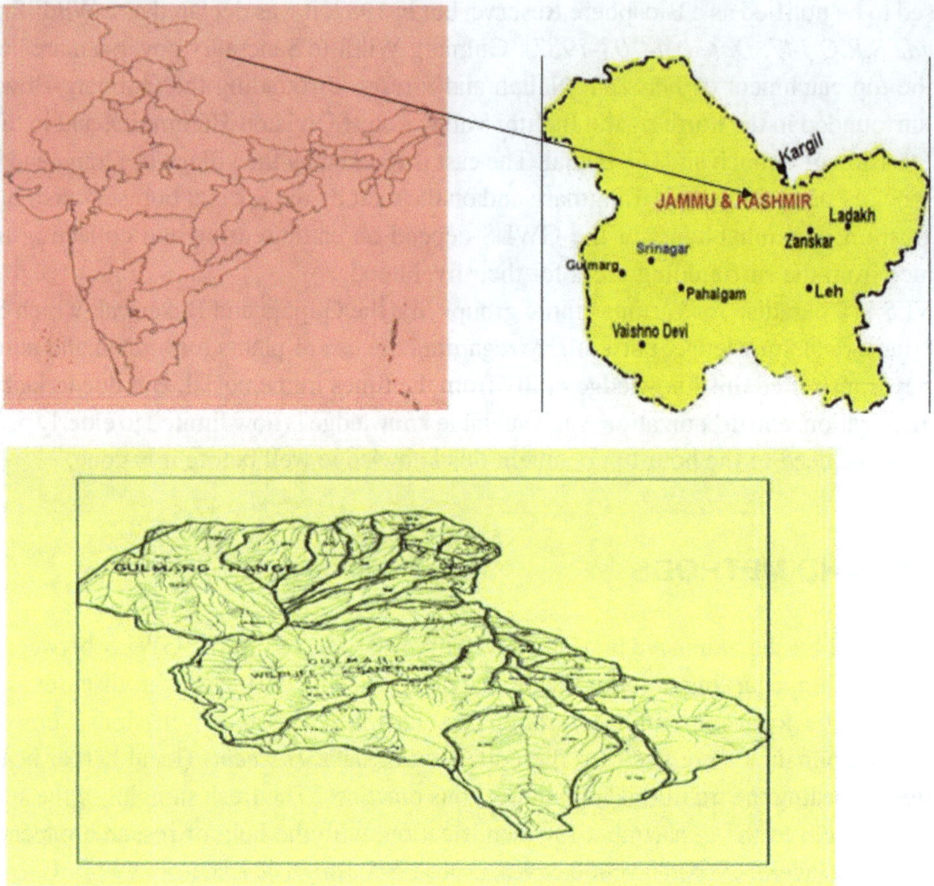

nerable) {(Fig. 9)}. An age-wise representation of respondents revealed that out of 220 respondents, 85 were male and 135 were female. A graphical representation of the same has been presented (Fig.10).

This study provides comprehensive information on the indigenous uses and traditional practices of the plants used to cure various diseases. Herbal medicines are accounted as safe, cheap and easily accessible health care approaches; that is why, about 80% of the world population depends on the traditional system of healthcare and relies on the use of traditional medicine of plant origin (Ahmad, 1999; Rai, et al., 2000). These days such practices are getting great attention due to the concept of herbal medicine, as they have no side effects, and, further documentation of traditional knowledge has gained a wide recognition due to an escalating faith in herbal medicine (Kaul, 1995). Therefore, there is an urgent need to document such kinds of information, as it is transmitted orally from one generation to others since time immemorial. Further, such information for other parts of Himalaya should also be documented so that a comprehensive database of plants used in indigenous practices could be made, otherwise this valuable treasure will disappear due to the advent of modernization.

Figure 2. Familywise distribution of plant species in GWLS

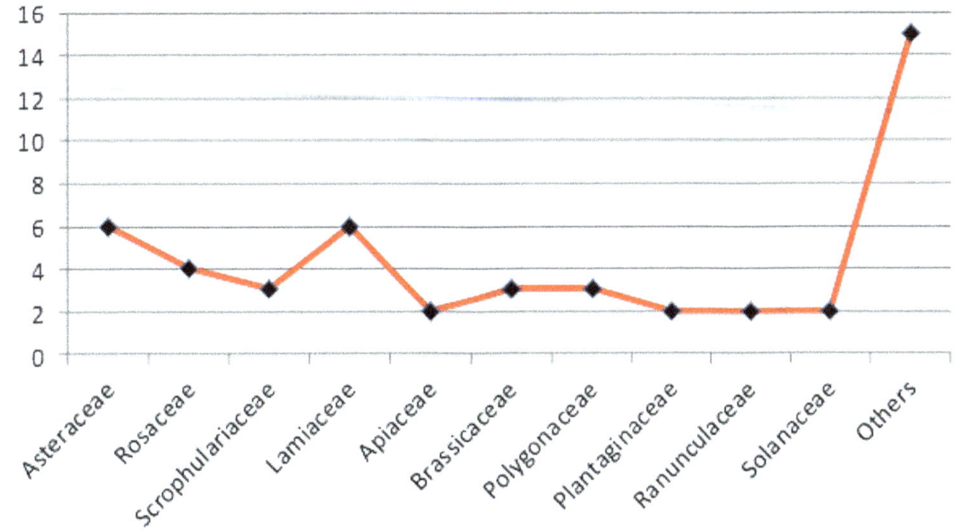

Figure 3. Life form of the plant species

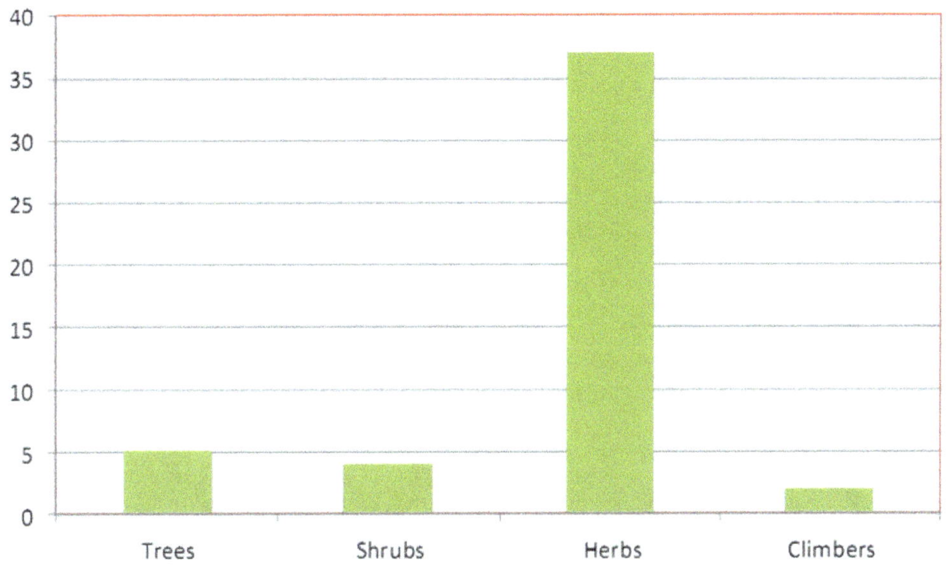

Figure 4. Altitudinal distribution of plants in GWLS

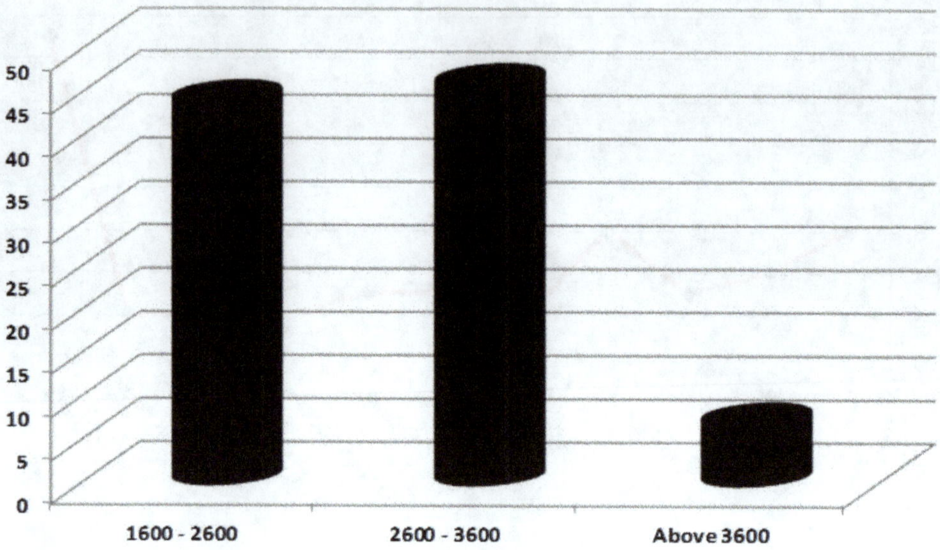

Figure 5. Disease wise utilization pattern of the species

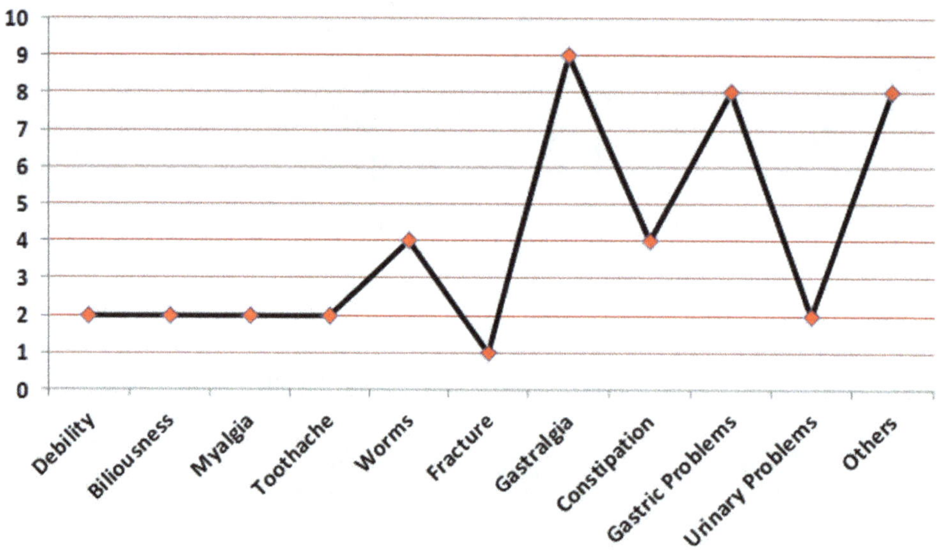

Figure 6. Utilization pattern of plant parts used in GWLS

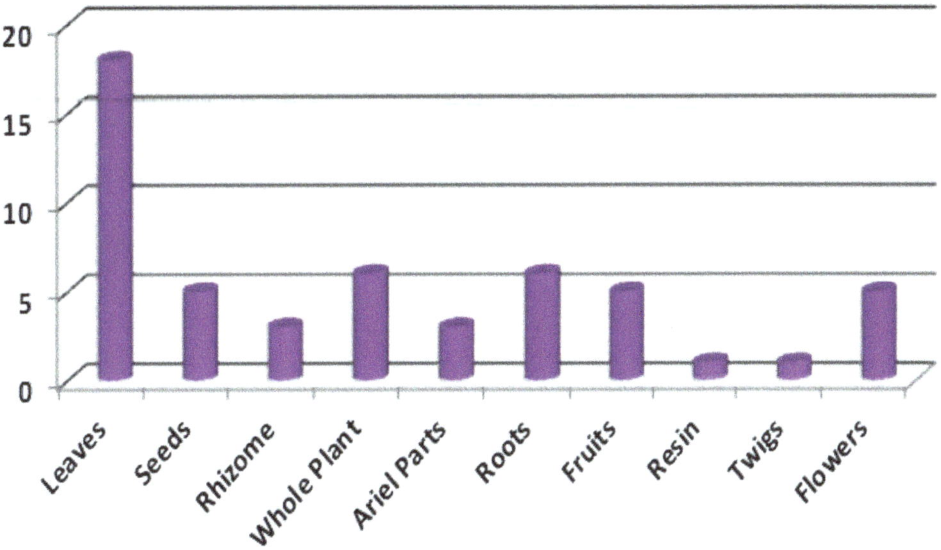

Figure 7. Native and non-nativeness in GWLS

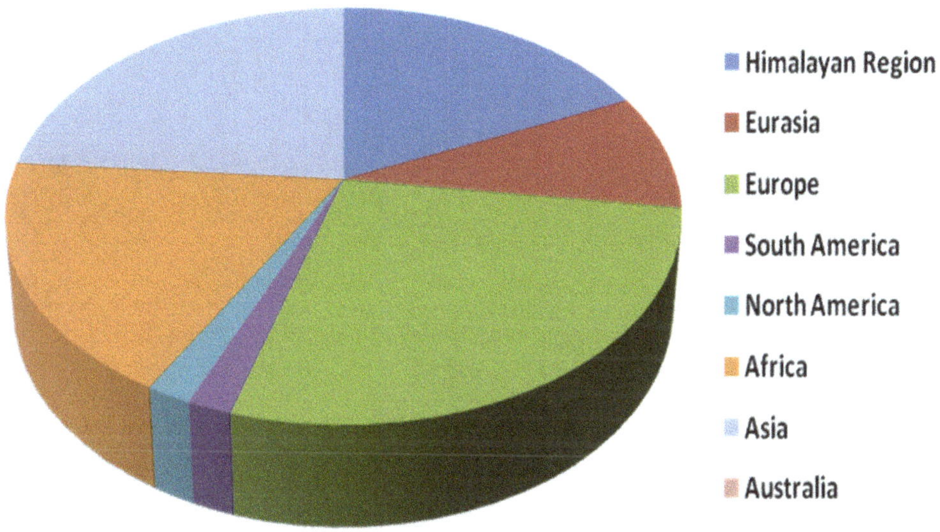

Figure 8. Mode of introduction of non-native species

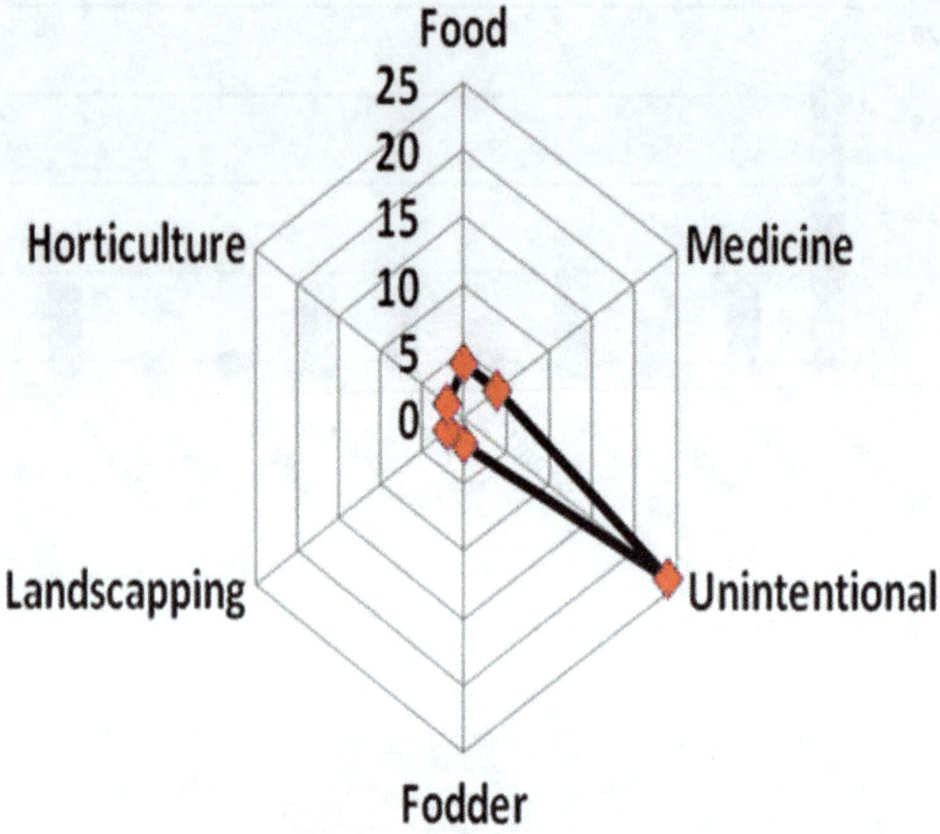

Figure 9. Diversity of threatened plant species in GWLS

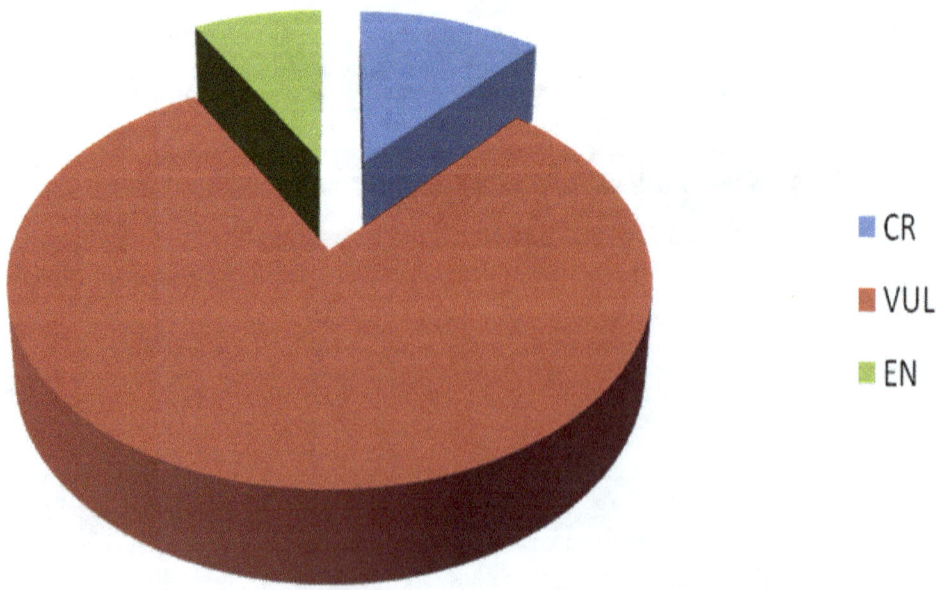

Figure 10. Graph showing gender-wise age structure of the participants

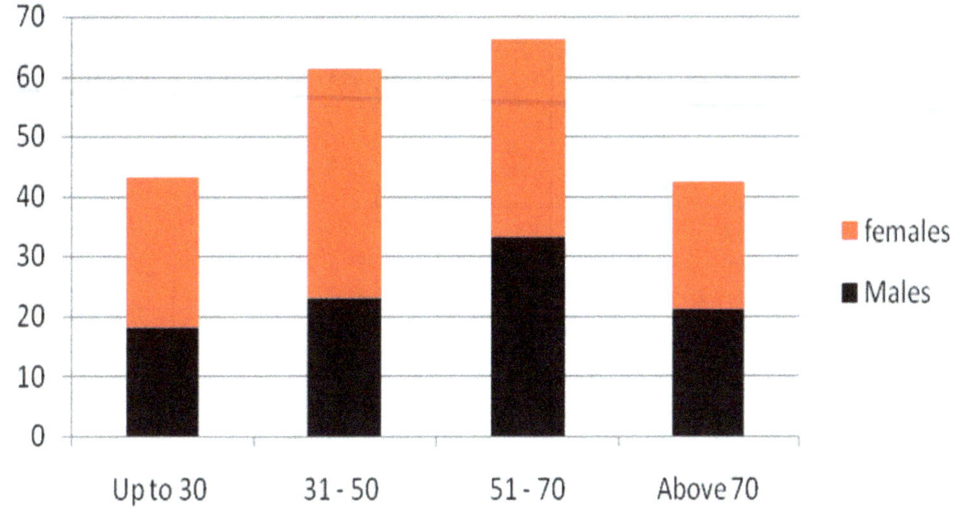

Table 1. Diversity, distribution, and indigenous practices in the GWLS

S. No	Taxa	Family	Local Name	Habit	Range of Altitude	Nativity	Part(s) used	Disease/ Ailment	Mode of Administration
01	*Angelica glauca* Edgew. **(CR)**	Apiaceae	Chora	H	2000 - 3800	Himalayan Region	L	Weakness / Debility	Curry prepared from the leaves is used to overcome the weakness after childbirth in women
02	*Foeniculum vulgare (L.)*Mill.**(1)**	Apiaceae	Badiyaan	H	1600 - 2400	Europe	S	Indigestion / Biliousness	Seeds are eaten raw to cure indigestion
03	*Arisaema jacquemontii* Blume **(VUL)**	Araceae	Hapat makai	H	1800 – 2400	Himalayan Region	Rh	Muscular pain/ Myalgia	Sap obtained by crushing the rhizome is used to massage the body to relieve muscular pain
04	*Achillea millefolium* L.**(2)** **(EN)**	Asteraceae	Pahel geseh	H	2200 - 3100	Asia	L	Toothache	Chewing of fresh leaves is a good remedy against toothache
05	*Artemisia absinthium* L. **(2)** **(EN)**	Asteraceae	Tethwan	H	1800 - 2800	Europe	WP	Worms	Juice obtained from the plant is taken orally to treat worms
07	*Conyza canadensis* (L.) Cronq. **(3)**	Asteraceae	Shallut	H	1600 - 2500	North America	L	Gases	Fresh leaves are carminative so are eaten raw to cure stomach gases
07	*Cotula anthemoides* (L.) **(3) (VUL)**	Asteraceae	Thul bobul	H	1800 - 2500	Asia Africa	AP	Fracture	Paste of the plant is tied on the fractured bone with a cloth for a week
08	*Inula racemose* Hook. f. **(3) (VUL)**	Asteraceae	Poshkar	H	2900 - 3200	Asia	L	Stomachache / Gastralgia	To cure abdominal pain, extract of the leaves is taken orally
09	*Taraxacum officinale* Wigg. **(3) (VUL)**	Asteraceae	Maidan Haend	H	1600 - 2500	Asia Australia	L	Weakness / Debility	Curry prepared from the leaves is used to overcome the weakness after childbirth in women
10	*Berberis lyceum* Royle	Berberidaceae	Kaw daech	S	1700 - 2600	Himalayan Region	Fr	Constipation	Fruits are eaten raw.

continued on following page

Table 1. Continued

S. No	Taxa	Family	Local Name	Habit	Range of Altitude	Nativity	Part(s) used	Disease/ Ailment	Mode of Administration
11	*Capsella bursa pastoris* (L.) Medic. **(3)**	Brassicaceae	Kral mound	H	1800 - 3000	Asia Africa	L	Intestinal infection	Fresh leaves are eaten raw to cure intestinal infection.
12	*Descurainia sophia* (L.) Webb. ex. Prantl. **(3)**	Brassicaceae	Khaempa	H	1700 - 2900	Africa	S	Stomachache / Gastralgia	Seeds are crushed, mixed with water, and are taken orally to cure abdominal pain
13	*Nasturtium officinale* (L.) R. Br. **(3)**	Brassicaceae	Kul Haekh	H	1600 - 2100	Asia	L	Stomachache / Gastralgia	Juice is extracted from the leaves, which is taken orally to cure stomachache
14	*Lychnis coronaria* (L.) Desr. **(3)** **(VUL)**	Caryophylaceae	Shosh Gaseh	H	1600 – 2700	Europe	F	Respiratory problems	Extract of flowers is taken orally
15	*Colchicum luteum* Baker **(VUL)**	Colchicaceae	Virkim Posh	H	1900 - 2700	Himalayan Region	WP	Wounds	Paste of the plant is applied on wounds
16	*Solena amplexicaulis* (Lam.) Gandhi **(3)**	Cucurbitaceae	Khakri	C	1600 - 1900	Asia Australia	Rh	Diarrhoea	The rhizome is boiled in water and the water is used to cure diarrhoea
17	*Dioscorea deltoidea* Wall. ex Kunth**(2)** **(VUL)**	Dioscoriaceae	Kreath	C	2000 - 2800	Asia	R	Respiratory problem	Extract of root is taken orally
18	*Indigofera hetrrantha* Wall. ex. Brandis	Fabaceae	Zand	S	1700 - 2600	Himalayan Region	Rh	Gastric problems	The rhizome of the plant are chewed to cure gastric disorders
19	*Geranium wallichianum* D. Don ex Sweet **(3)** **(EN)**	Geraniaceae	Kaw ashud	H	2000 - 3200	Africa & Europe	R	Gastric problems	The roots of the plant are chewed to cure gastric disorders
20	*Juglans regia* L. **(VUL)**	Juglandaceae	Doon	T	1600 - 1800	Himalayan Region	L	Worms	Leaves are boiled in water and the water is taken orally to cure worms
21	*Ajuga bracteosa* Wall. **(3)** **(VUL)**	Lamiaceae	Jaan e adam	H	1600 - 2800	Asia - Africa	L	Hair care	Leaves are boiled in water and the water is used to cure dandruff and lice
22	*Lamium album* L. **(3)** **(VUL)**	Lamiaceae	Zakhmi Dawah	H	1800 - 2800	Europe	L	Wounds	Paste of leaves is applied on wounds for quick healing
23	*Mentha arvensis* L. **(1)**	Lamiaceae	Pudni	H	1600 - 2600	Europe Africa	L	Biliousness	To cure indigestion, leaves are eaten raw or mixed with milk or curd
24	*Nepata cetaria* L. **(3)** **(VUL)**	Lamiaceae	Ghandi soi	H	2500 - 3200	Eurasia	L	Nausea	Fresh leaves are chewed in case of nausea
25	*Prunella vulgaris* L. **(2)**	Lamiaceae	Kalveuth	H	1600 - 2600	Europe Africa	F	Headache/ Cephalalgia	Paste of flower tops is used to massage on forehead to cure headache
26	*Thymus serphyllum* L.**(4)** **(EN)**	Lamiaceae	Javend	S	2000 - 3400	Europe	S	Worms	Seeds in the form of powder are used as antihelminthic
27	*Lavatera kashmeriana* Malt.	Malvaceae	Saz posh	H	1800 - 2500	Himalayan Region	F	Throat infection	A paste is prepared from the flowers, which is then tied on the throat with a cloth to cure infection of throat
28	*Oxalis corniculata* (L.) Kuntze **(3)**	Oxalidaceae	Amrul	H	1600 - 2500	Eurasia	L	Stomachache / Gastralgia	To cure abdominal pain, fresh leaves are eaten raw or in the form of decoction

continued on following page

Table 1. Continued

S. No	Taxa	Family	Local Name	Habit	Range of Altitude	Nativity	Part(s) used	Disease/ Ailment	Mode of Administration
29	*Pinus walliachiana* Jackson **(VUL)**	Pinnaceae	Kayur	T	2300 - 2800	Himalayan Region	Rn	Cracks	Resin obtained from the stem of the plant is applied on the heels.
30	*Plantago lanceolate* L. **(3)**	Plantaginaceae	Gull	H	1600 - 2700	Africa Europe	AP	Kidney problems	Cooked plant acts as diuretic and stimulates flow of urine, this maintains kidney function.
31	*Plantago major* L. **(3)**	Plantaginaceae	Bud gul	H	1600 - 2800	Eurasia	WP	Gastritis	To cure acidity, decoction of the plant is taken orally
32	*Zea mays* L. **(1)**	Poaceae	Makai	H	1600 - 2600	South America	F	Kidney stones/ Nephrplithiasis	A soup is prepared from the styles of the flowers, which is taken early in the morning for a month to get relief from kidney stones
33	*Sinopodophyllum hexandrum* (Royle) Ying **(EN)**	Polygonaceae	Wan wangun	H	2800 – 4000	Himalayan Region	Fr	Gastritis	Fruits are eaten to cure burning sensation of stomach
34	*Polygonum hydropiper* L. **(1)**	Polygonaceae	Choki chin	H	1600 – 3000	Europe	L	Gastritis	To cure acidity, decoction prepared from the leaves is taken orally
35	*Rumex acetosa* L. **(3)**	Polygonaceae	Abej	H	1700 - 2500	Eurasia	AP	Stomachache / Gastralgia	The ariel portion is cooked and is taken to cure stomachache
36	*Aconitum heterophyllum* Wall. ex Royle **(3) (CR)**	Ranunculaceae	Patris	H	2400 – 3700	Asia	R	Stomachache / Gastralgia	Extract of the roots are taken orally to cure abdominal pain
37	*Ranunculus arvensis* L. **(3)**	Ranunculaceae	–	H	1600 – 3200	Africa Europe	WP	Constipation	The plant is cooked and is taken orally to cure constipation.
38	*Crataegus songarica* Koch. **(5)**	Rosaceae	Ring kul	S	1600 - 1800	Europe Africa	F	Hypertension	Fruits are eaten to cure hypertension
39	*Geum elatum* Wall. ex G. Don **(VUL)**	Rosaceae	Shoonkar	H	2300 - 3600	Himalayan Region	R	Constipation	Roots are crushed and the paste is taken orally to cure constipation
40	*Prunus armeniaca* L. **(6)**	Rosaceae	Chaire	T	1600 - 2300	Asia	S	Constipation	Seed powder mixed with milk is taken before going to bed to cure constipation.
41	*Prunus persica* (L.) Batson *(6)*	Rosaceae	Chinun	T	1600 - 2300	Asia	L	Stomachache / Gastralgia	To cure abdominal pain, decoction of leaves is taken
42	*Salix acmophylla* Boiss. **(4)**	Salicaceae	Krill vir	T	1700 - 2200	Asia	Tw	Gastric problems	The twigs of the plant are chewed to cure gastric problems
43	*Bergenia ligulata* Engl.	Saxifragaceae	Pashanbeda	H	1800 – 3800	Himalaya Region	R	Gastric problems	Roots are crushed, mixed with water, and is taken orally to cure gastric disorders
44	*Verbascum thapsus* L. **(3)**	Scrophulariaceae	Wan tamook	H	1600 - 2800	Europe	R	Inflammation	Paste prepared by crushing the roots of the plant is applied to reduce the swelling due to fracture

continued on following page

Table 1. Continued

S. No	Taxa	Family	Local Name	Habit	Range of Altitude	Nativity	Part(s) used	Disease/ Ailment	Mode of Administration
45	*Verbena officinalis* L. (3)	Scrophulariaceae	Hutmool	H	1700 - 2300	Europe	WP	Worms	The plant is crushed and then mixed with hot water and is then taken orally to cure abdominal worms.
46	*Veronica persica* Poir. (3)	Scrophulariaceae	Pout keach	H	1600 - 3100	Asia	WP	Indigestion	To cure indigestion, extract of the plant is taken orally
47	*Hyoscyamus niger* L. (3) (VUL)	Solanaceae	Bazaar Bhang	H	1800 - 2900	Africa Europe	S	Toothache	Seeds are burned and the ash is kept in the mouth for some time to cure toothache
48	*Solanum nigrum* L. (3)	Solanaceae	Kambai	H	1700 - 2400	Eurasia	L	Stomachache / Gastralgia	To cure stomachache, decoction of leaves is taken orally

Abbreviations Used: Status - CR = Critically Endangered, E = Endangered and VUL = Vulnerable; Mode of introduction - 1 = Food, 2 = Medicinal, 3 = Unintentional, 4 = Fodder, 5 = Landscaping and 6 = Horticulture

REFERENCES

Adhikari, B. S., Babu, M. M., Saklani, P. L., & Rawat, G. S. (2010). Medicinal plants diversity and their conservation status in Wildlife Institute of India (WII) campus, Dehradun. *Ethnobotanical Leaflets, 1*, 69.

Akash, & Navneet. (2018). Eco-tourism as a viable option for conservation of wildlife in protected areas under Shivalik hills of the outer Himalaya, India. In R. Sharma, & R. Prakash, (Eds.), *Environmental Impact of Tourism in Developing Nations*. (1-348). IGI Global.

Akash, & Navneet. (2019). Impact of climatic changes on earth's survivals- an analysis of changing environment and its consequences at global level. In A. Karmaoui (Ed.), *Climate Change and Its Impact on Ecosystem Services and Biodiversity in Arid and Semi-Arid Zones* (1-408). IGI Global.

Akash, Navneet, & Bhandari, B. S. (2018a). Phytosociological studies, biodiversity conservation in a subtropical moist deciduous forest of Rajaji tiger reserve. Uttarakhand, India. *International Journal of Research and Analytical Review, 5*(3), 39-51.

Akash, Navneet, & Bhandari, B. S. (2018b). Phytosociological investigation, biodiversity conservation and life form pattern in a *Holeptelea integrifolia* community under Rajaji tiger reserve, Uttarakhand, India. *International Research Journal of Biological Science, 7*(7), 1-8.

Akash, Navneet, & Bhandari, B. S. (2018c). Tree diversity, stand structure, and community composition in tropical forest of Rajaji tiger reserve, Northern India. *Journal of Applied and Natural Sciences, 10*(3), 945–953.

Akash, Navneet, & Bhandari, B. S. (2019). A community analysis of woody species in a tropical forest of Rajaji tiger reserve. *Environment and Ecology, 37*(1), 48-55.

Akash, Navneet, Bhandari, B. S., & Bijalwan, K. (2019). Phytodiversity, stand structure, and biodiversity conservation in a tropical forest community under Rajaji tiger reserve, Uttarakhand, India. *ENVIS Bulletin Himalayan Ecology*, 26, 7-16.

Ara, S., & Naqshi, A. R. (1992). Ethnobotanical studies in Gureiz Valley. *Journal of Economic and Taxonomic Botany*, *17*(3).

Baba, I. A., Dubey, S., Alia, A., Saxena, R. C., Itoo, A. A., & Powar, A. (2012). Ethnobotanical survey of medicinal plants used by the people of District Ganderbal Jammu and Kashmir. *Research Journal of Pharmaceutical, Biological, and Chemical Sciences*, *3*, 549–556.

Balick, M. J. (1996). Transforming ethnobotany for the new millennium. *Annals of the Missouri Botanical Garden*, *83*(1), 58–66. doi:10.2307/2399968

Bhat, F., Mahajan, D. M., Sayyed, M. R. G., & Bhat, A. (2014). Ethno-medicinal survey of North Kashmir Himalaya – A case study of Lolab valley (J&K), India, *Ecology Environmental Conservation*, *20*(1), 59–71.

Bhat, N. A. (1984). Floristic Composition of Gulmarg (Baramulla) (Unpublished doctoral dissertation). University of Kashmir, Srinagar.

Bhat, T. A., Nigam, G., & Majaz, M. (2012a). Traditional use of medicinal plants by Gujjar and Bakarwal tribes in Pir Panjal Range of the Shopion District, Kashmir. *International Journal of Research and Development in Pharmacy and Life Sciences*, *1*, 160–166.

Bhat, T. A., Nigam, G., & Majaz, M. (2012b). Study of some medicinal plants of the Shopion district, Kashmir (India) with emphasis on their traditional use by Gujjar and Bakerwal tribes. *Asian Journal of Pharmaceutical and Clinical Research*, *5*(2), 94–98.

Chak, I., Agarwal, R. K., & Kak, A. M. (2009). Ethno medicinal study of some important plants used in the treatment of hair and boils in district Pulwama of Kashmir. *Annals of Forestry (Dehra Dun)*, *17*(1), 101–107.

Cox, P. A. (2000). Will tribal knowledge survive the millennium? *Science*, *287*(5450), 44–45. doi:10.1126cience.287.5450.44 PMID:10644221

Dad, J. M., & Khan, A. B. (2011). Threatened medicinal plants of Gurez valley, Kashmir Himalayas: Distribution pattern and current conservation status. *The International Journal of Biodiversity Science, Ecosystem Services & Management*, *7*(1), 20–26. doi:10.1080/21513732.2011.602646

Dar, G. H., Bhagat, R. C., & Khan, M. A. (2002). *Biodiversity of Kashmir Himalaya*. Srinagar, India: Valley Book House.

Fabricant, D. S., & Farnsworth, N. R. (2001). The value of plants used in traditional medicine for drug discovery. *Environmental Health Perspectives*, *109*, 69–75. PMID:11250806

Hassan, G. A., Ahmad, T. A., & Mohi-ud-din, R. A. (2013). An Ethnobotanical study in Budgam of Kashmir Valley: An attempt to explore and document traditional knowledge of the area. *International Research Journal of Pharmacy*, *4*(1), 201–204.

Hussain, M. (2001). *Geography of Jammu and Kashmir*. New Delhi, India: Rajesh Publications.

Jan, R. A., & Khare, N. (2013). Ethnopharmacological uses of plants among tribal and rural folks of Shopion Forest Area of Kashmir. *International Journal of Science and Research*, *4*(5), 232–234.

Javeid, G. N. (1968). Flora of Srinagar, Kashmir. *Science*, *5*, 59–71.

Kapahi, B. K., Srivastava, T. N., & Sarin, Y. K. (1993). Traditional medicinal of Gurez (Kashmir) – An ethnobotanical study. *Ancient Science of Life*, 119–124. PMID:22556637

Kaul, M. K. (2010). High altitude botanicals in integrative medicine – Case study from Northwest Himalaya. *Indian Journal of Traditional Knowledge*, *9*(1), 18–25.

Khan, S., Kamili, A., & Gupta, R. C. (2016). Economic and medicinal properties of some medicinal plants found in Kashmir Himalaya. *Journal of Medicinal Plant Studies*, *4*(3), 38–44.

Khan, Z. S., Khuroo, A. A., & Dar, G. H. (2004). Ethnomedicinal survey of Uri, Kashmir Himalaya. *Indian Journal of Traditional Knowledge*, *3*(4), 351–357.

Khuroo, A. A., Malik, A. H., Dar, A. R., Dar, G. H., & Khan, Z. S. (2007). Ethno veterinary medicinal uses of some plant species by the Gujjar tribe of Kashmir Himalayas. *Asian Journal of Plant Sciences*, *6*(1), 148–152. doi:10.3923/ajps.2007.148.152

Khuroo, A. A., Rashid, I., Reshi, Z., Dar, G. H., & Wafai, B. A. (2007). The Alien Flora of Kashmir Himalaya. *Biological Invasions*, *9*(3), 262–292. doi:10.100710530-006-9032-6

Lone, A. H., & Pandit, A. K. (2007). Medicinal plant wealth of Langate Forest Division in Kashmir Himalayas. *Journal of Himalayan Ecology and Sustainable Development*, *2*, 73–78.

Lone, F. A. (2003). Folklore medicinal system of Uri sector Kashmir valley, India, In proceeding of the 2nd world congress on Biotechnology Development of Herbal Medicine, Feb. 20-22, (India). Academic Press.

Lone, P., Bhardwaj, A. K., Shah, K. W., & Tabasum, S. (2014). Ethnobotanical survey of some threatened medicinal plants of Kashmir Himalaya, India. *Journal of Medicinal Plants Research*, *8*(47), 1362–1373.

Lone, P. A., & Bhardwaj, A. K. (2013a). Traditional herbal based disease treatment in some rural areas of Bandipora District of Jammu and Kashmir, India. *Asian Journal of Pharmaceutical and Clinical Research*, *6*(4), 162–171.

Lone, P. A., & Bhardwaj, A. K. (2013b). Ethnomedicinal uses of certain locally available plants of Bandipora district of Jammu & Kashmir, India, *International Journal of Medicinal and Aromatic Plants*, (3), 470–485.

Mala, F. A., Lone, M. A., Lone, F. A., & Arya, N. (2012). Ethno medicinal survey of Kajinaag Range of Kashmir, Himalaya, India. *International Journal of Pharma and Bio Sciences*, (3): 442–449.

Malik, A. R., Siddique, M. A. A., Sofi, P. A., & Butola, J. S. (2011). Ethnomedicinal practices and conservation status of medicinal plants of North Kashmir Himalayas. *Research Journal of Medicinal Plant*, *5*(5), 515–530. doi:10.3923/rjmp.2011.515.530

Mir, G. M., & John, S. A. (2014). Ethno medicinal study of Pulwama Tehsil, Jammu & Kashmir. *Journal of Medicinal Plant Studies*, *2*(4), 5–8.

Mir, M. Y. (2014a). Ethno medicinal survey of plants from Kupwara, J & K, India. *International Journal of Advanced Research*, *2*(1), 846–857.

Mir, M. Y. (2014b). Ethnoveterinary studies in tribals of Kupwara, J & K, India. *International Journal of Multidisciplinary Research and Development*, *1*(6), 92–96.

Mir, M. Y. (2014c). Documentation and ethnobotanical survey of wild edible plants used by the tribals of Kupwara, J & K, India. *International Journal of Herbal Medicine*, *2*(4), 11–18.

Mir, M. Y. (2014d). Indigenous knowledge of using medicinal plants in treating skin disease by tribal's of Kupwara, J & K, India. *International Journal of Herbal Medicine*, *1*(6), 62–68.

Mittermeier, R. A., Gil, P. R., Hoffman, M., Pilgrim, J., Brooks, T., Mittermeier, C. G., ... Da Fonseca, G. A. B. (2005). *Hotspots revised: Earth's biologically richest and most threatened terrestrial ecoregions*. Conservation International.

Navchoo, I. A., & Kachroo, P. (1995). *Flora of Pulwama, Kashmir*. Dehradun: Bishen Singh and Mahendra Pal Singh.

Noman, A. (2003). Influence of different doses of nitrogen fertilizer on ajwain. (Unpublished master's thesis). University of Agriculture. Faisalabad. Pakistan.

Rajoriya, C. M., Choudhary, R. G., Shah, I. A., Rawat, R. S., & Jat, B. L. (2016). Ethno – medicinal survey of North Kashmir with special reference to Bandipora. *International Journal for Research in Applied Science and Engineering Technology*, *4*(11), 95–106.

Ramakrishnappa, K. (2002). *Impact of cultivation and gathering of medicinal plants on Biodiversity: Case study from India. Forestry and Fisheries*. Rome: FAO.

Samant, S. S., Dhar, U., & Palni, L. M. S. (1998). *Medicinal plants of Indian Himalaya: Diversity distribution potential values*. Gyanodaya Prakashan, Nanital.

Shah, A., Karim, A., Ahmad, J., & Sharma, M. P. (2015). Medicinal shrubs used by the Gujjar-Bakerwal tribes against various tribes against various non-communicable diseases in Rajouri district, (J&K), India. *Indian Journal of Traditional Knowledge*, *14*(3), 466–473.

Shapoo, G. A., Kaloo, Z. A., Ganie, A. H., & Singh, S. (2013). Ethnobotanical survey and documentation of some Orichid species of Kashmir Himalaya, J&K, India. *International Journal of Pharmaceutical and Biological Research*, *4*, 32–40.

Singh, B., & Bedi, Y. S. (2017). Eating from raw wild plants from Himalaya: Traditional Knowledge Documentary on *Sheena* Tribe in Kashmir. *Indian Journal of Natural Products and Resources*, *8*(3), 269–275.

Singh, J. B., & Kachroo, P. (1994). Forest flora of Pir Panjal Range (northwestern Himalaya). Dehradun, India. Singh, V. (2002). Herbal remedies in traditional medicines of the local valley in Kashmir Himalayas, India, round progress in medicinal plants. *Ethnomedicine and Pharmacology*, *1*, 63–71.

Wagay, N. A. (2014). Medicinal flora and ethno-botanical knowledge of Baramulla Tehsil in Jammu and Kashmir, India. *International Journal of Advanced Biotechnology and Research*, *5*(3), 539–546.

Wani, Z. A., & Kumar, N., & Akash. (2015). Addition to the ethnomedicinal plants used by the local people of Gulmarg and its allied areas, Kashmir, Jammu and Kashmir, India. *International Journal of Current Research in Bioscience in Plant Biology*, *2*(10), 18–22.

Wani, Z. A., Kumar, N., & Akash. (2016). Ethnobotanical study of some threatened plants in district Baramulla, Kashmir, Jammu and Kashmir, India. *International Journal of Current Research in Bioscience in Plant Biology*, *3*(2), 58–64. doi:10.20546/ijcrbp.2016.302.007

Compilation of References

Abbiw, D. K. (1990). Useful plants of Ghana, West African uses of wild and cultivated plants. England, Ltd., (28-34). Practical Action Publishing.

Abdel-Kader, M. S., Bahler, B. D., Malone, S., Werkhoven, M. C., van Troon, F., D, ... Kingston, D. G. I. (1998). DNA-damaging steroidal alkaloids from *Eclipta alba* from the Suriname rainforest. *Journal of Natural Products*, *61*(10), 1202–1208. doi:10.1021/np970561c PMID:9784152

Abdullah, S., Abidin, S. A. Z., Murad, N. A., Makpol, S., & Ngah, W. Z. W., & Yusof, Y. A. M. (2010). Ginger extract (*Zingiber officinale*) triggers apoptosis and G0/G1 cells arrest in HCT 116 and HT 29 colon cancer cell lines. *African Journal of Biochemistry Research*, *4*(4), 134–142.

Abebe, T., Wiersum, K., & Bongers, F. (2010). Spatial and temporal variation in crop diversity in agroforestry homegardens of southern Ethiopia. *Agroforestry Systems*, *78*(3), 309–322. doi:10.100710457-009-9246-6

Abid, M. J., Kumar, K., Khan, R., Ali, S., Chandra, P., Rani, R. & Khan, N. A. (2016). Assessment of Leucorrhea diseases in female students. *Journal of Scientific and Innovative Research*, *5*(4), 116–118.

Abiraamasri, B. L., & Lakshmi, T. (2016). In vitro anti-arthritic activity of *Hemidesmus indicus* root extract. *International Journal of Pharmaceutical Sciences Review and Research*, *41*(2), 15–17.

Abo El-Maati, M. F., Labib, S. M., Al-Gaby, A. M. A., & Ramadan, M. F. (2012). Antioxidant properties of different extracts from five medicinal plants. *Zagazig Journal of Agricultural Research*, *39*(4), 1–13.

Adhikari, B. S., Babu, M. M., Saklani, P. L., & Rawat, G. S. (2010). Medicinal plants diversity and their conservation status in Wildlife Institute of India (WII) campus, Dehradun. *Ethnobotanical Leaflets*, *1*, 6–9.

Adjanohoun, E. J. (1988). Médecine Traditionnelle et Pharmacopée: Contribution aux études ethnobotaniques et floristiques en République Populaire du Congo. Paris (France Agence de Cooperation Culturelle et Technique. Agarwal, V. S. (1997). Drug plants of India. (1-656). New Delhi, India: Kalyani Publishers.

Agarwal, M., Walia, S., Dhingra, S., & Khambay, B. P. S. (2001). Insect growth inhibition, antifeedant and antifungal activity of compounds isolated/derived from *Zingiber officinale* Roscoe (ginger) rhizomes. *Pest Management Science*, *57*(3), 289–300. doi:10.1002/ps.263 PMID:11455660

Agarwal, P., Fatima, A., & Singh, P. P. (2012). Herbal medicine scenario in India and European countries. *Journal of Pharmacognosy & Phytochemistry*, *1*(4), 88–93.

Agelet, A., Bonet, M. À., & Vallès, J. (2000). Homegardens and their role as a main source of medicinal plants in mountain regions of Catalonia (Iberian Peninsula). *Economic Botany*, *54*(3), 295–309. doi:10.1007/BF02864783

Aggarwal, B. B. (2010). Targeting inflammation-induced obesity and metabolic diseases by curcumin and other nutraceuticals. *Annual Review of Nutrition, 30*(1), 173–199. doi:10.1146/annurev.nutr.012809.104755 PMID:20420526

Agnihotri, R., Sharma, S., Joshi, M., & Palni, L. (2004). Crop diversity in the home gardens of the Kumaun region of central Himalaya, India. *Plant Genetic Resources Newsletter.*

Agnihotri, N., & Gupta, A. K. (2013). Folkloremedicines for cuts and wounds in Kalyanpur block of Kanpur District, Uttar Pradesh, India. *Ph Tech Med, 2*, 381–386.

Agra, M. F., Baracho, G. S., Nurit, K., Basílio, I. J. L. D., & Coelho, V. P. M. (2007). Medicinal and poisonous diversity of the flora of "Cariri Paraibano". *Brazilian Journal of Ethnopharmacology, 111*(2), 383–395. doi:10.1016/j.jep.2006.12.007 PMID:17236731

Ahlawat, K. S., & Khatkar, B. S. (2011). Processing Food Applications and Safety of Aloe vera Products: A Review. *Journal of Food Science and Technology, 48*(5), 525–533. doi:10.100713197-011-0229-z PMID:23572784

Ahluwalia, K. S. (1968). Medicinal plants of Kerala-V. *Nagarjun, 11*(7), 363–369.

Ahmad, M., Rookh, M., Rehman, A. B., Muhammad, N., Younus, M., & Wazir, A. (2014). Assessment of anti-inflammatory, anti-ulcer and neuro-pharmacological activities of *Cyperus rotundus* Linn. *Pakistan Journal of Pharmaceutical Sciences, 27*(6). PMID:26045387

Ahmad, S. S., Khan, S. M., & Butt, A. (2015). Exploring the medicinal plants wealth: An assessment of traditional medico-botanical knowledge of local communities in Changa Manga forest, Pakistan. *Pharmaceutical Chemistry Journal, 2*(3), 24–32.

Ahn, J. H., Lee, T. W., Kim, K. H., Byun, H., Ryu, B., Lee, K. T., ... Choi, J. H. (2015). 6-Acetoxy Cyperene, a Patchoulane-type Sesquiterpene Isolated from *Cyperus rotundus* Rhizomes Induces Caspase-dependent Apoptosis in Human Ovarian Cancer cells. *Phytotherapy Research, 29*(9), 1330–1338. doi:10.1002/ptr.5385 PMID:26062076

Ajith, T. A. (2010). Ameliorating reactive oxygen species-induced in vitro lipid peroxidation in brain, liver, mitochondria and DNA damage by *Zingiber officinale* Roscoe. *Indian Journal of Clinical Biochemistry, 25*(1), 67–73. doi:10.100712291-010-0014-1 PMID:23105887

Akan, H., & Bakır Sade, Y. (2015). Kâhta (adıyaman) merkezi ve narince köyü'nün etnobotanik açıdan araştırılması. *BEÜ Fen Bilimleri Dergisi, 4*(2), 219–224.

Akash, & Navneet. (2018). Eco-tourism as a viable option for conservation of wildlife in protected areas under Shivalik hills of the outer Himalaya, India. In R. Sharma, & R. Prakash, (Eds.), *Environmental Impact of Tourism in Developing Nations.* (1-348). IGI Global.

Akash, & Navneet. (2019). Impact of climatic changes on earth's survivals- an analysis of changing environment and its consequences at global level. In A. Karmaoui (Ed.), *Climate Change and Its Impact on Ecosystem Services and Biodiversity in Arid and Semi-Arid Zones* (1-408). IGI Global.

Akash, Navneet, & Bhandari, B. S. (2018a). Phytosociological studies, biodiversity conservation in a subtropical moist deciduous forest of Rajaji tiger reserve. Uttarakhand, India. *International Journal of Research and Analytical Review, 5*(3), 39-51.

Akash, Navneet, & Bhandari, B. S. (2018b). Phytosociological investigation, biodiversity conservation and life form pattern in a *Holeptelea integrifolia* community under Rajaji tiger reserve, Uttarakhand, India. *International Research Journal of Biological Science, 7*(7), 1-8.

Akash, Navneet, & Bhandari, B. S. (2018c). Tree diversity, stand structure, and community composition in tropical forest of Rajaji tiger reserve, Northern India. *Journal of Applied and Natural Sciences, 10*(3), 945–953.

Akash, Navneet, & Bhandari, B. S. (2019). A community analysis of woody species in a tropical forest of Rajaji tiger reserve. *Environment and Ecology, 37*(1), 48-55.

Akash, Navneet, Bhandari, B. S., & Bijalwan, K. (2019). Phytodiversity, stand structure, and biodiversity conservation in a tropical forest community under Rajaji tiger reserve, Uttarakhand, India. *ENVIS Bulletin Himalayan Ecology*, 26, 7-16.

Akaydin, G., Şimşek, I., Arituluk, Z. C., & Yeşilada, E. (2013). An ethnobotanical survey in selected towns of the Mediterranean subregion (Turkey). *Turkish Journal of Biology, 37*(2), 230–247.

Akgul, A., Akgul, A., Senol, S. G., Yildirim, H., Secmen, O., & Dogan, Y. (2018). An ethnobotanical study in Midyat (Turkey), a city on the silk road where cultures meet. *Journal of Ethnobiology and Ethnomedicine, 14*(12), 1–18. PMID:29415748

Akinyele, B. O., & Odiyi, A. C. (2007). Comparative study of vegetative morphology and the existing taxonomic status of Aloe vera L. *Journal of Plant Sciences, 2*(5), 558–563. doi:10.3923/jps.2007.558.563

Akkol, E. K., Koca, U., Pesin, I., & Yilmazer, D. (2011). Evaluation of the wound healing potential of *Achillea biebersteinii* Afan.(Asteraceae) by in vivo excision and incision models. *Evidence-Based Complementary and Alternative Medicine, 2011*, 1–7. doi:10.1093/ecam/nep039 PMID:19546149

Aktan, T. (2011). *Yenişehir (bursa) köylerinin etnobotanik özellikleri* (Unpublished master's thesis). Celal Bayar Üniversitesi, Turkey.

Alam, M. (2011). Tropical homegardens in Bangladesh: Characteristics and sustainability. In Alternative Farming Systems, Biotechnology, Drought Stress, and Ecological Fertilisation (pp. 245–262). Springer.

Alam, M. I., Auddy, B., & Gomes, A. (1994). Isolation, purification and partial characterization of viper venom inhibiting factor from the root extract of the Indian medicinal plant sarsaparilla (*Hemidesmus indicus* R. Br.). *Toxicon, 32*(12), 1551–1557. doi:10.1016/0041-0101(94)90314-X PMID:7725324

Alam, M. I., & Gomes, A. (1998a). Viper venom-induced inflammation and inhibition of free radical formation by pure compound (2-hydroxy-4-methoxy benzoic acid) isolated and purified from anantamul (*Hemidesmus indicus* R. BR) root extract. *Toxicon, 36*(1), 207–215. doi:10.1016/S0041-0101(97)00070-6 PMID:9604294

Alam, M. I., & Gomes, A. (1998b). Adjuvant effects and antiserum action potentiation by a (herbal) compound 2-hydroxy-4-methoxy benzoic acid isolated from the root extract of the Indian medicinal plant 'sarsaparilla' (*Hemidesmus indicus* R. Br.). *Toxicon, 36*(10), 1423–1431. doi:10.1016/S0041-0101(98)00076-2 PMID:9723840

Albuquerque, U. P., Andrade, L., & Caballero, J. (2005). Structure and floristics of homegardens in northeastern Brazil. *Journal of Arid Environments, 62*(3), 491–506. doi:10.1016/j.jaridenv.2005.01.003

Albuquerque, U. P., Ramos, M. A., de Lucena, R. F. P., & Alencar, N. L. (2014). Methods and techniques used to collect ethnobiological data. In *Methods and techniques in Ethnobiology and Ethnoecology* (pp. 15–37). Springer. doi:10.1007/978-1-4614-8636-7_2

Alcorn, J. B. (1989). Process as resource. *Advances in Economic Botany, 7*, 63–77.

Alessandro, B., & Stefano, S. (2005). *The great Aloe book, history, botany, composition, and pharmacological aspects of this legendary plant*. Trento: Ziccari Pty Ltd.

Alhassan, A. M., & Ahmed, Q. U. (2016). Averrhoa bilimbi Linn.: A review of its ethnomedicinal uses, phytochemistry, and pharmacology. *Journal of Pharmacy & Bioallied Sciences*, *8*(4), 265–271. doi:10.4103/0975-7406.199342 PMID:28216948

Ali, A. M. S. (2005). Homegardens in smallholder farming systems: Examples from Bangladesh. *Human Ecology*, *33*(2), 245–270. doi:10.100710745-005-2434-8

Ali, M., Ravinder, E., & Ramachandran, R. (2001). Anticoagulant activity of *Tridax procumbens*. *Fitoterapia*, *72*, 313–315. doi:10.1016/S0367-326X(00)00296-3 PMID:11295316

Ali, S. I., Gopalakrishnan, B., & Venkatesalu, V. (2017). Pharmacognosy, phytochemistry and pharmacological properties of *Achillea millefolium* L.: A review. *Phytotherapy Research*, *31*(8), 1140–1161. doi:10.1002/ptr.5840 PMID:28618131

Alonso-Castro, A. J., Domínguez, F., Zapata-Morales, J. R., & Carranza-Álvarez, C. (2015). Plants used in the traditional medicine of Mesoamerica (Mexico and Central America) and the Caribbean for the treatment of obesity. *Journal of Ethnopharmacology*, *175*(June 2016), 335–345. . 2015.09.029 doi:10.1016/j.jep

Alsalhi, M. S., Devanesan, S., Alfuraydi, A. A., Vishnubalaji, R., Munusamy, M. A., Murugan, K., ... Benelli, G. (2016). Green synthesis of silver nanoparticles using *Pimpinella anisum* seeds: Antimicrobial activity and cytotoxicity on human neonatal skin stromal cells and colon cancer cells. *International Journal of Nanomedicine*, *11*, 4439–4449. doi:10.2147/IJN.S113193 PMID:27660438

Altundağ, E. (2009). *Iğdır ilinin (doğu anadolu bölgesi) doğal bitkilerinin halk tarafından kullanımı* (Unpublished doctoral dissertation), İstanbul Üniversitesi, Turkey.

Altundağ, E. (2009). *Iğdır ilinin (doğu anadolu bölgesi) doğal bitkilerinin halk tarafından kullanımı* (Unpublished doctoral dissertation). İstanbul Üniversitesi, İstanbul.

Altundag, E., & Ozturk, M. (2011). Ethnomedicinal studies on the plant resources of east Anatolia, Turkey. *Procedia: Social and Behavioral Sciences*, *19*, 756–777. doi:10.1016/j.sbspro.2011.05.195

Amsalu, N., Bezie, Y., Fentahun, M., Alemayehu, A., & Amsalu, G. (2018). Use and conservation of medicinal plants by indigenous people of Gozamin Wereda, East Gojjam Zone of Amhara region, Ethiopia: An ethnobotanical approach. *Evidence-Based Complementary and Alternative Medicine*, *2018*, 1–23. doi:10.1155/2018/2973513 PMID:29743921

Ananthi, R., Chandra, N., & Santhiya, S. T. (2010a). Protective effect of *Hemidesmus indicus* R. Br. root extract against cisplatin-induced cytogenetic damage in mouse bone marrow cells. *Genetics and Molecular Biology*, *33*(1), 182–185. doi:10.1590/S1415-47572010005000011 PMID:21637624

Ananthi, R., Chandra, N., Santhiya, S. T., & Ramesh, A. (2010b). Genotoxic and antigenotoxic effects of *Hemidesmus indicus* R. Br. root extract in cultured lymphocytes. *Journal of Ethnopharmacology*, *127*(2), 558–560. doi:10.1016/j.jep.2009.10.034 PMID:19896526

Andersson, G. B. J. (1997). The epidemiology of spinal disorders. In J. W. Frymoyer (Ed.), *The Adult Spine: Principles and Practice* (2nd ed., pp. 93–141). New York, NY: Raven Press.

Andrade, J. M., Faustino, C., García, C., Ladeiras, D., Reis, C. P., & Rijo, P. (2018). *Rosmarinus officinalis* L.: An update review of its phytochemistry and biological activity. *Future Science OA*, *4*(4).

Andrewartha, H. G., & Birch, L. C. (1954). *The distribution and abundance of animals*. Chicago, IL: University of Chicago Press.

Anonymous. (1885). The wealth of India: A dictionary of Indian raw materials and industrial products. New Delhi, India: A. Publication and Information Directorate CSIR. *Raw Materials*, *1*.

Aqil, F., Zahin, M., & Ahmad, I. (2008). Antimutagenic activity of methanolic extracts of four ayurvedic medicinal plants. *Indian Journal of Experimental Biology*, *46*(9), 668–672. PMID:18949897

Arambewela, L. S., Arawwawala, M., Owen, N. L., & Jarvis, B. (2007). Volatile oil of *Alpinia galanga* willd. of Sri Lanka. *The Journal of Essential Oil Research*, *19*(5), 455–456. doi:10.1080/10412905.2007.9699950

Ara, S., & Naqshi, A. R. (1992). Ethnobotanical studies in Gureiz Valley. *Journal of Economic and Taxonomic Botany*, *17*(3).

Archana, S. J., Paul, R., & Tiwari, A. (2011). Indian medicinal plants: A rich source of natural immuno-modulator. *International Journal of Pharmacology*, *7*(2), 198–205. doi:10.3923/ijp.2011.198.205

Ardestani, A., & Yazdanparast, R. (2007). *Cyperus rotundus* suppresses AGE formation and protein oxidation in a model of fructose-mediated protein glycoxidation. *International Journal of Biological Macromolecules*, *41*(5), 572–578. doi:10.1016/j.ijbiomac.2007.07.014 PMID:17765965

Arı, S. (2014). *Afyonkarahisar ve civarında halk tarafından kullanılan bazı bitkilerin etnobotanik özellikleri* (Unpublished doctoral dissertation) Afyon Kocatepe Üniversitesi, Afyon.

Arıturuk, Z. C. (2010). *Tefenni (burdur) ilçesinin florası ve halk ilaçları* (Unpublished master's thesis), Hacettepe Üniversitesi, Turkey.

Arora, R. (1987). Ethnobotany and its role in the domestication and conservation of native plant genetic resources. In A Manual of Ethnobotany (pp. 94–102).

Arora, R. K., & Anjula, P. (1996). *Wild edible plants of India: Diversity conservation and use*. New Delhi, India: ICAR.

Arqueta, A. (1994). *Atlas de las plantas de la medicina tradicional mexicana* (Vol. 3). Universidad Autónoma de Chiapas, Consorcio de Ciencias Agropecuarias, Instituto de Estudios Indígenas.

Arroyo-Johnson, C., & Mincey, K. D. (2016). Obesity epidemiology trends by race/ethnicity, gender, and education. *Gastroenterology Clinics of North America*, *45*(4), 571–579. doi:10.1016/j.gtc.2016.07.012 PMID:27837773

Asem, S. D., & Laitonjam, W. S. (2012). Investigation of the structure-nonlinearity relationship of zederone from the rhizomes of *Curcuma caesia* Roxb. *Indian Journal of Chemistry-Part B Organic Including Medicinal, 51*(12), 1738-1742.

Asmat, U., Abad, K., & Ismail, K. (2016). Diabetes mellitus and oxidative stress – A concise review. *Saudi Pharmaceutical Journal*, *24*(5), 547–553. doi:10.1016/j.jsps.2015.03.013 PMID:27752226

Asmelash, T. G. (2018). The Types of Aloe Species and Their Multi-Functions in Hammer District of Southern Ethiopia. *Annals of Reviews and Research*, *1*(2), 555–559.

Asolkar, L.V., Kakkar, K. K., & Chakra, O. J. (1992). Second supplement to glossary of Indian medicinal plants with active principles part-1 (A-K) 1965-1981. NISC, CSIR, New Delhi, India, pp. 265-266.

Athanasakis, K. (2017). The Socioeconomic effects of uncontrolled hypertension. *Current Vascular Pharmacology*, *16*(1), 5–9. doi:10.2174/1570161115666170413145125 PMID:28412912

Atkinson, E. T. (1882). *The Himalayan gazetteer*. New Delhi, India: Cosmo.

Austin, A. (2008). A review on Indian sarsaparilla, *Hemidesmus indicus* (L.) R. Br. *The Journal of Biological Sciences*, *8*(1), 1–12. doi:10.3923/jbs.2008.1.12

Austin, A., Jagadeesan, M., & Gowrishankar, R. (2003). Antimicrobial activity of *Hemidesmus indicus* var. *indicus* R. Br. against human isolates of *Helicobacter pylori*. *Natural Product Sciences*, *9*(1), 1–3.

Austin, A., & Jegadeesan, M. (2002). Acute and sub-acute toxicity studies of *Hemidesmus indicus* var. pubescens R. Br. *Ancient Science of Life*, *22*(2), 34–41. PMID:22557084

Austin, A., & Jegadeesan, M. (2003). Biochemical studies on the anti-ulcerogenic potential of *Hemidesmus indicus* R. Br. var. *indicus*. *Journal of Ethnopharmacology*, *84*(2-3), 149–156. doi:10.1016/S0378-8741(02)00291-X PMID:12648808

Avelino-Flores, M. D. C., Cruz-López, M. D. C., Jiménez-Montejo, F. E., & Reyes-Leyva, J. (2015). Cytotoxic activity of the methanolic extract of *Turnera diffusa* Willd. on breast cancer cells. *Journal of Medicinal Food*, *18*(3), 299–305. doi:10.1089/jmf.2013.0055 PMID:25299247

Ayala Garay, A., Rivas Valencia, P., Cortes-Espinoza, L., de la Olán, M., Escobedo-López, D., & Espitia Rangel, E. (2014). La rentabilidad del cultivo de amaranto (*Amaranthus* spp.) en la región centro de México. *CIENCIA ergo-sum*, *21*(1), 47–54.

Ayensu, E. S. (1981). Medicinal plants of the West Indies. Reference Publications, Inc.

Azimi, A., Ghaffari, S. M., Riazi, G. H., Arab, S. S., Tavakol, M. M., & Pooyan, S. (2016). α-Cyperone of *Cyperus rotundus* is an effective candidate for reduction of inflammation by destabilization of microtubule fibers in brain. *Journal of Ethnopharmacology*, *194*, 219–227. doi:10.1016/j.jep.2016.06.058 PMID:27353867

Babaei, A., Manafi, M., & Tarafi, H. (2013). A study on Effects of Aloe vera leaf Extract on growth of *Aspergillus flavus*. *Annual Research & Review in Biology*, *3*, 1091–1097.

Baba, I. A., Dubey, S., Alia, A., Saxena, R. C., Itoo, A. A., & Powar, A. (2012). Ethnobotanical survey of medicinal plants used by the people of District Ganderbal Jammu and Kashmir. *Research Journal of Pharmaceutical, Biological, and Chemical Sciences*, *3*, 549–556.

Baby, J., & Justin, S. R. (2010). Pharmacognostic and phytochemical properties of *Aloe vera* -an overview. *International Journal of Pharmaceutical Sciences Review and Research*, *4*, 106.

Baby, S., Dan, M., Thaha, A. R., Johnson, A. J., Kurup, R., Balakrishnapillai, P., & Lim, C. K. (2009). High content of zerumbone in volatile oils of *Zingiber zerumbet* from southern India and Malaysia. *Flavour and Fragrance Journal*, *24*(6), 301–308. doi:10.1002/ffj.1940

Badgujar, S. B., & Bandivdekar, A. H. (2015). Evaluation of a lactogenic activity of an aqueous extract of *Cyperus rotundus* Linn. *Journal of Ethnopharmacology*, *163*, 39–42. doi:10.1016/j.jep.2015.01.019 PMID:25625349

Bağcı, Y. (2000). Aladağlar (yahyalı, kayseri) ve çevresinin etnobotanik özellikleri. *Ot Sistematik Botanik Dergisi*, *7*(1), 89–94.

Baharvand-Ahmadi, B., & Asadi-Samani, M. (2016). A mini-review on the most important effective medicinal plants to treat hypertension in ethnobotanical evidence of Iran. *Journal of Nephropharmacology*, *6*(1), 3–8. PMID:28197520

Baheti, J. R., Goyal, R. K., & Shah, G. B. (2006). Hepatoprotective activity of *Hemidesmus indicus* R. Br. in rats. *Indian Journal of Experimental Biology*, *44*, 399–402. PMID:16708894

Bai, A., Ara, T., Deva, A. S., Madan, J., & Sharma, S. (2013). Preparation and Evaluation of Novel Aloe vera Gel Beads. *Journal of Global Biosciences*, *2*(6), 206–216.

Bajpay, A., Nainwal, R. C., Singh, D., & Tewari, S. K. (2018). Medicinal value of *Cyperus rotundus* Linn: An updated review. *Medicinal Plant*, *10*, 165–170.

Baker, J. T., Borris, R. P., Carté, B., Cordell, G. A., Soejarto, D. D., Cragg, G. M., & Tyler, V. E. (1995). Natural product drug discovery and development: New perspectives on international collaboration. *Journal of Natural Products*, *58*(9), 1325–1357. doi:10.1021/np50123a003 PMID:7494142

Balami, N. P. (2004). Ethnomedicinal uses of plants among the Newar community of Pharping village of Kathmandu district, Nepal. *Tribhuvan University Journal*, *24*(1), 13–19. doi:10.3126/tuj.v24i1.251

Balamurugan, S., Vijayakumar, S., Prabhu, S., & Yabesh, J. E. M. (2018). Traditional plants used for the treatment of gynaecological disorders in Vedaranyam Taluk, South India: An ethnomedicinal survey. *Journal of Traditional and Complementary Medicine*, *8*(2), 308–323. doi:10.1016/j.jtcme.2017.06.009 PMID:29736387

Balasubramanian, J., & Narayanan, N. (2013). Natures Gift. *Species*, *2*(1), 11–13.

Balick, M. J. (1996). Transforming ethnobotany for the new millennium. *Annals of the Missouri Botanical Garden*, *83*(1), 58–66. doi:10.2307/2399968

Balick, M. J., & Cox, P. A. (1966). *Plants, people, and culture: The science of ethnobotany*. NewYork: Scientific American Library, W. H. Freeman & Co.

Balick, M. J., & Cox, P. A. (1997). *Plants, people, and culture: The Science of ethnobotany*. New York, NY: Scientific American Library.

Balick, M. J., & Cox, P. A. R. (1996). *Plants, people, and culture: the science of ethnobotany* (p. 219). New York: Scientific American Library.

Balunas, M. J., & Kinghorn, A. D. (2005). Drug discovery from medicinal plants. *Life Sciences*, *78*(5), 431–441. doi:10.1016/j.lfs.2005.09.012 PMID:16198377

Banerjee, A. K., & Banerjee, I. (1986). A survey of medicinal plants in Shevaroy hills. *Journal of Economic and Taxonomic Botany*, *8*, 271–290.

Banerjee, S. K., & Maulik, S. K. (2002). Effect of garlic on cardiovascular disorders: A review. *Nutrition Journal*, *1*(1), 4. doi:10.1186/1475-2891-1-4 PMID:12537594

Banjerdpongchai, R., Punyati, P., Nakrob, A., Pompimon, W., & Kongtawelert, P. (2011). 4¢-Hydroxycinnamaldehyde from *Alpinia galanga* (Linn.) induces human leukemic cell apoptosis via mitochondrial and endoplasmic reticulum stress pathways. *Asian Pacific Journal of Cancer Prevention*, *12*(3), 593–598. PMID:21627350

Ban, N., & Coomes, O. T. (2004). Home gardens in Amazonian Peru: Diversity and exchange of planting material. *Geographical Review*, *94*(3), 348–367. doi:10.1111/j.1931-0846.2004.tb00177.x

Baqueiro-Peña, I., & Guerrero-Beltrán, J. A. (2014). Uses of *Justicia spicigera* in medicine and as a source of pigments. *Functional Foods in Health and Disease*, *4*(9), 401–414. doi:10.31989/ffhd.v4i9.150

Barbhuiya, A. R., Sahoo, U. K., & Upadhyaya, K. (2016). Plant diversity in the indigenous home gardens in the eastern Himalayan region of Mizoram, northeast India. *Economic Botany*, *70*(2), 115–131. doi:10.100712231-016-9349-8

Bargali, K. (2015). Comparative participation of rural women in agroforestry home gardens in Kumaun Himalaya, Uttarakhand, India. *Asian Journal of Agricultural Extension. Economia e Sociologia (Evora, Portugal)*, *6*(1), 16–22.

Bargali, K. (2016). Traditional homegardens as a sustainable ecosystem for maintenance of biodiversity: A case study from Kumaun Himalaya, India. *Journal of Biodiversity*, *7*(2), 88–100. doi:10.1080/09766901.2016.11884761

Barr, K. P., & Harrast, M. A. (2007). Low back pain. In: Braddom RL, editor. Physical medicine and rehabilitation. Philadelphia, CN: Saunder & Elsevier, 883-929.

Bartolome, A. P., Villaseñor, I. M., & Yang, W. C. (2013). *Bidens pilosa* L.(Asteraceae): Botanical properties, traditional uses, phytochemistry, and pharmacology. *Evidence-Based Complementary and Alternative Medicine, 2013*, 1–5. doi:10.1155/2013/340215 PMID:23935661

Başer, K. H. C., Honda, G., & Miki, W. (1986). Herb drugs and herbalists in turkey. Institute for the Study of Languages and Cultures of Asia and Africa, Studia Culturae Islamicae, 27, Tokyo.

Batoro, J., & Siswanto, D. (2017). Ethnomedicinal survey of plants used by local society in Poncokusumo district, Malang, East Java Province, Indonesia. *Asian Journal of Medical and Biological Research, 3*(2), 158–167. doi:10.3329/ajmbr.v3i2.33563

Bawankar, R., Singh, P., & Subramanian, B. (2014). Bioactive compounds and medicinal properties of Aloe vera L.: An update. *Journal of Plant Sciences, 2*(3), 102–107.

Bayram, E., Kırıcı, S., Tansı, S., Yılmaz, G., Arabacı, O., Kızıl, S., & Telci, İ. (2010). Tıbbi ve aromatik bitkiler üretiminin arttırılması olanakları. In Türkiye ziraat mühendisliği vii. teknik kongresi bildiriler kitabı 1 (pp. 437-456). Ankara, Turkey.

Baytop, T. (1984). Türkiye'de bitkiler ile tedavi. İstanbul, Turkey: İ. Ü. Yayınları. 3255 Ecz. Fak. No: 40.

Baytop, T. (1999). *Türkiye'de bitkiler ile tedavi, 2. baskı*. İstanbul, Turkey: Nobel Tıp Kitapevleri.

Baytop, T. (1999). *Türkiye'de bitkiler ile tedavi: geçmişte ve bugün*. İstanbul, Turkey: Nobel Tıp Kitabevleri.

Beck, S., Mathison, H., Todorov, T., Calderon-Juarez, E. A., & Kopp, O. R. (2018). A review of medicinal uses and pharmacological activities of *Tridax procumbens* (L.). *Journal of Plant Studies, 7*(1), 19–35. doi:10.5539/jps.v7n1p19

Behar, N., Tiwari, K. L., Jadhav, S. K., Vairappan, C. S., Elias, U. M., Ramachandram, T. R., ... Haldar, P. K. (2013). Comparative phytochemical screening of bioactive compounds in *Curcuma caesia* Roxb. and *Curcuma longa*. *Research Journal of Medicinal Plant, 7*(2), 113–118. doi:10.3923/rjmp.2013.113.118

Behura, S., & Srivastava, V. K. (2004). Essential oils of leaves of *Curcuma* species. *The Journal of Essential Oil Research, 16*(2), 109–110. doi:10.1080/10412905.2004.9698664

Bekalo, T. H., Woodmatas, S. D., & Woldemariam, Z. A. (2009). An ethnobotanical study of medicinal plants used by local people in the lowlands of Konta Special Woreda, southern nations, nationalities and peoples regional state, Ethiopia. *Journal of Ethnobiology and Ethnomedicine, 5*(1), 26. doi:10.1186/1746-4269-5-26 PMID:19775482

Bell, T. W. (1890). Medical notes on New Zealand. *New Zealand Medical Journal, 3*(65-83), 129-145.

Bergenudd, H., & Nilsson, B. O. (1988). Back pain in middle age: Occupational workload and psychologic factors: An epidemiologic survey. *Spine, 13*(1), 58–60. doi:10.1097/00007632-198801000-00014 PMID:2967994

Berkes, F. (1993). Traditional ecological knowledge in perspective. In Traditional ecological knowledge: concepts and cases (Vol. 1). Canadian Museum of Nature/International Development Research Centre, Ottawa.

Berkes, F. (1999). *Sacred ecology: traditional ecological knowledge and resource management*. Philadelphia, PA: Taylor and Francis.

Berkes, F., Colding, J., & Folke, C. (2000). Rediscovery of traditional ecological knowledge as adaptive management. *Ecological Applications, 10*(5), 1251–1262. doi:10.1890/1051-0761(2000)010[1251:ROTEKA]2.0.CO;2

Berkes, F., Folke, C., & Gadgil, M. (1994). Traditional ecological knowledge, biodiversity, resilience and sustainability. In *Biodiversity conservation* (pp. 269–287). Dordrecht, The Netherlands: Springer.

Berlin, E. A., & Berlin, B. (2005). Some Field Methods in Medical Ethnobiology. *Field Methods*, *17*(3), 235–268. doi:10.1177/1525822X05277532

Bersani, F. S., Coviello, M., Imperatori, C., Francesconi, M., Hough, C. M., Valeriani, G., . . . Corazza, O. (2015). Adverse psychiatric effects associated with herbal weight-loss products. *BioMed Research International, 2015*, 10 pages. doi:10.1155/2015/120679

Best, E. (1904). Maori medical lore. Notes on sickness and disease among the Maori people of New Zealand, and their treatment of the sick; together with some account of various beliefs, superstitions and rites pertaining to sickness, and the treatment thereof, as collected from the Tuhoe tribe. Part I. *The Journal of the Polynesian Society (NZ)*, *13*(52), 213–237.

Bhandari, P. R. (2015). *Crocus sativus* L. (saffron) for cancer chemoprevention: A mini review. *Journal of Traditional and Complementary Medicine*, *5*(2), 81–87. doi:10.1016/j.jtcme.2014.10.009 PMID:26151016

Bharati, K. A., & Sharma, B. (2012). Plants used as ethnoveterinary medicines in Sikkim Himalayas. *Ethnobotany Research and Applications*, *10*, 339–356. doi:10.17348/era.10.0.339-356

Bhaskara Rao, K. V., Nidhi, H., Dipankar, D., Garima, D., Kumar, G., & Karthik, L. (2015). Phytochemical profile, in vitro antioxidant property and HPTLC analysis of methanol extract of *Bougainvillea glabra* (*Nyctaginaceae*). *International Journal of Pharmaceutical Sciences Review and Research*, *31*(2), 235–241.

Bhat, N. A. (1984). Floristic Composition of Gulmarg (Baramulla) (Unpublished doctoral dissertation). University of Kashmir, Srinagar.

Bhat, A., Satpathy, G., & Gupta, R. K. (2015). Evaluation of Nutraceutical properties of *Amaranthus hypochondriacus* L. grains and formulation of value-added cookies. *Journal of Pharmacognosy and Phytochemistry*, *3*(5), 51–54.

Bhat, F., Mahajan, D. M., Sayyed, M. R. G., & Bhat, A. (2014). Ethno-medicinal survey of North Kashmir Himalaya – A case study of Lolab valley (J&K), India, *Ecology Environmental Conservation*, *20*(1), 59–71.

Bhati, N., Jaidka, S., & Somani, R. (2015). Evaluation of antimicrobial efficacy of Aloe vera and Meswak containing dentifrices with ñouridated dentifrice: An in vivo study. *Journal of International Society of Preventive & Community Dentistry*, *5*(5), 394–399. doi:10.4103/2231-0762.165924 PMID:26539392

Bhat, T. A., Nigam, G., & Majaz, M. (2012a). Traditional use of medicinal plants by Gujjar and Bakarwal tribes in Pir Panjal Range of the Shopion District, Kashmir. *International Journal of Research and Development in Pharmacy and Life Sciences*, *1*, 160–166.

Bhat, T. A., Nigam, G., & Majaz, M. (2012b). Study of some medicinal plants of the Shopion district, Kashmir (India) with emphasis on their traditional use by Gujjar and Bakerwal tribes. *Asian Journal of Pharmaceutical and Clinical Research*, *5*(2), 94–98.

Bhattacharjee, K., & Kumar, S. (2000). *Handbook of medicinal plants*. India: Aavishkar Publishers.

Bhattacharya, M., Malik, S., & Singh, A. (2011). Aloe vera barbedensis: A review on its Ethnopharmacological value. *Journal of Pharmacy Research*, *4*(1), 4507–4510.

Bhatt, V. P., & Negi, G. C. S. (2006). Ethnomedicinal plant resources of Jaunsari tribe of Garhwal Himalaya, Uttaranchal. *Indian Journal of Traditional Knowledge*, *5*(3), 331–335.

Bhuiyan, M. N. I., Chowdhury, J. U., & Begum, J. (2008). Chemical investigation of the leaf and rhizome essential oils of *Zingiber zerumbet* (L.) Smith from Bangladesh. *Bangladesh Journal of Pharmacology*, *4*(1), 9–12. doi:10.3329/bjp.v4i1.845

Bhutya, R. K. (2011). *Ayurvedic medicinal plants of India* (Vol. 1-2). Jodhpur, Rajasthan: Scientific Publishers.

Bidinotto, L. T., Spinardi-Barbisan, A. L. T., Rocha, N. S., Salvadori, D. M. F., & Barbisan, L. F. (2006). Effects of ginger (*Zingiber officinale* Roscoe) on DNA damage and development of urothelial tumors in a mouse bladder carcinogenesis model. *Environmental and Molecular Mutagenesis*, *47*(8), 624–630. doi:10.1002/em.20248 PMID:16878317

Biering-Sørensen, F., Thomsen, C. E., & Hilden, J. (1989). Risk indicators for low back trouble. *Scandinavian Journal of Rehabilitation Medicine*, *21*(3), 151–157. PMID:2529630

Bisht, A. K., Bhatt, A., & Dhar, U. (2006). Prioritization and conservation of Himalayan medicinal plants: Angelica glauca as a case study. [Not cited in text.]. *Ethnobotany Research and Applications*, *4*, 11–23. doi:10.17348/era.4.0.11-24

Bisht, A., Bisht, G. R. S., Singh, M., Gupta, R., & Singh, V. (2011). Chemical composition and antimicrobial activity of essential oil of tubers of *Cyperus rotundus* Linn. collected from Dehradun (Uttarakhand). *International Journal of Research in Pharmaceutical and Biomedical Sciences*, *2*, 661–665.

Blanckaert, I., Swennen, R. L., Flores, M. P., López, R. R., & Saade, R. L. (2004). Floristic composition, plant uses and management practices in homegardens of San Rafael Coxcatlán, valley of Tehuacán-Cuicatlán, Mexico. *Journal of Arid Environments*, *57*(2), 179–202. doi:10.1016/S0140-1963(03)00100-9

Blumenthal, M., Goldberg, A., & Brinckmann, J. (2000). Herbal Medicine. Expanded Commission E monographs. Integrative Medicine Communications. *American Botanical Council*, *3*, 78–83.

Boericke, W. (1988). Pocket Manual of Homoeopathic Materia Medica. New Delhi, India. Homeopathic publication.

Bopanna, K. N., Bhagyalakshmi, N., Rathod, S. P., Balaraman, R., & Kannan, J. (1997). Cell culture derived *Hemidesmus indicus* in the prevention of hypercholesterolemia in normal and hyperlipidemic rats. *Indian Journal of Pharmacology*, *29*(2), 105–109.

Borges, C. C., Matos, T. F., Moreira, J., Rossato, A. E., Zanette, V. C., & Amaral, P. A. (2013). *Bidens pilosa* L.(Asteraceae): Traditional use in a community of southern Brazil. *Revista Brasileira de Plantas Medicinais*, *15*(1), 34–40. doi:10.1590/S1516-05722013000100004

Bose, M. F. J. N., Aron, S., & Mehalingam, P. (2014). An ethnobotanical study of medicinal plants used by the Paliyars aboriginal community in Virudhunagar district, Tamil Nadu, India. *Indian Journal of Traditional Knowledge*, *13*(3), 613–618.

Bouasla, A., & Bouasla, I. (2017). Ethnobotanical survey of medicinal plants in northeastern of Algeria. *Phytomedicine*, *36*(September), 68–81. doi:10.1016/j.phymed.2017.09.007 PMID:29157830

Boudreau, M. D., Beland, F. A., Nichols, J. A., & Pogribna, M. (2013). Toxicology and carcinogenesis studies of a non-colorized whole leaf extract of Aloe barbadensis Miller (Aloe vera) in rats and mice (drinking water study). *National Toxicology Program Technical Report Series*, *577*(577), 1–266. PMID:24042237

Bowen, G. A. (2009). Document Analysis as a Qualitative Research Method. *Qualitative Research Journal*, *9*(2), 45–65. doi:10.3316/QRJ0902027

Bower, A., Marquez, S., & de Mejía, E. G. (2016). The health benefits of selected culinary herbs and spices found in the traditional Mediterranean diet. *Critical Reviews in Food Science and Nutrition*, *56*(16), 2728–2746. doi:10.1080/10408398.2013.805713 PMID:25749238

Bozyel, M. E., Merdamert Bozyel, E., Canli, K., & Altuner, E. M. (2019). Anticancer Uses of Medicinal Plants in Turkish Traditional Medicine. *KSU J. Agric Nat*, *22*(Suppl 2), 465–484.

Bozyel, M. E., & Merdamert, E. (2018). Antiurolithiatic activity of medicinal plants in turkey. In R. Efe (Ed.), *Science, ecology and engineering research in the globalizing world* (pp. 152–167). Sofia, Bulgaria: St. Kliment Ohridski University Press.

Bozyel, M. E., & Merdamert, E. (2018). Antiurolithiatic activity of medicinal plants in turkey. In R. Efe (Ed.), *Science, ecology, and engineering research in the globalizing world* (pp. 152–167). Sofia: St. Kliment Ohridski University Press.

Bozyel, M. E., Şenturan, M., Benek, A., Merdamert Bozyel, E., Canli, K., & Altuner, E. M. (2019). In vitro antimicrobial activity screening of heliotropium europaeum against wide range of microorganisms and multi drug resistant (mdr) bacteria. *European Journal of Biomedical and Pharmaceutical Sciences*, *6*(3), 113–117.

Bozzi, A., Perrin, C., Austin, S., & Arce, V. F. (2007). Quality and authenticity of commercial Aloe vera gel power. *Food Chemistry*, *103*(1), 22–30. doi:10.1016/j.foodchem.2006.05.061

Braatz, S. M. (1992). *Conserving biological diversity: a strategy for protected areas in the Asia-Pacific region* (Vol. 23). Washington, DC: World Bank Publications.

Bray, F., Ferlay, J., Soerjomataram, I., Siegel, R. L., Torre, L. A., & Jemal, A. (2018). Global cancer statistics 2018: GLOBOCAN estimates of incidence and mortality worldwide for 36 cancers in 185 countries. *CA: a Cancer Journal for Clinicians*, *68*(6), 394–424. doi:10.3322/caac.21492 PMID:30207593

Bremer, K. R. (1994). *Asteraceae: cladistics & classification.* Portland, OR: Timber Press.

Brown, C. H. (1984). *Language and living things: Uniformities in folk classification and naming.* New Brunswick, NJ: Rutgers University Press.

Bucker, A., Falcao-Bucker, N. C., Nunez, C. V., Pinheiro, C. C. D. S., & Tadei, W. P. (2013). Evaluation of larvicidal activity and brine shrimp toxicity of rhizome extracts of *Zingiber zerumbet* (L.) Smith. *Revista da Sociedade Brasileira de Medicina Tropical*, *46*(3), 377–380. doi:10.1590/0037-8682-1309-2013 PMID:23856868

Buck, M., & Hamilton, C. (2011). The Nagoya protocol on access to genetic resources and the fair and equitable sharing of benefits arising from their utilization to the convention on biological diversity. *Review of European Community & International Environmental Law*, *20*(1), 47–61. doi:10.1111/j.1467-9388.2011.00703.x

Budak, E. Ç., & Bozkurt, M. R. (2013). Vertebra lomber disklerde meydana gelen bozulmaların manyetik rezonans görüntüleme (mrg) ile analizi. *AJIT-e: Online Academic Journal of Information Technology*, *4*(12), 125–144.

Bulut, G., & Tuzlaci, E. (2013). An ethnobotanical study of medicinal plants in turgutlu (Manisa-turkey). *Journal of Ethnopharmacology*, *149*(3), 633–647. doi:10.1016/j.jep.2013.07.016 PMID:23933313

Bulut, G., Tuzlacı, E., Doğan, A., & Şenkardeş, İ. (2014). An ethnopharmacological review on the Turkish apiaceae species. *Journal of Faculty Pharmacy of Istanbul University*, *44*(2), 163–179.

Butler, M. S. (2004). The role of natural product chemistry in drug discovery. *Journal of Natural Products*, *67*(12), 2141–2153. doi:10.1021/np040106y PMID:15620274

Calvet-Mir, L., Gómez-Baggethun, E., & Reyes-García, V. (2012). Beyond food production: Ecosystem services provided by home gardens. A case study in Vall Fosca, Catalan Pyrenees, northeastern Spain. *Ecological Economics*, *74*, 153–160. doi:10.1016/j.ecolecon.2011.12.011

Camacho, M. R., Chávez, D., Mata, R., & Palacios-Ríos, M. (1991). Chemical studies on Mexican plants used in traditional medicine XXII: Constituents of *Equisetum myriochaetum* Schlecht. et Cham. *Fitoterapia1*, *63*(5), 471–472.

Camara-Leret, R., Paniagua-Zambrana, N., Balslev, H., & Macia, M. J. (2014). Ethnobotanical knowledge in vastly under-documented in Northwestern South America. *PLoS One*, *9*(1). doi:10.1371/journal.pone.0085794 PMID:24416449

Canli, K., Bozyel, M. E., & Altuner, E. M. (2017c). In vitro antimicrobial activity screening of maclura pomifera fruits against wide range of microorganisms. *International Journal of Pharmaceutical Science Invention*, *6*(8), 19–22.

Canli, K., Simsek, O., Yetgin, A., & Altuner, E. M. (2017d). Determination of the chemical composition and antimicrobial activity of frankenia hirsuta. *Bangladesh Journal of Pharmacology*, *12*(4), 463–469. doi:10.3329/bjp.v12i4.33652

Canli, K., Yetgin, A., Akata, I., & Altuner, E. M. (2016). In vitro antimicrobial screening of aquilaria agallocha roots. *African Journal of Traditional, Complementary, and Alternative Medicines*, *13*(5), 178–181. PMID:28487909

Canli, K., Yetgin, A., Akata, I., & Altuner, E. M. (2017a). Antimicrobial activity and chemical composition screening of epilobium montanum root. *Indian Journal of Pharmaceutical Education and Research*, *51*(3s), 239–243. doi:10.5530/ijper.51.3s.21

Canli, K., Yetgin, A., Akata, I., & Altuner, E. M. (2017b). Antimicrobial activity and chemical composition screening of anacyclus pyrethrum root. *Indian Journal of Pharmaceutical Education and Research*, *51*(3s), 244–248. doi:10.5530/ijper.51.3s.22

Canli, K., Yetgin, A., Benek, A., Bozyel, M. E., & Altuner, E. M. (2019). In vitro antimicrobial activity screening of ethanol extract of lavandula stoechas and investigation of its biochemical composition. *Advances in Pharmacological Sciences, 2019. Article ID, 3201458*, 1–6. doi:10.1155/2019/3201458

Canter, P. H., Thomas, H., & Ernst, E. (2005). Bringing medicinal plants into cultivation: Opportunities and challenges for biotechnology. *Trends in Biotechnology*, *23*(4), 180–185. doi:10.1016/j.tibtech.2005.02.002 PMID:15780709

Cardellina, J. H. (2002). Challenges and opportunities confronting the botanical dietary supplement industry. *Journal of Natural Products*, *65*(7), 1073–1084. doi:10.1021/np0200515 PMID:12141880

Carlquist, S. (1976). Tribal interrelationships and phylogeny of the Asteraceae. Aliso: *A Journal of Systematic and Evolutionary Botany, 8*(4), 465-492.

Carvalho, A. R. Jr, Diniz, R. M., Suarez, M. A., Figueiredo, C. S. S. S., Zagmignan, A., Grisotto, M. A. G., ... da Silva, L. C. N. (2018). Use of some Asteraceae plants for the treatment of wounds: From ethnopharmacological studies to scientific evidences. *Frontiers in Pharmacology*, *9*, 1–11. doi:10.3389/fphar.2018.00784 PMID:30186158

Casas, A., Blancas, J., & Lira, R. (2016). Ethnobotany of Mexico, (April). doi:10.1007/978-1-4614-6669-7

Cavalcanti, B. C., Ferreira, J. R., Cabral, I. O., Magalhães, H. I., de Oliveira, C. C., Rodrigues, F. A., ... Moraes, M. O. (2012). Genetic toxicology evaluation of essential oil of *Alpinia zerumbet* and its chemoprotective effects against H2O2-induced DNA damage in cultured human leukocytes. *Food and Chemical Toxicology*, *50*(11), 4051–4061. doi:10.1016/j.fct.2012.03.038 PMID:22446814

CBD. (1992). Convention on Biological Diversity, Rio de Janeiro, Brazil. Available from http://www.cbd.int/convention

Cetkovic, G. S., Djilas, S. M., Canadanović-Brunet, J. M., & Tumbas, V. T. (2004). Antioxidant properties of marigold extracts. *Food Research International*, *37*(7), 643–650. doi:10.1016/j.foodres.2004.01.010

Ceylan, S., Azal, Ö., Taşlipinar, A., Türker, T., Açikel, C. H., & Gulec, M. (2009). Complementary and alternative medicine use among Turkish diabetes patients. *Complementary Therapies in Medicine*, *17*(2), 78–83. doi:10.1016/j.ctim.2008.07.003 PMID:19185265

Chadha, Y. R. (1995). The wealth of India, raw materials. *Publications and Information Directorate, 10*(Sp-W), 591.

Chak, I., Agarwal, R. K., & Kak, A. M. (2009). Ethno medicinal study of some important plants used in the treatment of hair and boils in district Pulwama of Kashmir. *Annals of Forestry (Dehra Dun)*, *17*(1), 101–107.

Chakraborty, A. K., Gaikwad, A. V., & Singh, K. B. (2012). Phytopharmacological review on *Acanthospermum hispidum*. *Journal of Applied Pharmaceutical Science*, *2*(1), 144–148.

Chandra, P., Sachan, N., & Pal, D. (2015). Protective effect of *Dalbergia sissoo* Roxb. ex DC. (Family: Fabaceae): Leaves against experimentally induced diarrhoea and peristalsis in mice. *Toxicology and Industrial Health*, *3*(12), 1229–1235. doi:10.1177/0748233713491815 PMID:23788395

Chandra, R., Deepak, D., & Khare, A. (1994). Pregnane glycosides from *Hemidesmus indicus*. *Phytochemistry*, *35*(6), 1545–1548. doi:10.1016/S0031-9422(00)86891-0

Chan, E. W., Ng, V. P., Tan, V. V., & Low, Y. Y. (2011). Antioxidant and antibacterial properties of *Alpinia galanga, Curcuma longa*, and *Etlingera elatior* (Zingiberaceae). *Pharmacognosy Journal*, *3*(22), 54–61. doi:10.5530/pj.2011.22.11

Chane-Ming, J., Vera, R., & Chalchat, J. C. (2003). Chemical composition of the essential oil from rhizomes, leaves and flowers of *Zingiber zerumbet* Smith from Reunion Island. *The Journal of Essential Oil Research*, *15*(3), 202–205. doi:10.1080/10412905.2003.9712114

Chang, C. J., Tzeng, T. F., Liou, S. S., Chang, Y. S., & Liu, I. M. (2012). Absence of genotoxic and mutagenic effects of *Zingiber zerumbet* (L.) Smith (Zingiberaceae) extract. *Evidence-Based Complementary and Alternative Medicine*, *2012*, 1–7. PMID:22844331

Chatterjee, I., Chakravarty, A. K., & Gomes, A. (2006). *Daboia russellii* and *Naja kaouthia* venom neutralization by lupeol acetate isolated from the root extract of Indian sarsaparilla *Hemidesmus indicus* R. Br. *Journal of Ethnopharmacology*, *106*(1), 38–43. doi:10.1016/j.jep.2005.11.031 PMID:16426782

Chatterjee, P., C.hakraborty, B., & Nandy, S. (2013). Aloe vera plant: Review with significant pharmacological activities. *Mintage Journal of Pharmaceutical and Medical Sciences*, *2*(3), 20–24.

Chatterjee, R. C., & Bhattacharya, B. K. (1955). A note on the isolation of ß-sitosterol from *Hemidesmus indicus*. *Journal of the Indian Chemical Society*, *32*, 485–486.

Cheah, P. B., & Abu Hasim, N. H. (2000). Natural antioxidant extract from galangal (*Alpinia galanga*) for minced beef. *Journal of the Science of Food and Agriculture*, *80*(10), 1565–1571. doi:10.1002/1097-0010(200008)80:10<1565::AID-JSFA677>3.0.CO;2-7

Cheah, P. B., & Gan, S. P. (2000). Antioxidative/antimicrobial effects of galangal and α-tocopherol in minced beef. *Journal of Food Protection*, *63*(3), 404–407. doi:10.4315/0362-028X-63.3.404 PMID:10716573

Chen, C. C., Hsu, J. D., Wang, S. F., Chiang, H. C., Yang, M. Y., Kao, E. S., ... Wang, C. J. (2003). *Hibiscus sabdariffa* extract inhibits the development of atherosclerosis in cholesterol-fed rabbits. *Journal of Agricultural and Food Chemistry*, *51*(18), 5472–5477. doi:10.1021/jf030065w PMID:12926900

Chen, Y., Zhao, Y. Y., Wang, X. Y., Liu, J. T., Huang, L. Q., & Peng, C. S. (2011). GC-MS analysis and analgesic activity of essential oil from fresh rhizoma of *Cyperus rotundus*. *Zhong Yao Cai*, *34*, 1225–1229. PMID:22233038

Chepkosgey, M. P., & Jerotich, C. G. (2016). Trends for application of indigenous knowledge in natural resource management among Nandi people, Kenya. *Journal of Environmental Science. Computer Science and Engineering Technology.*, *5*(3), 513–527.

Chetri, B. K., Ghalley, L., Penjor, D., Dechen, K., & Gyeltshen, T. (2018). Ethnobotanical study on wealth of homegardens in Gosiling Gewog of Tsirang District. *Asian Journal of Research in Botany*.

Chetri, B. K. (2019). *Ethnobotanical Study of South Eastern Foothills of Bhutan*. Asian Plant Research Journal.

Chetri, K., Zangpo, T., Tshering, P., Rai, S., Tenzin, P., & Roy, S. (2016). Ethnobotanical study of the southern foothills of the Himalayas in Bhutan. Shrub Deonte. *The Research Journal of Sherubtse College*, *12*(1 & 2), 63–70.

Chewonarin, T., Kinouchi, T., Kataoka, K., Arimochi, H., Kuwahara, T., Vinitketkumnuen, U., & Ohnishi, Y. (1999). Effects of roselle (*Hibiscus sabdariffa* Linn.), a Thai medicinal plant, on the mutagenicity of various known mutagens in *Salmonella typhimurium* and on formation of aberrant crypt foci induced by the colon carcinogens Azoxymethane and 2-Amino-1-methyl-6-ph. *Food and Chemical Toxicology*, *37*(6), 591–601. doi:10.1016/S0278-6915(99)00041-1 PMID:10478827

Chiang, L. C., Chang, J. S., Chen, C. C., Ng, L. T., & Lin, C. C. (2003). Anti-Herpes simplex virus activity of *Bidens pilosa* and *Houttuynia cordata*. *The American Journal of Chinese Medicine*, *31*(3), 355–362. doi:10.1142/S0192415X03001090 PMID:12943167

Chidrawar, V. R., Ushir, Y. V., Sudarshan, S., Patel, K. N., Patel, N. J., & Vadalia, K. R. (2009). Possible role of natural nephroprotective; *Hemidesmus indicus* in congestive heart failure. *Pharmacognosy Research*, *1*(6), 367–374.

Chien, T. Y., Chen, L. G., Lee, C. J., Lee, F. Y., & Wang, C. C. (2008). Anti-inflammatory constituents of *Zingiber zerumbet*. *Food Chemistry*, *110*(3), 584–589. doi:10.1016/j.foodchem.2008.02.038

Choi, H. J., Lee, J. H., & Jung, Y. S. (2014). (+)-Nootkatone inhibits tumor necrosis factor α/interferon γ-induced production of chemokines in HaCaT cells. *Biochemical and Biophysical Research Communications*, *447*(2), 278–284. doi:10.1016/j.bbrc.2014.03.121 PMID:24704449

Choi, S., & Chung, M. H. (2003). A review on the relationship between Aloe vera components and their biologic effects. *Seminar in integrative. Medicine*, *1*(1), 53–62.

Chopra, R. N., Nayar, S. L., & Chopra, I. C. (2002). *Glossary of Indian Medicinal Plants*. New Delhi, India: Niscair publisher.

Çilden, E. (2011). *Paşayaylası (aydın) florası ve etnobotanik özellikleri* (Unpublished master's thesis), Hacettepe Üniversitesi, Turkey.

Clarke, L. W., Li, L., Jenerette, G. D., & Yu, Z. (2014). Drivers of plant biodiversity and ecosystem service production in home gardens across the Beijing municipality of china. *Urban Ecosystems*, *17*(3), 741–760. doi:10.100711252-014-0351-6

Clements, F. E., & Shelford, V. E. (1939). *Bio-ecology*. New York: John Wiley & Sons.

Colding, J. (1998). Analysis of hunting options by the use of general food taboos. *Ecological Modelling*, *110*(1), 5–17. doi:10.1016/S0304-3800(98)00038-6

Colding, J., & Folke, C. (2001). Social taboos: 'invisible' systems of local resource management and biological conservation. *Ecological Applications*, *11*, 584–600.

Colding, J., Folke, C., & Elmqvist, T. (2003). Social institutions in ecosystem management and biodiversity conservation. *Tropical Ecology*, *44*(1), 25–41.

Çömlekçioğlu, N., & Karaman, Ş. (2008). The medicinal plants found in the local herbal markets in the city of kahramanmaraş in turkey. *KSU Journal of Science and Engineering*, *11*(1), 23–32.

Çömlekçioğlu, N., & Karaman, Ş. (2008). The medicinal plants found in the local herbal markets in the city of Kahramanmaraş in Turkey. *KSU Journal of Science and Engineering*, *11*(1), 23–32.

Committee of Experts on Cosmetic Products. (2008). *Aloe extracts with anthraquinones. Active ingredients used in cosmetics: safety survey*. Strasbourg, France: Council of Europe Publishing.

Coopoosamy, R. M., & Magwa, M. L. (2006). Antibacterial activity of Aloe emodin and aloin A isolated from *Aloe excelsa*. *African Journal of Biotechnology*, *5*(11), 1092–1094.

Córdova Villalobos, J. Á. (2016). La obesidad: La verdadera pandemia del siglo XXI. *Cirugia y Cirujanos*, *84*(5), 351–355. doi:10.1016/j.circir.2016.08.001 PMID:27601182

Coronado, G. D., Thompson, B., Tejeda, S., & Godina, R. (2004). Attitudes and Beliefs among Mexican Americans about type II diabetes. *Journal of Health Care for the Poor and Underserved*, *15*(4), 576–588. doi:10.1353/hpu.2004.0057 PMID:15531816

Cotton, C. M., & Wilkie, P. (1996). *Ethnobotany: Principles and applications*. Chichester, UK: John Wiley & Sons.

Cousins, S. R., & Witkowski, E. T. F. (2012). African aloe ecology, a review. *Journal of Arid Environments*, *85*(1), 1–17. doi:10.1016/j.jaridenv.2012.03.022

Cox, P. A. (1994). The ethnobotanical approach to drug discovery: Strengths and limitations. *Ciba Foundation Symposium*, *185*, 25–41. PMID:7736859

Cox, P. A. (2000). Will tribal knowledge survive the millennium? *Science*, *287*(5450), 44–45. doi:10.1126cience.287.5450.44 PMID:10644221

Cragg, G. M., & Boyd, M. (1996). Drug discovery and development at the National Cancer Institute: The role of natural products of plant origin. In M. J. Balick, E. Elisabetsky, & S. A. Laird (Eds.), Medicinal Plant Resources of the Tropical Forest (101–136). New York: Columbia University Press.

Cragg, G. M., Schepartz, S. A., Suffness, M., & Grever, M. R. (1993). The taxol supply crisis: New NCI policies for handling the large-scale production of novel natural product anticancer and anti-HIV agents. *Journal of Natural Products*, *56*(10), 1657–1668. doi:10.1021/np50100a001 PMID:7903979

Creswell, J. W. (2009). *Research Design* (3rd ed.). United States: Sage.

Cunningham, A. B. (1993). *African Medicinal Plants setting priorities at the interface between conservation and primary health care*. Working Paper 1. Paris, France: UNESCO.

Cunningham, A. B. (2008). People's science: Southern African ethnobotany in global perspective. *South African Journal of Botany*, *2*(74), 357–359. doi:10.1016/j.sajb.2008.01.014

Da Rocha Fernandes, J., Ogurtsova, K., Linnenkamp, U., Guariguata, L., Seuring, T., Zhang, P., ... Makaroff, L. E. (2016). IDF Diabetes Atlas estimates of 2014 global health expenditures on diabetes. *Diabetes Research and Clinical Practice*, *117*, 48–54. doi:10.1016/j.diabres.2016.04.016 PMID:27329022

Da Silva Lima, N., De Paula Numata, E., De Souza Mesquita, L. M., Dias, P. H., Vilegas, W., Gambero, A., & Ribeiro, M. L. (2017). Modulatory effects of guarana (*Paullinia cupana*) on adipogenesis. *Nutrients*, *9*(6), 635. doi:10.3390/nu9060635 PMID:28632199

Dabaghian, F. H., Hashemi, M., Entezari, M., Movassaghi, S., Goushegir, S. A., Kalantari, S., ... Sharifi, Z. N. (2015). Effect of *Cyperus rotundus* on ischemia-induced brain damage and memory dysfunction in rats. *Iranian Journal of Basic Medical Sciences.*, *18*, 199–204. PMID:25825638

Dad, J. M., & Khan, A. B. (2011). Threatened medicinal plants of Gurez valley, Kashmir Himalayas: Distribution pattern and current conservation status. *The International Journal of Biodiversity Science, Ecosystem Services & Management*, *7*(1), 20–26. doi:10.1080/21513732.2011.602646

Dafallah, A. A., & Al-Mustafa, Z. (2004). Investigation of the anti-inflammatory activity of *Acacia nilotica* and *Hibiscus sabdariffa*. *The American Journal of Chinese Medicine*, *24*(03n04), 263–269. doi:10.1142O192415x96000323

Dağlı, M. (2015). *Şanlıurfa merkez ve bağlı köylerde etnobotanik bir araştırma* (Unpublished master's thesis). Harran Üniversitesi, Şanlıurfa.

Dai, D. N., Thang, T. D., Chau, L. T., & Ogunwande, I. A. (2013). Chemical Constituents of the Root Essential Oils of *Zingiber rubens* Roxb., and *Zingiber zerumbet* (L.) Smith. *American Journal of Plant Sciences*, *4*(1), 7–10. doi:10.4236/ajps.2013.41002

Daimei, P., & Kumar, Y. (2014). Ethnobotanical uses of gingers in Tamenglong district, Manipur, Northeast India. *Genetic Resources and Crop Evolution*, *61*(1), 273–285. doi:10.100710722-013-0054-5

Dandekar, S. P., Bajpai, N. D., & Sakharwade, S. N. (2018). A multifunctional *Hemidesmus indicus* as cosmetic agent: A review. *International Journal of Scientific Development and Research*, *3*(10), 62–69.

Dang, G. K., Parekar, R. R., Kamat, S. K., Scindia, A. M., & Rege, N. N. (2011). Antiinflammatory activity of *Phyllanthus emblica, Plumbago zeylanica* and *Cyperus rotundus* in acute models of inflammation. *Phytotherapy Research*, *25*(6), 904–908. doi:10.1002/ptr.3345 PMID:21132843

Danielsen, F., Burgess, N. D., Balmford, A., Donald, P. F., Funder, M., Jones, J. P., & Child, B. (2009). Local participation in natural resource monitoring: A characterization of approaches. *Conservation Biology*, *23*(1), 31–42. doi:10.1111/j.1523-1739.2008.01063.x PMID:18798859

Darekar, R. S., Khetre, A. B., Sinha, P. K., Jeswani, R. M., & Damle, M. C. (2008). Quantitation of lupeol octacosanoate in *Hemidesmus indicus* R. Br. root powder by HPTLC. *Journal of Planar Chromatography - Modern TLC*, *22*(6), 453–456. doi:10.1556/JPC.22.2009.6.13

Dar, G. H., Bhagat, R. C., & Khan, M. A. (2002). *Biodiversity of Kashmir Himalaya*. Srinagar, India: Valley Book House.

Das, A. K., Dutta, B., & Sharma, G. (2008). *Medicinal plants used by different tribes of Cachar district*. Assam.

Das, A. P., & Pandey, A. K. (Eds.). (2007). *Advances in ethnobotany*. Dehradun, India: Bishen Singh Mahendra Pal Singh.

Das, P. C., Joshi, P. C., & Mandal, S., Das, A., Chatterjee, A., & Banerji, A. (1992). New Coumarino-Lignoids from *Hemidesmus indicus* R. Br. *ChemInform*, *23*(37), 342–345.

Das, S., Choudhury, M. D., Mandal, S. C., & Talukdar, A. D. (2012). Traditional knowledge of ethnomedicinal hepatoprotective plants used by certain ethnic communities of Tripura State. *Indian Journal of Fundamental and Applied Life Sciences*, *2*, 84–97.

Das, S., Dash, S. K., & Padhy, S. N. (2003a). Ethno-medicinal informations from Orissa State, India, A review. *Journal of Human Ecology (Delhi, India)*, *14*(3), 165–227. doi:10.1080/09709274.2003.11905616

Das, S., & Devaraj, S. N. (2006). Glycosides derived from *Hemidesmus indicus* R. Br. root inhibit adherence of *Salmonella typhimurium* to host cells: Receptor mimicry. *Phytotherapy Research*, *20*(9), 784–793. doi:10.1002/ptr.1963 PMID:16807882

Das, S., & Devaraj, S. N. (2007). Protective role of *Hemidesmus indicus* R. Br. root extract against *Salmonella typhimurium*-induced cytotoxicity in Int 407 cell line. *Phytotherapy Research*, *21*(12), 1209–1216. doi:10.1002/ptr.2241 PMID:17685386

Das, S., & Devaraj, S. N. (2008). Effect of *Hemidesmus indicus* R. Br. root extract against *Salmonella enterica* serovar *Typhimurium*-induced apoptosis in murine macrophage cell line (P388D1). *The Indian Journal of Medical Research*, *128*(5), 647–657. PMID:19179687

Das, S., Prakash, R., & Devaraj, S. N. (2003b). Antidiarrhoeal effects of methanolic root extract of *Hemidesmus indicus* (Indian sarsaparilla)-an in vitro and in vivo study. *Indian Journal of Experimental Biology*, *41*, 363–366. PMID:15255649

Das, S., & Singh Bisht, S. (2012). The bioactive and therapeutic potential of *Hemidesmus indicus* R. Br.(Indian Sarsaparilla) root. *Phytotherapy Research*, *27*(6), 791–801. doi:10.1002/ptr.4788 PMID:22887725

Das, T., & Das, A. K. (2005). Inventorying plant biodiversity in homegardens: A case study in Barak valley, Assam, north east India. *Current Science*, 155–163.

Daswani, P. G., Brijesh, S., Tetali, P., & Birdi, T. J. (2011). Studies on the activity of *Cyperus rotundus* Linn. tubers against infectious diarrhea. *Indian Journal of Pharmacology*, *43*(3), 340–344. doi:10.4103/0253-7613.81502 PMID:21713044

Davis, P. H. (1965-1985). *Flora of Turkey and the East Aegean Islands* (Vols. 1-9). Edinburgh, UK: Edinburgh at the University Press.

Davis, P. H. (1965-1985). *Flora of turkey and the east Aegean islands*. Edinburgh at the University Press, Vol. 1-9, Edinburgh.

De Santayana, M. P., Pieroni, A., & Kuri, R. K. (2010). *Ethnobotany in the new Europe: People, health, and wild plant resources* (Vol. 14; Studies In). Berghahn Books.

Deepak, D., Srivastav, S., & Khare, A. (1995b). *Structural studies of oligosaccharide segment of hemisine and desmisine.* Paper presented at the Trends in Carbohydrate Chemistry Developed From a Carbohydrate Conference, Lucknow, India.

Deepak, D., Srivastav, S., & Khare, A. (1995a). Indicusin-A pregnane diester triglycoside from *Hemidesmus indicus* R. Br. *Natural Product Letters*, *6*(2), 81–86. doi:10.1080/10575639508044094

Demirci, S., & Özhatay, N. (2012). An ethnobotanical study in kahramanmaraş (turkey); wild plants used for medicinal purpose in andirin, kahramanmaraş. *Turk J Pharm Sci*, *9*(1), 75–92.

Demirhan Erdemir, A. (2007). *Şifalı bitkiler: doğal ilaçlarla geleneksel tedaviler*. İstanbul, Turkey: Alfa Yayınları.

Denzin, N. K., & Lincoln, Y. S. (1994). Introduction: Entering the Field of Qualitative Research. In: N. K. Denzin, & Y. S. Lincoln (Eds.), Handbook of Qualitative Research. Thousand Oaks, CA: Sage.

Desai, S., Babaria, P., Nakarani, M., Shah, K., & Paranjape, A. (2017). Antiosteoporotic effect of *Hemidesmus indicus* Linn. on ovariectomised rats. *Journal of Ethnopharmacology*, *199*, 1–8. doi:10.1016/j.jep.2017.01.031 PMID:28119100

Devi, H. P., Mazumder, P. B., & Devi, L. P. (2015). Antioxidant and Antimutagenic activity of *Curcuma caesia* Roxb. rhizome extracts. *Toxicology Reports*, *2*, 423–428. doi:10.1016/j.toxrep.2014.12.018 PMID:28962377

Devi, N. B., Das, A. K., & Singh, P. K. (2017). Ethnobotanical importance of *Zingiber* in the Imphal Valley District, Manipur with *Zingiber chrysanthum* Roscoe as a new record for Manipur. *Journal of Pharmaceutical and Biological Sciences*, *12*(3), 29–32.

Deyo, R. A., & Phillips, W. R. (1996). Low back pain: A primary care challenge. *Spine*, *21*(24), 2826–2832. doi:10.1097/00007632-199612150-00003 PMID:9112706

Deyo, R. A., & Tsui-Wu, Y. J. (1987). Descriptive epidemiology of low-back pain and its related medical care in the United States. *Spine*, *12*(3), 264–268. doi:10.1097/00007632-198704000-00013 PMID:2954221

Dhall, K., & Dogra, M. (1988). Phase I and II clinical trials with *Vicoa indica* (Banjauri): Herbal medicine, as an antifertility agent. *Contraception*, *37*(1), 75–84. doi:10.1016/0010-7824(88)90150-3 PMID:3365984

Dhal, N. K., Panda, S. S., & Muduli, S. D. (2015). Traditional uses of medicinal plants by native people in Nawarangpur district, Odisha, India. *Asian Journal of Plant Science and Research*, *5*(2), 27–33.

Dhanam, S., & Elayaraj, B. (2014). Ethnomedicinal aspects of some weeds from paddy fields of Villupuram district in Tamil Nadu, India. *International Letters of Natural Sciences*, *19*, 1–10. doi:10.18052/www.scipress.com/ILNS.19.1

Dharmananda, S. (2013). A Popular Remedy Escapes Notice of Western Practitioners. Retrieved from http://www.itmonline.org/arts/bidens.htm

Dhar, U. (2000). Prioritization of conservation sites in the timberline zone of west Himalaya. In S. Singh, A. R. K. Sastry, R. Mehta, & V. Uppal (Eds.), *Setting biodiversity conservation priorities for India* (pp. 193–211). New Delhi, India: WWF India.

Dhar, U., Manjkhola, S., Joshi, M., Bhatt, A., Bisht, A. K., & Joshi, M. (2002). Current status and future strategy for development of medicinal plants sector in Uttaranchal, India. *Current Science*, *83*, 956–964.

Dhasmana, H. (1985). Medicinal plants of Pauri town (Garhwal) and adjacent forest regions (part 1: Family compositae). *Journal of Scientific Research in Plants and Medicines*, *6*, 45–47.

Díaz-Reviriego, I., González-Segura, L., Fernández-Llamazares, Á., Howard, P. L., Molina, J., & Reyes-García, V. (2016). Social organization influences the exchange and species richness of medicinal plants in Amazonian home gardens. *Ecology and Society*, *21*(1), 1. doi:10.5751/ES-07944-210101 PMID:27668001

Dickson, M., & Gagnon, J. P. (2004). Key factors in the rising cost of new drug discovery and development. *Nature Reviews. Drug Discovery*, *3*(5), 417–429. doi:10.1038/nrd1382 PMID:15136789

Dimo, T., Azay, J., Tan, P. V., Pellecuer, J., Cros, G., Bopelet, M., & Serrano, J. J. (2001). Effects of the aqueous and methylene chloride extracts of *Bidens pilosa* leaf on fructose-hypertensive rats. *Journal of Ethnopharmacology*, *76*(3), 215–221. doi:10.1016/S0378-8741(01)00229-X PMID:11448541

Ding, D., Lawson, K. D., Kolbe-Alexander, T. L., Finkelstein, E. A., Katzmarzyk, P. T., van Mechelen, W., & Pratt, M. (2016). The economic burden of physical inactivity: A global analysis of major non-communicable diseases. *Lancet*, *388*(10051), 1311–1324. doi:10.1016/S0140-6736(16)30383-X PMID:27475266

Dobson, T. (1992). Loss of biodiversity: An international environmental policy perspective. *North Carolina Journal of International Law and Commercial Regulation*, *17*, 277.

Doğanoğlu, Ö., Gezer, A., & Yücedağ, C. (2006). Göller bölgesi-yenişarbademli yöresi'nin önemli bazı tıbbi ve aromatik bitki taksonları üzerine araştırmalar. *Süleyman Demirel Üniversitesi Fen Bilimleri Enstitüsü Dergisi*, *10*(1), 66–73.

Dokuparthi, S. K., & Manikanta, P. (2015). Phytochemical and pharmacological studies on *Chrysanthemum coronarium* L.: A review. *Journal of Drug Discovery and Therapeutics*, *3*(27), 11–16.

Duarte, M. C., Figueira, G. M., Sartoratto, A., Rehder, V. L., & Delarmelina, C. (2005). Anti-*Candida* activity of Brazilian medicinal plants. *Journal of Ethnopharmacology*, *97*(2), 305–311. doi:10.1016/j.jep.2004.11.016 PMID:15707770

Dũng, N. X., Chính, T. D., & Leclercq, P. A. (1995). Chemical investigation of the aerial parts of *Zingiber zerumbet* (L.) Sm. from Vietnam. *The Journal of Essential Oil Research*, *7*(2), 153–157. doi:10.1080/10412905.1995.9698490

Durodola, J. I. (1977). Antibacterial property of crude extracts from a herbal wound healing remedy–*Ageratum conyzoides*, L. *Planta Medica*, *32*(08), 388–390. doi:10.1055-0028-1097620 PMID:594219

Dutta, A. T., Ghosh, S., & Chopra, R. N. (1938). Chemische Untersuchung der Wurzeln von *Hemidesmus Indicus*. I. Teil. *Archiv der Pharmazie*, *276*(6), 333–340. doi:10.1002/ardp.19382760602

Dutta, M. K., Sen, T. K., & Sikdar, S. (1982). Some preliminary observations on anti-inflammatory properties of *Hemidesmus indicus* in rat. *Indian Journal of Pharmacology*, *14*, 78.

Dutta, M. L. (2017). Plants used as ethnomedicine by the Thengal Kacharies of Assam, India. *Asian Journal of Plant Science and Research*, *7*, 7–8.

Efterpi, V. C., & Panagita, C. F. P. (2010). Aloe vera, a plant for many uses. *Journal of Food Agriculture and Environment*, *8*(2), 245–249.

Egesie, U. G., Chima, K. E., & Galam, N. Z. (2011). Anti-inflammatory and Analgesic Effects of Aqueous Extract of Aloe Vera (*Aloe barbadensis*) in Rats. *African Journal of Biomedical Research*, *14*(1), 209–212.

Eichemberg, M. T., Amorozo, M. C. M., & Moura, L. C. (2009). Species composition and plant use in old urban homegardens in Rio Claro, Southeast of Brazil. *Acta Botanica Brasílica*, *23*(4), 1057–1075. doi:10.1590/S0102-33062009000400016

El-Ghorab, A. H., Nauman, M., Anjum, F. M., Hussain, S., & Nadeem, M. (2010). A comparative study on chemical composition and antioxidant activity of ginger (*Zingiber officinale*) and cumin (*Cuminum cyminum*). *Journal of Agricultural and Food Chemistry*, *58*(14), 8231–8237. doi:10.1021/jf101202x PMID:20590154

El-Gohary, H. M. A. (2004). Study of essential oils of the tubers of *Cyperus rotundus* L. and *Cyperus alopecuroides* Rottb. *Bulletin of Faculty of Pharmacy, Cairo University*, *42*, 157–164.

El-Hawary, S. S., Kassem, H. A., Motaal, A. A., Tawfik, W. A., Hassanein, H. M., & El-Shamy, S. S. (2013). GC-MS Analysis of the essential oil of *Alpinia zerumbet* (Pers.) BL and in vitro hepatoprotection and cytotoxicity study. MPC-4. *Planta Medica*, *79*(13), PI106. doi:10.1055-0033-1352195

Elujoba, A., Odeleye, O., & Ogunyemi, C. (2005). Traditional medicine development for medical and dental primary health care delivery system in Africa. *African Journal of Traditional, Complementary, and Alternative Medicines*, *2*(1), 46–61. doi:10.4314/ajtcam.v2i1.31103

Elzaawely, A. A., Xuan, T. D., & Tawata, S. (2007). Essential oils, kava pyrones and phenolic compounds from leaves and rhizomes of *Alpinia zerumbet* (Pers.) BL Burtt. & RM Sm. and their antioxidant activity. *Food Chemistry*, *103*(2), 486–494. doi:10.1016/j.foodchem.2006.08.025

Emias, D., Bisrat, D., Alvaro, V., & Ben-Erik, V. W. (2000). Chemistry of Aloe species. *Current Organic Chemistry*, *4*(10), 1055–1078. doi:10.2174/1385272003375932

Emre, A. (2012). *Şifalı bitkiler ve vitaminler*. İstanbul, Turkey: Alfa Yayınları.

Endo, K., Kanno, E., & Oshima, Y. (1990). Structures of antifungal diarylheptenones, gingerenones A, B, C and isogingerenone B, isolated from the rhizomes of *Zingiber officinale*. *Phytochemistry*, *29*(3), 797–799. doi:10.1016/0031-9422(90)80021-8

Erik, S., & Tarikahya, B. (2004). Türkiye florası üzerine. *Kebikeç*, *17*, 139–163.

Erol, M. K. (1995). *Eğirdir (ısparta) yöresinin geleneksel halk ilacı olarak kullanılan bitkileri* (Unpublished master's thesis), Marmara Üniversitesi, Turkey.

Ersöz, T. (2012). Bitkisel ilaçlar ve gıda takviyeleri ile ilgili genel yaklaşım ve sorunlar. *MİSED*, *27-28*, 9–19.

Ertuğ, F., Tümen, G., & Çelik, A. (2003). Buldan (denizli) etnobotanik alan araştırma raporu 2002 yılı çalışması. *Türkiye Bilimler Akademisi, TÜBA-TÜKSEK Türkiye Kültür Envanteri Pilot Bölge Çalışmaları*, 76-87.

Eryaşar, P. (1998). *Gönen (balıkesir) yöresinin geleneksel halk ilacı olarak kullanılan bitkileri* (Unpublished master's thesis). Marmara Üniversitesi, Turkey.

Eşen, B. (2008). *Aydınlar köyü ve çevresinin (erdemli/mersin) etnobotanik özellikleri* (Unpublished master's thesis), Selçuk Üniversitesi, Turkey.

Eshun, K., & He, Q. (2004). Aloe vera: A valuable ingredient for the food, pharmaceutical and cosmetic industries: A Review. *Critical Reviews in Food Science and Nutrition*, *44*(2), 91–96. doi:10.1080/10408690490424694 PMID:15116756

Espinosa, L. N. (2016). Tratamiento de la obesidad con productos naturales. *Enfermería Investiga, Investigación, Vinculación. Docencia y Gestión*, *1*(4), 164–168.

Etkin, N. L. (1993). Anthropological methods in ethnopharmacology. *Journal of Ethnopharmacology*, *38*(2-3), 93–104. doi:10.1016/0378-8741(93)90003-N PMID:8510473

Etkin, N. L. (2006). *Edible medicines: An ethnopharmacology of food*. Tucson, AZ: The University of Arizona Press.

Evans, D. A., Rajasekharan, S., & Subramoniam, A. (2004). Enhancement in the absorption of water and electrolytes from rat intestine by *Hemidesmus indicus* R. Br. root (water extract). *Phytotherapy Research*, *18*(7), 511–515. doi:10.1002/ptr.1402 PMID:15305307

Eyzaguirre, P., & Watson, J. (2002). Home gardens and agrobiodiversity: An overview across regions. *Home gardens and in Situ Conservation of Plant Genetic Resources in Farming Systems, 10*.

Fabricant, D. S., & Farnsworth, N. R. (2001). The value of plants used in traditional medicine for drug discovery. *Environmental Health Perspectives, 109*(1), 69-75.

Fabricant, D. S., & Farnsworth, N. R. (2001). The value of plants used in traditional medicine for drug discovery. *Environmental Health Perspectives*, *109*(1), 69–75. PMID:11250806

Farook, S. M., Atlee, W. C., Kannan, S., Kumar, S., & Davey, M. S. (2011). Assessment of analgesic, anti-pyretic and anti-inflammatory activity of hydro-alcoholic fraction of *Hemidesmus indicus* root in experimental animals. *Scholars Research Library Der Pharmacia Lettre*, *3*(1), 442–447.

Faruque, M. O., Uddin, S. B., Barlow, J. W., Hu, S., Dong, S., Cai, Q., ... Hu, X. (2018). Quantitative ethnobotany of medicinal plants used by indigenous communities in the Bandarban district of Bangladesh. *Frontiers in Pharmacology*, *9*(40), 1–12. doi:10.3389/fphar.2018.00040 PMID:29467652

Faydaoğlu, E., & Sürücüoğlu, M. S. (2011). Geçmişten günümüze tıbbi ve aromatik bitkilerin kullanılması ve ekonomik önemi. *Kastamonu Üniversitesi Orman Fakültesi Dergisi*, *11*(1), 52–67.

Fazal, H., Ahmad, N., Ullah, I., Inayat, H., Khan, L., & Abbasi, B. H. (2011). Antibacterial potential in *Parthenium hysterophorus, Stevia rebaudiana* and *Ginkgo biloba*. *Pakistan Journal of Botany*, *43*(2), 1307–1313.

Federation, I. D. (2017). *IDF Diabetes Atlas* (8th ed.). Brussels, Belgium: International Diabetes Federation; doi:10.1016/S0140-6736(16)31679-8

Fernandes, E. C., & Nair, P. R. (1986). An evaluation of the structure and function of tropical homegardens. *Agricultural Systems*, *21*(4), 279–310. doi:10.1016/0308-521X(86)90104-6

Ferrannini, E., & Cushman, W. C. (2012). Diabetes and hypertension: The bad companions. *Lancet*, *380*(9841), 601–610. doi:10.1016/S0140-6736(12)60987-8 PMID:22883509

Ferrara, L. (2014). The *Garcinia cambogia* in phytotreatment of obesity: Activities of the hydroxycitric acid. *European Scientific Journal*, *10*(21), 291–301.

Ferro, V., Bradbury, B., Cameron, P., Shakir, E., Rahman, S., & Stimson, W. (2003). In vitro susceptibility of *Shigella flexneri* and *Streptococcus pyogenes* to inner gel of Aloe barbadensis Miller. *Antimicrobial Agents and Chemotherapy*, *3*(1), 1137–1139. doi:10.1128/AAC.47.3.1137-1139.2003 PMID:12604556

Ferruzzi, L., Turrini, E., Burattini, S., Falcieri, E., Poli, F., Mandrone, M., ... Hrelia, P. (2013). *Hemidesmus indicus* induces apoptosis as well as differentiation in a human promyelocytic leukemic cell line. *Journal of Ethnopharmacology*, *147*(1), 84–91. doi:10.1016/j.jep.2013.02.009 PMID:23500881

Feyzioglu, G. C., & Tornuk, F. (2016). Development of chitosan nanoparticles loaded with summer savory (*Satureja hortensis* L.) essential oil for antimicrobial and antioxidant delivery applications. *Lebensmittel-Wissenschaft + Technologie*, *70*, 104–110. doi:10.1016/j.lwt.2016.02.037

Ffolliott, P. (1998). Multiple benefits of arid land agroforestry home gardens and riparian ecosystems. In *North American Conference on Enterprise Development Through Agroforestry*.

Fierascu, I., Dinu-Pirvu, C. E., Fierascu, R. C., Velescu, B. S., Anuta, V., Ortan, A., & Jinga, V. (2018). Phytochemical profile and biological activities of *Satureja hortensis* L.: A review of the last decade. *Molecules (Basel, Switzerland)*, *57*(24), 11557–11563. doi:10.3390/molecules23102458 PMID:30257512

Fimognari, C., Lenzi, M., Ferruzzi, L., Turrini, E., Scartezzini, P., Poli, F., ... Cantelli-Forti, G. (2011). Mitochondrial pathway mediates the antileukemic effects of *Hemidesmus indicus*, a promising botanical drug. *PLoS One*, *6*(6), 215–244. doi:10.1371/journal.pone.0021544 PMID:21738701

Fitwe, P., Rathod, V., Sarnaik, D. K., & Wakade, R. B. (2013). Sugandi the Healing-magical-spiritual dream herb. *Asian Journal of Biomedical and Pharmaceutical Sciences*, *3*(22), 6–8.

Flores-Flores, A., Hernández-Abreu, O., Rios, M. Y., León-Rivera, I., Aguilar-Guadarrama, B., Castillo-España, P., ... Estrada-Soto, S. (2016). Vasorelaxant mode of action of dichloromethane-soluble extract from *Agastache mexicana* and its main bioactive compounds. *Pharmaceutical Biology*, *54*(12), 2807–2813. doi:10.1080/13880209.2016.1184690 PMID:27252080

Folke, C., Berkes, F., & Colding, J. (1998). Ecological practices and social mechanisms for building resilience and sustainability. In F. Berkes, & C. Folke (Eds.), *Linking social and ecological systems: management practices and social mechanisms for building resilience* (pp. 414–436). Cambridge, UK: Cambridge University Press.

Fort, R. S., Trinidad Barnech, J. M., Dourron, J., Colazzo, M., Aguirre-Crespo, F. J., Duhagon, M. A., & Álvarez, G. (2018). Isolation and structural characterization of bioactive molecules on prostate cancer from Mayan traditional medicinal plants. *Pharmaceuticals*, *11*(3), 78. doi:10.3390/ph11030078 PMID:30110911

Foster, M., Hunter, L., & Samman, S. (2011). Evaluation of the nutritional and metabolic effects of Aloe vera. In I. F. F. Benzie, & S. Watchel-Galor (Eds.), Herbal Medicine: Biomolecular and Clinical Aspects (2nd ed.). Boca Raton, FL: CRC Press. doi:10.1201/b10787-4

Frankel, O. H., Brown, A. H., & Burdon, J. J. (1995). *The conservation of plant biodiversity*. Cambridge University Press.

Frantz, S., & Smith, A. (2003). New drug approvals for 2002. *Nature Reviews. Drug Discovery*, *2*(2), 95–96. doi:10.1038/nrd1014 PMID:12572538

Fritz, G. J. (2005). Paleoethnobotanical methods and applications. H. D. G. Maschner, & C. Chippindale (Eds.). Handbook of archaeological methods (pp. 771–832). Walnut Creek, CA: Altamira Press.

Funk, V. A., Bayer, R. J., Keeley, S., Chan, R., Watson, L., Gemeinholzer, B., & Jansen, R. K. (2005). Everywhere but Antarctica: Using a super tree to understand the diversity and distribution of the Compositae. *Biologiske Skrifter*, *55*, 343–374.

Furkan, M. K. (2016). *Adıyaman ilinde yetişen bazı bitkilerin etnobotanik özellikleri* (Unpublished master's thesis), Adıyaman Üniversitesi, Adıyaman.

Gadge, N. B., & Jalalpure, S. S. (2011). Natriuretic and saluretic effects of *Hemidesmus indicus* R. Br. root extracts in rats. *Indian Journal of Pharmacology*, *43*(6), 714–717. PMID:22144781

Gadgil, M., Berkes, F., & Folke, C. (1993). Indigenous knowledge for biodiversity conservation. *Ambio*, *22*, 151–156.

Gairola, S., Shariff, N. M., Bhatt, A., & .... (2010). Influence of climate change on production of secondary chemicals in high altitude medicinal plants: Issues needs immediate attention. *Journal of Medicinal Plants Research*, *4*(18), 1825–1829.

Galani, V. J., Patel, B. G., & Rana, D. G. (2010). *Sphaeranthus indicus* Linn.: A phytopharmacological review. *International Journal of Ayurveda Research*, *1*(4), 247–253. doi:10.4103/0974-7788.76790 PMID:21455454

Galati, E. M., Monforte, M. T., Forestieri, A. M., Miceli, N., Bade, A., & Trovato, A. (1999). *Salvadora persica* L.: Hypolipidemic activity on experimental hypercholesterolemia in rat. *Phytomedicine*, *6*(3), 181–185. doi:10.1016/S0944-7113(99)80007-5 PMID:10439483

Galhena, D. H., Freed, R., & Maredia, K. M. (2013). Home gardens: A promising approach to enhance household food security and wellbeing. *Agriculture & Food Security*, *2*(1), 8. doi:10.1186/2048-7010-2-8

Galhena, P. B., Samarakoon, S. R., Thabrew, M. I., Weerasinghe, G. A. K., Thammitiyagodage, M. G., Ratnasooriya, W. D., & Tennekoon, K. H. (2012). Anti-inflammatory activity is a possible mechanism by which the polyherbal formulation comprised of *Nigella sativa* (seeds), *Hemidesmus indicus* (root), and *Smilax glabra* (rhizome) mediates its antihepatocarcinogenic effects. *Evidence-Based Complementary and Alternative Medicine*, *2012*, 1–11. doi:10.1155/2012/108626 PMID:23243426

Galluzzi, G., Eyzaguirre, P., & Negri, V. (2010). Home gardens: Neglected hotspots of agro-biodiversity and cultural diversity. *Biodiversity and Conservation*, *19*(13), 3635–3654. doi:10.100710531-010-9919-5

Ganesan, S., Venkateshan, G., & Banumathy, N. (2006). Medicinal plants used by ethnic group Thottianaickans of Semmalai hills (reserved forest), Tiruchirappalli district, Tamil Nadu. *Indian Journal of Traditional Knowledge*, *5*(2), 245–252.

Gangwar, A., & Ramakrishnan, P. (1990). Ethnobiological notes on some tribes of Arunachal Pradesh, northeastern India. *Economic Botany*, *44*(1), 94–105. doi:10.1007/BF02861071

Gao, X. M., Zhang, T. M., Zhang, J. R., Guo, J. S., & Zhong, G. S. (2007). Chinese material medica. *Beijing: China Press of Traditional Chinese Medicine, 60, 323*–324.

García, J. G., García, M. L., Zamilpa, A., Almanza, J. C. P., Jasso, I. E., Ramos, R. R., & Alarcón-Aguilar, F. J. (2017). Chemical characterization of a hypoglycemic extract from *Cucurbita ficifolia* Bouche that induces liver glycogen accumulation in diabetic mice. *African Journal of Traditional, Complementary, and Alternative Medicines*, *14*(3), 218–230. doi:10.21010/ajtcam.v14i3.24 PMID:28480434

Gaur, R. D. (1999). *Flora of the District Garhwal, North West Himalaya*. Transmedia.

Gaur, R. D., & Sharma, J. (2011). Indigenous knowledge on the utilization of medicinal plant diversity in the Siwalik region of Garhwal Himalaya, Uttarakhand. *Journal of Forest and Environmental Science*, *27*(1), 23–31.

Gautam, R., Sthapit, B., & Shrestha, P. (2006). *Home gardens in Nepal: Proceeding of a workshop on enhancing the contribution of home garden to on-farm management of plant genetic resources and to improve the livelihoods of Nepalese farmers: Lessons learned and policy implications,* August 6–7, 2004. Pokhara, Nepal: LI-BIRD, Biodiversity International and SDC.

Gautam, A., & Batra, A. (2012). Ethnomedicinal plant product used by the local traditional practitioners in Mount Abu. *World Journal of Pharmaceutical Sciences*, *1*, 10–18.

Gautam, R. P., Rajkumar, S. D., Srivastava, S. K., & Singh, S. K. (2015). Folk medicinal uses of plants from Kusmi Forest, Uttar Pradesh, Gorakhpur, India. *International Journal of Current Microbiology and Applied Sciences*, *4*, 343–351.

Gautam, R., Sthapit, B., Subedi, A., Poudel, D., Shrestha, P., & Eyzaguirre, P. (2009). Home gardens management of key species in Nepal: A way to maximize the use of useful diversity for the well-being of poor farmers. *Plant Genetic Resources*, *7*(2), 142–153. doi:10.1017/S1479262108110930

Gautam, S., & Awasthi, P. (2007). Nutrient composition and physiochemical characteristics of *Aloe vera* (*Aloe barbadensis*) powder. *Journal of Food Science and Technology*, *44*(1), 224–225.

Gaziano, T. A., Bitton, A., Anand, S., & Weinstein, M. C. (2009). The global cost of nonoptimal blood pressure. *Journal of Hypertension*, *27*(7), 1472–1477. doi:10.1097/HJH.0b013e32832a9ba3 PMID:19474763

Gebremedhin, Y. (2018). Review on Medicinal Value of Aloe vera in Veterinary Practice. *Biomedical Journal of Science and Technical Research*, *6*(1), 1–6.

Gençay, A. (2007). *Cizre (şırnak)'nin etnobotanik özellikleri* (Unpublished master's thesis). Yüzüncü Yıl Üniversitesi, Van.

Gençler, Ö. A. M., & Koyuncu, M. (2005). Traditional medicinal plants used in pınarbaşı area (kayseri-türkiye). *Turkish J. Pharm. Sci.*, *2*(2), 63–82.

George, S., Tushar, K. V., Unnikrishnan, K. P., Hasim, K. M., & Balachandran, I. (2008). *Hemidesmus indicus* (L.) R. Br. A Review. *Journal of Plant Sciences*, *3*(2), 146–156. doi:10.3923/jps.2008.146.156

Ghannadi, A., Rabbani, M., Ghaemmaghami, L., & Malekian, N. (2012). Phytochemical screening and essential oil analysis of one of the Persian sedges; *Cyperus rotundus* L. *International Journal of Pharmaceutical Sciences and Research*, *3*, 424–427.

Ghasemzadeh, A., Jaafar, H. Z., & Rahmat, A. (2010). Antioxidant activities, total phenolics and flavonoids content in two varieties of Malaysia young ginger (*Zingiber officinale* Roscoe). *Molecules (Basel, Switzerland)*, *15*(6), 4324–4333. doi:10.3390/molecules15064324 PMID:20657444

Giday, M., & Ameni, G. (2003). An ethnobotanical survey of plants of veterinary importance in two woredas of southern Tigray, northern Ethiopia. *Sinet. An Ethiopian Journal of Science*, *26*(2), 123–136.

Gilgil, E., Kaçar, C., Bütün, B., Tuncer, T., Urhan, S., Yildirim, Ç., ... Dündar, Ü. (2005). Prevalence of low back pain in a developing urban setting. *Spine*, *30*(9), 1093–1098. doi:10.1097/01.brs.0000161007.46849.4c PMID:15864165

Girach, R. D., Aminuddin, Siddiqui, P. A., & Khan, S. A. (1994). Traditional plant remedies among the Kondh of District Dhenkanal (Orissa). *International Journal of Pharmacognosy*, *32*(3), 274–283. doi:10.3109/13880209409083005

Gnabre, J., Bates, R., & Huang, R. C. (2015). Creosote bush lignans for human disease treatment and prevention: Perspectives on combination therapy. *Journal of Traditional and Complementary Medicine*, *5*(3), 119–126. doi:10.1016/j.jtcme.2014.11.024 PMID:26151022

Gopi, G., Elumalai, A., & Jayasri, P. (2012). A concise review on *Tagetes erecta. International Journal of Phytopharmacy Research, 3*(1), 16–19.

Gopi, S., & Setty, O. H. (2010). Beneficial effect of the administration of *Hemidesmus indicus* against bromobenzene induced oxidative stress in rat liver mitochondria. *Journal of Ethnopharmacology, 127*(1), 200–203. doi:10.1016/j.jep.2009.09.043 PMID:19799985

Govea-Salas, M., Morlett-Chávez, J., Rodriguez-Herrera, R., & Ascacio-Valdés, J. (2017). Some Mexican plants used in traditional medicine. In *Aromatic and Medicinal Plants: Back to Nature* (pp. 191–200). BoD: Books on Demand.

Grace, O. M., Simmonds, M. S. J., Smith, G. F., & Van, W. (2009). Documented utility and biocultural value of Aloe L. (Asphodelaceae) a review. *Economic Botany, 63*(2), 167–178. doi:10.100712231-009-9082-7

Graul, A. I. (2001). The year's new drugs. *Drug News & Perspectives, 14*(1), 12–31. PMID:12819805

Greenwell, M., & Rahman, P. K. S. M. (2015). Medicinal plants: Their use in anticancer treatment. *International Journal of Pharmaceutical Sciences and Research, 6*(10), 4103–4112. doi:10.13040/IJPSR.0975-8232.6(10).4103-12 PMID:26594645

Grenier, L. (1998). *Working with indigenous knowledge: a guide for researchers*. IDRC.

Groom, Q. J., & Reynolds, T. (1987). Barbaloin in Aloe species. *Planta Medica, 53*(4), 345–348. doi:10.1055-2006-962735 PMID:17269040

Grzanna, R., Lindmark, L., & Frondoza, C. (2005). Ginger - A herbal medicinal product with broad anti-inflammatory actions. *Journal of Medicinal Food, 8*(2), 125–132. doi:10.1089/jmf.2005.8.125 PMID:16117603

Guerra, R. N., Pereira, H. A., Silveria, L. M., & Olea, R. S. (2003). Immunomodulatory property of AlternantheratenellaColla aqueous extracts in mice. *Braz. J. Med. Biol. Res.,* 36, 1,215-1,219.

Guerrini, A., Mancini, I., Maietti, S., Rossi, D., Poli, F., Sacchetti, G., ... Borgatti, M. (2014). Expression of pro-inflammatory interleukin-8 is reduced by ayurvedic decoctions. *Phytotherapy Research, 28*(8), 1173–1181. doi:10.1002/ptr.5109 PMID:24395450

Günbatan, T., Gürbüz, İ., & Özkan, A. M. G. (2016). The current status of ethnopharmacobotanical knowledge in çamlıdere (Ankara, Turkey). *Turkish Journal of Botany, 40*(3), 241–249. doi:10.3906/bot-1501-37

Güner, A., Aslan, S., Ekim, T., Vural, M., & Babaç, M. T. (Eds.). (2012). *Türkiye bitkileri listesi damarlı bitkiler*. İstanbul, Turkey: Nezahat Gökyiğit Botanik Bahçesi ve Flora Araştırmaları Derneği Yayını.

Güneş, S. (2010). *Karaisalı (adana) ve köylerinde halkın kullandığı doğal bitkilerin etnobotanik yönden araştırılması* (Unpublished master's thesis), Niğde Üniversitesi, Niğde.

Gupta, A. P., Gupta, M. M., & Kumar, S. (1999). Simultaneous determination of curcuminoids in *Curcuma* samples using high performance thin layer chromatography. *Journal of Liquid Chromatography & Related Technologies, 22*(10), 1561–1569. doi:10.1081/JLC-100101751

Gupta, M. M., Verma, R. K., & Misra, L. N. (1992). Terpenoids from *Hemidesmus indicus. Phytochemistry, 31*(11), 4036–4037. doi:10.1016/S0031-9422(00)97582-4

Gupta, N. S. (2006). *The ayurvedic system of indian medicine*. New Delhi, India: Bharatiya Kala Prakashan.

Gupta, P. N. (1981). Antileprotic action of an extract from "Anantamul" (*Hemidesmus Indicus* R. Br.). *Leprosy in India, 53*(3), 354–359. Retrieved from https://easyayurveda.com/2013/12/20/sariva-hemidesmus-indicus-benefits-usage-dose-side-effects/ PMID:7278140

Gutiérrez-Grijalva, E. P., Picos-Salas, M. A., Leyva-López, N., Criollo-Mendoza, M. S., Vazquez-Olivo, G., & Heredia, J. B. (2018). Flavonoids and phenolic acids from Oregano: Occurrence, biological activity and health benefits. *Plants*, *7*(1), 2. doi:10.3390/plants7010002 PMID:29278371

Güzel, Y., Güzelşemme, M., & Miski, M. (2015). Ethnobotany of medicinal plants used in antakya: A multicultural district in hatay province of turkey. *Journal of Ethnopharmacology*, *174*, 118–152. doi:10.1016/j.jep.2015.07.042 PMID:26239155

Habila, J. D., Bello, I. A., Dzikwi, A. A., Musa, H., & Abubakar, N. (2010). Total phenolics and antioxidant activity of *Tridax procumbens* Linn. *African Journal of Pharmacy and Pharmacology*, *4*(3), 123–126.

Haghgoo, R., Mehran, M., Zadeh, H. F., Afshari, E., & Zadeh, N. F. (2017). Comparison Between Antibacterial Effect of Chlorhexidine 0.2% and Different Concentrations of *Cyperus rotundus* Extract: An *In vitro* Study. *Journal of International Society of Preventive & Community Dentistry*, *7*, 242–246. PMID:29026695

Hahn, G. (1992). *Hypericum perforatum* (St. John's wort): A medicinal herb used in antiquity and still of interest today. *Journal of Naturopathic Medicines*, *3*, 94–96.

Haji Faraji, M., & Haji Tarkhani, A. H. (1999). The effect of sour tea (*Hibiscus sabdariffa*) on essential hypertension. *Journal of Ethnopharmacology*, *65*(3), 231–236. doi:10.1016/S0378-8741(98)00157-3 PMID:10404421

Hajra, P. K., & Jain, K. K. (1983). *A contribution to the botany of Nanda Devi National Park*. Howrah: Botanical Survey of India.

Hamilton, A. C. (2004). *Medicinal plants, conservation and livelihoods*. Kluwer Academic Publishers. Retrieved from http://www.cbd.int/doc/articles/2004/A

Hamilton, A. C. (2004). Medicinal plants, conservation and livelihoods. *Biodiversity and Conservation*, *13*(8), 1477–1517. doi:10.1023/B:BIOC.0000021333.23413.42

Hamman, J. (2008). Composition and application of Aloe vera leaf gel. *International Journal of Pharmacological Science*, *12*(8), 1599–1616.

Hammer, W. (1999). *Functional soft tissue examination and treatment by manual methods* (2nd ed.). Gaithersburg, MD: Aspen.

Hannan, M. T., & Freeman, J. (1989). *Organizational ecology*. Cambridge, MA: Harvard University Press.

Hansson, T. H., & Hansson, E. K. (2000). The effects of common medical interventions on pain, back function, and work resumption in patients with chronic low back pain: A prospective 2-year cohort study in six countries. *Spine*, *25*(23), 3055–3064. doi:10.1097/00007632-200012010-00013 PMID:11145817

Hany, Y. H., El-sayed, M., Magda, A. S., Ahmed, Z., & Magdy, A. A. (2004). Efficacy of Aloe vera gel leaves for treatment of skin infections among animals. In *Proceedings of the first anniversary conference of the Faculty of Veterinary Science*. Moshtohor: Semantic Scholar.

Harborne, J. B. (1962). Plant polyphenols. 5: Occurrence of azalein and related pigments in flowers of *Plumbago* and *Rhododendron* species. *Archives of Biochemistry and Biophysics*, *96*(1), 171–178. doi:10.1016/0003-9861(62)90467-8 PMID:13904580

Harris, E. M. (1995). Inflorescence and floral ontogeny in Asteraceae: A synthesis of historical and current concepts. *Botanical Review*, *61*(3), 93–131. doi:10.1007/BF02887192

Hasanein, P., Felehgari, Z., & Emamjomeh, A. (2016). Preventive effects of *Salvia officinalis* L. against learning and memory deficit induced by diabetes in rats: Possible hypoglycaemic and antioxidant mechanisms. *Neuroscience Letters*, *622*, 72–77. doi:10.1016/j.neulet.2016.04.045 PMID:27113201

Hassan, G. A., Ahmad, T. A., & Mohi-ud-din, R. A. (2013). An Ethnobotanical study in Budgam of Kashmir Valley: An attempt to explore and document traditional knowledge of the area. *International Research Journal of Pharmacy*, *4*(1), 201–204.

Hassan, S. U., Haddawy, P., & Zhu, J. (2014). A bibliometric study of the world's research activity in sustainable development and its sub-areas using scientific literature. [CrossRef]. *Scientometrics*, *99*(2), 549–579. doi:10.100711192-013-1193-3

Haunss, S. (2013). *Conflicts in the knowledge society: the contentious politics of intellectual property* (Vol. 20). Cambridge University Press; doi:10.1017/CBO9781139567633

Heber, D. (2007). *Physicians' Desk Reference for Herbal Medicines* (4th ed.). Montvale: Thompson Health Care.

Hefferman, C. (2013). *Qualitative Research Approach.* [Online] Available from http://www.explorable.com

Heinrich, M., & Teoh, H. L. (2004). Galanthamine from snowdrop: The development of a modern drug against Alzheimer's disease from local Caucasian knowledge. *Journal of Ethnopharmacology*, *92*(2-3), 147–162. doi:10.1016/j.jep.2004.02.012 PMID:15137996

Hemadri, K., & Rao, S. S. (1983). Leucorrhoea and Menorrhagia: Tribal medicine. *Ancient Science of Life*, *3*(1), 40–41. PMID:22557376

Hemanth Kumar, K., Tamatam, A., Pal, A., & Khanum, F. (2013). Neuroprotective effects of *Cyperus rotundus* on SIN-1 induced nitric oxide generation and protein nitration: Ameliorative effect against apoptosis mediated neuronal cell damage. *Neurotoxicology*, *34*, 150–159. doi:10.1016/j.neuro.2012.11.002 PMID:23174672

Hemmati, A. A., Arzi, A., & Amin, M. (2002). Effect of *Achillea millefolium* extract in wound healing of rabbit. *Journal of Natural Remedies*, *2*(2), 164–167.

Hertling, D., & Kessler, R. M. (2006). *Management of common musculoskeletal disorders: physical therapy principles and methods*. Lippincott Williams & Wilkins.

High, C., & Shackleton, C. M. (2000). The comparative value of wild and domestic plants in home gardens of a South African rural village. *Agroforestry Systems*, *48*(2), 141–156. doi:10.1023/A:1006247614579

Hoffman, B., & Gallaher, T. (2007). Importance indices in ethnobotany. *Ethnobotany Research and Applications*, *5*, 201–218. doi:10.17348/era.5.0.201-218

Hoogerbrugge, I., & Fresco, L. O. (1993). Homegarden systems: Agricultural characteristics and challenges. *IIED, 39.*

Horton, D., & Mackay, R. (2003). Using evaluation to enhance institutional learning and change: Recent experiences with agricultural research and development. *Agricultural Systems*, *78*(2), 127–142. doi:10.1016/S0308-521X(03)00123-9

Hossain, S., Kader, G., Nikkon, F., & Yeasmin, T. (2012). Cytotoxicity of the rhizome of medicinal plants. *Asian Pacific Journal of Tropical Biomedicine*, *2*(2), 125–127. doi:10.1016/S2221-1691(11)60205-0 PMID:23569881

Huai, H., & Hamilton, A. (2009). Characteristics and functions of traditional homegardens: A review. *Frontiers of Biology in China*, *4*(2), 151–157. doi:10.100711515-008-0103-1

Huang, B., He, D., Chen, G., Ran, X., Guo, W., Kan, X., ... Liu, J. (2018). α-Cyperone inhibits LPS-induced inflammation in BV-2 cells through activation of Akt/Nrf2/HO-1 and suppression of the NF-κB pathway. *Food & Function*, *9*(5), 2735–2743. doi:10.1039/C8FO00057C PMID:29667667

Huang, K. C. (1998). *The dried tuber of Cyperus rotundus L. The pharmacology of Chinese herbs* (2nd ed., pp. 320–321). Boca Raton, FL: CRC Press.

Hussain, J., Muhammad, Z., Ullah, R., Khan, F. U., Khan, I. U., Khan, N., & Jan, S. (2010). Evaluation of the chemical composition of *Sonchus eruca* and *Sonchus asper. The Journal of American Science*, *6*(9), 231–235.

Hussain, M. (2001). *Geography of Jammu and Kashmir*. New Delhi, India: Rajesh Publications.

Hussain, M., Waqas, H. M., Hussain, I., Majeed, A., Raza, S. M., & Janbaz, K. H. (2018). Pharmacological validation of the folkloric uses of *Cyperus rotundus* L. in different ailments: An *in vivo* and *in vitro* research. *Pakistan Journal of Pharmaceutical Sciences*, *31*, 95–102. PMID:29348090

Hutter, J. A., Salmon, S., Stavinoha, W. B., Satsangi, N., Williams, R. F., & Streeper, R. T. (1996). Anti-inflammatory C-Glycosylchromone from Aloe Barbadensis. *Journal of Natural Products*, *59*(1), 541–543. doi:10.1021/np9601519 PMID:8778246

Iddamaldeniya, S. S., Thabrew, M. I., Wickramasinghe, S. M. D. N., Ratnatunge, N., & Thammitiyagodage, M. G. (2006). A long-term investigation of the anti-hepatocarcinogenic potential of an indigenous medicine comprised of Nigella sativa, *Hemidesmus indicus* and Smilax glabra. *Journal of Carcinogenesis*, *5*(11), 1–7. PMID:16684351

Iddamaldeniya, S. S., Wickramasinghe, N., Thabrew, I., Ratnatunge, N., & Thammitiyagodage, M. G. (2003). Protection against diethylnitrosoamine-induced hepatocarcinogenesis by an indigenous medicine comprised of *Nigella sativa, Hemidesmus indicus* and *Smilax glabra*: A preliminary study. *Journal of Carcinogenesis*, *2*(6), 1–6. PMID:14613573

Idrisi, M. S., Badola, H. K., & Singh, R. (2010). Indigenous knowledge and medicinal use of plants by local communities in Rangit Valley, south Sikkim, India. *NeBio*, *1*, 34–45.

Ikram, M. (1981). A review on the medicinal plants. *Hamdard*, *24*(1/2), 102–129.

Imam, M. Z., & Sumi, C. D. (2014). Evaluation of antinociceptive activity of hydromethanol extract of *Cyperus rotundus* in mice. *BMC Complementary and Alternative Medicine*, *14*(1), 83. doi:10.1186/1472-6882-14-83 PMID:24589067

Imran, M., Hussain, A., Gurmani, Z. A., Zahid, M. S., & Khan, S. (2008). Cultivation of sunflower and its utilization in livestock. *Science Journal of Technology Dev*, *27*, 1–2.

Inta, A., Shengji, P., Balslev, H., Wangpakapattanawong, P., & Trisonthi, C. (2008). A comparative study on medicinal plants used in Akha's traditional medicine in China and Thailand: Cultural coherence or ecological divergence? *Journal of Ethnopharmacology*, *116*(3), 508–517. doi:10.1016/j.jep.2007.12.015 PMID:18280071

Ishii, Y., Tanizawa, H., & Takino, Y. (1994). Studies of Aloe vera: Mechanism of Cathartic Effect. *Biological & Pharmaceutical Bulletin*, *17*(1), 651–653. doi:10.1248/bpb.17.651 PMID:7920425

Islam, M. Z., Hossain, M. T., Hossen, F., Akter, M. S., & Mokammel, M. A. (2016). In-vitro antioxidant and antimicrobial activity of *Bougainvillea glabra* flower. *Research Journal of Medicinal Plant, 10*(3), 228–236. . 2016.228.236 doi:10.3923/rjmp

Ito, T., Endo, H., Shinohara, H., Oyama, M., Akao, Y., & Iinuma, M. (2012). Occurrence of stilbene oligomers in *Cyperus* rhizomes. *Fitoterapia*, *83*(8), 1420–1429. doi:10.1016/j.fitote.2012.08.005 PMID:23339254

Itrat, M., & Zarnigar, K. (2013). *Aloe vera*: A review of its clinical effectiveness. *International Research Journal of Pharmacy*, *4*(1), 75–79.

Izaola, O., de Luis, D., Sajoux, I., Domingo, J. C., & Vidal, M. (2015). Inflamación y obesidad (lipoinflamación). *Nutrición Hospitalaria*, *31*(6), 2352–2358. PMID:26040339

İzgi, M. N. (2017). Intent-to-treat medicinal plants found ın the flora of mardin province of turkey and neighborhood. In *International Symposium of Medicinal, Aromatic and Dye Plants (REYHAN 2017)* (pp. 1-8). Malatya. Academic Press.

İzgi, M. N. (2017). Intent-to-treat medicinal plants found ın the flora of mardin province of turkey and neighborhood. In *International Symposium of Medicinal, Aromatic and Dye Plants (REYHAN 2017) Proceedings Book* (pp. 1-8). Malatya.

Jahan, F. I., Hasan, M. R. U., Jahan, R., Seraj, S., Chowdhury, A. R., Islam, M. T., ... Rahmatullah, M. (2011). A comparison of medicinal plant usage by folk medicinal practitioners of two adjoining villages in Lalmonirhat district, Bangladesh. *American-Eurasian Journal of Sustainable Agriculture*, *5*, 46–66.

Jahan, R., Al-Nahain, A., Majumder, S., & Rahmatullah, M. (2014). Ethnopharmacological significance of *Eclipta alba* (L.) Hassk. (Asteraceae). *International Scholarly Research Notices*, *2014*, 1–22. doi:10.1155/2014/385969 PMID:27355071

Jain, A., & Basal, E. (2003). Inhibition of Propionibacterium acnes-induced mediators of inflammation by Indian herbs. *Phytomedicine*, *10*(1), 34–38. doi:10.1078/094471103321648638 PMID:12622461

Jain, S. K. (1991). *Dictionary of Indian folk medicine and ethnobotany*. New Delhi, India: Deep Publications.

Jain, S. K. (2004). Credibility of traditional knowledge: The criteria of multilocational and multiethnic use. *Indian Journal of Traditional Knowledge*, *3*(2), 137–153.

Jain, S. K. (2005). Dynamism in ethnobotany. *Ethnobotany*, *17*, 3–13.

Jain, S. K., & Jain, V. (2015). Evolution of ethnobotany in last six decades in India: An overview. *Ethnobotany*, *27*, 1–6.

Jain, S. K., & Jain, V. (2017). *Methods and approaches in ethnobotany (Concepts, practices and prospects)*. New Delhi, India: Deep Publications.

Jain, S. K., & Rao, R. R. (1976). *A handbook of field and herbarium methods*. New Delhi, India: Today and Tomorrow's Printers and Publishers.

Jain, S. K., & Rao, R. R. (1977). *A Handbook of field and herbarium methods*. New Delhi, India: Today and Tomorrow's Printers and Publishers.

Jain, V., & Jain, S. K. (2016). *Compendium of Indian folk medicine and ethnobotany (1991-2015)*. New Delhi, India: Deep Publications.

Jaju, S., Indurwade, N., Sakarkar, D., Fuloria, N., & Ali, M. (2009). Isolation of galangogalloside from rhizomes of *Alpinia galanga*. *International Journal of Green Pharmacy*, *3*(2), 144. doi:10.4103/0973-8258.54907

James, S. L., Abate, D., Abate, K. H., Abay, S. M., Abbafati, C., Abbasi, N., ... Murray, C. J. L. (2018). Global, regional, and national incidence, prevalence, and years lived with disability for 354 diseases and injuries for 195 countries and territories, 1990–2017: A systematic analysis for the Global Burden of Disease Study 2017. *Lancet*, *392*(10159), 1798–1858. doi:10.1016/S0140-6736(18)32279-7 PMID:30496104

Jang, D. S., Han, A. R., Park, G., Jhon, G. J., & Seo, E. K. (2004). Flavonoids and aromatic compounds from the rhizomes of *Zingiber zerumbet*. *Archives of Pharmacal Research*, *27*(4), 386–389. doi:10.1007/BF02980078 PMID:15180302

Jang, D. S., & Seo, E. K. (2005). Potentially bioactive two new natural sesquiterpenoids from the rhizomes of *Zingiber zerumbet*. *Archives of Pharmacal Research*, *28*(3), 294–296. doi:10.1007/BF02977795 PMID:15832816

Jan, R. A., & Khare, N. (2013). Ethnopharmacological uses of plants among tribal and rural folks of Shopion Forest Area of Kashmir. *International Journal of Science and Research*, *4*(5), 232–234.

Janssen, A. M., & Scheffer, J. J. (1985). Acetoxychavicol Acetate, an Antifungal Component of *Alpinia galanga*1. *Planta Medica*, *51*(6), 507–511. doi:10.1055-2007-969577 PMID:17345272

Jantan, I. B., Ahmad, F. B., & Ahmad, A. S. (2004). Constituents of the rhizome and seed oils of greater galangal *Alpinia galangal* (L.) Willd. from Malaysia. *The Journal of Essential Oil Research*, *16*(3), 174–176. doi:10.1080/10412905.2004.9698687

Jatoi, S. A., Kikuchi, A., Gilani, S. A., & Watanabe, K. N. (2007). Phytochemical, pharmacological and ethnobotanical studies in mango ginger (*Curcuma amada* Roxb.; Zingiberaceae). Phytotherapy Research. *An International Journal Devoted to Pharmacological and Toxicological Evaluation of Natural Product Derivatives*, *21*(6), 507–516. PMID:17397131

Javeid, G. N. (1968). Flora of Srinagar, Kashmir. *Science*, *5*, 59–71.

Jayaram, S., & Dharmesh, S. M. (2011). Assessment of antioxidant potentials of free and bound phenolics of *Hemidesmus indicus* (L) R. Br against oxidative damage. *Pharmacognosy Research*, *3*(4), 225–231. doi:10.4103/0974-8490.89741 PMID:22224044

Jebasingh, D., Devavaram, J. D., Venkataraman, S., Adeghate, E., & Starling, E. B. (2014). The protective effects of *Cyperus rotundus* on behavior and cognitive function in a rat model of hypoxia injury. *Pharmaceutical Biology*, *52*(12), 1558–1569. doi:10.3109/13880209.2014.908395 PMID:25026346

Jeeva, S., Lyndem, F. G., Sawian, J. T., Laloo, R. C., & Mishra, B. P. (2011). MyricaesculentaBuch.- Ham. ex D. Don. - a potential ethnomedicinal species in a subtropical forest of Meghalaya, northeast India. *Asian Pacific Journal of Tropical Biomedicine*, *1*(2), S174–S177. doi:10.1016/S2221-1691(11)60150-0

Jensen, G. M., Lenninger, J., Beaman, R., Taylor, K., & Benson, N. (2015). Support of Joint Function, Range of Motion and Physical Activity Levels by Consumption of a Water-Soluble Egg Membrane Hydrolyzate. *Journal of Medicinal Food*, *18*(9), 1042–1048. doi:10.1089/jmf.2015.0041 PMID:26355952

Jensen, M. C., Brant-Zawadzki, M. N., Obuchowski, N., Modic, M. T., Malkasian, D., & Ross, J. S. (1994). Magnetic resonance imaging of the lumbar spine in people without back pain. *The New England Journal of Medicine*, *331*(2), 69–73. doi:10.1056/NEJM199407143310201 PMID:8208267

Jeong, S. J., Miyamoto, T., Inagaki, M., Kim, Y. C., & Higuchi, R. (2000). Rotundines A-C, three novel sesquiterpene alkaloids from *Cyperus rotundus*. *Journal of Natural Products*, *63*(5), 673–675. doi:10.1021/np990588r PMID:10843585

Jiang, H., Sólyom, A. M., Timmermann, B. N., & Gang, D. R. (2005). Characterization of gingerol-related compounds in ginger rhizome (*Zingiber officinale* Rosc.) by high-performance liquid chromatography/electrospray ionization mass spectrometry. *Rapid Communications in Mass Spectrometry*, *19*(20), 2957–2964. doi:10.1002/rcm.2140 PMID:16189817

Jiménez-Arellanes, M. A., García-Martínez, I., & Rojas-Tomé, S. (2014). Biological potential of medicinal species of *Cnidoscolus* genus. *Revista Mexicana de Ciencias Farmacéuticas*, *45*(4), 1–6.

Jin, J. H., Lee, D. U., Kim, Y. S., & Kim, H. P. (2011). Anti-allergic activity of sesquiterpenes from the rhizomes of *Cyperus rotundus*. *Archives of Pharmacal Research*, *34*(2), 223–228. doi:10.100712272-011-0207-z PMID:21380805

Jitoe, A., Masuda, T., Tengah, I. G. P., Suprapta, D. N., Gara, I. W., & Nakatani, N. (1992). Antioxidant activity of tropical ginger extracts and analysis of the contained curcuminoids. *Journal of Agricultural and Food Chemistry*, *40*(8), 1337–1340. doi:10.1021/jf00020a008

Johannes, R. E. (1998). The case for data-less marine resource management: Examples from tropical nearshore finfisheries. *Trends in Ecology & Evolution*, *13*(6), 243–246. doi:10.1016/S0169-5347(98)01384-6 PMID:21238285

Johari, S., Joshi, C., & Gandhi, T. (2016). Effect of *Cyperus Rotundus* on Cytokine Gene Expression in Experimental Inflammatory Bowel Disease. *Iranian Journal of Medical Sciences*, *41*, 391–398. PMID:27582588

Jolad, S. D., Lantz, R. C., Chen, G. J., Bates, R. B., & Timmermann, B. N. (2005). Commercially processed dry ginger (*Zingiber officinale*): Composition and effects on LPS-stimulated PGE 2 production. *Phytochemistry*, *66*(13), 1614–1635. doi:10.1016/j.phytochem.2005.05.007 PMID:15996695

Jones, G. E. (1987). *The conservation of ecosystems and species*. New York: Croom Helm.

Joshi, A., & Chauhan, R. S. (2013). Phytochemical Analysis and Cytotoxicity Studies of *Curcuma amada* rhizomes in BHK-21 Cells. *International Journal of Scientific Research in Environmental Sciences*, *1*(12), 365–371. doi:10.12983/ijsres-2013-p365-371

Joshi, B., & Pant, S. C. (2012). Ethnobotanical study of some common plants used among the tribal communities of Kashipur, Uttarakhand. *Indian Journal of Natural Products and Resources*, *3*(2), 262–266.

Joshi, J., Vaidya, R., Shah, R., & Vaidya, A. B. (2008). The phytochemical and pharmacological basis of Ayurvedic plants used in treatment of Leucorrhoea and Menorrhagia. *The Journal of Research and Education in Indian Medicine*, *14*, 1–16.

Joshi, K. (2005). Ethnobotanical information on *Cyperus rotundus* L. in Nepal. *Bionotes*, *7*, 86–88.

Jung, S. H., Kim, S. J., Jun, B. G., Lee, K. T., Hong, S. P., Oh, M. S., ... Choi, J. H. (2013). α-Cyperone, isolated from the rhizomes of *Cyperus rotundus*, inhibits LPS-induced COX-2 expression and PGE2 production through the negative regulation of NFκB signalling in RAW 264.7 cells. *Journal of Ethnopharmacology*, *147*(1), 208–214. doi:10.1016/j.jep.2013.02.034 PMID:23500883

Juntachote, T., & Berghofer, E. (2005). Antioxidative properties and stability of ethanolic extracts of Holy basil and Galangal. *Food Chemistry*, *92*(2), 193–202. doi:10.1016/j.foodchem.2004.04.044

Kabir, M. E., & Webb, E. L. (2008). Can homegardens conserve biodiversity in Bangladesh? *Biotropica*, *40*(1), 95–103.

Kadel, C., & Jain, A. K. (2008). Folklore claims on snakebite among some tribal communities of central India. *Indian Journal of Traditional Knowledge*, *7*(2), 296–299.

Kainthla, R. P., Kashyap, R. S., Deopujari, J. Y., Purohit, H. J., Taori, G. M., & Daginawala, H. F. (2006). Effect of *Hemidesmus indicus* (Anantmool) extract on IgG production and adenosine deaminase activity of human lymphocytes in vitro. *Indian Journal of Pharmacology*, *38*(3), 190–193. doi:10.4103/0253-7613.25806

Kakarla, L., Katragadda, S. B., Tiwari, A. K., Kotamraju, K. S., Madhusudana, K., Kumar, D. A., & Botlagunta, M. (2016). Free radical scavenging, α-glucosidase inhibitory and anti-inflammatory constituents from Indian sedges, *Cyperus scariosus* R.Br and *Cyperus rotundus* L. *Pharmacognosy Magazine*, *12*(47), S488–S496. doi:10.4103/0973-1296.191467 PMID:27761080

Kala, C. P. (2009). Aboriginal uses and management of ethnobotanical species in deciduous forests of Chhattisgarh state in India. *Journal of Ethnobiology and Ethnomedicine*, *5*(1), 20–25. doi:10.1186/1746-4269-5-20 PMID:19653889

Kala, C. P. (2010). Home gardens and management of key species in the Pachmarhi biosphere reserve of India. *Journal of Biodiversity*, *1*(2), 111–117. doi:10.1080/09766901.2010.11884722

Kala, C. P., & Ratajc, P. (2012). High altitude biodiversity of the Alps and the Himalayas: Ethnobotany, plant distribution and conservation perspective. *Biodiversity and Conservation*, *21*(4), 1115–1126. doi:10.100710531-012-0246-x

Kamala, A., Middha, S. K., Gopinath, C., Sindhura, H. S., & Karigar, C. S. (2018). *In vitro* Antioxidant Potentials of *Cyperus rotundus* L. Rhizome Extracts and Their Phytochemical Analysis. *Pharmacognosy Magazine*, *14*(54), 261–267. doi:10.4103/pm.pm_228_17 PMID:29720842

Kandari L. S., & Maikhuri, R. K. (2009). *Artemisia maritime (Purchu): Option to improve local economy of higher Himalaya. A report* (46–47). Kurukshetra, Hariyana: Publisher Name.

Kandari, L. S. (2005). *Eco-physiological and socio-economic studies of some rhizomatous medicinal and aromatic plants species,* 1–153. (Doctoral dissertation). HNB Garhwal University, India.

Kandari, L. S., Phondani, P. C., Payal, K. C., Rao, K. S., & Maikhuri, R. K. (2012). Ethnobotanical study towards conservation of medicinal and aromatic plants in upper catchments of Dhauli Ganga in the Central Himalaya. *Journal of Mountain Science*, *9*(2), 286–296. doi:10.100711629-012-2049-7

Kandhari, K., Mishra, J. P. N., & Singh, R. P. (2019). Acacetin inhibits cell proliferation, survival, and migration in human breast cancer cells. *International Journal of Pharmacy and Biological Sciences-IJPBS*, *9*(1), 443–452.

Kandikattu, H. K., Deep, S. N., Razack, S., Amruta, N., Prasad, D., & Khanum, F. (2017). Hypoxia induced cognitive impairment modulating activity of *Cyperus rotundus. Physiology & Behavior*, *175*, 56–65. doi:10.1016/j.physbeh.2017.03.035 PMID:28351559

Kapahi, B. K., Srivastava, T. N., & Sarin, Y. K. (1993). Traditional medicinal of Gurez (Kashmir) – An ethnobotanical study. *Ancient Science of Life*, 119–124. PMID:22556637

Kapoor, S. L., & Kapoor, L. D. (1980). Medicinal plant wealth of the Karimnagar district of Andhra Pradesh. *Bulletin of Medical Ethnobotany and Research*, *1*, 120–144.

Kapur, S. K. (1993). Ethno-medico plants of Kangra valley (Himachal Pradesh). *Journal of Economic and Taxonomic Botany*, *17*, 395–408.

Karadeniz, C., Pınarlı, F. G., Oğuz, A., Gürsel, T., & Canter, B. (2007). Complementary/alternative medicine use in a pediatric oncology unit in Turkey. *Pediatric Blood & Cancer*, *48*(5), 540–543. doi:10.1002/pbc.21012 PMID:16900485

Karakurt, E. (2014). *Kelkit (gümüşhane) ilçesinin etnobotanik özellikleri* (Unpublished master's thesis), Erzincan Üniversitesi, Erzincan.

Karakurt, E. (2014). *Kelkit (gümüşhane) ilçesinin etnobotanik özellikleri* (Unpublished master's thesis), Erzincan Üniversitesi, Turkey.

Karcı, E., Gürbüz, İ., Akaydın, G., & Günbatan, T. (2017). Folk medicines of bafra (Samsun-Turkey). *Turkish Journal of Biochemistry*, *42*(4), 381–399. doi:10.1515/tjb-2017-0172

Karkala, M., & Bhushan, B. (2014). Aloe vera: A wonder plant its history, cultivation and medicinal uses. *Journal of Pharmacognosy and Phytochemistry*, *2*(5), 85–88.

Karmakar, I., Dolai, N., Saha, P., Sarkar, N., Bala, A., & Haldar, P. K. (2011). Scavenging activity of *Curcuma caesia* rhizome against reactive oxygen and nitrogen species. *Oriental Pharmacy and Experimental Medicine*, *11*(4), 221–228. doi:10.100713596-011-0030-6

Karthikumar, S., Vigneswari, K., & Jegatheesan, K. (2007). Screening of antibacterial and antioxidant activities of leaves of *Eclipta prostrata* (L). *Scientific Research and Essays*, (4): 101–104.

Kasiram, K., Sakharkar, P. R., & Patil, A. T. (2000). Antifungal activity of *Calendula officinalis. Indian Journal of Pharmaceutical Sciences*, *62*(6), 464–466.

Kasturi, T. R., Thomas, M., & Abraham, E. M. (1973). Essential oil of *Ageratum conyzoides*: Isolation and structure of two new constituents. *Indian Journal of Chemistry*, 91–95.

Kaul, M. K. (2010). High altitude botanicals in integrative medicine – Case study from Northwest Himalaya. *Indian Journal of Traditional Knowledge*, *9*(1), 18–25.

Kaur, A., Singh, R., Dey, C. S., Sharma, S. S., Bhutani, K. K., & Singh, I. P. (2010). Antileishmanial phenylpropanoids from *Alpinia galanga* (Linn.) Willd. *Indian Journal of Experimental Biology*, *48*, 314–317. PMID:21046987

Kaur, A., Singh, S., Shirwaikar, A., & Setty, M. (2011). Effect of ethanolic extract of *Hemidesmus indicus* roots on Cisplatin induced Nephrotoxicity in rats. *Journal of Pharmacy Research*, *4*(8), 2523–2525.

Kaur, R., & Dogra, N. K. (2014). A review on traditional uses, chemical constituents and pharmacology of *Ageratum conyzoides* L.(Asteraceae). *International Journal of Pharmaceutical and Biological Archives*, *5*(5), 33–45.

Kaushik, N. K., Bagavan, A., Rahuman, A. A., Mohanakrishnan, D., Kamaraj, C., Elango, G., ... Sahal, D. (2013). Antiplasmodial potential of selected medicinal plants from eastern Ghats of South India. *Experimental Parasitology*, *134*(1), 26–32. doi:10.1016/j.exppara.2013.01.021 PMID:23399920

Kawlni, L., Bora, M., Upadhyay, S. N., Mukherjee, K., & Hazra, J. (2017). Pharmacological and therapeutic profile of Anantamula (*Hemidesmus indicus* (L.) R. Br.: A comprehensive review. *International Journal of Ayurveda and Pharma Research*, *5*(11), 49–57.

Kearney, P. M., Whelton, M., Reynolds, K., Muntner, P., Whelton, P. K., & He, J. (2005). Global burden of hypertension: Analysis of worldwide data. *Lancet*, *365*(9455), 217–223. doi:10.1016/S0140-6736(05)17741-1 PMID:15652604

Kebede, A., Ayalew, S., Mesfin, A., & Mulualem, G. (2017). An ethnoveterinary study of medicinal plants used for the management of livestock ailments in selected kebeles of Dire Dawa administration, eastern Ethiopia. *Journal of Plant Sciences*, *5*(1), 34–42.

Kehlenbeck, K., Arifin, H. S., & Maass, B. L. (2007). Plant diversity in homegardens in a socio-economic and agro-ecological context. In *Stability of Tropical Rainforest Margins* (pp. 295–317). Springer. doi:10.1007/978-3-540-30290-2_15

Kehlenbeck, K., & Maass, B. L. (2004). Crop diversity and classification of homegardens in central Sulawesi, Indonesia. *Agroforestry Systems*, *63*(1), 53–62. doi:10.1023/B:AGFO.0000049433.95038.25

Kennedy, D. O., & Wightman, E. L. (2011). Herbal extracts and phytochemicals: Plant secondary metabolites and the enhancement of human brain function. *Advances in Nutrition*, *2*(1), 32–50. doi:10.3945/an.110.000117 PMID:22211188

Khajuria, A. K., & Bisht, N. S. (2017). Ethnomedicinal plants used to treat Nephrolithiasis: A case study Pauri (PAURI Garhwal), Uttarakhand. *International Journal of Herbal Medicine*, *5*, 10–13.

Khan, B. A., Abraham, A., & Leelamma, S. (1996). Biochemical response in rats to the addition of curry leaf (*Murraya koenigii*) and mustard seeds (*Brassica juncea*) to the diet. *Plant Foods for Human Nutrition (Dordrecht, The Netherlands)*, *49*(4), 295–299. doi:10.1007/BF01091978 PMID:8983055

Khan, B. A., Abraham, A., & Leelamma, S. (1998). Influence of spices: *Murraya koenigii* and *Brassica juncea* on rats fed atherogenic diet. *Journal of Food Science and Technology (Mysore), 35(1), 66-68.*

Khandelwal, V. K. M., Balaraman, R., Ondrejcakova, M., Pancza, D., & Ravingerova, T. (2010). Effect of *Hemidesmus indicus* on ischemia-reperfusion injury in the isolated rat heart. *Pharmaceutical Biology*, *48*(6), 611–614. doi:10.3109/13880200903218943 PMID:20645732

Khandelwal, V. K. M., Balaraman, R., Pancza, D., & Ravingerova, T. (2011). *Hemidesmus indicus* and *Hibiscus rosa-sinensis* affect ischemia reperfusion injury in isolated rat hearts. *Evidence-Based Complementary and Alternative Medicine*, *2011*, 1–8. doi:10.1155/2011/802937 PMID:20953394

Khanna, V. G., & Kannabiran, K. (2008). Antimicrobial activity of saponin fraction from the roots of *Hemidesmus indicus*. *Research Journal of Medicinal Plant*, *2*(1), 39–42. doi:10.3923/rjmp.2008.39.42

Khanna, V. G., & Kannabiran, K. (2009). Anticancer-cytotoxic activity of saponins isolated from the leaves of *Gymnema sylvestre* and *Eclipta prostrata* on Hela cells. *International Journal of Green Pharmacy*, *3*(3), 227–229. doi:10.4103/0973-8258.56280

Khan, S. U., Khan, R. U., Mehmood, S., Khan, A., Ullah, I., & Bokhari, T. Z. (2013). Medicinal plants used to cure diarrhea and dysentery by the local inhabitants of district Bannu, Khyber Pakhtoon Khwa, Pakistan. *Advances in Pharmaceutical and Ethnomedicines*, *1*, 15–18.

Khan, S., Kamili, A., & Gupta, R. C. (2016). Economic and medicinal properties of some medicinal plants found in Kashmir Himalaya. *Journal of Medicinal Plant Studies*, *4*(3), 38–44.

Khan, Z. S., Khuroo, A. A., & Dar, G. H. (2004). Ethnomedicinal survey of Uri, Kashmir Himalaya. *Indian Journal of Traditional Knowledge*, *3*(4), 351–357.

Khare, C. P. (2007). Indian Medicinal Plants - An Illustrated Dictionary, Springer Science+Business Media, 733-734.

Khare, C. P. (2004). *Encyclopedia of indian medicinal plants: rational western therapy, ayurvedic and other traditional usage, botany*. New York, NY: Springer.

Khare, C. P. (2008). *Indian medicinal plants: an illustrated dictionary*. Springer Science & Business Media.

Khattak, S., Shah, H. U., Ahmad, W., & Ahmad, M. (2005). Biological effects of indigenous medicinal plants *Curcuma longa* and *Alpinia galanga*. *Fitoterapia*, *76*(2), 254–257. doi:10.1016/j.fitote.2004.12.012 PMID:15810156

Khuroo, A. A., Malik, A. H., Dar, A. R., Dar, G. H., & Khan, Z. S. (2007). Ethno veterinary medicinal uses of some plant species by the Gujjar tribe of Kashmir Himalayas. *Asian Journal of Plant Sciences*, *6*(1), 148–152. doi:10.3923/ajps.2007.148.152

Khuroo, A. A., Rashid, I., Reshi, Z., Dar, G. H., & Wafai, B. A. (2007). The Alien Flora of Kashmir Himalaya. *Biological Invasions*, *9*(3), 262–292. doi:10.100710530-006-9032-6

Khwaja, S., Mahmood, T., & Siddiqui, H. H. (2016). Effect of ethanolic extract of *Cyperus rotundus* L. against isoprenaline induced cardiotoxicity. *Indian Journal of Experimental Biology*, *54*, 670–675. PMID:30084568

Kianbakht, S., Khalighi-Sigaroodi, F., & Dabaghian, F. H. (2013). Improved glycemic control in patients with advanced type 2 diabetes mellitus taking *Urtica dioica* leaf extract: A randomized double-blind placebo-controlled clinical trial. *Clinical Laboratory*, *59*(9–10), 1071–1076. PMID:24273930

Kilani-Jaziri, S., Neffati, A., Limem, I., Boubaker, J., Skandrani, I., Sghair, M. B., ... Chekir-Ghedira, L. (2009). Relationship correlation of antioxidant and antiproliferative capacity of *Cyperus rotundus* products towards K562 erythroleukemia cells. *Chemico-Biological Interactions*, *181*(1), 85–94. doi:10.1016/j.cbi.2009.04.014 PMID:19446539

Kilani, S., Ben Sghaier, M., Limem, I., Bouhlel, I., Boubaker, J., Bhouri, W., ... Chekir-Ghedira, L. (2008). *In vitro* evaluation of antibacterial, antioxidant, cytotoxic and apoptotic activities of the tubers infusion and extracts of *Cyperus rotundus*. *Bioresource Technology*, *99*(18), 9004–9008. doi:10.1016/j.biortech.2008.04.066 PMID:18538563

Kim, H. & Song, M. J. (2013). Ethnomedicinal practices for treating liver disorders of local communities in the southern regions of Korea and alternative medicine. doi:10.1155/2013/869176

Kim, H. S., Kacew, S., & Lee, B. M. (1999). In vitro chemopreventive effects of plant polysaccharides (*Aloe barbadensis Miller, Lentinusedodes, Ganoderma Lucidum and CariolusVesicolor*). *Carcinogenesis*, *20*(8), 1637–1871. doi:10.1093/carcin/20.8.1637 PMID:10426820

Kim, H. S., & Lee, B. M. (1997). Inhibition of benzopyrene-DNA adduct formation by *Aloe* barbadensis Miller. *Carcinogenesis*, *18*(1), 771–776. doi:10.1093/carcin/18.4.771 PMID:9111213

Kim, J. S., Im Lee, S., Park, H. W., Yang, J. H., Shin, T. Y., Kim, Y. C., ... Kim, D. K. (2008). Cytotoxic components from the dried rhizomes of *Zingiber officinale* Roscoe. *Archives of Pharmacal Research*, *31*(4), 415–418. doi:10.100712272-001-1172-y PMID:18449496

Kim, K. J., & Jansen, R. K. (1995). ndhF sequence evolution and the major clades in the sunflower family. *Proceedings of the National Academy of Sciences of the United States of America*, *92*(22), 10379–10383. doi:10.1073/pnas.92.22.10379 PMID:7479788

Kim, W., & Egan, J. M. (2008). The role of incretins in glucose homeostasis and diabetes treatment. *Pharmacological Reviews*, *60*(4), 470–512. doi:10.1124/pr.108.000604 PMID:19074620

Kinghorn, A. D. (2001). Pharmacognosy in the 21st century. *Journal of Pharma and Pharmacology, 53*, 135–148.

Kıran, Ö. (2006). *Kozan yöresindeki tıbbi bitkiler ve bunların halk arasında kullanılışı* (Unpublished master's thesis), Çukurova Üniversitesi, Adana.

Kiranmai, M., & Ibrahim, M. (2012). Antibacterial potential of different extracts of *Tagetes erecta* Linn. *International Journal of Pharmaceutics*, *2*(1), 90–96.

Kirtikar, K. R. (1987). *Indian Medicinal Plants. Lalit Mohan Basu, Allahabad. 2, (146)*. New Delhi, India: Jayyd Press.

Kirtikar, K. R., & Basu, B. D. (1981). ICS. Indian medicinal plants. Allahabad, India. *Lalit Mohan Basu*, *2*(2), 1346–1348.

Kızılarslan, Ç. (2008). *İzmit körfezi'nin güney kesiminde etnobotanik bir araştırma* (Unpublished master's thesis), İstanbul Üniversitesi, Turkey.

Knekt, P., Kumpulainen, J., Järvinen, R., Rissanen, H., Heliövaara, M., Reunanen, A., ... Aromaa, A. (2002). Flavonoid intake and risk of chronic diseases. *The American Journal of Clinical Nutrition*, *76*(3), 560–568. doi:10.1093/ajcn/76.3.560 PMID:12198000

Koehn, F. E., & Carter, G. T. (2005). The evolving role of natural products in drug discovery. *Nature Reviews. Drug Discovery*, *4*(3), 206–220. doi:10.1038/nrd1657 PMID:15729362

Koleva, V., Dragoeva, A., Nanova, Z., Koynova, T., & Dashev, G. (2015). An ethnobotanical study on current status of some medicinal plants used in Bulgaria. *International Journal of Current Microbiology and Applied Sciences*, *4*(4), 297–305.

Komakech, R., Matsabisa, M. G., & Kang, Y. (2019). The wound healing potential of *Aspilia africana* (Pers.) CD Adams (Asteraceae). *Evidence-Based Complementary and Alternative Medicine*, *2019*, 1–12. doi:10.1155/2019/7957860 PMID:30800171

Korkmaz, M., & Karakurt, E. (2014). Kelkit (gümüşhane) aktarlarında satılan tıbbi bitkiler. *Journal of Natural and Applied Science*, *18*(3), 60–80.

Kostova, I. (2005). Synthetic and natural coumarins as cytotoxic agents. *Current Medicinal Chemistry. Anti-Cancer Agents*, *5*(1), 29–46. doi:10.2174/1568011053352550 PMID:15720259

Kotnis, M. S., Patel, P., Menon, S. N., & Sane, R. T. (2004). Renoprotective effect of *Hemidesmus indicus*, a herbal drug used in gentamicin-induced renal toxicity. *Nephrology (Carlton, Vic.)*, *9*(3), 142–152. doi:10.1111/j.1440-1797.2004.00247.x PMID:15189175

Kposowa, A. J., & Tsunokai, G. T. (2002). Searching for relief: Racial differences in treatment of patients with back pain. *Race and Society*, *5*(2), 193–223. doi:10.1016/j.racsoc.2004.01.004

Krebs, C. J. (1985). *Ecology: experimental analysis of distribution and abundance* (3rd ed.). New York: Harper & Row.

Krishnaraj, M., Manibhushanrao, K., & Mathivanan, N. (2010). A comparative study of phenol content and antioxidant activity between non-conventional *Curcuma caesia* Roxb. and *Curcuma amada* Roxb. *International Journal of Plant Production*, *4*(3), 169–174.

Kubitzki, K., Kadereit, J. W., & Jeffrey, C. (2007). The families and genera of vascular plants: *Flowering plants. Eudicots*, *8*, 1–645.

Kubota, K., Nakamura (Murayam), K., Kobayashi, A., & Amaike, M. (1998). Acetoxy-1, 8-cineoles as Aroma Constituents of *Alpinia galanga* Willd. *Journal of Agricultural and Food Chemistry*, *46*(12), 5244–5247. doi:10.1021/jf9804239

Kültür, Ş. (2007). Medicinal plants used in Kırklareli Province (Turkey). *Journal of Ethnopharmacology*, *111*(2), 341–364. doi:10.1016/j.jep.2006.11.035 PMID:17257791

Kumar, G. S., Jayaveera, K. N., Kumar, A. C. K., Bharathi, T. B. M., Umachigi, P. S., & Vrushabendra, S. (2008). Evaluation of antioxidant and antiacne properties of terpenoidal fraction of *Hemidesmus indicus* (Indian sarsaparilla). *The Internet Journal of Aesthetic and Antiaging Medicine*, *1*(1).

Kumar, A., Agarwal, S., Singh, A., & Deepak, D. (2012). Medicobotanical study of some weeds growing in Moradabad District of Western Uttar Pradesh in India. *Indian Journal of Social Research*, *3*(1), 107–111.

Kumar, G. S., Jayaveera, K. N., Kumar, C. K., Sanjay, U. P., Swamy, B. M., & Kumar, D. V. (2007). Antimicrobial effects of Indian medicinal plants against acne-inducing bacteria. *Tropical Journal of Pharmaceutical Research*, *6*(2), 717–723. doi:10.4314/tjpr.v6i2.14651

Kumari, P., Joshi, G. C., & Tiwari, L. M. (2011). Diversity and status of ethnomedicinal plants of Almora district in Uttarakhand, India. *International Journal of Biodeversity and Conservation*, *3*(7), 298–326.

Kumar, K. H., & Khanum, F. (2013). Hydroalcoholic extract of *Cyperus rotundus* ameliorates H_2O_2-induced human neuronal cell damage via its anti-oxidative and anti-apoptotic machinery. *Cellular and Molecular Neurobiology*, *33*(1), 5–17. doi:10.100710571-012-9865-8 PMID:22869350

Kumar, M., Bussmann, R. W., Mukesh, J., & Kumar, P. (2011). Ethnomedicinal uses of plants close to rural habitation in Garhwal Himalaya, India. *Journal of Medicinal Plants Research*, *5*(11), 2252–2260.

Kumar, M., Paul, Y., & Anand, V. K. (2009). An ethnobotanical study of medicinal plants used by the locals in Kishtwar, Jammu and Kashmir, India. *Ethnobotanical Leaflets*, *13*, 1240–1256.

Kumar, S., & Yadav, J. P. (2014). Ethnobotanical and pharmacological properties of *Aloe vera*: A Review. *Journal of Medicinal Plants Research*, *8*(4), 1387–1398.

Lachhiramka, P., & Patil, S. (2016). Cholesterol lowering property of garlic (*Allium sativum*) on patients with hypercholesterolemia. *International Journal of Medical Science and Public Health*, *5*(11), 2249. doi:10.5455/ijmsph.2016.28032016449

Lakshman, H. C., Yeasmin, T., & Gabriel, K. P. (2014). Herbs of Asteraceae and their ethano medicinal uses in dermatological problems. *Journal of Bio-Science*, *22*, 127–129. doi:10.3329/jbs.v22i0.30016

Lakshman, K., Jayaprakash, B., & Joshi, H. (2005). Comparative anti-inflammatory activity studies of four species of Sariva. *Indian Journal of Pharmaceutical Sciences*, *67*(4), 481–482.

Lakshman, K., Shivaprasad, H. N., Jaiprakash, B., & Mohan, S. (2006). Anti-inflammatory and antipyretic activities of *Hemidesmus indicus* root extract. *African Journal of Traditional, Complementary, and Alternative Medicines*, *3*(1), 90–94.

Lama, Y. C., Ghimire, S. K., & Aumeeruddy, T. Y. (2001). *Medicinal plants of Dolpo: Amchis' knowledge and conservation*. Kathmandu: WWF Nepal Program.

Lamont, S. R., Eshbaugh, W. H., & Greenberg, A. M. (1999). Species composition, diversity, and use of homegardens among three Amazonian villages. *Economic Botany*, *53*(3), 312–326. doi:10.1007/BF02866644

Lawal, O. A., & Oyedeji, A. O. (2009). Chemical composition of the essential oils of *Cyperus rotundus* L. from South Africa. *Molecules (Basel, Switzerland)*, *14*(8), 2909–2917. doi:10.3390/molecules14082909 PMID:19701133

Lawand, R. V., & Gandhi, S. V. (2013). Comparison of *Curcuma caesia* Roxb. with other Commonly Used *Curcuma* Species by HPTLC. *Journal of Pharmacognosy and Phytochemistry*, *2*(4), 126–131.

Lawless, J., & Allen, J. (2000). *Aloe vera-Natural wonder care*. Hammersmith: Harper Collins Publishers.

Lechat-Vahirua, I., François, P., Menut, C., Lamaty, G., & Bessiere, J. M. (1993). Aromatic plants of French Polynesia. I. Constituents of the essential oils of rhizomes of three Zingiberaceae: *Zingiber zerumbet* Smith, *Hedychium coronarium* Koenig and *Etlingera cevuga* Smith. *The Journal of Essential Oil Research*, *5*(1), 55–59. doi:10.1080/10412905.1993.9698170

Lee, C. C., & Houghton, P. (2005). Cytotoxicity of plants from Malaysia and Thailand used traditionally to treat cancer. *Journal of Ethnopharmacology*, *100*(3), 237–243. doi:10.1016/j.jep.2005.01.064 PMID:15888378

Lee, C. H., Hwang, D. S., Kim, H. G., Oh, H., Park, H., Cho, J. H., ... Oh, M. S. (2010). Protect.ive effect of Cyperi rhizoma against 6-hydroxydopamine-induced neuronal damage. *Journal of Medicinal Food*, *13*(3), 564–571. doi:10.1089/jmf.2009.1252 PMID:20521982

Lee, P., Helewa, A., Goldsmith, C. H., Smythe, H. A., & Stitt, L. W. (2001). Low back pain: Prevalence and risk factors in an industrial setting. *The Journal of Rheumatology*, *28*(2), 346–351. PMID:11246674

Lemaure, B., Touché, A., Zbinden, I., Moulin, J., Courtois, D., Macé, K., & Darimont, C. (2007). Administration of *Cyperus rotundus* tubers extract prevents weight gain in obese Zucker rats. *Phytotherapy Research*, *21*(8), 724–730. doi:10.1002/ptr.2147 PMID:17444573

Lepofsky, D., Heyerdahl, E., Lertzman, K., Schaepe, D., & Mierendorf, B. (2003). Historical meadow dynamics in southwest British Columbia: A multidisciplinary analysis. *Conservation Ecology*, *7*(3), 5–9. doi:10.5751/ES-00559-070305

Liebenson, C. (2007). *Rehabilitation of the spine: a practitioner's manual*. Lippincott Williams & Wilkins.

Linger, E. (2014). Agro-ecosystem and socio-economic role of homegarden agroforestry in Jabithenan district, northwestern Ethiopia: Implication for climate change adaptation. *SpringerPlus*, *3*(1), 154. doi:10.1186/2193-1801-3-154 PMID:24790810

Lin, S. Q., Zhou, Z. L., & Li, C. Y. (2018). Cyprotuoside C and Cyprotuoside D, Two New Cycloartane Glycosides from the Rhizomes of *Cyperus rotundus*. *Chemical & Pharmaceutical Bulletin*, *66*(1), 96–100. doi:10.1248/cpb.c17-00608 PMID:29118309

Lin, T. L., Lin, H. H., Chen, C. C., Lin, M. C., Chou, M. C., & Wang, C. J. (2007). *Hibiscus sabdariffa* extract reduces serum cholesterol in men and women. *Nutrition Research (New York, N.Y.)*, *27*(3), 140–145. doi:10.1016/j.nutres.2007.01.007

Little, K. (1965). *West African urbanization: a study of voluntary associations in social change*. CUP Archive.

Liu, Z., & Zhao, X. (2010). Effects of different extraction methods on extraction rate of polysaccharide from *Sonchus oleraceus* L. *Medicinal Plant*, *1*(5), 31–35.

Logeeswari, K., & Sripathi, S. K. (2012). Wound healing medicinal plants: A review. *International Journal of Chemical Environmental and Pharmaceutical Research*, *3*(3), 199–218.

Lombardino, J. G., & Lowe, J. A. III. (2004). A guide to drug discovery: The role of the medicinal chemist in drug discovery—then and now. *Nature Reviews. Drug Discovery*, *3*(10), 853–862. doi:10.1038/nrd1523 PMID:15459676

Lone, F. A. (2003). Folklore medicinal system of Uri sector Kashmir valley, India, In proceeding of the 2nd world congress on Biotechnology Development of Herbal Medicine, Feb. 20-22, (India). Academic Press.

Lone, P. A., & Bhardwaj, A. K. (2013b). Ethnomedicinal uses of certain locally available plants of Bandipora district of Jammu & Kashmir, India, *International Journal of Medicinal and Aromatic Plants*, (3), 470–485.

Lone, A. H., & Pandit, A. K. (2007). Medicinal plant wealth of Langate Forest Division in Kashmir Himalayas. *Journal of Himalayan Ecology and Sustainable Development*, *2*, 73–78.

Lone, P. A., & Bhardwaj, A. K. (2013a). Traditional herbal based disease treatment in some rural areas of Bandipora District of Jammu and Kashmir, India. *Asian Journal of Pharmaceutical and Clinical Research*, *6*(4), 162–171.

Lone, P., Bhardwaj, A. K., Shah, K. W., & Tabasum, S. (2014). Ethnobotanical survey of some threatened medicinal plants of Kashmir Himalaya, India. *Journal of Medicinal Plants Research*, *8*(47), 1362–1373.

Long, A. N., & Dagogo-Jack, S. (2011). Comorbidities of diabetes and hypertension: Mechanisms and approach to target organ protection. *Journal of Clinical Hypertension*, *13*(4), 244–251. doi:10.1111/j.1751-7176.2011.00434.x PMID:21466619

López-Romero, P., Pichardo-Ontiveros, E., Avila-Nava, A., Vázquez-Manjarrez, N., Tovar, A. R., Pedraza-Chaverri, J., & Torres, N. (2014). The effect of nopal (*Opuntia Ficus Indica*) on postprandial blood glucose, incretins, and antioxidant activity in Mexican patients with type 2 diabetes after consumption of two different composition breakfasts. *Journal of the Academy of Nutrition and Dietetics*, *114*(11), 1811–1818. doi:10.1016/j.jand.2014.06.352 PMID:25132122

Luna-Vázquez, F. J., Ibarra-Alvarado, C., Del Rayo Camacho-Corona, M., Rojas-Molina, A., Rojas-Molina, J. I., García, A., & Bah, M. (2018). Vasodilator activity of compounds isolated from plants used in Mexican traditional medicine. *Molecules (Basel, Switzerland)*, *23*(6), 1474. doi:10.3390/molecules23061474 PMID:29912156

Lunsford, K. E., Bodzin, A. S., Reino, D. C., Wang, H. L., & Busuttil, R. W. (2016). Dangerous dietary supplements: *Garcinia cambogia*-associated hepatic failure requiring transplantation. *World Journal of Gastroenterology*, *22*(45), 10071–10076. doi:10.3748/wjg.v22.i45.10071 PMID:28018115

Madhu, A., Gupta, G., Arali, B. K., Chellappan, D., & Dua, K. (2017). Anti-psychotic activity of aqueous root extract of *Hemidesmus indicus*: A time bound study in rats. *Recent Patents on Drug Delivery & Formulation*, *11*(1), 36–41. PMID:27993107

Madhu, A., Keerthi, P. H. V., Singh, J., & Shivalinge, G. K. P. (2009). To evaluate the antiepileptic activity of aqueous root extract of *Hemidesmus indicus* in rats. *Archives of Pharmaceutical Sciences and Research*, *1*(1), 43–47.

Maenthaisong, R., Chaiyakunapruk, N., Niruntraporn, S., & Kongkaew, C. (2007). The efficacy of *Aloe vera* for burn wound healing: A systematic review. *Burns*, *33*(1), 713–718. doi:10.1016/j.burns.2006.10.384 PMID:17499928

Magee, D. J. (2013). *Orthopedic physical assessment*. Elsevier Health Sciences.

Magurran, A. E. (2013). *Measuring biological diversity*. John Wiley & Sons.

Mahae, N., & Chaiseri, S. (2009). Antioxidant activities and antioxidative components in extracts of *Alpinia galanga* (L.) Sw. *Witthayasan Kasetsat Witthayasat, 43*(2), 358–369.

Mahalingam, G., & Kannabiran, K. (2008). Hypoglycemic activity of *Hemidesmus indicus* R. Br. on streptozotocin-induced diabetic rats. *International Journal of Diabetes in Developing Countries, 28*(1), 6–10. doi:10.4103/0973-3930.41979 PMID:19902032

Mahalingam, G., & Kannabiran, K. (2009a). *Hemidesmus indicus* root extract ameliorates diabetes-mediated metabolic changes in rats. *International Journal of Green Pharmacy, 3*(4), 314–318. doi:10.4103/0973-8258.59739

Mahalingam, G., & Kannabiran, K. (2009b). Antidiabetic activity of 2-hydroxy 4-methoxy benzoic acid isolated from the roots of *Hemidesmus indicus* on streptozotocin-induced diabetic rats. *International Journal of Diabetes and Metabolism, 17*(2), 53–57. doi:10.1159/000497673

Maikhuri, R. K., Nautiyal, S., Rao, K. S., & Saxena, K. G. (1998). Medicinal plant cultivation and biosphere reserve management: A case study from the Nanda Devi Biosphere Reserve, Himalaya. *Current Science, 74*(2), 157–163.

Maikhuri, R. K., Rao, K. S., Chauhan, K., Kandari, L. S., Prasad, P., & Rajasekaran, C. (2003). Development of marketing of medicinal plants and other forest products: Can it be a pathway for effective management and conservation? *Indian Forester, 129*, 169–178.

Maikhuri, R., & Gangwar, A. (1993). Ethnobiological notes on the Khasi and Garo tribes of Meghalaya, northeast India. *Economic Botany, 47*(4), 345–357. doi:10.1007/BF02907348

Maiti, S., & Chandra, R. (2002). *Cultivation of Aloe vera. National Research Centre for Medicinal and Aromatic plants*. Gujarat: Anand Press.

Mala, F. A., Lone, M. A., Lone, F. A., & Arya, N. (2012). Ethno medicinal survey of Kajinaag Range of Kashmir, Himalaya, India. *International Journal of Pharma and Bio Sciences*, (3): 442–449.

Malathi, M., & Maharani, B. (2011). Evaluation of anticonvulsant activity of ethanolic extract of roots of *Hemidesmus indicus* using adult albino rats. *Journal of Pharmacy Research, 4*(10), 3345–3347.

Malek, V. G., Abbasian, A., Karimi, A., Khanlarkhani, N., & Parvari, S. (2017). 5 medicinal plants used in iranian traditional medicine for low back pain: A quick review. *European Journal of Experimental Biology, 7*(4, 24), 1-4.

Malik, A. H., Khuroo, A. A., Dar, G. H., & Khan, Z. S. (2011). Ethnomedicinal uses of some plants in the Kashmir Himalaya. *Indian Journal of Traditional Knowledge, 10*(2), 362–366.

Malik, A. R., Siddique, M. A. A., Sofi, P. A., & Butola, J. S. (2011). Ethnomedicinal practices and conservation status of medicinal plants of North Kashmir Himalayas. *Research Journal of Medicinal Plant, 5*(5), 515–530. doi:10.3923/rjmp.2011.515.530

Malik, I. Z., & Zarnigar. (2013). Aloe vera, a review article on its clinical effectiveness. *International Research Journal of Pharmacy, 4*(8), 75–79. doi:10.7897/2230-8407.04812

Mallavarapu, G. R., Rao, L., Ramesh, S., Dimri, B. P., Rajeswara Rao, B. R., Kaul, P. N., & Bhattacharya, A. K. (2002). Composition of the volatile oils of *Alpinia galanga* rhizomes and leaves from India. *The Journal of Essential Oil Research, 14*(6), 397–399. doi:10.1080/10412905.2002.9699900

Mallikarjun, N., Mesta, S. C., Sudharshan, S. J., & Vinayaka, K. S. (2010). Antimosquito (Insecticidal) activity of extracts of *Hemidesmus indicus* and *Swertia chirata* against *Aedes aegypti* mosquito larvae-A Comparative Study. *Drug Invent Today, 2*, 106–108.

Malyer, H., Aydın, Ö. A., Tümen, G., & Er, S. (2004). Tekirdağ ve çevresindeki aktarlarda satılan bazı bitkiler ve tıbbi kullanım özellikleri. *Dumlupınar Üniversitesi Fen Bilimleri Enstitüsü Dergisi, 7*, 103–112.

Mandal, S., Das, P. C., Joshi, P. C., Das, A., & Chatterjee, A. (1991). Hemidesminine-A New Coumarino-Lignoid from *Hemidesmus indicus* R. Br. *ChemInform, 22*(37), 712–713.

Manicad, G. (1996). Biodiversity conservation and development: the collaboration of formal and non-formal institutions. *Biotechnology and development monitor, 26*, 15-17.

Mannarreddy, P., Denis, M., Munireddy, D., Pandurangan, R., Thangavelu, K. P., & Venkatesan, K. (2017). Cytotoxic effect of *Cyperus rotundus* rhizome extract on human cancer cell lines. *Biomedicine and Pharmacotherapy, 95*, 1375–1387. doi:10.1016/j.biopha.2017.09.051 PMID:28946185

Mantle, D., Gok, M. A., & Lennard, T. W. (2001). Adverse and beneficial effects of plant extracts on skin and skin disorders. *Adverse Drug Reactions and Toxicological Reviews, 20*(2), 89–103. PMID:11482001

Marchand, F., & Ahmed, A. M. (1990, May). Investigation of the laminate structure of lumbar disc anulus fibrosus. *Spine, 15*(5), 402–410. doi:10.1097/00007632-199005000-00011 PMID:2363068

Marcia, S., Bellini, M., Hirsch, J. A., Chandra, R. V., Piras, E., Marras, M., ... Saba, L. (2018). Efficacy of an ethyl alcohol gel in symptomatic disc herniation. *European Journal of Radiology, 109*, 101–107. doi:10.1016/j.ejrad.2018.10.029 PMID:30527290

Marsh, D. M., & Trenham, P. C. (2008). Current trends in plant and animal population monitoring. *Conservation Biology, 22*(3), 647–655. doi:10.1111/j.1523-1739.2008.00927.x PMID:18445076

Marsh, R. (1998). Building on traditional gardening to improve household food security. *Food, Nutrition and Agriculture*, 4–14.

Martínez Salvador, L. (2016). Seguridad alimentaria, autosuficiencia y disponibilidad del amaranto en México. *Problemas del Desarrollo, 47*(186), 107–132. doi:10.1016/j.rpd.2016.08.004

Martin, G. J. (2010). *Ethnobotany: A methods manual*. Routledge. doi:10.4324/9781849775854

Martín, M., Herrero, A., & Echavarría, I. (2015). El cáncer de mama. *Arbor, 191*(773), a234. doi:10.3989/arbor.2015.773n3004

Martins, N., Barros, L., Santos-Buelga, C., Henriques, M., Silva, S., & Ferreira, I. C. F. R. (2014). Decoction, infusion and hydroalcoholic extract of *Origanum vulgare* L.: Different performances regarding bioactivity and phenolic compounds. *Food Chemistry, 158*, 73–80. doi:10.1016/j.foodchem.2014.02.099 PMID:24731316

Mary, N. K., Achuthan, C. R., Babu, B. H., & Padikkala, J. (2003). In vitro antioxidant and antithrombotic activity of *Hemidesmus indicus* (L) R. Br. *Journal of Ethnopharmacology, 87*(2-3), 187–191. doi:10.1016/S0378-8741(03)00119-3 PMID:12860306

Mathew, N. S., & Negi, P. S. (2017). Traditional uses, phytochemistry and pharmacology of wild banana (*Musa acuminata* Colla): A review. *Journal of Ethnopharmacology, 196*(20), 124–140. 2016.12.009

Matsuda, H., Pongpiriyadacha, Y., Morikawa, T., Ochi, M., & Yoshikawa, M. (2003). Gastroprotective effects of phenylpropanoids from the rhizomes of *Alpinia galanga* in rats: Structural requirements and mode of action. *European Journal of Pharmacology, 471*(1), 59–67. doi:10.1016/S0014-2999(03)01785-0 PMID:12809953

Maurya, S. K., Kushwaha, A. K., & Seth, A. (2015). Ethnomedicinal review of Usnakantaka (*Echinops echinatus* Roxb.). *Pharmacognosy Reviews*, *9*(18), 149–154. doi:10.4103/0973-7847.162138 PMID:26392713

Mavi, S., & Shava, S. (1997). *Traditional methods of conserving medicinal plants in Zimbabwe*. BGCNews 2. Journal archive.

Mayachiew, P., & Devahastin, S. (2008). Antimicrobial and antioxidant activities of Indian gooseberry and galangal extracts. *Lebensmittel-Wissenschaft + Technologie*, *41*(7), 1153–1159. doi:10.1016/j.lwt.2007.07.019

Mayer, F. L., Wilson, D., & Hube, B. (2013). *Candida albicans* pathogenicity mechanisms. *Virulence*, *4*(2), 119–128. doi:10.4161/viru.22913 PMID:23302789

Mazumder, P. M., Das, S., Das, S., & Das, M. K. (2010). Cytotoxic activity of methanolic extracts of *Berberis aristata* DC and *Hemidesmus indicus* R.Br. in MCF7 Cell Line. *Journal of Current Pharmaceutical Research*, *1*(1), 12–15.

Mazzio, E. A., & Soliman, K. F. A. (2009). In Vitro screening for the tumoricidal properties of international medicinal herbs. *Phytotherapy Research*, *23*(3), 385–398. doi:10.1002/ptr.2636 PMID:18844256

Mbajiuka, C. S., Obeagu, E. I., & Okwandu, G. E. (2014). Antimicrobial effects of Aloe vera on some human pathogens. *International Journal of Current Microbiology and Applied Sciences*, *3*(3), 1022–1028.

McFarlane, S. I., Muniyappa, R., Shin, J. J., Bahtiyar, G., & Sowers, J. R. (2004). Osteoporosis and cardiovascular disease: Brittle bones and boned arteries, is there a link? *Endocrine*, *23*(1), 1–10. doi:10.1385/ENDO:23:1:01 PMID:15034190

McNaughton, S. J. (1989). Ecosystems and conservation in the twenty-first century. In D. Western, & M. C. Pearl (Eds.), Conservation for the twenty-first century (pp. 109-120). New York: Oxford University Press.

Meena, K. L., & Yadav, B. L. (2011). Some ethnomedicinal plants used by the *Garasia* tribe of district Sirohi, Rajasthan. *Indian Journal of Traditional Knowledge*, *10*, 354–357.

Mehdizadeh, M., Hashem Dabaghian, F., Shojaee, A., Molavi, N., Taslimi, Z., Shabani, R., & Soleimani Asl, S. (2017). Protective Effects of *Cyperus Rotundus* Extract on Amyloid β-Peptide (1-40)-Induced Memory Impairment in Male Rats: A Behavioral Study. *Basic and Clinical Neuroscience*, *8*(3), 249–254. doi:10.18869/nirp.bcn.8.3.249 PMID:28781732

Mehta, A., Sethiya, N. K., Mehta, C., & Shah, G. B. (2012). Anti-arthritis activity of roots of *Hemidesmus indicus* R. Br. (Anantmul) in rats. *Asian Pacific Journal of Tropical Medicine*, *5*(2), 130–135. doi:10.1016/S1995-7645(12)60011-X PMID:22221757

Mena, P., Cirlini, M., Tassotti, M., Herrlinger, K. A., Dall'Asta, C., & Del Rio, D. (2016). Phytochemical profiling of flavonoids, phenolic acids, terpenoids, and volatile fraction of a rosemary (*Rosmarinus officinalis* L.) extract. *Molecules (Basel, Switzerland)*, *21*(11), 1576. doi:10.3390/molecules21111576 PMID:27869784

Mesfin, F., Demissew, S., & Teklehaymanot, T. (2009). An ethnobotanical study of medicinal plants in Wonago Woreda, SNNPR, Ethiopia. *Journal of Ethnobiology and Ethnomedicine*, *5*(1), 28. doi:10.1186/1746-4269-5-28 PMID:19821994

Metin, A. (2009). *Mut ve çevresinde yetişen bitkilerin (mersin) etnobotanik özellikleri* (Unpublished master's thesis), Selçuk Üniversitesi, Turkey.

Miller, J. A., Schmatz, C., & Schultz, A. B. (1988). Lumbar disc degeneration: Correlation with age, sex, and spine level in 600 autopsy specimens. *Spine*, *13*(2), 173–178. doi:10.1097/00007632-198802000-00008 PMID:3406837

Minnis, P. E. (2000). Ethnobotany: A reader, (1–325). Norman, OK: University of Oklahoma Press.

Miraj, S., & Kiani, S. (2016). Study of therapeutic effects of *Cynara scolymus* L.: A review. *Der Pharmacia Lettre*, *8*(9), 168–173.

Mir, G. M., & John, S. A. (2014). Ethno medicinal study of Pulwama Tehsil, Jammu & Kashmir. *Journal of Medicinal Plant Studies*, *2*(4), 5–8.

Mir, M. Y. (2014). Indigenous knowledge of using medicinal plants in treating skin diseases by tribal's of Kupwara, J&K, India. *International Journal of Herbal Medicine*, *1*(6), 62–68.

Mir, M. Y. (2014a). Ethno medicinal survey of plants from Kupwara, J & K, India. *International Journal of Advanced Research*, *2*(1), 846–857.

Mir, M. Y. (2014b). Ethnoveterinary studies in tribals of Kupwara, J & K, India. *International Journal of Multidisciplinary Research and Development*, *1*(6), 92–96.

Mir, M. Y. (2014c). Documentation and ethnobotanical survey of wild edible plants used by the tribals of Kupwara, J & K, India. *International Journal of Herbal Medicine*, *2*(4), 11–18.

Mir, M. Y. (2014d). Indigenous knowledge of using medicinal plants in treating skin disease by tribal's of Kupwara, J & K, India. *International Journal of Herbal Medicine*, *1*(6), 62–68.

Misael, P. C. E., De Guadalupe, T. M. B. M., Paola, D. C. S. G. M., & Carlos, R. L. J. (2014). Effect of *Cynara scolymus* (artichoke) in homeopathic doses on body mass index in obese and overweight patients. *Biomedical & Pharmacology Journal*, *7*(2), 525–533. doi:10.13005/bpj/520

Misri, J., Fatema, H., Brishti, M., & Hoque, M. (2014). Aloe vera gel as a novel edible coating for fresh fruits: A Review. *American Journal of Food Science and Technology*, *2*(3), 93–97. doi:10.12691/ajfst-2-3-3

Mithila, M. V., & Khanum, F. (2015). Effectual comparison of quinoa and amaranth supplemented diets in controlling appetite; a biochemical study in rats. *Journal of Food Science and Technology*, *52*(10), 6735–6741. doi:10.100713197-014-1691-1 PMID:26396423

Mitra, R., & Gupta, R. C. (1997). Punarnava: An Ayurvedic drug of repute. *Applied Botany Abstracts*, *17*(3), 209–227.

Mitsui, S., Kobayashi, S., Nagahori, H., & Ogiso, A. (1976). Constituents from seeds of *Alpinia galanga* Wild. and their anti-ulcer activities. *Chemical & Pharmaceutical Bulletin*, *24*(10), 2377–2382. doi:10.1248/cpb.24.2377 PMID:1017082

Mittermeier, R. A., Gil, P. R., Hoffman, M., Pilgrim, J., Brooks, T., Mittermeier, C. G., ... Da Fonseca, G. A. B. (2005). *Hotspots revised: Earth's biologically richest and most threatened terrestrial ecoregions*. Conservation International.

Mohamadi, N., Sharififar, F., Koohpayeh, A., & Daneshpajouh, M. (2015). Traditional and Ethnobotanical uses of medicinal plants by ancient populations in Khabr and Rouchon of Iran. *Journal of Applied Pharmaceutical Science*, *5*(11), 101–107. doi:10.7324/JAPS.2015.501117

Mohamed-Ibrahim, S. R., Mohamed, G. A., Abdullah Khayat, M. T., Zayed, M. F., & Soliman El-Kholy, A. A. (2018). Anti-inflammatory terpenoids from *Cyperus rotundus* rhizomes. *Pakistan Journal of Pharmaceutical Sciences*, *31*, 1449–1456. PMID:30058534

Moideen, M. M., Varghese, R., Kumar, E. K., & Dhanapal, C. K. (2011). Wound healing activity of ethanolic extract of *Hemidesmus indicus* (Linn) R.Br leaves in rats. *Research Journal of Pharmaceutical, Biological and Chemical Sciences*, *2*(3), 643–651.

Moliner, C., Barros, L., Dias, M. I., López, V., Langa, E., Ferreira, I. C. F. R., & Gómez-Rincón, C. (2018). Edible flowers of *Tagetes erecta* L. as functional ingredients: Phenolic composition, antioxidant and protective effects on *Caenorhabditis elegans*. *Nutrients*, *10*(12), 2002. doi:10.3390/nu10122002 PMID:30567311

Monica, K. C., Mark, K., & Paul, O. (2016). Traditional controls of harvesting and conserving medicinal plants in Keiyo south sub-county, Kenya. *International Journal of Humanities and Social Science*, *6*(11), 54–63.

Monroy-Ortíz, C., & Castillo-España, P. (2007). *Medicinal plants used in the State of Morelos*. Morelos, México: Centro de Investigaciones Biológicas, Universidad Autónoma del Estado de Morelos.

Montagnini, F. (2006). Homegardens of Mesoamerica: biodiversity, food security, and nutrient management. In *Tropical Homegardens* (pp. 61–84). Springer. doi:10.1007/978-1-4020-4948-4_5

Morales-Sánchez, A., & Fuentes-Pananá, E. M. (2014). Human viruses and cancer. *Viruses*, *6*(10), 4047–4079. doi:10.3390/v6104047 PMID:25341666

Moreira-Muñoz, A. (2007). The Austral floristic realm revisited. *Journal of Biogeography*, *34*(10), 1649–1660. doi:10.1111/j.1365-2699.2007.01757.x

Morita, H., & Itokawa, H. (1986). New diterpenes from *Alpinia galanga* Wild. *Chemistry Letters*, *15*(7), 1205–1208. doi:10.1246/cl.1986.1205

Motaleb, M. A. (2011). *Selected Medicinal plants of Chittagong Hill Tracts*. Bangladesh: International Union for Conservation of Nature.

Mpalantinos, M. A., Soares de Moura, R., Parente, J. P., & Kuster, R. M. (1998). Biologically active flavonoids and kava pyrones from the aqueous extract of *Alpinia zerumbet*. *Phytotherapy Research*, *12*(6), 442–444. doi:10.1002/(SICI)1099-1573(199809)12:6<442::AID-PTR320>3.0.CO;2-Y

Muangnoi, P., Lu, M., Lee, J., Thepouyporn, A., Mirzayans, R., Le, X. C., ... Changbumrung, S. (2007). Cytotoxicity, apoptosis and DNA damage induced by *Alpinia galanga* rhizome extract. *Planta Medica-Natural Products and Medicinal Plant Research*, *73*(8), 748–754. PMID:17611930

Mukhopadhyay, M. J., & Mukherjee, A. (2000). Clastogenic effect of ginger rhizome in mice. *Phytotherapy Research*, *14*(7), 555–557. doi:10.1002/1099-1573(200011)14:7<555::AID-PTR659>3.0.CO;2-J PMID:11054851

Muley, B. P., Khadabadi, S. S., & Banarase, N. B. (2009). Phytochemical constituents and pharmacological activities of *Calendula officinalis* Linn (Asteraceae): A review. *Tropical Journal of Pharmaceutical Research*, *8*(5), 455–465. doi:10.4314/tjpr.v8i5.48090

Mundy, C., & Kirkpatrick, P. (2004). Tiotropium bromide. *Nature Reviews. Drug Discovery*, *3*(8), 643–645. doi:10.1038/nrd1472 PMID:15317149

Murakami, S., Li, W., Matsuura, M., Satou, T., Hayashi, S., & Koike, K. (2009). Composition and seasonal variation of essential oil in *Alpinia zerumbet* from Okinawa Island. *Journal of Natural Medicines*, *63*(2), 204–208. doi:10.100711418-008-0306-4 PMID:19067113

Murshed, S., Begum, R., & Nahar, N. (2005). Hypoglycemic and hypolipidemic effect of *Hemidesmus indicus* root on diabetic model rats. *Diabetes Research (Edinburgh, Lothian)*, *39*, 15–23.

Murti, P. B. R., & Seshadri, T. R. (1941). A study of the chemical components of the roots of *Decalepis Hamiltonii* (Makali Veru). In *Proceedings of the Indian Academy of Sciences-Section A*. India: Springer.

Mussarat, S. (2014). AbdEl-Salam, N. M., Tariq, A., Wazir, S. M., Ullah, R., & Adnan, M. (2014). Use of ethnomedicinal plants by the people living around Indus River. *Evidence-Based Complementary and Alternative Medicine*, 212634–212634. PMID:24778701

Mustafa, A., Ali, M., & Khan, N. Z. (2005). Volatile Oil Constituents of the Fresh Rhizomes of *Curcuma amada* Roxb. *The Journal of Essential Oil Research, 17*(5), 490–491. doi:10.1080/10412905.2005.9698974

Mutawakil, M. H. (2012). Antigenotoxic Effects of *Zingiber officinale* Extract on *Saccharomyces cerevisiae* D7. *Researcher, 4*(2), 51–55.

Nadaf, M., Joharchi, M. R., & Amiri, M. S. (2019). Ethnomedicinal uses of plants for the treatment of nervous disorders at the herbal markets of Bojnord, North Khorasan Province, Iran. *Avicenna Journal of Phytomedicine, 9*, 153–163. PMID:30984580

Nadkarni, K. H. (2007). Indian Materica Medica. (1-739). Bombay Popular Publisher, Bombay.

Nagarajan, S., Rao, J. M. L., & Gurudutt, K. N. (2001). Chemical composition of the volatiles of *Hemidesmus indicus* R. Br. *Flavour and Fragrance Journal, 16*(3), 212–214. doi:10.1002/ffj.985

Najafian, J., Abdar-Esfahani, M., Arab-Momeni, M., & Akhavan-Tabib, A. (2014). Safety of herbal medicine in treatment of weight loss. *ARYA Atherosclerosis, 10*(1), 55–58. PMID:24963315

Nakatani, N., Jitoe, A., Masuda, T., & Yonemori, S. (1991). Flavonoid constituents of *Zingiber zerumbet* Smith. *Agricultural and Biological Chemistry, 55*(2), 445–460. doi:10.1080/00021369.1991.10870577

Nam, J. H., & Lee, D. U. (2015). Inhibitory effect of oleanolic acid from the rhizomes of *Cyperus rotundus* on transient receptor potential vanilloid 1 channel. *Planta Medica, 81*, 20–25. PMID:25402944

Nam, J. H., Nam, D. Y., & Lee, D. U. (2016). Valencene from the Rhizomes of *Cyperus rotundus* Inhibits Skin Photoaging-Related Ion Channels and UV-Induced Melanogenesis in B16F10 Melanoma Cells. *Journal of Natural Products, 79*(4), 1091–1096. doi:10.1021/acs.jnatprod.5b01127 PMID:26967731

Nampoothiri, S. V., Esakkidurai, T., & Pitchumani, K. (2015). Identification and Quantification of Phenolic Compounds in *Alpinia galanga* and *Alpinia calcarata* and its Relation to Free Radical Quenching Properties: A Comparative Study. *Journal of Herbs, Spices, & Medicinal Plants, 21*(2), 140–147. doi:10.1080/10496475.2014.923358

Nandal, U., & Bhardwaj, R. L. (2012). *Aloe vera*: A valuable wonder plant for food, medicine and cosmetic use - A review. *International Journal of Pharmaceutical Science Research, 13*(1), 59–67.

National Cancer Institute. (2019). Types of cancer treatment. Retrieved August 8, 2019, from Methanolic extract of *Moringa oleifera* leaves improves glucose tolerance, glycogen synthesis and lipid metabolism in alloxan-induced diabetic rats. *Journal of Basic and Clinical Physiology and Pharmacology, 26*(6), 585–593. doi:10.1515/jbcpp-2014-0129

National Center for Complementary and Integrative Health. (2016). *National toxicology program.* Available from https://nccih.nih.gov/health/aloevera

Nautiyal, M. C., & Nautiyal, B. P. (2004). Collaboration between farmers, research institutions and industry: Experiences of Picrorhiza kurrooa cultivation at Gheshe village in Chamoli district, Uttaranchal. In G. Alam, & J. Belt (Eds.) Searching synergy: stakeholder views on developing a sustainable medicinal plant chain in Uttaranchal, India. Amsterdam, The Netherlands: KIT Publishers, KIT (Bulletin no. 359 pp. 63–71).

Nautiyal, S., Maikhuri, R. K., Semwal, R. L., & Rao, K. S. (1998). Conservation through cultivation: A case study of medicinal plants in buffer zone village of NDBR. In Research for mountain development: Some initiatives and accomplishments. Gyanodaya Prakashan, Nainital: Himvikas Publication. (vol. 12. pp. 357–374).

Nautiyal, S., Maikhuri, R. K., Rao, K. S., & Saxena, K. G. (2001). Medicinal plant resources in Nanda Devi Biosphere Reserve in the Central Himalaya, India. *Herbs Spices & Medicinal Plants, 8*(4), 47–64. doi:10.1300/J044v08n04_06

Nautiyal, S., Rajan, K. S., & Shibasaki, R. (2004). Environmental conservation vs. compensation: Explorations from the Uttaranchal Himalaya. *Environmental Informatics Archives*, *2*, 24–35.

Navchoo, I. A., & Kachroo, P. (1995). *Flora of Pulwama, Kashmir*. Dehradun: Bishen Singh and Mahendra Pal Singh.

Naveh, Z., & Lieberman, A. S. (2013). *Landscape ecology: theory and application*. Springer Science & Business Media.

Nayar, M. P., & Shastry, A. R. K. (1990). *Red data book of Indian plants* (Vol. I). Calcutta, India: Botanical Survey of India.

Negi, K. S., Tiwari, J. K., & Gaur, R. D. (1987). A contribution to the flora of Khatling glacier in the Garhwal Himalaya (District-Tehri). *U. P. J. Bombay Nat. Hist. Soc*, *84*, 585–598.

Negi, V. M., & Chauhan, N. S. (2009). Medicinal and aromatic plants wealth of a tribal district Kinnaur in Himachal Himalayas. *Indian Forester*, *135*(6), 838–852.

Negi, V. S., Maikhuri, R. K., & Vashishtha, D. P. (2011). Traditional healthcare practices among the villages of Rawain Valley, Uttarkashi, Uttarakhand, India. *Indian Journal of Traditional Knowledge*, *10*(3), 533–537.

Neogi, B., Prasad, M. N. V., & Rao, R. R. (1989). Ethnobotany of some weeds of Khasi and Garo hills, Meghalaya, Northeastern India. *Economic Botany*, *43*(4), 471–479. doi:10.1007/BF02935921

Newman, D. J., Cragg, G. M., & Snader, K. M. (2003). Natural products as sources of new drugs over the period 1981−2002. *Journal of Natural Products*, *66*(7), 1022–1037. doi:10.1021/np030096l PMID:12880330

Nigam, N., Bhui, K., Prasad, S., George, J., & Shukla, Y. (2009). [6]-Gingerol induces reactive oxygen species regulated mitochondrial cell death pathway in human epidermoid carcinoma A431 cells. *Chemico-Biological Interactions*, *181*(1), 77–84. doi:10.1016/j.cbi.2009.05.012 PMID:19481070

Njoroge, G. N., Kaibui, I. M., Njenga, P. K., & Odhiambo, P. O. (2010). Utilization of priority traditional medicinal plants and local people's knowledge on their conservation status in arid lands of Kenya (Mwingi district). *Journal of Ethnobiology and Ethnomedicine*, *6*(1), 1–22. doi:10.1186/1746-4269-6-22 PMID:20712897

Noman, A. (2003). Influence of different doses of nitrogen fertilizer on ajwain. (Unpublished master's thesis). University of Agriculture. Faisalabad. Pakistan.

Oberai, K., Khare, M. P., & Khare, A. (1985). A pregnane ester diglycoside from *Hemidesmus indicus*. *Phytochemistry*, *24*(10), 2395–2397. doi:10.1016/S0031-9422(00)83049-6

Oboh, G., Akinyemi, A. J., & Ademiluyi, A. O. (2012). Antioxidant and inhibitory effect of red ginger (*Zingiber officinale* var. Rubra) and white ginger (*Zingiber officinale* Roscoe) on Fe2+ induced lipid peroxidation in rat brain *in vitro*. *Experimental and Toxicologic Pathology*, *64*(1), 31–36. doi:10.1016/j.etp.2010.06.002 PMID:20598871

Oh, G. S., Yoon, J., Lee, G. G., Kwak, J. H., & Kim, S. W. (2015). The Hexane Fraction of *Cyperus rotundus* Prevents Non-Alcoholic Fatty Liver Disease Through the Inhibition of Liver X Receptor α-Mediated Activation of Sterol Regulatory Element Binding Protein-1c. *The American Journal of Chinese Medicine*, *43*(03), 477–494. doi:10.1142/S0192415X15500305 PMID:25967664

Oksuz, E. (2006). Prevalence, risk factors, and preference-based health states of low back pain in a Turkish population. *Spine*, *31*(25), E968–E972. doi:10.1097/01.brs.0000247787.25382.3c PMID:17139213

Okunade, A. L. (2002). *Ageratum conyzoides* L. (Asteraceae). *Fitoterapia*, *73*(1), 1–16. doi:10.1016/S0367-326X(01)00364-1 PMID:11864757

Ong, T. P., Moreno, F. S., & Ross, S. A. (2012). Targeting the epigenome with bioactive food components for cancer prevention. *Journal of Nutrigenetics and Nutrigenomics*, *4*(5), 275–292. doi:10.1159/000334585 PMID:22353664

Oral, D. Ç. (2007). *Konya ilinde kullanılan halk ilaçları üzerinde etnobotanik araştırmalar* (Unpublished master's thesis), Gazi Üniversitesi, Turkey.

Osmaston, A. E. (1978). *Forest flora for Kumaun*. Dehra Dun, India: International Book Distributors.

Oyebode, O., Kandala, N. B., Chilton, P. J., & Lilford, R. J. (2016). Use of traditional medicine in middle-income countries: A WHO–SAGE study. *Health Policy and Planning*, *31*(8), 984–991. doi:10.1093/heapol/czw022 PMID:27033366

Özgen, U., Kaya, Y., & Houghton, P. (2012). Folk medicines in the villages of Ilıca District (Erzurum, Turkey). *Turkish Journal of Biology*, *36*(1), 93–106.

Ozturk, M., & Altay, V. (2017). Endemic medicinal and aromatic plants of turkiye -conservation and sustainable use. In *1st International Congress on Medicinal and Aromatic Plants: "Natural and Healthy Life" Proceedings Book* (pp. 11-39). Konya, Turkey.

Padhy, S. N., Mahato, S. B., & Dutta, N. L. (1973). Triterpenoids from the roots of *Hemidesmus indicus*. *Phytochemistry*, *12*(1), 217–218. doi:10.1016/S0031-9422(00)84656-7

Padulosi, S., Leaman, D., & Quek, P. (2002). Challenges and opportunities in enhancing the conservation and use of medicinal and aromatic plants. *Journal of Herbs, Spices, & Medicinal Plants*, *9*(4), 243–267. doi:10.1300/J044v09n04_01

Paembonan, S., Millang, S., Dassir, M., & Ridwan, M. (2018). Species variation in home garden agroforestry system in south Sulawesi, Indonesia and its contribution to farmer's income. In *IOP Conference Series: Earth and Environmental Science* (Vol. 157, pp. 012004). IOP Publishing. 10.1088/1755-1315/157/1/012004

Paksoy, M. Y., Selvi, S., & Savran, A. (2016). Ethnopharmacological survey of medicinal plants in Ulukışla (Niğde-Turkey). *Journal of Herbal Medicine*, *6*(1), 42–48. doi:10.1016/j.hermed.2015.04.003

Pala, N. A., Sarkar, B. C., Shukla, G., Chettri, N., Deb, S., Bhat, J. A., & Chakravarty, S. (2019). Floristic composition and utilization of ethnomedicinal plant species in home gardens of the eastern Himalaya. *Journal of Ethnobiology and Ethnomedicine*, *15*(1), 14. doi:10.118613002-019-0293-4 PMID:30782184

Pal, D., Dutta, S., & Sarkar, A. (2009). Evaluation of CNS activities of ethanol extract of roots and rhizomes of *Cyperus rotundus* in mice. *Acta Poloniae Pharmaceutica*, *66*, 535–541. PMID:19894649

Panahi, Y., Sharif, M. R., Sharif, A., Beiraghdar, F., Zahiri Z, Amirchoopani, G., Eisar, T. M., & Sahebkar, A. (2012). A randomized comparative trial on the therapeutic efficacy of topical Aloe vera and *Calendula officinalis* on diaper dermatitis in children. *Scientific World Journal, 8*(10), 22-34.

Panchal, G. A., Panchal, S. J., & Patel, J. A. (2009). *Hemidesmus indicus*: A review. *Pharmacologyonline*, *2*, 758–771.

Panda, A., & Misra, M. K. (2011). Ethnomedicinal survey of some wetland plants of South Orissa and their conservation. *Indian Journal of Traditional Knowledge*, *10*(2), 296–303.

Pande, H. C., & Pande, P. C. (2002). *An illustrated fern flora of the Kumaon Himalaya* (Vol. II). Dehra Dun, India: Bishen Singh Mahendra Pal Singh.

Pandey, A. K., & Chowdhury, A. R. (2003). Volatile constituents of the rhizome oil of *Curcuma caesia* Roxb. from central India. *Flavour and Fragrance Journal*, *18*(5), 463–465. doi:10.1002/ffj.1255

Pandey, A., & Singh, S. (2016). Aloe vera: A Systematic Review of Its Industrial and Ethnomedical Efficacy. *International Journal of Pharmaceutical Research and Allied Sciences*, *5*(1), 21–33.

Pandey, G. S. (2013). *Bhavprakash Nighantu*. Varanasi, India: Chaukhamba Bharti Academy.

Panero, J. L., & Funk, V. A. (2002). Toward a phylogenetic subfamilial classification for the Compositae (Asteraceae). *Proceedings of the Biological Society of Washington, 115*, 909–922.

Panero, J. L., & Funk, V. A. (2008). The value of sampling anomalous taxa in phylogenetic studies: Major clades of the Asteraceae revealed. *Molecular Phylogenetics and Evolution, 47*(2), 757–782. doi:10.1016/j.ympev.2008.02.011 PMID:18375151

Pan, S. Y., Zhou, S. F., Gao, S. H., Yu, Z. L., Zhang, S. F., Tang, M. K., ... Ko, K. M. (2013). New perspectives on how to discover drugs from herbal medicines: CAM'S outstanding contribution to modern therapeutics. *Evidence-Based Complementary and Alternative Medicine, 2013*, 1–25. doi:10.1155/2013/627375 PMID:23634172

Panthong, A., Kanjanapothi, D., Taesotikul, T., & Taylor, W. C. (1991). Ethnobotanical review of medicinal plants from Thai traditional books, part II: Plants with antidiarrheal, laxative and carminative properties. *Journal of Ethnopharmacology, 31*(2), 121–156. doi:10.1016/0378-8741(91)90001-T PMID:2023424

Pant, S., & Samant, S. S. (2010). Ethnobotanical observation in the Mornaula Reserve Forest of Kumoun, West Himalaya, India. *Ethnobotanical Leaflets, 14*, 193–217.

Pant, S., Samant, S. S., & Arya, S. C. (2009). Diversity and indigenous household remedies of the inhabitants surrounding Mornaula reserve forest in West Himalaya. *Indian Journal of Traditional Knowledge, 8*(4), 606–610.

Panyadee, P., Balslev, H., Wangpakapattanawong, P., & Inta, A. (2019). Medicinal plants in homegardens of four ethnic groups in Thailand. *Journal of Ethnopharmacology, 239*, 111927. doi:10.1016/j.jep.2019.111927 PMID:31055002

Papp, N., Birkás-Frendl, K., Farkas, Á., & Pieroni, A. (2013). An ethnobotanical study on home gardens in a Transylvanian Hungarian csángó resources village (Romania). *Genetic and Crop Evolution, 60*(4), 1423–1432. doi:10.100710722-012-9930-7

Paranjape, P. (2001). *Indian medicinal plants. Forgotten healer: a guide to Ayurvedic herbal medicine*. Delhi, India: Chaukhamba Sanskrit Pratisthan.

Pardeshi, M. D., & Tapar, K. K. (2016). Formulation and Evaluation of Aloe vera Gel with Active Salt and Alum: As a New Dentifrice. *International Journal of Research in Engineering and Social Sciences, 6*(5), 297–310.

Park, S. E., Shin, W. T., Park, C., Hong, S. H., Kim, G. Y., Kim, S. O., ... Choi, Y. H. (2014). Induction of apoptosis in MDA-MB-231 human breast carcinoma cells with an ethanol extract of *Cyperus rotundus* L. by activating caspases. *Oncology Reports, 32*(6), 2461–2470. doi:10.3892/or.2014.3507 PMID:25241797

Parthiban, R., Vijayakumar, S., Prabhu, S., & Yabesh, J. G. E. M. (2016). Quantitative traditional knowledge of medicinal plants used to treat livestock diseases from Kudavasal Taluk of Thiruvarur district, Tamil Nadu, India. [Brazilian Journal of Pharmacognosy]. *Revista Brasileira de Farmacognosia, 26*(1), 109–121. doi:10.1016/j.bjp.2015.07.016

Parvez, M. K., Al-Dosari, M. S., Arbab, A. H., & Niyazi, S. (2019). The *in vitro* and *in vivo* anti-hepatotoxic, anti-hepatitis B virus and hepatic CYP450 modulating potential of *Cyperus rotundus. Saudi Pharmaceutical Journal, 27*(4), 558–564. doi:10.1016/j.jsps.2019.02.003 PMID:31061625

Patel, D. M., Kanzariya, N. R., Patel, N. J., Mehta, H. R., & Modh, H. A. (2010). Evaluation of the efficacy of methanolic extract of *Hemidesmus indicus* in urolithiasis in wistar rats. *Inventi Rapid: Ethnopharmacology*. Retrieved from http://inventi.in/journal/art_reader/3/9176/rapid/ethnopharmacology/pi

Patel, D. K. (2012). Study on medicinal plants with special reference to family Asteraceae, Fabaceae and Solanaceae in GGV-Campus, Bilaspur (CG) in central India. *Current Biotica, 3*(4), 34–38.

Pathak, A. K., Saraf, S., & Dixit, V. K. (1991). Hepatoprotective activity potential of *Tridax procumbens. BMC Complementary and Alternative Medicine, 10*, 10–15.

Patwardhan, B., Mutalik, G., & Tillu, G. (2015). Drug discovery and Ayurveda. In Patwardhan & ... (Eds.), *Integrative approaches for health: Biomedical research, Ayurveda and Yoga* (pp. 229–258). Amsterdam, The Netherlands: Academic Press, Elsevier. doi:10.1016/B978-0-12-801282-6.00009-7

Paul, S., Devi, N., & Sarma, G. C. (2013). Ethnobotanical utilization of some medicinal plants by Bodo people of Manas Biosphere Reserve in the treatment of malaria. *International Research Journal of Pharmacy, 4*(6), 102–105. doi:10.7897/2230-8407.04622

Paul, S., Rahman, A. N. M. A., Al Mahmud, M. A., Uzzal, A. R., & Das, Z. C. (2017). Wound healing by marigold (*Calendula officinalis*) and turmeric (*Curcuma longa*) paste: A comparative approach. *Journal of Advanced Veterinary and Animal Research, 4*(4), 333–342. doi:10.5455/javar.2017.d229

Paulsen, E., Korsholm, L., & Brandrup, F. (2005). A double-blind, placebo-controlled study of a commercial Aloe vera gel in the treatment of slight to moderate psoriasis vulgaris. *Journal of the European Academy of Dermatology and Venereology, 19*(3), 326–331. doi:10.1111/j.1468-3083.2004.01186.x PMID:15857459

Payal, K. (2007). Eco-physiological and socio-economic studies of some multipurpose Allium species of higher Himalaya. (Ph.D. Thesis), HAPPRC, H.N.B. Garhwal University. Srinagar Garhwal Central University, India.

Payal, K., Maikhuri, R. K., Rao, K. S., & Kandari, L. S. (2014). Effect of temperature, photoperiod, gibberellic acid and water based pre-soaking treatments on seeds germination of *Allium stracheyi* Baker: An endangered alpine species of Central Himalaya, India. *Journal Plant Biosystems—An International Journal Dealing with all Aspects of Plant Biology, 148*(6), 1075–1084.

Peng, F., Tao, Q., Wu, X., Dou, H., Spencer, S., Mang, C., ... Hao, X. (2012). Cytotoxic, cytoprotective and antioxidant effects of isolated phenolic compounds from fresh ginger. *Fitoterapia, 83*(3), 568–585. doi:10.1016/j.fitote.2011.12.028 PMID:22248534

Pereira, R. L., Ibrahim, T., Lucchetti, L., da Silva, A. J. R., & de Moraes, V. L. G. (1999). Immunosuppressive and anti-inflammatory effects of methanolic extract and the polyacetylene isolated from *Bidens pilosa* L. *Immunopharmacology, 43*(1), 31–37. doi:10.1016/S0162-3109(99)00039-9 PMID:10437654

Peshkin, A. (1993). The Goodness of Qualitative Research. *Educational Researcher, 22*(2), 23–29. doi:10.3102/0013189X022002023

Petlevski, R., Hadžija, M., Slijepčevič, M., & Juretič, D. (2001). Effect of "antidiabetis" herbal preparation on serum glucose and fructosamine in NOD mice. *Journal of Ethnopharmacology, 75*(2), 181–184. doi:10.1016/S0378-8741(01)00177-5 PMID:11297848

Petrie, J. R., Guzik, T. J., & Touyz, R. M. (2018). Diabetes, hypertension, and cardiovascular disease: Clinical insights and vascular mechanisms. *The Canadian Journal of Cardiology, 34*(5), 575–584. doi:10.1016/j.cjca.2017.12.005 PMID:29459239

Peyre, A., Guidal, A., Wiersum, K. F., & Bongers, F. (2006). Dynamics of homegarden structure and function in Kerala, India. *Agroforestry Systems, 66*(2), 101–115. doi:10.100710457-005-2919-x

Phillips, O., & Gentry, A. H. (1993). The useful plants of Tambopata, Peru: I. Statistical hypotheses tests with a new quantitative technique. *Economic Botany, 47*(1), 15–32. doi:10.1007/BF02862203

Picavet, H. S. J., Schouten, J. S. A. G., & Smit, H. A. (1999). Prevalence and consequences of low back problems in the netherlands, working vs non-working population, the MORGEN-study. *Public Health*, *113*(2), 73–77. doi:10.1016/S0033-3506(99)00122-5 PMID:10355306

Pieroni, A., & Price, L. (2006). Eating and healing: Traditional food as medicine (1–237). Binghamton, NY: Haworth Press.

Pieroni, A. (2001). Evaluation of the cultural significance of wild food botanicals traditionally consumed in Northwestern Tuscany. *Italy. Journal of Ethnobiology*, *21*(1), 89–104.

Pillai, M. K., Asmawi, M. Z., Choon, T. S., Sasidharan, S., Lachumy, S. J., & Ismail, R. (2012). Evaluation of radical scavenging activity of rhizome extracts of *Alpinia galanga* and *Zingiber officinale. Journal of Medicinal Plants Research*, *6*(41), 5406–5411. doi:10.5897/JMPR11.084

Pirbalouti, A. G., Koohpayeh, A., & Karimi, I. (2010). The wound healing activity of flower extracts of *Punica granatum* and *Achillea kellalensis* in Wistar rats. *Acta Poloniae Pharmaceutica*, *67*(1), 107–110. PMID:20210088

Pirttilä, T., Wilcock, G., Truyen, L., & Damaraju, C. V. (2004). Long-term efficacy and safety of galantamine in patients with mild-to-moderate Alzheimer's disease: Multicenter trial. *European Journal of Neurology*, *11*(11), 734–741. doi:10.1111/j.1468-1331.2004.00885.x PMID:15525294

Polat, R. (2010). *Havran ve burhaniye (balıkesir) çevresinde tarımsal biyoçeşitlilik ve etnobotanik araştırmaları* (Unpublished doctoral dissertation), Balıkesir Üniversitesi, Balıkesir.

Polat, R., Cakilcioglu, U., & Satıl, F. (2013). Traditional uses of medicinal plants in Solhan (Bingöl-Turkey). *Journal of Ethnopharmacology*, *148*(3), 951–963. doi:10.1016/j.jep.2013.05.050 PMID:23770029

Polat, R., Güner, B., Babacan, E. Y., & Çakılcıoğlu, U. (2017). Survey of wild food plants for human consumption in Bingöl (Turkey). *Indian Journal of Traditional Knowledge*, *16*(3), 378–384.

Polat, R., Satıl, F., & Çakılcıoğlu, U. (2011). Medicinal plants and their use properties of sold in herbal market in bingöl (turkey) district. *Biological Diversity and Conservation*, *4*(3), 25–35.

Polat, R., Satıl, F., & Çakılcıoğlu, U. (2011). Medicinal plants and their use properties of sold in herbal market in Bingöl (Turkey) district. *Biological Diversity and Conservation*, *4*(3), 25–35.

Prabakan, M., Anandan, R., & Devaki, T. (2000). Protective effect of *Hemidesmus indicus* against rifampicin and isoniazid-induced hepatotoxicity in rats. *Fitoterapia*, *71*(1), 55–59. doi:10.1016/S0367-326X(99)00120-3 PMID:11449471

Prabhakkar, A. R., Karuna, Y. M., Yavagal, C., & Deepak, B. M. (2015). Cavity disinfection in minimally invasive dentistry – comparative evaluation of Aloe vera and propolis: A randomized clinical trial. *Contemporary Clinical Dentistry*, *6*(1), 24–31. doi:10.4103/0976-237X.152933 PMID:25821369

Prajapati, N. D., Purohit, S. S., Sharma, A. K., & Kumar, T. (2003). *A Handbook of Medicinal Plants* (pp. 1–1155). Jodhpur, India: Agrobios.

Prakash, K., Sethi, A., Deepak, D., Khare, A., & Khare, M. P. (1991). Two pregnane glycosides from *Hemidesmus indicus. Phytochemistry*, *30*(1), 297–299. doi:10.1016/0031-9422(91)84141-E

Prance, G. T., Baleé, W., Boom, B. M., & Carneiro, R. L. (1987). Quantitative ethnobotany and the case for conservation in Ammonia. *Conservation Biology*, *1*(4), 296–310. doi:10.1111/j.1523-1739.1987.tb00050.x

Prasad, S., Kashyap, R. S., Deopujari, J. Y., Purohit, H. J., Taori, G. M., & Daginawala, H. F. (2007). Effect of *Fagonia arabica* (Dhamasa) on in vitro thrombolysis. *BMC Complementary and Alternative Medicine*, *7*(1), 36–41. doi:10.1186/1472-6882-7-36 PMID:17986325

Pratibha, J., & Sahu, T. R. (1993). An ethnobotanical study of Noradehi Sanctuary Park of Madhya Pradesh, India: Native plant remedies for scorpion sting and snake bite. *Journal of Economic and Taxonomic Botany*, *17*(2), 315–328.

Prema, D., Kamaraj, M., Achiraman, S., & Udayakumar, R. (2014). *In vitro* antioxidant and cytotoxicity studies of *Curcuma amada* Roxb. (Mango ginger). *International Journal of Scientific and Research Publications*, *4*(4), 1–6.

Previati, M., Corbacella, E., Astolfi, L., Catozzi, M., Khan, M. T. H., Lampronti, I., ... Martini, A. (2007). Ethanolic extract from *Hemidesmus indicus* (Linn) displays otoprotectant activities on organotypic cultures without interfering on gentamicin uptake. *Journal of Chemical Neuroanatomy*, *34*(3-4), 128–133. doi:10.1016/j.jchemneu.2007.05.009 PMID:17640851

Pringle, R. M. (2017). Upgrading protected areas to conserve wild biodiversity. *Nature*, *546*(7656), 91–99. doi:10.1038/nature22902 PMID:28569807

PROTA. (2017). *Plant Resources of Tropical Africa.* Available from https://www.prota4u.org/

Purohit, A., Maikhuri, R. K., Rao, K. S., & Nautiyal, S. (2001). Impact of bark removal on Taxus baccata L. (Himalayan Yew) in Nanda Devi Biosphere Reserve, Garhwal Himalaya, India. *Current Science*, *81*(5), 586–590.

Pusalkar, P. K., & Srivastava, S. K. (2018). *Flora of Uttarakhand* (Vol. 1). Kolkata, India: Botanical Survey of India.

Pushparaj, P., Tan, C. H., & Tan, B. K. H. (2000). Effects of *Averrhoa bilimbi* leaf extract on blood glucose and lipids in streptozotocin-diabetic rats. *Journal of Ethnopharmacology*, *72*(1–2), 69–76. doi:10.1016/S0378-8741(00)00200-2 PMID:10967456

Quinlan, M. (2005). Considerations for collecting freelists in the field: Examples from ethnobotany. *Field Methods*, *17*(3), 219–234. doi:10.1177/1525822X05277460

Quinn, P. J., Carter, M. E., Markey, B. K., & Carter, R. C. (1994). *Clinical veterinary microbiology* (1st ed.). United States: Mosby.

Radha, S., Borra, S. K., Krishna, S., & Gowrinath, R. M. (2011). Anti-ulcer effects of Aloe vera in non-steroidal and anti-inflammatory drug induced peptic ulcers in rats. *African Journal of Pharmacy and Pharmacology*, *5*(1), 1867–1871.

Raghupathy, L. (2001). Conservation and sustainable use of medicinal plants: Current issues. In S. Samant, U. Dhar, & L. M. S. Palni (Eds.), Himalayan medicinal plants: Potential and prospects (415–426). Nainital: Gyanodaya Prakashan.

Rahman, M. (2013). A study on exploration of ethnobotanical knowledge of rural community in Bangladesh: basis for biodiversity conservation. *ISRN Biodiversity*, 2013.

Rahman, K., & Lowe, G. M. (2006). Garlic and cardiovascular disease: A critical review. *The Journal of Nutrition*, *136*(3), 736S–740S. doi:10.1093/jn/136.3.736S PMID:16484553

Rai, S., Mukherjee, K., Mal, M., Wahile, A., Saha, B. P., & Mukherjee, P. K. (2006). Determination of 6-gingerol in ginger (*Zingiber officinale*) using high-performance thin-layer chromatography. *Journal of Separation Science*, *29*(15), 2292–2295. doi:10.1002/jssc.200600117 PMID:17120812

Rajamma, A. G., Bai, V., & Nambisan, B. (2012). Antioxidant and antibacterial activities of oleoresins isolated from nine *Curcuma* species. *Phytopharmacology*, *2*, 312–317.

Rajar, U. D., Majeed, R., Parveen, N., Sheikh, I., & Sushel, C. (2008). Efficacy of Aloe vera gel in the treatment of vulval lichen planus. *Journal of the College of Physicians and Surgeons--Pakistan*, *18*(10), 612–614. PMID:18940117

Rajeswari, R., Umadevi, M., Sharmila, C. R., Pushpa, R., Selvavenkadesh, S., Sampath Kumar, K. P., & Debjit, B. (2018). Aloe vera: The Miracle Plant Its Medicinal and Traditional Uses in India. *Journal of Pharmacognosy and Phytochemistry, 1*(4), 118–124.

Rajkumari, R., Nirmala, R., Singh, P., Das, A. K., Dutta, B., & Pinokiyo, A. (2014). Ethnoveterinary plants used by the chiru tribes of Manipur, northeast India.

Rajoriya, C. M., Choudhary, R. G., Shah, I. A., Rawat, R. S., & Jat, B. L. (2016). Ethno – medicinal survey of North Kashmir with special reference to Bandipora. *International Journal for Research in Applied Science and Engineering Technology, 4*(11), 95–106.

Raju, V. S., & Reddy, K. N. (2005). Ethnomedicine for dysentery and diarrhoea from Khammam district of Andhra Pradesh. *Indian Journal of Traditional Knowledge, 4*(4), 443–447.

Ramachandran, V. S., & Nair, N. C. (1981). Ethnobotanical observations on irulars of Tamil Nadu (India). *Journal of Economic and Taxonomic Botany, 2*, 183–190.

Ramakrishnappa, K. (2002). *Impact of cultivation and gathering of medicinal plants on Biodiversity: Case study from India. Forestry and Fisheries*. Rome: FAO.

Ramana, M. V. (2008). Ethnomedicinal and ethnoveterinary plants from Boath, Adilabad district, Andhra Pradesh, India. *Ethnobotanical Leaflets, 2008*(1), 46.

Ramírez, G., Zavala, M., Pérez, J., & Zamilpa, A. (2012). In vitro screening of medicinal plants used in Mexico as antidiabetics with glucosidase and lipase inhibitory activities. *Evidence-Based Complementary and Alternative Medicine, 2012*, 1–6. doi:10.1155/2012/701261 PMID:23082084

Ramirez-Sosa, C. R. (2006). Quantitative ethnobotany in El Salvador, Central America: A model to study ethnobotanical knowledge dynamics. (2005). In *Proceedings of the 4th International Congress of Ethnobotany (ICEB 2005)*, (557–564). Central America.

Rana, C. S., Ballabha, R., Sharma, A., Dangwal, L. R., & Tiwari, J. K. (2013). Herbal remedies for leucorrhoea: A study from the Garhwal Himalaya, India. *Global Journal of Research on Medicinal Plants and Indigenous Medicine, 2*(10), 685–691.

Rana, P., Tewari, S. K., Kumar, V., & Kumar, A. (2016). Floristic structure, composition and functional characteristics of homegardens in Garhwal region, Uttarakhand, India. *International Journal of Agriculture Environment and Biotechnology, 9*(6), 1045. doi:10.5958/2230-732X.2016.00133.9

Rana, V. S., Verdeguer, M., & Blazquez, M. A. (2012). Chemical composition of the essential oil of *Zingiber zerumbet* var. darcyi. *Natural roduct. Communications, 7*(10), 1369–1370. PMID:23157013

Ranjan, S., Jadon, V. S., Sharma, N., Singh, K., Parcha, V., Gupta, S., & Bhatt, J. P. (2010). Anti-inflammatory and analgesic potential of leaf extract of *Allium stracheyi.* [Not cited in text.]. *Journal of Applied Sciences Research, 6*(2), 139–143.

Rao, G. M., Venkateswararao, C., Rawat, A. K. S., Pushpangadan, P., & Shirwaikar, A. (2005). Antioxidant and antihepatotoxic activities of *Hemidesmus indicus* R. Br. *Acta Pharmaceutica Turcica, 47*(2), 107–113.

Rao, S., Rao, G. M. N., & Murthy, P. P. (2016). Diversity and indigenous uses of some ethno medicinal plants in Papikondalu wildlife Sanctuary, Eastern Ghats of Andhra Pradesh, India. *American Journal of Ethnomedicine, 3*, 6–25.

Raskin, I., Ribnicky, D. M., Komarnytsky, S., Ilic, N., Poulev, A., Borisjuk, N., ... Fridlender, B. (2002). Plants and human health in the twenty-first century. *Trends in Biotechnology, 20*(12), 522–531. doi:10.1016/S0167-7799(02)02080-2 PMID:12443874

Rau, M. A. (1975). *High altitude flowering plants of West Himalaya.* Howrah: Botanical Survey of India - Indian Botanic Garden.

Ravishankara, M. N., Shrivastava, N., Padh, H., & Rajani, M. (2002). Evaluation of antioxidant properties of root bark of *Hemidesmus indicus* R. Br. (Anantmul). *Phytomedicine*, *9*(2), 153–160. doi:10.1078/0944-7113-00104 PMID:11995949

Redl, K., Breu, W., Davis, B., & Bauer, R. (1994). Anti-inflammatory active polyacetylenes from *Bidens campylotheca. Planta Medica*, *60*(1), 58–62. doi:10.1055-2006-959409 PMID:8134419

Rehecho, S., Uriarte-Pueyo, I., Calvo, J., Vivas, L. A., & Calvo, M. I. (2011). Ethnopharmacological survey of medicinal plants in Nor-Yauyos, a part of the Landscape Reserve Nor-Yauyos-Cochas, Peru. *Journal of Ethnopharmacology*, *133*(1), 75–85. doi:10.1016/j.jep.2010.09.006 PMID:20837126

Resende, J. H. C., De Aquino, G. S. T., Do Nascimento, F. R. F., Aguiar, M. M., & Fiorelli, R. K. A. (2015). Oral use of an infusion of leaves of *Solanum paniculatum* L., *Jacaranda brasiliensis* and *Sonchus oleraceus* for Treatment of Vitiligo. Journal of Cosmetics. *Dermatological Sciences and Applications*, *5*(4), 317–331. doi:10.4236/jcdsa.2015.54039

Reuter, J., Jocher, A., Stump, J., Grossjohann, B., Franke, G., & Schempp, C. M. (2008). Investigation of the anti-inñammatory potential of Aloe vera gel (97.5%) in the ultraviolet erythema test. *Skin Pharmacology and Physiology*, *21*(2), 106–110. doi:10.1159/000114871 PMID:18253066

Reyes-García, V., Aceituno-Mata, L., Calvet-Mir, L., Garnatje, T., Gómez-Baggethun, E., Lastra, J. J., ... Pardo-de-Santayana, M. (2014). Resilience of traditional knowledge systems: The case of agricultural knowledge in home gardens of the Iberian Peninsula. *Global Environmental Change*, *24*, 223–231. doi:10.1016/j.gloenvcha.2013.11.022

Reyes-García, V., Huanca, T., Vadez, V., Leonard, W., & Wilkie, D. (2006). Cultural, practical, and economic value of wild plants: A quantitative study in the Bolivian Amazon. *Economic Botany*, *60*(1), 62–74. doi:10.1663/0013-0001(2006)60[62:CPAEVO]2.0.CO;2

Reynolds, T. (2004). Aloe chemistry. In T. Reynolds (Ed.), Aloes: The genus Aloe (38th ed). Boca Raton, FL: CRC Press.

Rhama, S., & Madhavan, S. (2011). Antibacterial activity of the flavonoid, patulitrin isolated from the flowers of *Tagetes erecta* L. *International Journal of Pharm Tech Research*, *3*(3), 1407–1409.

Ried, K., Toben, C., & Fakler, P. (2013). Effect of garlic on serum lipids: An updated meta-analysis. *Nutrition Reviews*, *71*(5), 282–299. doi:10.1111/nure.12012 PMID:23590705

Rikhari, H. C., Negi, G. C. S., & Singh, S. P. (1993). Species and community diversity pattern in an alpine meadow of central Himalaya. In: US. Dhar (Ed.), Himalayan biodiversity: Conservation strategies. (pp. 205–218). Kosi, Almora: G. B. Pant Institute of Himalayan Environment and Development.

Rivera-Madrid, R., Aguilar-Espinosa, M., Cárdenas-Conejo, Y., & Garza-Caligaris, L. E. (2016). Carotenoid derivates in Achiote (*Bixa orellana*) seeds: Synthesis and health promoting properties. *Frontiers in Plant Science*, *7*, 1406. doi:10.3389/fpls.2016.01406 PMID:27708658

Rizos, C. V., & Elisaf, M. S. (2014). Antihypertensive drugs and glucose metabolism. *World Journal of Cardiology*, *6*(7), 517–530. doi:10.4330/wjc.v6.i7.517 PMID:25068013

Robbins, P. (2011). *Political ecology: A critical introduction* (Vol. 16). John Wiley & Sons.

Robinson, H. (2004). New supertribes Helianthodae and Senecionodae, for the subfamily Asteroideae (Asteraceae). *Phytologia*, *86*(3), 116–120. doi:10.5962/bhl.part.28428

Robinson, H. (2005). Validation of the supertribe Asterodae. *Phytologia*, *87*, 72–88.

Rodriguez-Fragoso, L., Reyes-Esparza, J., Burchiel, S. W., Herrera-Ruiz, D., & Torres, E. (2008). Risks and benefits of commonly used herbal medicines in Mexico. *Toxicology and Applied Pharmacology, 227*(1), 125–135. . 2007.10.005 doi:10.1016/j.taap

Rogers, M. (2015). Herbalpedia: Herbs and everything about them, Sunflower. Retrieved from https://www.herbalpedia.com/blog

Ro, J. Y., Lee, B., Kim, J. Y., Chung, Y., Chung, M. H., Lee, S. K., ... Park, Y. I. (2000). Inhibitory mechanism of aloe single component (Alprogen) on mediator release in guinea pig lung mast cells activated with specific antigen-antibody reactions. *Journal of Pharmacological Experiment Therapeutics*, *292*(1), 21–114. PMID:10604937

Rosenthal, J. (2002). Curtain has fallen on hopes of legal bioprospecting. *Nature*, *416*(6876), 15–19. doi:10.1038/416015a PMID:11882865

Rout, O. P., Acharya, R., & Mishra, S. K. (2011). In-Vitro Antioxidant potentials in leaves of Coleus aromaticus Benth and rhizomes of *Zingiber zerumbet* (L.) SM. *Journal of Applied Pharmaceutical Science*, *1*(8), 194–198.

Royal Botanical Garden Kew. (2013). *World Checklist of Aloe vera Species*. Available from www.londonexplorepass.com

Roy, S. K., Ali, M., Sharma, M. P., & Ramachandram, R. (2001). New pentacyclic triterpenes from the roots of *Hemidesmus indicus. Die Pharmazie*, *56*(3), 244–246. PMID:11265594

Roy, S. K., Ali, M., Sharma, M. P., & Ramachandram, R. (2002). Phytochemical investigation of *Hemidesmus indicus* R. Br. roots. *Indian Journal of Chemistry, Section B. Organic Chemistry Including Medicinal Chemistry*, *41B*(11), 2390–2394.

Rubina, L., Priyanka, T., & Ebenezer, J. (2009). Isolation, purification and evaluation of antibacterial agents from Aloe vera. *Brazilian Journal of Microbiology*, *40*(4), 906–915. doi:10.1590/S1517-83822009000400023 PMID:24031440

Rugalema, G., Johnsen, F., Okting'Ati, A., & Minjas, A. (1994). The homegarden agro-forestry system of Bukoba district, north-western Tanzania. 3. An economic appraisal of possible solutions to falling productivity. *Agroforestry Systems*, *28*(3), 227–236. doi:10.1007/BF00704758

Rulz, M. A., Pleguezuelos, M., Murioz, M., & Gallardo, V. (2004). Moisturizing Capacity of Aloe vera Gel in Skin Creams with Silicon-Based and Olive Oil-Based Latex Preparations. *Journal of Applied Cosmetology*, *22*(1), 25–33.

Ruslay, S., Abas, F., Shaari, K., Zainal, Z., Sirat, H., Israf, D. A., & Lajis, N. H. (2007). Characterization of the components present in the active fractions of health gingers (*Curcuma xanthorrhiza* and *Zingiber zerumbet*) by HPLC–DAD–ESIMS. *Food Chemistry*, *104*(3), 1183–1191. doi:10.1016/j.foodchem.2007.01.067

Rybalchenko, N. P., Prykhodko, V. A., Nagorna, S. S., Volynets, N. N., Ostapchuk, A. N., Klochko, V. V., ... Avdeeva, L. V. (2010). In vitro antifungal activity of phenylheptatriyne from *Bidens cernua* L. against yeasts. *Fitoterapia*, *81*(5), 336–338. doi:10.1016/j.fitote.2009.10.007 PMID:19837144

Saal, J. A. (1996). Natural history and nonoperative treatment of lumbar disc herniation. *Spine*, *21*(24Suppl), 2–9. doi:10.1097/00007632-199612151-00002 PMID:9112320

Sabale, P., Bhimani, B., Prajapati, C., & Sabale, V. (2012). An overview of medicinal plants as wound healers. *Journal of Applied Pharmaceutical Science*, *2*(11), 143–150.

Sabran, S. F., Mohamed, M., & Bakar, M. F. A. (2016). Ethnomedical knowledge of plants used for the treatment of tuberculosis in Johor, Malaysia. *Evidence-Based Complementary and Alternative Medicine*, *2016*, 2850845–2850845. doi:10.1155/2016/2850845 PMID:26881002

Saday, H. (2009). *Güzeloluk köyü ve çevresinin (erdemli/mersin) etnobotanik özellikleri* (Unpublished doctoral dissertation). Selçuk Üniversitesi, Turkey.

Sadıkoğlu, E. (2003). *Koçarlı (aydın) yöresinin geleneksel halk ilacı olarak kullanılan bitkileri* (Unpublished master's thesis). Marmara Üniversitesi, Turkey.

Sadıkoğlu, N., & Alpınar, K. (2001). Etnobotanik açıdan bartın. In XIII. bitkisel ilaç hammaddeleri toplantısı bildiri kitabı (pp. 87-100). İstanbul, Turkey.

Sadruddin. (1997). Medicinal plants of Bhutan: A conspectus. *Sherub Doenme, 3*(1 & 2), 32–55.

Safdar, W., Majeed, H., Naveed, I., Kayani, W. K., Ahmed, H., Hussain, S., & Kamal, A. (2010). Pharmacognostical study of the medicinal plant *Calendula officinalis* L. (family Compositae). *International Journal of Cell Science & Molecular Biology, 1*, 108–116.

Sağında, A. (2014). *Silifke bölgesindeki bitkisel halk ilaçlarının araştırılması* (Unpublished doctoral dissertation). Selçuk Üniversitesi, Turkey.

Sağıroğlu, M., Topuz, T., Ceylan, K., & Turna, M. (2013). An ethnobotanical survey from yahyalı (kayseri) and tarsus (mersin). *Sakarya University J Sci and Lit, 2*, 13–37.

Saha, J. C., Savini, E. C., & Kasinathan, S. (1961). Ecbolic properties of Indian medicinal plants. Part 1. *The Indian Journal of Medical Research, 49*, 130–151. PMID:14496025

Saheb, T. S., Zakia, S., Swamulu, M., & Mahesh, M. (2017). Medicinal plants used against gynaecological disorders by the tribal people of Nallamalais Andhra Pradesh. *Journal of Pharmacognosy and Phytochemistry, 6*(2), 242–244.

Sahu, C. R., Nayak, R. K., & Dhal, N. K. (2013). Traditional herbal remedies for various diseases used by tribals of Boudh district, Odisha, India for sustainable development. *International Journal of Herbal Medicine, 1*(1), 12–20.

Sahu, R. K., Singh, H., & Roy, A. (2009). Antioxidative characteristics of ethanol and aqueous extracts of *Curcuma amada* rhizomes. *Research Journal of Pharmacognosy and Phytochemistry, 1*(1), 41–43.

Sahu, T. R. (1984). Less known uses of weeds as medicinal plants. *Ancient Science of Life, 3*(4), 245–249. PMID:22557414

Saikia, P., & Khan, M. (2014). Homegardens of upper Assam, northeast India: A typical example of on farm conservation of agarwood (*Aquilaria malaccensis* Lam.). *The International Journal of Biodiversity Science, Ecosystem Services & Management, 10*(4), 262–269. doi:10.1080/21513732.2014.973449

Saikia, P., & Khan, M. L. (2011). Diversity of medicinal plants and their uses in home-gardens of upper Assam, northeast India. [AJPBR]. *Asian Journal of Pharmaceutical & Biological Research, 1*(3).

Sakinoğlu Oruç, F. Ç., & Oruç, S. H. (2017). Ethnobotonical aspects of some species in düzce and its vicinity. In *1st International Congress on Medicinal and Aromatic Plants: "Natural and Healthy Life" Proceedings Book* (pp. 391-399). Konya, Turkey.

Salah, S. H., Abdouh, S., & Booles, H. (2012). Effect of *Zingiber Officinale* on paracetamol-induced genotoxicity in male rats. *Journal of Medicinal Plants Research, 6*(41), 5425–5434. doi:10.5897/JMPR12.218

Saleem, H., Zengin, G., Ahmad, I., Lee, J. T. B., Htar, T. T., Mahomoodally, F. M., ... Ahemad, N. (2019). Multidirectional insights into the biochemical and toxicological properties of *Bougainvillea glabra* (Choisy.) aerial parts: A functional approach for bioactive compounds. *Journal of Pharmaceutical and Biomedical Analysis, 170*(5), 132–138. doi:10.1016/j.jpba.2019.03.027 PMID:30921647

Salehi, B., Valussi, M., Flaviana Bezerra Morais-Braga, M., Nalyda Pereira Carneiro, J., Linkoln Alves Borges Leal, A., Douglas Melo Coutinho, H., ... Sharifi-Rad, J. (2018). *Tagetes* spp. Essential oils and other extracts: Chemical characterization and biological activity. *Molecules (Basel, Switzerland)*, *23*(11), 2847. doi:10.3390/molecules23112847 PMID:30388858

Samant, S. S., & Dhar, U. (1997). Diversity, endemism and economic potential of wild edible plants of Indian Himalaya. *International Journal of Sustainable Development and World Ecology*, *4*(3), 179–191. doi:10.1080/13504509709469953

Samant, S. S., Dhar, U., & Palni, L. M. S. (1998). *Medicinal plants of Indian Himalaya: Diversity distribution potential values*. Gyanodaya Prakashan, Nanital.

Samarakoon, S. R., Thabrew, I., Galhena, P. B., De Silva, D., & Tennekoon, K. H. (2010). A comparison of the cytotoxic potential of standardized aqueous and ethanolic extracts of a polyherbal mixture comprised of *Nigella sativa* (seeds), *Hemidesmus indicus* (roots) and *Smilax glabra* (rhizome). *Pharmacognosy Research*, *2*(6), 335–342. doi:10.4103/0974-8490.75451 PMID:21713135

Samarakoon, S. R., Thabrew, I., Galhena, P. B., & Tennekoon, K. H. (2012). Modulation of apoptosis in human hepatocellular carcinoma (HepG2 cells) by a standardized herbal decoction of *Nigella sativa* seeds, *Hemidesmus indicus* roots and *Smilax glabra* rhizomes with anti-hepatocarcinogenic effects. *BMC Complementary and Alternative Medicine*, *12*(1), 25. doi:10.1186/1472-6882-12-25 PMID:22458551

Samarghandian, S., & Borji, A. (2014). Anticarcinogenic effect of saffron (*Crocus sativus* L.) and its ingredients. *Pharmacognosy Research*, *6*(2), 99–107. doi:10.4103/0974-8490.128963 PMID:24761112

Samuelsson, G. (2004). Drugs of natural origin: A textbook of pharmacognosy, (1–350). Stockholm, Sweden: Swedish Pharmaceutical Press.

Samy, R. P., Thwin, M. M., Gopalakrishnakone, P., & Ignacimuthu, S. (2008). Ethnobotanical survey of folk plants for the treatment of snakebites in Southern part of Tamilnadu, India. *Journal of Ethnopharmacology*, *115*(2), 302–312. doi:10.1016/j.jep.2007.10.006 PMID:18055146

Sandeep, D., & Nair, C. K. K. (2010). Amelioration of cisplatin-induced nephrotoxicity by extracts of *Hemidesmus indicus* and *Acorus calamus*. *Pharmaceutical Biology*, *48*(3), 290–295. doi:10.3109/13880200903116048 PMID:20645815

Sanjit, K. K., & Tushar, K. B. (2018). Phytochemical constituents of *Aloe vera* and their Multifunctional properties: A comprehensive review. *International Journal of Pharmaceutical Sciences*, *9*(4), 1416–1423.

Santana, A. P. M. (2009) Safety and genotoxicity evaluation of the *Alpinia zerumbet* (pers.) Burtt & Smith tea on healthy volunteers. *Cadernos de Agroecologia, 8*(2).

Santana-Gálvez, J., Cisneros-Zevallos, L., & Jacobo-Velázquez, D. A. (2017). Chlorogenic acid: Recent advances on its dual role as a food additive and a nutraceutical against metabolic syndrome. *Molecules (Basel, Switzerland)*, *22*(3), 358. doi:10.3390/molecules22030358 PMID:28245635

Saraf, S., Pathak, A. K., & Dixit, V. K. (1991). Hair growth promoting activity of *Tridax procumbens*. *Fitoterapia*, *62*, 495–498.

Saravanan, N., & Nalini, N. (2007a). Antioxidant effect of *Hemidesmus indicus* on ethanol-induced hepatotoxicity in rats. *Journal of Medicinal Food*, *10*(4), 675–682. doi:10.1089/jmf.2006.110 PMID:18158840

Saravanan, N., & Nalini, N. (2007b). Effect of 2-hydroxy 4-methoxy benzoic acid on an experimental model of hyperlipidaemia, induced by chronic ethanol treatment. *The Journal of Pharmacy and Pharmacology*, *59*(11), 1537–1542. doi:10.1211/jpp.59.11.0011 PMID:17976265

Saravanan, N., Rajasankar, S., & Nalini, N. (2007). Antioxidant effect of 2-hydroxy-4-methoxy benzoic acid on ethanol-induced hepatotoxicity in rats. *The Journal of Pharmacy and Pharmacology*, *59*(3), 445–453. doi:10.1211/jpp.59.3.0015 PMID:17331349

Sargın, S. A. (2013). *Alaşehir ve çevresinde (manisa) tarımsal biyoçeşitlilik ve etnobotanik araştırmaları* (Unpublished doctoral dissertation). Balıkesir Üniversitesi, Balıkesir.

Sargın, S. A. (2013). *Alaşehir ve çevresinde (manisa) tarımsal biyoçeşitlilik ve etnobotanik araştırmaları* (Unpublished doctoral dissertation). Balıkesir Üniversitesi, Turkey.

Sargin, S. A. (2015). Ethnobotanical survey of medicinal plants in Bozyazı District of Mersin, Turkey. *Journal of Ethnopharmacology*, *173*, 105–126. doi:10.1016/j.jep.2015.07.009 PMID:26190351

Sargın, S. A., Selvi, S., & López, V. (2015). Ethnomedicinal plants of Sarigöl District (Manisa), Turkey. *Journal of Ethnopharmacology*, *171*, 64–84. doi:10.1016/j.jep.2015.05.031 PMID:26026370

Sarı, A. O., Oğuz, B., Bilgiç, A., Tort, N., Güvensen, A., & Şenol, S. G. (2010). Ege ve güney marmara bölgelerinde halk ilacı olarak kullanılan bitkiler. *Anadolu Ege Tarimsal Arastirma Enstitüsü Dergisi*, *20*(2), 1–21.

Sari, I. P., Nurrochmad, N., & Setiawan, I. M. (2013). Indonesian herbals reduce cholesterol levels in diet-induced hypercholesterolemia through lipase inhibition. *Malaysian Journal of Pharmaceutical Sciences*, *11*(1), 13–20.

Sarkar, P. R. (1983). *Yogic treatment and natural remedies. Tiljala*. Kolkata, India: AMPS Publications.

Sasikumar, B. (2005). Genetic resources of Curcuma: Diversity, characterization and utilization. *Plant Genetic Resources*, *3*(2), 230–251. doi:10.1079/PGR200574

Sastri, K. (2013). *Charak Samhita*. Varanasi, India: Chaukhambha Bharti Academy.

Sathe, V. (1985). *Culture and related corporate realities*. Homewood, IL: Irwin.

Satou, T., Murakami, S., Matsuura, M., Hayashi, S., & Koike, K. (2010). Anxiolytic effect and tissue distribution of inhaled *Alpinia zerumbet* essential oil in mice. *Natural Product Communications*, *5*(1), 143–146. doi:10.1177/1934578X1000500133 PMID:20184040

Sayed, H. M., Mohamed, M. H., Farag, S. F., Mohamed, G. A., Omobuwajo, O. R., & Proksch, P. (2008). Fructose-amino acid conjugate and other constituents from *Cyperus rotundus* L. *Natural Product Research*, *22*(17), 1487–1497. doi:10.1080/14786410802038556 PMID:19023813

Sayed, H. M., Mohamed, M. H., Farag, S. F., Mohamed, G. A., & Proksch, P. (2007). A new steroid glycoside and furochromones from *Cyperus rotundus* L. *Natural Product Research*, *21*(4), 343–350. doi:10.1080/14786410701193056 PMID:17479423

Schippmann, U., Leaman, D. J., & Cunningham, A. (2002). *Impact of cultivation and gathering of medicinal plants on biodiversity: Global trends and issues*. Biodiversity and the Ecosystem Approach in Agriculture, Forestry and Fisheries.

Schmeller, D. S. (2008). European species and habitat monitoring: Where are we now? *Biodiversity and Conservation*, *17*(14), 3321–3326. doi:10.100710531-008-9514-1

Schmink, M., Redford, K. H., & Padoch, C. (1992). Traditional peoples and the biosphere: framing the issues and defining the terms. In K. H. Redford, & C. Padoch (Eds.), *Conservation of neotropical forests: working from traditional resource use* (pp. 3–13). New York: Columbia University Press.

Selvam, J. P., & Durai, M. (2018). An ethnobotanical study of medicinal plants used for the dengue in Tiruchirappalli district. *International Journal of Recent Scientific Research*, *9*, 27108–27112.

Selvi, S., Satıl, F., Polat, R., & Çakılcıoğlu, U. (2012). Kazdağlarından (balıkesir-edremit) toplanan ve bölgedeki aktarlarda satılan tıbbi bitkiler üzerine bir araştırma. In Kazdağları iii. ulusal sempozyumu bildiri kitabı (pp. 505-513). Edremit, Balıkesir.

Selvi, S., Satıl, F., Polat, R., & Çakılcıoğlu, U. (2012). Kazdağlarından (balıkesir-edremit) toplanan ve bölgedeki aktarlarda satılan tıbbi bitkiler üzerine bir araştırma. In Kazdağları III. Ulusal Sempozyumu Bildiri Kitabı (pp. 505-513). Edremit, Balıkesir.

Semwal, D. P., Saradhi, P. P., Kala, C. P., & Sajwan, B. S. (2010). Medicinal plants used by local Vaidyas in Ukhimath block, Uttarakhand. *Indian Journal of Traditional Knowledge*, *9*(3), 480–485.

Sen, U. K. (2016). *Botanical and socio-cultural studies on some sacred groves of West Midnapore district, West Bengal.* (Ph.D. Thesis), Vidyasagar University, India.

Senan, V. P. (2014). Aloe vera: A miracle plant with biological action. *World Journal of Pharmaceutical Sciences*, *3*(11), 209–218.

Sengupta, S. D., & Paul, R. (2016). Exploration of knowledge on new ethno-botanical value of *Justicia spicigera* SCHLTDL. *Global Journal of Research on Medicinal Plants & Indigenous*, *5*(10), 261–266.

Sen, S. K., & Behera, L. M. (2008). Ethnomedicinal plants used by the tribals of Bargarh district to cure diarrhoea and dysentery. *Indian Journal of Traditional Knowledge*, *7*(3), 425–428.

Senthilkumar, M., Gurumoorthi, P., & Janardhanan, K. (2006). Some medicinal plants used by Irular, the tribal people of Marudhamalai hills, Coimbatore, Tamil Nadu. *Natural Product Radiance*, *5*(5), 382–388.

Seo, E. J., Lee, D. U., Kwak, J. H., Lee, S. M., Kim, Y. S., & Jung, Y. S. (2011). Antiplatelet effects of *Cyperus rotundus* and its component (+)-nootkatone. *Journal of Ethnopharmacology*, *135*(1), 48–54. doi:10.1016/j.jep.2011.02.025 PMID:21354294

Seo, W. G., Pae, H. O., Oh, G. S., Chai, K. Y., Kwon, T. O., Yun, Y. G., ... Chung, H. T. (2001). Inhibitory effects of methanol extract of *Cyperus rotundus* rhizomes on nitric oxide and superoxide productions by murine macrophage cell line, RAW 264.7 cells. *Journal of Ethnopharmacology*, *76*(1), 59–64. doi:10.1016/S0378-8741(01)00221-5 PMID:11378282

Seo, Y. J., Jeong, M., Lee, K. T., Jang, D. S., & Choi, J. H. (2016). Isocyperol, isolated from the rhizomes of *Cyperus rotundus*, inhibits LPS-induced inflammatory responses via suppression of the NF-κB and STAT3 pathways and ROS stress in LPS-stimulated RAW 264.7 cells. *International Immunopharmacology*, *38*, 61–69. doi:10.1016/j.intimp.2016.05.017 PMID:27240136

Sethi, A., Bhatia, A., Srivastava, S., Bhatia, G., Khan, M. M., Khanna, A. K., & Saxena, J. K. (2010). Pregnane glycoside from *Hemidesmus indicus* as a potential anti-oxidant and anti-dyslipidemic agent. *Natural Product Research*, *24*(15), 1371–1378. doi:10.1080/14786410802265084 PMID:20169502

Sethi, A., Srivastav, S. S., & Srivastav, S. (2006). Pregnane glycoside from *Hemidesmus indicus* R. Br. *Indian Journal of Heterocyclic Chemistry*, *16*(2), 191–192.

Seuring, T., Archangelidi, O., & Suhrcke, M. (2015). The economic costs of type 2 diabetes: A global systematic review. *PharmacoEconomics*, *33*(8), 811–831. doi:10.100740273-015-0268-9 PMID:25787932

Sezik, E., Yeşilada, E., Honda, G., Takaishi, Y., Takeda, Y., & Tanaka, T. (2001). Traditional medicine in Turkey x. folk medicine in Central Anatolia. *Journal of Ethnopharmacology*, *75*(2-3), 95–115. doi:10.1016/S0378-8741(00)00399-8 PMID:11297840

Shah, A., Karim, A., Ahmad, J., & Sharma, M. P. (2015). Medicinal shrubs used by the Gujjar-Bakerwal tribes against various tribes against various non-communicable diseases in Rajouri district, (J&K), India. *Indian Journal of Traditional Knowledge, 14*(3), 466–473.

Shaikh, P. Z. (2011). Study of anti-inflammatory activity of ethanolic extract of *Hemidesmus indicus* roots in acute, subchronic & chronic inflammation in experimental animals. *International Journal of Pharmacy & Life Sciences, 2*(10), 1154–1173.

Shan, B., Cai, Y. Z., Sun, M., & Corke, H. (2005). Antioxidant capacity of 26 spice extracts and characterization of their phenolic constituents. *Journal of Agricultural and Food Chemistry, 53*(20), 7749–7759. doi:10.1021/jf051513y PMID:16190627

Shapoo, G. A., Kaloo, Z. A., Ganie, A. H., & Singh, S. (2013). Ethnobotanical survey and documentation of some Orichid species of Kashmir Himalaya, J&K, India. *International Journal of Pharmaceutical and Biological Research, 4*, 32–40.

Sharma, J., Gairola, S., Gaur, R. D., & Painuli, R. M. (2011). Medicinal plants used for primary healthcare by Tharu tribe of Udham Singh Nagar, Uttarakhand, India. *International Journal of Medicinal and Aromatic Plants, 1*(3), 228–233.

Sharma, J., Gairola, S., Sharma, Y. P., & Gaur, R. D. (2014). Ethnomedicinal plants used to treat skin diseases by Tharu community of district Udham Singh Nagar, Uttarakhand, India. *Journal of Ethnopharmacology, 158*, 140–206. doi:10.1016/j.jep.2014.10.004 PMID:25448505

Sharma, L., & Khandelwal, S. (2010). Weeds of Rajasthan and their ethno-botanical importance. *Ethno Med, 4*(2), 75–79. doi:10.1080/09735070.2010.11886363

Sharma, P. C., Yelne, M. B., Dennis, T. J., Joshi, A., & Billore, K. V. (2000). *Database on medicinal plants used in Ayurveda*. New Delhi, India: Central Council for Research in Ayurveda and Siddha.

Sharma, P. V. (1998). *Dravyaguna vijnana*. Varanasi, India: Chaukhabha Bharti Academy.

Sharma, P., & Sharma, G. (1998). *Dhanvantari Nighantu*. Varanasi, India: Chaukhamba Oriental.

Sharma, S., & Gupta, N. (2012). Antimicrobial potential of a weed plant *Parthenium hysterphorus*: An in vivo study. *International Journal of Pharmaceutical Research Development, 4*, 112–118.

Sharrif, M. M., & Verma, S. K. (2011). Aloe vera, their chemicals composition and applications: A review. *International Journal of Biology and Medical Research, 2*(1), 466–471.

Sheeja, D. A., & Nair, M. S. (2012). Phytochemical constituents of *Curcuma amada. Biochemical Systematics and Ecology, 44*, 264–266. doi:10.1016/j.bse.2012.06.008

Shende, V., & Telrandhe, R. (2017). Formulation and Evaluation of Tooth Gel from Aloe vera Leaves Extract. *International Pharmaceutics and Drug Analysis, 5*(10), 394–398.

Shete, R. V., & Bodhankar, S. L. (2009). Neuropharmacology of ethanolic extract *of Hemidesmus indicus. Electronic Journal of Pharmacology and Therapy, 2*, 63–70.

Shete, R. V., & Bodhankar, S. L. (2010). *Hemidesmus indicus*: Evaluation of its nootropic effect in mice. *International Journal of Pharma and Bio Sciences, 1*(3), 1–10.

Shetty, T. K., Satav, J. G., & Nair, C. K. K. (2005). Radiation protection of DNA and membrane in vitro by extract of *Hemidesmus indicus. Phytotherapy Research, 19*(5), 387–390. doi:10.1002/ptr.1470 PMID:16106384

Shin, J. S., Hong, Y., Lee, H. H., Ryu, B., Cho, Y. W., Kim, N. J., ... Lee, K. T. (2015). Fulgidic Acid Isolated from the Rhizomes of *Cyperus rotundus* Suppresses LPS-Induced iNOS, COX-2, TNF-α, and IL-6 Expression by AP-1 Inactivation in RAW264.7 Macrophages. *Biological & Pharmaceutical Bulletin*, *38*(7), 1081–1086. doi:10.1248/bpb.b15-00186 PMID:26133719

Shi, Q., Li, L., Huo, C., Zhang, M., & Wang, Y. (2010). Study on natural medicinal chemistry and new drug development. *Zhongcaoyao-Chinese Traditional and Herbal Drugs*, *41*(10), 1583–1589.

Shojaii, A., & Abdollahi Fard, M. (2012). Review of pharmacological properties and chemical constituents of *Pimpinella anisum. ISRN Pharmaceutics*, *2012*, 1–8. doi:10.5402/2012/510795 PMID:22848853

Shrestha, P., Gautam, R., Rana, R. B., & Sthapit, B. (2001). Home gardens in Nepal: Status and scope for research and development. *Home gardens and in situ conservation of plant genetic resources in farming systems*, 17–19.

Shrivastava, R. J., & Heinen, J. T. (2005). Migration and home gardens in the Brahmaputra Valley, Assam, India. *Journal of Ecological Anthropology*, *9*(1), 20–34. doi:10.5038/2162-4593.9.1.2

Shukla, A. N., Srivastava, S., & Rawat, A. K. S. (2010). An ethnobotanical study of medicinal plants of Rewa district, Madhya Pradesh. *Indian Journal of Traditional Knowledge*, *9*(1) 191–202. Singh, G. & Rawat, G. S. (2011). Ethnomedicinal survey of Kedarnath Wildlife Sanctuary in Western Himalaya, India. *Indian Journal of Fundamental and Applied Life Sciences*, *1*(1), 35–46.

Shukla, A. N., Srivastava, S., & Rawat, A. K. S. (2010). An ethnobotanical study of medicinal plants of Rewa district, Madhya Pradesh. *Indian Journal of Traditional Knowledge*, *9*(1), 191–202.

Siddaraju, M. N., & Dharmesh, S. M. (2007). Inhibition of gastric H+, K+-ATPase and Helicobacter pylori growth by phenolic antioxidants of *Curcuma amada. Journal of Agricultural and Food Chemistry*, *55*(18), 7377–7386. doi:10.1021/jf070719r PMID:17725316

Siddhiqui, M. A. A., John, A. Q., & Paul, T. M. (1995). Status of some important medicinal and aromatic plants of Kashmir Himalaya. *Advances in Plant Sciences (Muzaffarnagar)*, *8*, 134–139.

Siddique, N. A., Bari, M. A., Naderuzzaman, A. T. M., Khatun, N., Rahman, M. H., Sultana, R. S., ... Rahman, M. M. (2004). Collection of indigenous knowledge and identification of endangered medicinal plants by questionnaire survey in lin rind tract of Bangladesh. *The Journal of Biological Sciences*, *4*(1), 72–80. doi:10.3923/jbs.2004.72.80

Siddiqui, M. B., Alam, M. M., & Husain, W. (1989). Traditional treatment of skin diseases in Uttar Pradesh, India. *Economic Botany*, *43*(4), 480–486. doi:10.1007/BF02935922

Sieber, S. S., da Silva, T. C., Campos, L. Z. O., Zank, S., & Albuquerque, U. P. (2014). *Participatory methods in ethnobiological and ethnoecological research* (pp. 39–58). New York, NY: Springer New York. doi:10.1007/978-1-4614-8636-7_3

Sigler, P., Saksena, R., Deepak, D., & Khare, A. (2000). C21 steroidal glycosides from *Hemidesmus indicus. Phytochemistry*, *54*(8), 983–987. doi:10.1016/S0031-9422(99)00611-1 PMID:11014302

Silori, C. S., & Badola, R. (2000). Medicinal plant cultivation and sustainable development: A case study in the buffer zone of the Nanda Devi Biosphere Reserve, Western Himalaya, India. *Mountain Research and Development*, *20*(3), 272–279. doi:10.1659/0276-4741(2000)020[0272:MPCASD]2.0.CO;2

Sim, Y., Choi, J. G., Gu, P. S., Ryu, B., Kim, J. H., Kang, I., ... Oh, M. S. (2016). Identification of Neuroactive Constituents of the Ethyl Acetate Fraction from Cyperi Rhizoma Using Bioactivity-Guided Fractionation. *Biomolecules & Therapeutics*, *24*(4), 438–445. doi:10.4062/biomolther.2016.091 PMID:27350341

Singh, A. K., Arunachalam, A., Ngachan, S. V., Mohapatra, K. P., & Dagar, J. C. (2014). *From shifting cultivation to integrating farming: Experience of agroforestry development in the northeastern Himalayan Region* (pp. 57–86). New Delhi, India: Springer India.

Singh, A., Nautiyal, M. C., Kunwar, R. M., & Bussmann, R. W. (2017). Ethnomedicinal plants used by local inhabitants of Jakholi block, Rudraprayag District, Western Himalaya, India. *Journal of Ethnobiology and Ethnomedicine*, *13*(1), 49. doi:10.118613002-017-0178-3 PMID:28836997

Singh, B., & Bedi, Y. S. (2017). Eating from raw wild plants from Himalaya: Traditional Knowledge Documentary on *Sheena* Tribe in Kashmir. *Indian Journal of Natural Products and Resources*, *8*(3), 269–275.

Singh, G., Kapoor, I. P. S., Singh, P., de Heluani, C. S., de Lampasona, M. P., & Catalan, C. A. (2008). Chemistry, antioxidant and antimicrobial investigations on essential oil and oleoresins of *Zingiber officinale*. *Food and Chemical Toxicology*, *46*(10), 3295–3302. doi:10.1016/j.fct.2008.07.017 PMID:18706468

Singh, J. B., & Kachroo, P. (1994). Forest flora of Pir Panjal Range (northwestern Himalaya). Dehradun, India. Singh, V. (2002). Herbal remedies in traditional medicines of the local valley in Kashmir Himalayas, India, round progress in medicinal plants. *Ethnomedicine and Pharmacology*, *1*, 63–71.

Singh, K., Maikhuri, R., Rao, K., & Saxena, K. (2008). Characterizing land-use diversity in village landscapes for sustainable mountain development: A case study from Indian Himalaya. *The Environmentalist*, *28*(4), 429–445. doi:10.100710669-008-9164-6

Singh, L. R., Singh, K., & Singh, D. (2018). Wild underutilized medicinal plant; motha (*Cyperus rotundus* Linn.). *International Journal of Current Research*, *10*, 64617–64619.

Singh, P., & Attri, B. L. (2014). Survey on traditional uses of medicinal plants of Bageshwar Valley (Kumaun Himalaya) of Uttarakhand, India. *International Journal of Conservation Science*, *5*(2), 223–234.

Singh, P., Khosa, R. L., Mishra, G., & Jha, K. K. (2015). Antidiabetic activity of ethanolic extract of *Cyperus rotundus* rhizomes in streptozotocin-induced diabetic mice. *Journal of Pharmacy & Bioallied Sciences*, *7*(4), 289–292. doi:10.4103/0975-7406.168028 PMID:26681885

Singh, S. D. J., Krishna, V., Mankani, K. L., Manjunatha, B. K., Vidya, S. M., & Manohara, Y. N. (2005). Wound healing activity of the leaf extracts and deoxyelephantopin isolated from *Elephantopus scaber* Linn. *Indian Journal of Pharmacology*, *37*(4), 238–242. doi:10.4103/0253-7613.16570

Sinha, R. K. (1991). Herbal remedies prescribed by the street herbal vendors (tribal medicineman) in the treatment of some common human ecological disorders (gynaecological disorders in women). *Ancient Science of Life*, *10*(3), 207–210. PMID:22556535

Sini, S., & Malathy, N. S. (2005). Antimicrobial properties of roots of medicinal plants. *Ancient Science of Life*, *25*, 62–65. PMID:22557193

Sivaraman, D., Kumar, P. S., Muralidharan, P., Rahman, H., Nagar, I., & Thoraipakkam, C. (2012). Effect of *Hemidesmus indicus* on cerebral infract Ischemia-reperfusion injured rats by four vessel occlusion method. *Pharmacologia*, *3*(4), 91–102. doi:10.5567/pharmacologia.2012.91.102

Šivel, M., Kráčmar, S., Fišera, M., Klejdus, B., & Kubáň, V. (2014). Lutein content in marigold flower (*Tagetes erecta* L.) concentrates used for production of food supplements. *Czech Journal of Food Sciences*, *32*(6), 521–525. doi:10.17221/104/2014-CJFS

Sjølie, A. N. (1998). The epidemiology of low back pain in the rest of the world. *Spine*, *23*(8), 960–961. PMID:9580967

Skouta, R., Morán-Santibañez, K., Valenzuela, C. A., Vasquez, A. H., & Fenelon, K. (2018). Assessing the antioxidant properties of *Larrea tridentata* extract as a potential molecular therapy against oxidative stress. *Molecules (Basel, Switzerland), 23*(7), 1826. doi:10.3390/molecules23071826 PMID:30041415

Smith, G. F., & Van, W. B. (2009). The Conservation Status of Aloe in Southern Africa: An Updated Synopsis. *Bothalia, 30*(2), 206–211. doi:10.4102/abc.v30i2.560

Sobrinho, A. C. N., de Morais, S. M., de Souza, E. B., & dos Santos Fontenelle, R. O. (2017). The genus *Eupatorium* L.(Asteraceae): A review of their antimicrobial activity. *Journal of Medicinal Plants Research, 11*(3), 43–57. doi:10.5897/JMPR2016.6313

Soejarto, D. D., Gyllenhaal, C., Fong, H. H. S., Xuan, L. T., Hiep, N. T., Hung, N. V., ... Pezzuto, J. M. (2004). The UIC ICBG (University of Illinois at Chicago International Cooperative Biodiversity Group) Memorandum of Agreement: A model of benefit-sharing arrangement in natural products drug discovery and development. *Journal of Natural Products, 67*(2), 294–299. doi:10.1021/np0304363 PMID:14987071

Soemarwoto, O. (1987). Homegardens: A traditional agroforestry system with a promising future. *Agroforestry: A decade of development*, 157–170.

Sofowora, A., Ogunbodede, E., & Onayade, A. (2013). The role and place of medicinal plants in the strategies for disease prevention. *African Journal of Traditional, Complementary, and Alternative Medicines: AJTCAM/African Networks on Ethnomedicines, 10*(5), 210–229.

Solowey, E., Lichtenstein, M., Sallon, S., Paavilainen, H., Solowey, E., & Lorberboum-Galski, H. (2014). Evaluating medicinal plants for anticancer activity. *The Scientific World Journal, 2014*, 1–12. doi:10.1155/2014/721402 PMID:25478599

Someya, Y., Kobayashi, A., & Kubota, K. (2001). Isolation and identification of trans-2-and trans-3-hydroxy-1, 8-cineole glucosides from *Alpinia galanga*. *Bioscience, Biotechnology, and Biochemistry, 65*(4), 950–953. doi:10.1271/bbb.65.950 PMID:11388479

Sonwa, M. M., & König, W. A. (2001). Chemical study of the essential oil of *Cyperus rotundus*. *Phytochemistry, 58*(5), 799–810. doi:10.1016/S0031-9422(01)00301-6 PMID:11672746

Soran, H., Adam, S., & Durrington, P. N. (2017). Are recent statin recommendations to employ fixed doses and abandon targets effective for treatment of hypercholesterolaemia? Investigation based on number needed to treat. *European Journal of Preventive Cardiology, 24*(1), 76–83. doi:10.1177/2047487316667785 PMID:27609614

Soumaya, K. J., Dhekra, M., Fadwa, C., Zied, G., Ilef, L., Kamel, G., & Leila, C. G. (2013). Pharmacological, antioxidant, genotoxic studies and modulation of rat splenocyte functions by *Cyperus rotundus* extracts. *BMC Complementary and Alternative Medicine, 13*(1), 28. doi:10.1186/1472-6882-13-28 PMID:23388107

Soumaya, K. J., Zied, G., Nouha, N., Mounira, K., Kamel, G., Genviève, F. D., & Leila, G. C. (2014). Evaluation of *in vitro* antioxidant and apoptotic activities of *Cyperus rotundus*. *Asian Pacific Journal of Tropical Medicine, 7*(2), 105–112. doi:10.1016/S1995-7645(14)60004-3 PMID:24461522

Sowmia, C., & Kokilavani, R. (2007). Antidiabetic and antihypercholesterolemic effect of *Hemidesmus indicus* Linn. R. root in Alloxan induced diabetic rats. *Ancient Science of Life, 26*(4), 4. PMID:22557243

Sreevani, K. H., Puranaik, G. S. J., & Naidu, M. M. (2013). Studies on antioxidant activity of *Zingiber zerumbet* spent and its constituents through *in vitro* models. *Wudpecker Journal of Food Technology, 1*(13), 48–55.

Srinivasan, K., Natarajan, D., Mohanasundari, C., Venkatakrishnan, C., & Nagamurugan, N. (2007). Antibacterial, preliminary phytochemical and pharmacognostical screening on the leaves of *Vicoa indica* (L.) DC. *Iranian Journal of Pharmacology and Therapeutics*, *6*(1), 109–110.

Srivastava, A. K., Srivastava, S. K., & Shah, N. C. (2001). Constituents of the rhizome essential oil of *Curcuma amada* Roxb. from India. *The Journal of Essential Oil Research*, *13*(1), 63–64. doi:10.1080/10412905.2001.9699608

Srivastava, T. N., Badola, D. P., Shah, D. C., & Gupta, O. P. (1984). Ethno-medico-botanical exploration of Figurez Valley Kashmir. *Bulletin of Medical Ethnobotany and Research*, *5*, 15–54.

Srivastava, T. N., Pathak, N. N., & Hussain, S. M. S. (1980). Medicinal plants of Kailash (J & K). *Bulletin of Ethno Medicine and Botanical Research*, *2*(2), 145–166.

Stanly, C., Bhatt, A., Ali, H. M. D., Keng, C. L., & Lim, B. P. (2011). Evaluation of free radical scavenging activity and total phenolic content in the petiole-derived callus cultures of *Zingiber zerumbet* Smith. *Journal of Medicinal Plants Research*, *5*(11), 2210–2217.

Stewart, B. W., & Wild, C. P. (2014). World cancer report 2014. *World Health Organization.*

Sthapit, B., Gautam, R., & Eyzaguirre, P. (2006). The value of home gardens to small farmers. In *Home gardens in Nepal: Proceedings of the workshop on enhancing the contribution of home garden to on-farm management of plant genetic resources and to improve the livelihoods of Nepalese farmers: Lessons learned and policy implications,* August 6–7, 2004, Pokhara, Nepal. LI-BIRD, Bioversity International and SDC. Local Initiatives for Biodiversity, 324, 8.

Stoilova, I., Krastanov, A., Stoyanova, A., Denev, P., & Gargova, S. (2007). Antioxidant activity of a ginger extract (*Zingiber officinale*). *Food Chemistry*, *102*(3), 764–770. doi:10.1016/j.foodchem.2006.06.023

Subedi, A., Suwal, R., Gautam, R., Sunwar, S., & Shrestha, P. (2006). Status and composition of plant genetic diversity in Nepalese home gardens. In *Home gardens in Nepal: Proceedings of the workshop on enhancing the contribution of home garden to on-farm management of plant genetic resources and to improve the livelihoods of Nepalese farmers: Lessons learned and policy implications, August 6–7, 2004, Pokhara, Nepal. LI-BIRD, Bioversity International and SDC. Local Initiatives for Biodiversity, 324*, 7.

Subramanian, S. S., & Nair, A. G. R. (1968). Flavonoids of some asclepiadaceous plants. *Phytochemistry*, *7*(9), 1703–1704. doi:10.1016/S0031-9422(00)88630-6

Sujatha, S., & Anusha, J. R. (2010). In vitro antibacterial activity on human pathogenic bacteria and larvicidal effect of root from *Hemidesmus indicus* (Linn.) on *Culex qinquifasciatus*. *International Journal of Phytomedicine*, *2*(4), 418–424.

Suksamerkun, W., Thongsomchitt, S., Wongkrajang, Y., Temsiririrkkul, R., Kitphati, W., & Thongpraditchote, S. (2013). Screening of antioxidant activity of vegetables in Thailand. *Journal of Asian Association of Schools of Pharmacy*, *2*, 254–261.

Sultana, S., Alam, A., Khan, N., & Sharma, S. (2003a). Inhibition of cutaneous oxidative stress and two-stage skin carcinogenesis by *Hemidesmus indicus* (L.) in Swiss albino mice. *Indian Journal of Experimental Biology*, *41*(12), 1416–1423. PMID:15320495

Sultana, S., Khan, M. A., Ahmad, M., & Zafar, M. (2006). Indigenous Knowledge of Folk Herbal Medicines by the Women of District Chakwal, Pakistan. *Ethnobotanical Leaflets*, *10*(1), 243–253.

Sultana, S., Khan, N., Sharma, S., & Alam, A. (2003b). Modulation of biochemical parameters by *Hemidesmus indicus* in cumene hydroperoxide-induced murine skin: Possible role in protection against free radicals-induced cutaneous oxidative stress and tumor promotion. *Journal of Ethnopharmacology*, *85*(1), 33–41. doi:10.1016/S0378-8741(02)00360-4 PMID:12576200

Sundriyal, M., & Sharma, B. (2016). Status of biodiversity in Central Himalaya. *Applied Ecology and Environmental Sciences*, *4*(2), 37–43.

Sunil, A. G., Kesavanarayanan, K. S., Kalaivani, P., Sathiya, S., Ranju, V., Priya, R. J., ... Babu, C. S. (2011). Total oligomeric flavonoids of *Cyperus rotundus* ameliorates neurological deficits, excitotoxicity and behavioral alterations induced by cerebral ischemic-reperfusion injury in rats. *Brain Research Bulletin*, *84*(6), 394–405. doi:10.1016/j.brainresbull.2011.01.008 PMID:21272618

Suntar, I. (2014). The medicinal value of Asteraceae family plants in terms of wound healing activity. FABAD. *Journal of Pharmaceutical Sciences*, *39*(1), 21–31. PMID:24450391

Suo, M., & Yang, J. (2014). Ceramides isolated from *Helianthus annuus* L. *Helvetica Chimica Acta*, *97*(3), 355–360. doi:10.1002/hlca.201300194

Suresh, J., Singh, A., Vasavi, A., Ihsanullah, M., & Mary, S. (2011). Phytochemical and pharmacological properties of *Artemisia pallens*. *International Journal of Pharmaceutical Sciences and Research*, *2*(12), 3081.

Sutalangka, C., & Wattanathorn, J. (2017). Neuroprotective and cognitive-enhancing effects of the combined extract of *Cyperus rotundus* and *Zingiber officinale*. *BMC Complementary and Alternative Medicine*, *17*(1), 135. doi:10.118612906-017-1632-4 PMID:28253877

Swathi, S., Amareshwari, P., & Venkatesh, K., R., R., A. (2019). Phytochemical and pharmacological benefits of *Hemidesmus indicus*: An updated review. *Journal of Pharmacognosy and Phytochemistry*, *8*(1), 256–262.

Szakiel, A., Paczkowski, C., & Henry, M. (2011). Influence of environmental abiotic factors on the content of saponins in plants. *Phytochemistry Reviews*, *10*(4), 471–491. doi:10.100711101-010-9177-x

Tabassum, Das. S., & Mazumder, A. (2015). Ethnobotanical review of *Hemidesmus indicus* R.Br. *International Journal of Current Research*, *7*(12), 24251-24256.

Tachakittirungrod, S., & Chowwanapoonpohn, S. (2007). Comparison of antioxidant and antimicrobial activities of essential oils from *Hyptis suaveolens* and *Alpinia galanga* growing in Northern Thailand. *CMU Journal of Natural Science*, *6*(1), 31–41.

Tadesse, B., Mulugeta, G., Fikadu, G., Sultan, A., & Nekemte, E. (2014). Survey on ethno-veterinary medicinal plants in selected woredas of east Wollega zone, western Ethiopia. *Journal of Biology. Agriculture and Healthcare*, *4*(17), 97–105.

Tan, P. V., Dimo, T., & Dongo, E. (2000). Effects of methanol, cyclohexane and methylene chloride extracts of *Bidens pilosa* on various gastric ulcer models in rats. *Journal of Ethnopharmacology*, *73*(3), 415–421. doi:10.1016/S0378-8741(00)00290-7 PMID:11090994

Tariq, A., Mussarat, S., & Adnan, M., Abd_Allah, E. F., Hashem, A., Alqarawi, A. A., & Ullah, R. (2015). Ethnomedicinal evaluation of medicinal plants used against gastrointestinal complaints. *BioMed Research International*, 892947. PMID:26114117

Tayade, S. K., & Patil, D. A. (2005). Ethnomedicinal traditions of tribals of Nandurbar District (Maharashtra). *Journal of Phytological Research*, *18*(2), 251–254.

Teklehaymanot, T. (2017). An ethnobotanical survey of medicinal and edible plants of Yalo Woreda in Afar regional state, Ethiopia. *Journal of Ethnobiology and Ethnomedicine*, *13*(1), 40–65. doi:10.118613002-017-0166-7 PMID:28679438

Tekle, Y. (2014). An ethno-veterinary botanical survey of medicinal plants in Kochore district of Gedeo zone, Southern Nations Nationalities and Peoples Regional State (SNNPRS). *Journal of Scientific and Innovative Research*, *3*(4), 433–445.

Thabrew, M. I., Mitry, R. R., Morsy, M. A., & Hughes, R. D. (2005). Cytotoxic effects of a decoction of *Nigella sativa, Hemidesmus indicus* and *Smilax glabra* on human hepatoma HepG2 cells. *Life Sciences*, *77*(12), 1319–1330. doi:10.1016/j.lfs.2005.01.022 PMID:15916774

Thatoi, H., & Patra, J. K. (2011). Biotechnology and pharmacological evaluation of medicinal plants: An overview. *Journal of Herbs, Spices, & Medicinal Plants*, *17*(3), 214–248. doi:10.1080/10496475.2011.602471

The Plant List. A working list of all plant species. Retrieved from www.theplantlist.org

Thomas, D., Govindhan, S., Baiju, E. C., Padmavathi, G., Kunnumakkara, A. B., & Padikkala, J. (2015). *Cyperus rotundus* L. prevents non-steroidal anti-inflammatory drug-induced gastric mucosal damage by inhibiting oxidative stress. *Journal of Basic and Clinical Physiology and Pharmacology*, *26*(5), 485–490. doi:10.1515/jbcpp-2014-0093 PMID:25915081

Thomas, E., Vandebroek, I., & Van Damme, P. (2007). What works in the field? A comparison of different interviewing methods in ethnobotany with special reference to the use of photographs. *Economic Botany*, *61*(4), 376–384. doi:10.1663/0013-0001(2007)61[376:WWITFA]2.0.CO;2

Thorne, R. F., & Reveal, J. L. (2007). An updated classification of the class Magnoliopsida ("Angiospermae"). *Botanical Review*, *73*(2), 67–181. doi:10.1663/0006-8101(2007)73[67:AUCOTC]2.0.CO;2

Tirumani, P., Venu, S., Sridhar, G., Praveen Kumar, M., Rajashekhar, A. V., & Naga Raju, T. (2018). Delaying of cataract through intervention of *Hemidesmus indicus* in STZ induced diabetic rats. *Natural Product Research*, *32*(11), 1295–1298. doi:10.1080/14786419.2017.1333991 PMID:28580798

Tiwari, J. K., Ballabha, R., & Tiwari, P. (2010). Diversity and present status of medicinal plants in and around Srinagar hydroelectric power project in Garhwal Himalaya, India: Needs for conservation. *Researcher*, *2*(2), 50–60.

Tiwari, J. K., Dangwal, L. R., Rana, C. S., Tiwari, P., & Ballabha, R. (2010). Indigenous uses of plant species in Nanda Devi Biosphere Reserve, Uttarakhand, India. *Rep. Opin*, *2*, 67–70.

Tiwari, V. J. (2018). Validity of Ethnopharmacological uses of *Vicoa indica* (L.) DC., Family–Asteraceae. *Research Journal of Pharmacognosy and Phytochemistry*, *10*(3), 207–210. doi:10.5958/0975-4385.2018.00033.X

Tobgay, S. (2006). Agriculture diversification in Bhutan.

Tongco, M. D. C. (2007). Purposive sampling as a tool for informant selection. *Ethnobotany Research and Applications*, *5*, 147–158. doi:10.17348/era.5.0.147-158

Torlak, H., Vural, M., & Aytaç, Z. (2010). *Türkiye'nin endemik bitkileri*. Ankara, Turkey: Kültür ve Turizm Bakanlığı.

Tran, H. H., Nguyen, M. C., Le, H. T., Nguyen, T. L., Pham, T. B., Chau, V. M., ... Nguyen, T. D. (2014). Inhibitors of α-glucosidase and α-amylase from *Cyperus rotundus*. *Pharmaceutical Biology*, *52*(1), 74–77. doi:10.3109/13880209.2013.814692 PMID:24044731

Travis, D. (2016). *Desk Research: The What, Why, and How.* Available from https//:www.userfocus.co.uk

Tremmel, M., Gerdtham, U. G., Nilsson, P. M., & Saha, S. (2017). Economic burden of obesity: A systematic literature review. *International Journal of Environmental Research and Public Health*, *14*(4), 1–18. doi:10.3390/ijerph14040435 PMID:28422077

Trinh, L. N., Watson, J. W., Hue, N. N., De, N. N., Minh, N., Chu, P., ... Eyzaguirre, P. B. (2003). Agrobiodiversity conservation and development in Vietnamese home gardens. *Agriculture, Ecosystems, & Environment*, *97*(1–3), 317–344. doi:10.1016/S0167-8809(02)00228-1

Tripathi, I. (2003). *Raj Nighantu*. Varanasi, India: Bharati Acadamy.

Trollope-Kumar, K. (2001). Cultural and biomedical meanings of the complaint of leukorrhea in South Asian women. *Tropical Medicine & International Health*, *6*(4), 260–266. doi:10.1046/j.1365-3156.2001.00699.x PMID:11348516

Tsoyi, K., Jang, H. J., Lee, Y. S., Kim, Y. M., Kim, H. J., Seo, H. G., ... Chang, K. C. (2011). (+)-Nootkatone and (+)-valencene from rhizomes of *Cyperus rotundus* increase survival rates in septic mice due to heme oxygenase-1 induction. *Journal of Ethnopharmacology*, *137*(3), 1311–1317. doi:10.1016/j.jep.2011.07.062 PMID:21843620

Tsuji, H., Hirano, N., Ohshima, H., Ishihara, H., Terahata, N., & Motoe, T. (1993). Structural variation of the anterior and posterior anulus fibrosus in the development of human lumbar intervertebral disc. A risk factor for intervertebral disc rupture. *Spine*, *18*(2), 204–210. doi:10.1097/00007632-199302000-00006 PMID:8441935

Tulukçu, E., & Sağdıç, O. (2011). Konya'da aktarlarda satılan tıbbi bitkiler ve kullanılan kısımları. *Erciyes Üniversitesi Fen Bilimleri Enstitüsü Dergisi*, *27*(4), 304–308.

Turner, N. J. (1995). Ethnobotany today in Northwestern North America Portland. In R. E. Schultes, & R. S. Von (Eds.), Ethnobotany: Evolution of a discipline (pp. 264–283). Portland, OR: Dioscorides Press.

Turner, N. J., Marshall, A., & Thompson, J. C. (2008). Ebb and flow- transmitting environmental knowledge in a contemporary aboriginal community. In J. S. Lutz, & B. Neis (Eds.), *Making and moving knowledge. Interdisciplinary and community-based research in a world on the edge* (pp. 45–63). Montreal, Canada: McGill-Queen's University Press.

Turrini, E., Calcabrini, C., Tacchini, M., Efferth, T., Sacchetti, G., Guerrini, A., ... Fimognari, C. (2018). In vitro study of the cytotoxic, cytostatic, and antigenotoxic profile of *Hemidesmus indicus* (L.) R. Br. (Apocynaceae) crude drug extract on t lymphoblastic cells. *Toxins*, *10*(2), 70. doi:10.3390/toxins10020070 PMID:29415441

Turrini, E., Ferruzzi, L., Guerrini, A., Gotti, R., Tacchini, M., Teti, G., ... Fimognari, C. (2015). In vitro anti-angiogenic effects of *Hemidesmus indicus* in hypoxic and normoxic conditions. *Journal of Ethnopharmacology*, *162*, 261–269. doi:10.1016/j.jep.2014.12.010 PMID:25560668

Tuzlacı, E. (2005). *Bodrum'da bitkiler ve yaşam*. İstanbul, Turkey: Güzel Sanatlar Matbaası.

Tuzlacı, E. (2006). *Şifa niyetine, türkiye'nin bitkisel halk ilaçları*. İstanbul, Turkey: Alfa Yayıncılık.

Tuzlacı, E. (2016). *Türkiye bitkileri geleneksel ilaç rehberi*. İstanbul, Turkey: İstanbul Tıp Kitabevleri.

Tuzlacı, E., & Erol, M. K. (1999). Turkish folk medicinal plants. part II: Eğirdir (ısparta). *Fitoterapia*, *70*(6), 593–610. doi:10.1016/S0367-326X(99)00074-X

Uddin, M. N., Rahman, M. A., Ahmed, N. U., Rana, M. S., Akter, R., & Chowdhury, A. M. A. (2010). Antioxidant, cytotoxic, and antimicrobial properties of *Eclipta alba* ethanol extract. *International Journal of Biological and Medical Research*, *1*(4), 341–346.

Uddin, S. J., Mondal, K., Shilpi, J. A., & Rahman, M. T. (2006). Antidiarrhoeal activity of *Cyperus rotundus*. *Fitoterapia*, *77*(2), 134–136. doi:10.1016/j.fitote.2004.11.011 PMID:16376024

Ugulu, I. (2011). Traditional ethnobotanical knowledge about medicinal plants used for external therapies in Alasehir, Turkey. *International Journal of Medicine and Aromatic Plants.*, *2*(1), 101–106.

Uğurlu, E., & Seçmen, Ö. (2008). Medicinal plants popularly used in the villages of Yunt Mountain (Manisa-Turkey). *Fitoterapia*, *79*(2), 126–131. doi:10.1016/j.fitote.2007.07.016 PMID:17878061

Ukiya, M., Akihisa, T., Yasukawa, K., Koike, K., Takahashi, A., Suzuki, T., & Kimura, Y. (2007). Triterpene glycosides from the flower petals of sunflower (*Helianthus annuus*) and their anti-inflammatory activity. *Journal of Natural Products*, *70*(5), 813–816. doi:10.1021/np0780021 PMID:17480100

Ukiya, M., Akihisa, T., Yasukawa, K., Tokuda, H., Suzuki, T., & Kimura, Y. (2006). Anti-inflammatory, anti-tumor-promoting, and cytotoxic activities of constituents of marigold (*Calendula officinalis*) flowers. *Journal of Natural Products*, *69*(12), 1692–1696. doi:10.1021/np068016b PMID:17190444

Ulbricht, C., Armstrong, J., Basch, E., Basch, S., Bent, S., Dacey, C., ... Weissner, W. (2007). An evidence-based systematic review of Aloe vera by the natural standard research collaboration. *Journal of Herbal Pharmacotherapy*, *7*(3-4), 279–323. doi:10.1080/15228940802153339 PMID:18928148

Umadevi, S., Mohanta, G. P., Kalaichelvan, V. K., & Manavalan, R. (2006). Studies on wound healing effect of *Flaveria trinervia* leaf in mice. *Indian Journal of Pharmaceutical Sciences*, *68*(1), 106–108. doi:10.4103/0250-474X.22979

Umair, M., Altaf, M., & Abbasi, A. M. (2017). An ethnobotanical survey of indigenous medicinal plants in Hafizabad district, Punjab-Pakistan. *PLoS One*, *12*(6). doi:10.1371/journal.pone.0177912 PMID:28574986

Umapriya, T., Rajendran, A., Aravindhan, V., Thomas, B., & Maharajan, M. (2011). Ethnobotany of Irular tribe in Palamalai Hills, Coimbatore, Tamil Nadu. *Natural Product Research*, *2*(2), 250–255.

UNESCO. (2002). Analytical report to governments and international partners on the follow-up to the World Conference on Science; June 26–July 1, 1999; Budapest, Hungary. Paris, France: UNESCO.

Uniyal, B., & Shiva, V. (2005). Traditional knowledge on medicinal plants among rural women of the Garhwal Himalaya, Uttaranchal. *Indian Journal of Traditional Knowledge*, *4*(3), 259–266.

Uniyal, S. K., Singh, K. N., Jamwal, P., & Lal, B. (2006). Traditional use of medicinal plants among the tribal communities of ChhotaBhangal, Western Himalaya. *Journal of Ethnobiology and Ethnomedicine*, *2*(1), 14. doi:10.1186/1746-4269-2-14 PMID:16545146

Upadhyay, B., Singh, K. P., & Kumar, A. (2011). Ethno-veterinary uses and informants consensus factor of medicinal plants of Sariska region, Rajasthan, India. *Journal of Ethnopharmacology*, *133*(1), 14–25. doi:10.1016/j.jep.2010.08.054 PMID:20817085

Upreti, K., Jalal, J. S., Tewari, L. M., Joshi, G. C., Pangtey, Y. P. S., & Tewari, G. (2009). Ethnomedicinal uses of pteridophytes of Kumaun Himalaya, Uttarakhand, India. *The Journal of American Science*, *5*, 167–170.

Uprety, Y., Asselin, H., Boon, E. K., Yadav, S., & Shrestha, K. K. (2010). Indigenous use and bio-efficacy of medicinal plants in the Rasuwa district, central Nepal. *Journal of Ethnobiology and Ethnomedicine*, *6*(1), 3. doi:10.1186/1746-4269-6-3 PMID:20102631

Urban, J. P. G., & McMullin, J. F. (1985). Swelling pressure of the inervertebral disc: Influence of proteoglycan and collagen contents. *Biorheology*, *22*(2), 145–157. doi:10.3233/BIR-1985-22205 PMID:3986322

USDA. (2012). *Aloe barbadensis Miller. Beltville*. National Germplasm Resources Laboratory.

Uysal, G. (2008). *Köyceğiz (muğla) ilçesinin etnobotaniği* (Unpublished master's thesis). Muğla Üniversitesi, Turkey.

Valsan, A., & Regi Raphael, K. (2016). Pharmacognostic Profile of *Averrhoa bilimbi* Linn. leaves. *South Indian Journal of Biological Sciences*, *2*(1), 75–80. doi:10.22205ijbs/2016/v2/i1/100347

Van Agtmael, M. A., Eggelte, T. A., & van Boxtel, C. J. (1999). Artemisinin drugs in the treatment of malaria from medicinal herb to registered medication. *Trends in Pharmacological Sciences*, *20*(5), 199–205. doi:10.1016/S0165-6147(99)01302-4 PMID:10354615

Van Tulder, M. W., Koes, B. W., & Bouter, L. M. (1995). A cost-of-illness study of back pain in The Netherlands. *Pain*, *62*(2), 233–240. doi:10.1016/0304-3959(94)00272-G PMID:8545149

Varenna, M., Manara, M., Galli, L., Binelli, L., Zucchi, F., & Sinigaglia, L. (2013). The association between osteoporosis and hypertension: The role of a low dairy intake. *Calcified Tissue International*, *93*(1), 86–92. doi:10.100700223-013-9731-9 PMID:23652773

Varzakas, T., Zakynthinos, G., & Verpoort, F. (2016). Plant food residues as a source of nutraceuticals and functional foods. *Foods*, *5*(4), 88. doi:10.3390/foods5040088 PMID:28231183

Vendruscolo, G. S., & Mentz, A. Z. (2006). Ethnobotanical survey of the medicinal plants used by the community of Ponta Grossa neighborhood, Porto Alegre, Rio Grande do Sul, Brazil. *Iheringia Botânica. Series*, *61*, 83–103.

Verma, P. R., Joharapurkar, A. A., Chatpalliwar, V. A., & Asnani, A. J. (2005). Antinociceptive activity of alcoholic extract of *Hemidesmus indicus* R. Br. in mice. *Journal of Ethnopharmacology*, *102*(2), 298–301. doi:10.1016/j.jep.2005.05.039 PMID:16081231

Veteto, J. R., & Skarbø, K. (2009). Sowing the seeds: Anthropological contributions to agrobiodiversity studies. *Culture & Agriculture*, *31*(2), 73–87. doi:10.1111/j.1556-486X.2009.01022.x

Vianna-da-Silva., N. M. (2003). Moraes ram, Da silva, A. J. M., Costa, P. R. R., Melo, P. A. Antivenom effect of a new synthetic coumestan analog of wedelolactone. *The Journal of Venomous Animals and Toxins*, *9*(2), 381.

Victório, C. P. (2011). Therapeutic value of the genus *Alpinia*, Zingiberaceae. *Revista Brasileira de Farmacognosia*, *21*(1), 194–201. doi:10.1590/S0102-695X2011005000025

Victório, C. P., Alviano, D. S., Alviano, C. S., & Lage, C. L. (2009). Chemical composition of the fractions of leaf oil of *Alpinia zerumbet* (Pers.) BL Burtt & RM Sm. and antimicrobial activity. *Revista Brasileira de Farmacognosia*, *19*(3), 697–701. doi:10.1590/S0102-695X2009000500008

Victório, C. P., Leitão, S. G., & Lage, C. L. (2010). Chemical composition of the leaf oils of *Alpinia zerumbet* (Pers.) Burtt et Smith and *A. purpurata* (Vieill) K. Schum. from Rio de Janeiro, Brazil. *The Journal of Essential Oil Research*, *22*(1), 52–54. doi:10.1080/10412905.2010.9700264

Vienna, C., Graz, R. B., Hohenhem, R. C., Milano, D. T., Trieste, A. T., & Wilen, T. (2007). Study of the assessment of plants/herbs, plant/herb extracts and their naturally or synthetically produced components as "additives" for use in animal production. *EFSA Supporting Publications*, *4*(1), 140–154.

Viladrich, A. (2007). Between bellyaches and lucky charms: Revealing Latinos' plant-healing knowledge and practices in New York City. In A. Pieroni, & I. Vandebroek (Eds.), Travelling cultures and plants: The ethnobiology and ethnopharmacy of migrations, (64–85). New York: Berghahn Books.

Vilar, D. D. A., Vilar, M. S. D. A., Moura, T. F. A. D. L. E., Raffin, F. N., De Oliveira, M. R., Franco, C. F. D. O., ... Barbosa-Filho, J. M. (2014). Traditional uses, chemical constituents, and biological activities of *Bixa orellana* L.: A review. *The Scientific World Journal*, 2014. PMID:25050404

Villarruel-López, A., López-de la Mora, D. A., Vázquez-Paulino, O. D., Puebla-Mora, A. G., Torres-Vitela, M. R., Guerrero-Quiroz, L. A., & Nuño, K. (2018). Effect of Moringa oleifera consumption on diabetic rats. *BMC Complementary and Alternative Medicine*, *18*(1), 127. doi:10.118612906-018-2180-2 PMID:29636032

Vishali, K., Kavitha, K. N. V., Rajesh, V., & Perumal, P. (2011). Anti-ulcer activity of *Hemidesmus indicus* root extract on Indomethacin induced gastric ulcer in albino wistar rats. *Journal of Pharmacy Research, 4*(2), 391–392.

Vishnupriya, M., Nishaa, S., Sasikumar, J. M., & Gopalakrishnan, V. K. (2012). Antioxidant Activity And Hydroxyl Radical Induced DNA Damage Protection Effect of Aqueous Extract of *Curcuma amada* Roxb. *Research Journal of Pharmaceutical, Biological and Chemical Sciences, 3*(4), 89.

Vishnuvardhan, Z., Jyothirmayi, G. L. V., & Jyothi, D. (2018). Medicinal plants of tribal traditional system from Guntur District, Andhra Pradesh, India. *International Journal of Life Sciences (Kathmandu), 6*(1), 194–204.

Vlkova, M., Polesny, Z., Verner, V., Banout, J., Dvorak, M., Havlik, J., ... Krausova, J. (2011). Ethnobotanical knowledge and agrobiodiversity in subsistence farming: Case study of home gardens in Phong My commune, central Vietnam. *Genetic Resources and Crop Evolution, 58*(5), 629–644. doi:10.100710722-010-9603-3

Vlok, G. J., & Hendrix, M. R. G. (1991). The lumbar disc: Evaluating the causes of pain. *Orthopedics, 14*(4), 419–425. PMID:1828114

Vogl, C. R., & Vogl-Lukasser, B. (2003). Tradition, dynamics and sustainability of plant species composition and management in homegardens on organic and non-organic smallscale farms in alpine eastern Tyrol, Austria. *Biological Agriculture and Horticulture, 21*(4), 349–366. doi:10.1080/01448765.2003.9755278

Vogl, C. R., Vogl-Lukasser, B., & Puri, R. K. (2004). Tools and methods for data collection in ethnobotanical studies of homegardens. *Field Methods, 16*(3), 285–306. doi:10.1177/1525822X04266844

Vogl-Lukasser, B., & Vogl, C. R. (2016). No Title. *Ethnobotany Research and Applications.*

Volinn, E. (1997). The epidemiology of low back pain in the rest of the world: A review of surveys in low-and middle-income countries. *Spine, 22*(15), 1747–1754. doi:10.1097/00007632-199708010-00013 PMID:9259786

Waddell, G. (1998). *The back pain revolution.* London, UK: Churchill Livingstone.

Wagay, N. A. (2014). Medicinal flora and ethno-botanical knowledge of Baramulla Tehsil in Jammu and Kashmir, India. *International Journal of Advanced Biotechnology and Research, 5*(3), 539–546.

Wagenitz, G. (1976). Systematics and phylogeny of the Compositae (Asteraceae*). Plant Systematics and Evolution, 125*(1), 29–46. doi:10.1007/BF00986129

Walsh, K., Cruddas, M., & Coggon, D. (1992). Low back pain in eight areas of Britain. *Journal of Epidemiology and Community Health, 46*(3), 227–230. doi:10.1136/jech.46.3.227 PMID:1386619

Wang, B. B., Niu, F. G., & Duan, Y. F. (2011). Comparison of Antioxidant Activity of Total Flavonoids from *Alpinia officinarum* Hance and *Alpinia galangal* Willd [J]. *Food Science, 7*, 028.

Wang, C. J., Wang, J. M., Lin, W. L., Chu, C. Y., Chou, F. P., & Tseng, T. H. (2000). Protective effect of *Hibiscus anthocyanins* against tert-butyl hydroperoxide-induced hepatic toxicity in rats. *Food and Chemical Toxicology, 38*(5), 411–416. doi:10.1016/S0278-6915(00)00011-9 PMID:10762726

Wani, Z. A., & Kumar, N., & Akash. (2015). Addition to the ethnomedicinal plants used by the local people of Gulmarg and its allied areas, Kashmir, Jammu and Kashmir, India. *International Journal of Current Research in Bioscience in Plant Biology, 2*(10), 18–22.

Wani, Z. A., Kumar, N., & Akash. (2016). Ethnobotanical study of some threatened plants in district Baramulla, Kashmir, Jammu and Kashmir, India. *International Journal of Current Research in Bioscience in Plant Biology, 3*(2), 58–64. doi:10.20546/ijcrbp.2016.302.007

Warren, D. M. (1992). *Indigenous knowledge, biodiversity conservation and development.* Keynote address at the International Conference on Conservation of Biodiversity in Africa: Local Initiatives and Institutional Roles, August 30-September 3, 1992, Nairobi, Kenya.

Warrier, P. K., Nambiar, V. P. K., & Ganapathy, P. M. (2000). Some important medicinal plants of the western ghats, India: a profile. New Delhi, India: Medicinal and Aromatic Plants Program in Asia (MAPPA), IDRC, SARO.

Weenen, H., Nkunya, M. H., Bray, D. H., Mwasumbi, L. B., Kinabo, L. S., & Kilimali, V. A. (1990). Antimalarial activity of Tanzanian medicinal plants. *Planta Medica, 56*(4), 368–370. doi:10.1055-2006-960984 PMID:2236289

Weinstein, S. M., Herring, S. A., & Cole, A. J. (1998). Rehabilitation of the patient with spinal pain. In Rehabilitation medicine principles and practice. 3rd ed. Philadelphia, PA: Lippincot Williams & Wilkins, 1423-1453.

Weissner, W. (2014). Anantamul (*Hemidesmus indicus*) A Review of Biomedical Studies and U.S. Products. *Ayurveda Journal of Health, 12*(4), 40–52.

Wiart, C. (2000). *Medicinal Plants of Southeast Asia.* Kuala Lumpur: Pelanduk Publisher.

Williams, N. M., & Baines, G. (Eds.). (1993). *Traditional ecological knowledge: wisdom for sustainable development. Centre for Resource and Environmental Studies.* Canberra, Australia: Australian National University.

Wilson, E. O. (1992). *The diversity of life.* Cambridge, MA: Harvard University Press.

Wong, K. C., & Wong, S. N. (1995). Volatile constituents of *Averrhoa bilimbi* L. fruit. *The Journal of Essential Oil Research, 7*(6), 691–693. doi:10.1080/10412905.1995.9700533

Wong, L. F., Lim, Y. Y., & Omar, M. (2009). Antioxidant and antimicrobial activities of some *Alpina* species. *Journal of Food Biochemistry, 33*(6), 835–851. doi:10.1111/j.1745-4514.2009.00258.x

World Health Organisation (WHO). (2000). *General guidelines for methodologies on research and evaluation of traditional medicine.* Geneva, Switzerland: World Health Organisation.

World Health Organization. (1978). *Alma-Ata 1978: Primary health care, health for all series (No. 1)* (p. 47). Geneva, Switzerland: WHO.

World Health Organization. (2013). *WHO Traditional Medicine Strategy 2014-2023. World Health Organization (WHO).*

World Resources Institute. (1992). *The World Conservation Union, and the United Nations Environment Programme, Global Biodiversity Strategy: Policy-makers' Guide.* Baltimore, MD: WRI Publications.

Wu, S., Zhu, W., Thompson, P., & Hannun, Y. A. (2018). Evaluating intrinsic and non-intrinsic cancer risk factors. *Nature Communications.* doi:10.103841467-018-05467-z PMID:30154431

Xavier, T. F., Kannan, M., Lija, L., Auxillia, A., Rose, A. K. F., & Kumar, S. S. (2014). Ethnobotanical study of Kani tribes in Thoduhills of Kerala, South India. *Journal of Ethnopharmacology, 152*(1), 78–90. doi:10.1016/j.jep.2013.12.016 PMID:24393788

Xiao-Lu, Z. H. U., Ming-Hua, Y. A. N. G., Jian-Guang, L. U. O., Huang, X. F., & Ling-Yi, K. O. N. G. (2009). A new phenylpropanoid from *Alpinia galanga. Chinese Journal of Natural Medicines, 7*(1), 19–20.

Xie, J.-T., Chang, W.-T., Wang, C.-Z., Mehendale, S. R., Li, J., Ambihaipahar, R., ... Yuan, C.-S. (2006). Curry Leaf (*Murraya koenigii* Spreng.) reduces blood cholesterol and glucose levels in ob/ob mice. *The American Journal of Chinese Medicine, 34*(02), 279–284. doi:10.1142/S0192415X06003825 PMID:16552838

Xu, H. B., Geng, C. A., Zhang, X. M., Ma, Y. B., Huang, X. Y., & Chen, J. J. (2016). Chemical structure of cyperotundic acid from rhizomes of *Cyperus rotundus. Zhongguo Zhongyao Zazhi*, *41*, 1066–1069. PMID:28875671

Xu, H. B., Ma, Y. B., Huang, X. Y., Geng, C. A., Wang, H., Zhao, Y., ... Chen, J. J. (2015). Bioactivity-guided isolation of anti-hepatitis B virus active sesquiterpenoids from the traditional Chinese medicine: Rhizomes of *Cyperus rotundus. Journal of Ethnopharmacology*, *171*, 131–140. doi:10.1016/j.jep.2015.05.040 PMID:26051832

Xu, H. X., Dong, H., & Sim, K. Y. (1996). Labdane diterpenes from *Alpinia zerumbet. Phytochemistry*, *42*(1), 149–151. doi:10.1016/S0031-9422(97)00900-X

Xu, H. X., Hui, D., & Sim, K. Y. (1995). The isolation of a new labdane diterpene from the seeds of *Alpinia zerumbet. Natural Product Letters*, *7*(1), 29–34. doi:10.1080/10575639508043183

Xu, Y., Zhang, H. W., Wan, X. C., & Zou, Z. M. (2009). Complete assignments of (1)H and (13)C NMR data for two new sesquiterpenes from *Cyperus rotundus* L. *Magnetic Resonance in Chemistry*, *47*(6), 527–531. doi:10.1002/mrc.2416 PMID:19288546

Xu, Y., Zhang, H. W., Yu, C. Y., Lu, Y., Chang, Y., & Zou, Z. M. (2008). Norcyperone, a novel skeleton norsesquiterpene from *Cyperus rotundus* L. *Molecules (Basel, Switzerland)*, *13*(10), 2474–2481. doi:10.3390/molecules13102474 PMID:18923338

Xu, Z., McClure, S. T., & Appel, L. J. (2018). Dietary cholesterol intake and sources among U.S. adults: Results from national health and nutrition examination surveys (NHANES), 2001–2014. *Nutrients*, *10*(6), 771. doi:10.3390/nu10060771 PMID:29903993

Yang, Y. P. (2012). Anti-inflammatory activity and oral cancer cells cytotoxicity of extracts from *Alpinia galangal* and *Zingiber zerumbet*. (Master's Thesis).

Yang, J.-L., & Shi, Y.-P. (2012). Structurally diverse terpenoids from the rhizomes of *Cyperus rotundus* L. *Planta Medica*, *78*(01), 59–64. doi:10.1055-0031-1280216 PMID:21928169

Yang, L., Ahmed, S., Stepp, J. R., Mi, K., Zhao, Y., Ma, J., ... Xue, D. (2014). Comparative homegarden medical ethnobotany of Naxi healers and farmers in northwestern Yunnan, China. *Journal of Ethnobiology and Ethnomedicine*, *10*(1), 6. doi:10.1186/1746-4269-10-6 PMID:24410825

Yang, Q. Y., Yao, C. S., & Fang, W. S. (2010). A new triglucosylated naphthalene glycoside from Aloe vera. *L. Fitoterapia*, *81*(1), 59–62. doi:10.1016/j.fitote.2009.07.006 PMID:19638305

Yang, X., & Eilerman, R. G. (1999). Pungent principal of *Alpinia galangal* (L.) Swartz and its applications. *Journal of Agricultural and Food Chemistry*, *47*(4), 1657–1662. doi:10.1021/jf9808224 PMID:10564034

Yashin, A., Yashin, Y., Wang, J. Y., & Nemzer, B. (2013). Antioxidant and antiradical activity of coffee. *Antioxidants*, *2*(4), 230–245. doi:10.3390/antiox2040230 PMID:26784461

Yazdanparast, R., & Ardestani, A. (2007). *In vitro* antioxidant and free radical scavenging activity of *Cyperus rotundus. Journal of Medicinal Food*, *10*(4), 667–674. doi:10.1089/jmf.2006.090 PMID:18158839

Yesilada, E., Honda, G., Sezik, E., Tabata, M., Fujita, T., Tanaka, T., ... Takaishi, Y. (1995). Traditional medicine in Turkey. v. folk medicine in the inner taurus mountains. *Journal of Ethnopharmacology*, *46*(3), 133–152. doi:10.1016/0378-8741(95)01241-5 PMID:7564412

Yeşilada, E., Honda, G., Sezik, E., Tabata, M., Goto, K., & Ikeshiro, Y. (1993). Traditional medicine in Turkey iv. Folk medicine in the Mediterranean subdivision. *Journal of Ethnopharmacology*, *39*(1), 31–38. doi:10.1016/0378-8741(93)90048-A PMID:8331960

Yesilada, E., Üstün, O., Sezik, E., Takaishi, Y., Ono, Y., & Honda, G. (1997). Inhibitory effects of turkish folk remedies on inflammatory cytokines: Interleukin-1α, interleukin-1β and tumor necrosis factor α. *Journal of Ethnopharmacology*, *58*(1), 59–73. doi:10.1016/S0378-8741(97)00076-7 PMID:9324006

Yetgin, A., Canli, K., & Altuner, E. M. (2018). Comparison of antimicrobial activity of allium sativum cloves from China and Taşköprü, Turkey. *Advances in pharmacological sciences, 2018. Article ID, 9302840*, 1–5. doi:10.1155/2018/9302840

Ying, J., & Bing, X. (2016). Chemical constituents of *Cyperus rotundus* L. and their inhibitory effects on uterine fibroids. *African Health Sciences*, *16*, 1000–1006. PMID:28479892

Yisehak, T., Eyosyas, K., Cruz, C., & Getachew, G. (2014). Potential Antibacterial Activity of Crude Extracts from *Aloe vera, Zingiber officinale* and Vinca major Medicinal Plants. *International Journal of Microbiological Research*, *5*(3), 202–207.

Yoshikawa, M., Murakami, T., Kishi, A., Kageura, T., & Matsuda, H. (2001). Medicinal flowers. III. Marigold. (1): Hypoglycemic, gastric emptying inhibitory, and gastroprotective principles and new oleanane-type triterpene oligoglycosides, calendasaponins A, B, C, and D, from Egyptian *Calendula officinalis*. *Chemical & Pharmaceutical Bulletin*, *49*(7), 863–870. doi:10.1248/cpb.49.863 PMID:11456093

Yu, D., Suzuki, M., Xie, L., Morris-Natschke, S. L., & Lee, K. H. (2003). Recent progress in the development of coumarin derivatives as potent anti-HIV agents. *Medicinal Research Reviews*, *23*(3), 322–345. doi:10.1002/med.10034 PMID:12647313

Yu, H. H., Lee, D. H., Seo, S. J., & You, Y. O. (2007). Anticariogenic properties of the extract of *Cyperus rotundus*. *The American Journal of Chinese Medicine*, *35*(3), 497–505. doi:10.1142/S0192415X07005016 PMID:17597508

Yusuf, M., Wahab, M., Yousuf, M., Chowdhury, J. U., & Begum, J. (2007). Some tribal medicinal plants of Chittagong hill tracts, Bangladesh. *Bangladesh Journal of Plant Taxonomy*, *14*(2), 117–128. doi:10.3329/bjpt.v14i2.531

Zaeoung, S., Plubrukarn, A., & Keawpradub, N. (2005). Cytotoxic and free radical scavenging activities of Zingiberaceous rhizomes. *Songklanakarin Journal of Science and Technology*, *27*(4), 799–812.

Zahra, A. A., Kadir, F. A., & Mahmood, A. A., Al hadi, A. A., Suzy, S. M., Sabri, S. Z., Latif, I. I., & Ketuly, K. A. (2011). Acute toxicity study and wound healing potential of *Gynura procumbens* leaf extract in rats. *Journal of Medicinal Plants Research*, *5*(12), 2551–2558.

Zancan, K. C., Marques, M. O., Petenate, A. J., & Meireles, M. A. A. (2002). Extraction of ginger (*Zingiber officinale* Roscoe) oleoresin with CO_2 and co-solvents: A study of the antioxidant action of the extracts. *The Journal of Supercritical Fluids*, *24*(1), 57–76. doi:10.1016/S0896-8446(02)00013-X

Zarei, M., & Javarappa, K. K. (2012). Anticarcinogenic and cytotoxic potential of *Hemidesmus indicus* root extract against Ehrlich Ascites tumor. *Der Pharmacia Lettre*, *4*(3), 906–991.

Zarger, R., & Stepp, J. (2004). Persistence of botanical knowledge among Tzeltal Maya children. *Current Anthropology*, *45*(3), 413–418. doi:10.1086/420908

Zhai, Z., Haney, D. M., Wu, L., Solco, A. K., Murphy, P. A., Wurtele, E. S., & Cunnick, J. E. (2009). Alcohol extract of *Echinacea pallida* reverses stress-delayed wound healing in mice. *Phytomedicine*, *16*(6-7), 669–678. doi:10.1016/j.phymed.2009.02.010 PMID:19303756

Zhao, Y., Chen, B., Shen, J., Wan, L., Zhu, Y., Yi, T., & Xiao, Z. (2017). The beneficial effects of quercetin, curcumin, and resveratrol in obesity. *Oxidative Medicine and Cellular Longevity, 2017*, 8 pages. doi:10.1155/2017/1459497

Zhao, Z., Matsunami, K., Otsuka, H., Negi, N., Kumar, A., & Negi, D. S. (2013). A condensed phenylpropanoid glucoside and pregnane saponins from the roots of *Hemidesmus indicus*. *Journal of Natural Medicines*, *67*(1), 137–142. doi:10.100711418-012-0659-6 PMID:22456894

Zhou, Z. L., Lin, S. Q., & Yin, W. Q. (2016). New cycloartane glycosides from the rhizomes of *Cyperus rotundus* and their antidepressant activity. *Journal of Asian Natural Products Research*, *18*(7), 662–668. doi:10.1080/10286020.2016.1142976 PMID:26978535

Zhou, Z. L., Yin, W. Q., Yang, Y. M., He, C. H., Li, X. N., Zhou, C. P., & Guo, H. (2016). New Iridoid Glycosides with Antidepressant Activity Isolated from *Cyperus rotundus*. *Chemical & Pharmaceutical Bulletin*, *64*(1), 73–77. doi:10.1248/cpb.c15-00686 PMID:26726748

Zhou, Z., & Yin, W. (2012). Two novel phenolic compounds from the rhizomes of *Cyperus rotundus* L. *Molecules (Basel, Switzerland)*, *17*(11), 12636–12641. doi:10.3390/molecules171112636 PMID:23099619

Zhou, Z., Yin, W., Zhang, H., Feng, Z., & Xia, J. (2013). A new iridoid glycoside and potential MRB inhibitory activity of isolated compounds from the rhizomes of *Cyperus rotundus* L. *Natural Product Research*, *27*(19), 1732–1736. doi:10.1080/14786419.2012.750318 PMID:23356789

Zimik, L., Saikia, P., & Khan, M. (2012). Comparative study on homegardens of Assam and Arunachal Pradesh in terms of species diversity and plant utilization pattern. *Research Journal of Agricultural Science*, *3*(3), 611–618.

About the Contributors

Akash is working as an Assistant professor (Guest) in Department of Botany and Microbiology, Gurukul Kangri University, India. He did his Bachelor degree in Science and Master degree (in Botany) form HNB Garhwal University, Uttarakhand, India and Ph.D.(in Botany) from Gurukul Kangri University, Haridwar under the supervision of Professor Navneet. His research was on the Estimation of Phytodiversity of Rajaji Tiger reserve, Soil nutrients dynamics and soil microbiological analysis. He has published more than a dozen research paper on national and international journal. Currently the author is serving as an active reviewer for the IGI global.

Navneet is a Professor in the Department of Botany and Microbiology Gurukul Kangri University Haridwar, India. He earned M.Sc. and M.Phil. Degree in 1985 and 1987 respectively and Ph.D. in 1991 from Kurukshetra University, Kurukshetra. He joined Gurukul Kangri University Haridwar as Lecturer in 1992. He has been the Head of the Department of Botany and Microbiology from July 2013 to June 2016. He has guided 14 Ph.D. students and is supervising 6 research students on varied aspects of microbiology. He had been the Dean Student Welfare and Coordinator, National Service Scheme of Gurukul Kangri University. He is at present holding the position of Dean, Faculty of Life Sciences. He has published over 100 research papers and reviews in National and International journals. Dr. Navneet is a Fellow of the Indian Botanical Society, Fellow of Indian Phytopathological Society and life member of several academic societies.

B. S. Bhandari started his academic career with receiving Gold Medal in Master's Degree. Thereafter, joined Ph. D. programme in the global burning issue of Fire Ecology. He has prevailed the CSIR, Pool Officership and DST Young Scientist Award of Government of India. Presently, he is a senior faculty in the Department of Botany and Microbiology, HNB Garhwal University (A Central University), Uttarakhnad, India. He has a wide experience of teaching and research in the field of Himalayan Ecology, Environmental Sciences and Conservation Biology. He has published more than three dozen research paper and articles in the journals of national and international repute. So far, he has supervised five Ph. D. scholars leading to doctorate degree. He is an ardent trekker and photographer devoted to document the floral and faunal diversity. For the last one decade he is working on the floral and avifaunal relationships of the ecosystems.

* * *

Dickson Adom is a hardworking researcher scientist with the Department of Educational Innovations in Science and Technology, Kwame Nkrumah University of Science and Technology, Ghana. He holds a Bachelor of Arts, Master of Philosophy and Doctor of Philosophy degrees from the same institution as well as a postgraduate degree in Education from the University of Education, Winneba, Ghana. He is an expert in the use of Traditional Knowledge Systems and Cultural Anthropology issues in enhancing the conservation of the environment and its rich biological diversities. Also, he undertakes studies in Rural Community Participation in Environmental Development Schemes, International Conventions related to Biodiversity Conservation, Protected Areas and Ecosystem Management through Community Involvement Approaches and Strategies, Agricultural Anthropology, NFTPs in Biodiversity Management in Forest-Fringe Communities, Development of Biodiversity Conservation Strategies, Ecotourism Development, Cultural Practices in Biodiversity Management and other areas in this field. He is an expert in academic and research writing and teaches research and academic writing. He has authored a total of 62 publications with over 30 papers, book chapters and reports in General Agriculture, Agricultural Anthropology, Environmental Science, Academic writing, Education, History and Culture. He had the world championship 2018 award in the field of Agriculture (Ecosystem Conservation). He is an editor and reviewer for several internationally refereed journals.

Ram Kumar Bhakat, Associate Professor, is the ex Head of the Department of Botany and Forestry, Vidyasagar University. His area of expertise includes the taxonomy and biodiversity conservation by traditional methods in West Midnapore and adjacent districts. Contribution: Research questions and manuscript preparation.

Elif Merdamert Bozyel currently works as Independent researcher. Elif research in Moleculer Biology. Her most recent article is 'Anticancer Uses of Medicinal Plants in Turkish Traditional Medicine.' and conference paper is 'Recommendations for determination and conservation of Plant Biodiversity of Ladik Limestone Quarry'. Her most read chapter is 'Antiurolithiatic Activity of Medicinal Plants in Turkey'.

Mustafa Eray Bozyel currently works as Independent researcher. M. Eray research in Botany. His most recent article is 'Anticancer Uses of Medicinal Plants in Turkish Traditional Medicine.' and conference paper is 'Recommendations for determination and conservation of Plant Biodiversity of Ladik Limestone Quarry'. His most read chapter is 'Antiurolithiatic Activity of Medicinal Plants in Turkey'.

Kerem Canli currently works as Associate Professor in Dokuz Eylül University. Kerem research in Botany and Microbiology. His most recent article is 'Anticancer Uses of Medicinal Plants in Turkish Traditional Medicine.' and conference paper is 'In Vitro Antimicrobial Activity Screening of Achillea Phrygia and Determination of the Ethanol Extract Composition by Gas Chromatography/Mass Spectrometry'.

Bimal Chetri is a regular faculty member under Department of Environment and Life Sciences, Sherubtse College under Royal University of Bhutan. Holds M.Sc. in Botany with specialization in Seed Technology. He teaches a few core modules from plant science like Ecophysiology of Plants, Plant Diversity and Ethnobotany to undergrads and honors students. About Author: Teaching has always been his self impetus. The most important thing he likes about teaching is, he gets a complete sense of contentment and happiness while he is teaching subjects of his interest to students. Teaching is a profession he believes that provides learning to learn kind of environment. Teaching in the university provides me with all the resources and opportunity to progress into my career, carry out researches thereby enabling learning as continuous process. He believes so much in team spirit, hard work and dedication. He documents traditional knowledge (TK) about plants on ethnobotanical studies related to medicines, including plants with stress tolerant traits. These studies are possible only in rural settings. Ecophysiology, ethnoecology, ethnophenological studies of wetland plant communities interest me. He always tries to see things rather holistically, other entities of nature like plant, animal and physical entity, co-exist harmoniously in nature, except for human being. Man is a problem. Therefore, it is very important to understand role of human being in nature to work for better society. Awards: • ASEAN-India Science,Technology & Innovation Cooperation Research and Training Fellowship for Developing Country Scientists (RTF-DCS) scheme 2019 • Funded by Department of Science and Technology India Science Research Fellowship Programme Jan-June 2018. Awarded by India Science and Research Fellowship (ISRF)2017-18 Programme to pursue research in frontier/advanced areas of science including Medicine and Agriculture for a period of 3-6 months by CCSTDS, DST, India.

Maidul Islam obtained his BPharm degree from the Department of Pharmacy, University of Development Alternative. He is currently in the final stage of completion of his Masters degree in Biotechnology & Genetic Engineering at the same University. His research interest lies in Medicinal Plants of Bangladesh and he already has a number of publications in this area.

Rownak Jahan obtained her PhD degree from Kansas State University, USA. She is currently the Chairperson of two Departments, Department of Biotechnology & Genetic Engineering and Department of Molecular Medicine & Bioinformatics at the University of Development Alternative. Her interest lies in nutrigenomics, phytoremediation, and functional food plants.

Vartika Jain, Ph.D., FES, FAPT, FSMRIM, is Assistant Professor, at Department of Botany, Govt. Meera Girls' College, Udaipur, Rajasthan working on various dimensions of Ethnobotany, Ethnopharmacology, Phytochemistry and Plant taxonomy for last 14 years. She is a Gold Medalist in M.Sc. Botany, qualified CSIR-NET (JRF) as also International aptitude tests like GRE, AGRE and TOEFL. She has published 57 research and review articles in National and International Journals, 12 chapters in books and written more than 60 popular articles in magazines and newspapers. She has authored five reference books which include the gigantic book entitled "Compendium of Indian Folk Medicine and Ethnobotany" comprising data of more than 4500 plant species. She has worked extensively on medicinal plant Semal (Bombax ceiba) and also worked for its conservation and known as 'Semal Lady of India' as also received various Awards like Maharana Fateh Singh Award (1997 & 1999), Young Scientist Award (2010), Platinum Shakti Award (2015), Prof. P. Sensarma Medal (2016), Young Achiever Award in Ethnobiology (2017) and Samaj Shiromani Award (2019). She has participated and presented research papers in more than 40 National and International Conferences, delivered invited lectures in training courses and also organized various National and International scientific events at Udaipur. She is member of many Academic Societies including The National Academy of Sciences, India and International Society of Ethnobiology, USA as well as working as a Secretary, Society for Microvita Research and Integrated Medicine (SMRIM), Udaipur, and editor of a tri-annual Bulletin of Microvita Research and Integrated Medicine (BOMRIM).

Khoshnur Jannat is currently a Laboratory Instructor in the Department of Pharmacy, University of Development Alternative, Bangladesh. She obtained her first Masters degree in Pharmacy and her second Masters degree in Biotechnology & Genetic Engineering from the University of Development Alternative. Her interest lies in drug discovery from anti-diabetic and analgesic plants of Bangladesh.

Laxman Singh Kandari is presently working as an Associate Professor in Department of Forestry, Mizan Tepi University, Ethiopia. Dr. Kandari did his Ph.D. degree for the work entitled "Ecophysiological and Socioeconomic Studies of Some Rhizomatous Medicinal and Aromatic Plants Species" which was conducted in Nanda Devi Biosphere Reserve (NDBR), Uttarakhand, India. Dr. Kandari also worked briefly with Indian Institute of Forest Management, Bhopal in a project entitled 'Operational strategy for sustainable forestry development with community participation in India' which was sponsored by the International Tropical Timber Organization (ITTO). Dr. Kandari extended his work as a Postdoctoral Research Fellow at Research Centre for Plant Growth & Development, University or KwaZulu-Natal South Africa (2009-2011). He worked on conservation and multiplication of threatened wild medicinal plants. Dr. Kandari has significant contribution in the field of natural Science and forestry. He worked to natural science and social science for conservation and management of medicinal and aromatic plants together Dr. Kandari having more than 40 quality publications in both International and National research journals. He has been teaching undergraduate and post-graduate students in Ethiopia and has more than 13 years teaching and research experience.

Kapil Kharkwal is working as a Research Associate in Forest Research Institute, Dehradun, India.

Mohan Kumar is an agronomist with high interest in plant science and ethnomedicine in the Tamil Nadu Agricultural University in India.

R. Maikhuri is working as Scientist-G, Scientist In-Charge (Garhwal Regional Centre) in G.B. Pant National Institute of Himalayan Environment and Sustainable Development. His area of research is rural area development, natural resource management, restoration ecology, bioresources utilization, and conservation of medicinal plants.

Dalip Mansotra is working as an Assistant Professor in Department of Zoology, Gurukul Kangri University, India.

Shreekar Pant is a Senior Assistant Professor in the Department of Botany at Baba Ghulam Shah Badshah University, in Rajouri, J&K.

Kusum Payal is a PhD and currently works as Principal investigator of Project under Scheme for Young Scientist and Technologist (SYST) of SEED Division DST. Doing her research work from University School of Environment Management, Guru Gobind Singh Indraprastha University, New Delhi, India.

Taufiq Rahman's early academic training was in pharmaceutical sciences with a Bachelor and a Master of Pharmacy (B.Pharm and M.Pharm) degrees from University of Dhaka, Bangladesh. His earlier research interest included ethnopharmacology and chemistry of natural products. After spending a year in Dhaka as a lecturer of pharmacy, he pursued an MSc in Molecular Pharmacology in the University of Manchester under the Commonwealth Scholarship and Fellowship program in 2003. In 2005, he started a PhD at the Department of Pharmacology, University of Cambridge. After the completion of his PhD in 2008, he continued research in the same lab as a post doc and then from October 2009, he started his first independent position as a Drapers' Research Fellow at Pembroke College. In 2011, he received the Early Career Research Award in Signal Transduction from the Biochemical Society. From October 2012 till September 2017, he worked as a Royal Society University Research Fellow in the Department. Since 2016, he has been working as a University Lecturer based in the Department of Pharmacology, University of Cambridge.

Mohammed Rahmatullah is currently Dean, Faculty of Life Sciences at the University of Development Alternative, Dhaka, Bangladesh. He obtained his PhD from the University of Hong Kong and served in various capacities at Kansas State University, Auburn University and Weis Center for Research, USA.

Kottapalli S. Rao is working as Head and Professor in Department of Botany, University of Delhi, India. He is currently involved in academic work related to natural resource management and sustainable development in Himalaya and biodiversity assessment in critical ecosystems in hill and mountain regions. He is a fellow of Linnean Society of London, India and Indian Botanical Society.

Namgay Rinchen is a B.Sc. Life Science student currently studying in Sherubtse College under Royal University of Bhutan (RUB). His interest on ethnography motivated him to record the uses ethnomedicinal plants since schooling. He is an active early researcher in the field of ethnobotanical studies.

Paul Sekyere is a chemical biologist from the Department of Chemistry in the Kwame Nkrumah University of Science and Technology in Kumasi, Ghana. His research areas include Agriculture, Plant Science and Biochemistry.

Uday Kumar Sen, Ph.D., is a post doctoral researcher at the Department of Botany and Forestry, Vidyasagar University. His research focuses on the taxonomy, ethnobotany, traditional conservation practices and distribution of angiosperm and non-angiosperm plants of West Midnapore and adjacent districts. Contribution: General concept and research procedure, data with sample collection and identification, herbarium work for voucher specimen, morphological study, interpretation of results and manuscript preparation.

Nasrin Akter Shova obtained her BPharm degree from the Department of Pharmacy, University of Development Alternative. She is currently in the final stage of completion of her Masters degree in Biotechnology & Genetic Engineering at the same University. Her research interest lies in Natural Products and Drug Discovery and she already has a number of publications in this area.

Karma Wangchuck completed B.Sc. General from Sherubtse College (Affiliated to Delhi University). Completed M.Sc. in Biodiversity and Plant Taxonomy from the University of Edinburgh, Scotland, UK. Last phase doing PhD at Queensland University of Technology (QUT) Brisbane, Queensland, Australia. Teaching Botany and Plant Science modules at Sherubtse College since 2002.

Zishan Wani is a PhD Scholar in the Department of Botany at Baba Ghulam Shah Badshah University, in Rajouri, J&K.

Index

H

I

K

L

M

N

P

Q

S

T

www.ingramcontent.com/pod-product-compliance
Lightning Source LLC
LaVergne TN
LVHW060301011026
849161LV00011B/2753